Economic Botany: Principles and Practices

Economic Botany
Principles and Practices

by

GERALD E. WICKENS

KLUWER ACADEMIC PUBLISHERS
DORDRECHT / BOSTON / LONDON

Library of Congress Cataloging-in-Publication Data

Economic botany : principles and practices / edited by Gerald E. Wickens.
   p. cm.
   ISBN 0-7923-6781-2 (hardcover : alk. paper)
   1. Botany, Economic. I. Wickens, G. E.

   SB107 .E26 2001
   581.6--dc21

                             00-066289

ISBN 0-7923-6781-2

Published by Kluwer Academic Publishers,
P.O. Box 17, 3300 AA Dordrecht, The Netherlands.

Sold and distributed in North, Central and South America
by Kluwer Academic Publishers,
101 Philip Drive, Norwell, MA 02061, U.S.A.

In all other countries, sold and distributed
by Kluwer Academic Publishers,
P.O. Box 322, 3300 AH Dordrecht, The Netherlands.

Printed on acid-free paper

Printed in the Netherlands.

Foreword

The strength of this book is that it is written by someone who has spent a lifetime devoted to the science of economic botany. The author has brought together his vast experience in the field in Africa with his studies of arid land plants at the Royal Botanic Gardens, Kew. The result is an informative and reliable text that covers a vast range of topics. It is also firmly based upon the author's research and interest in plant taxonomy and therefore fully acknowledges the importance of correct naming and classification in the field of science of economic botany. The coverage is of economic botany in its broadest sense. I was delighted to find such topics as ecophysiology, plant breeding, the environment and conservation are included in the text. This gives the book a much more comprehensive coverage than most other texts on the subject. I was also glad to see that the book covers the use of various organisms that are no longer considered part of the plant kingdom such as various species of fungi and algae. It is indeed a broad ranging book that will be of use to many people interested in the uses of plants and fungi.

Economic botany is once again being given more prominence as a discipline because of its enormous relevance to both conservation and sustainable development. Those people involved in those topics should find this a most useful resource. The development of new sustainable agroforestry systems, and the search for new plant products both depend upon a good knowledge of economic botany. I wish this text had been available at the time I was setting up various new courses in economic botany.

Professor Sir Ghilean Prance, FRS, VMH
Formerly Director of the Royal Botanic Gardens, Kew

Contents

FIGURE

TABLES

Acknowledgements

I gratefully acknowledge the help given by colleagues at the Royal Botanic Gardens, Kew, especially Frances E.M. Cook, also Dr A. Abu-Rabia (Beer-Sheva), Professor L. Boulos (Cairo), P.W. Coward (The Hill Brush Company), Professor M.N. El Hadidi (Cairo University), Dr H.N. Le Houérou (Montpellier), Dr S.R. King (Shaman Pharmaceuticals), P. Scott (National Trust); also for the library facilities provided by FAO (Rome), Institute of Food Research (Norwich), John Innes Centre (Norwich), Linnean Society (London), Norfolk County Library (Aylsham Branch), Welcome Institute for the History of Medicine (London) and to Plant Health, Ministry of Agriculture, Fisheries and Food, and Plant Health, Forestry Commission; I am particularly grateful to Professor E.A. Bell, Bill Howard and Dr Peter Lapinskas for their critical reading of parts of the text.

I also wish to thank the following for permission to cite material: Market House Books Ltd. (*Penguin Dictionary of Botany* ©1984), Reader's Digest Assoc. Ltd. (*Reader's Digest Universal Dictionary* © 1986), John Wiley & Sons Ltd (Cotton, C.M. *Ethnobotany Principles and Applications* ©1996), Royal Society of Chemistry (D'Mello, J.P.F., Duffus, C.M. and Duffus, J.H. (eds.) *Toxic Substances in Crop Plants* ©1991), F. Rexen, F. (Rexen, F. and Munck, *Cereal Crops for Industrial Use in Europe* ©1984), FAO (*FAO Production Year Book 1993*, vol. 47 and *Non-Wood Forest Products* no.2: *Gum naval stores* ©1995, IDRC, Ottawa (The Crucible Group, *People, Plants and Patents* ©1994) Ericht Hoyt (Hoyt, E., *Conserving the World Relatives of Crops* ©1988), Addison, Wesley, Longman (*Penguin Dictionary of Chemistry*, 2nd edn. © 1990).

Finally, I am indebted to Arno Flier and Angela Timmerman of Kluwer Academic Publishers, Dordrecht for their forbearance with my incompetence in the preparation of camera-ready copy.

Preface

As a background to the rationale for this book, these ramblings of a retired economic botanist are intended to explain my attitude to the many facets of economic botany. I first became interested in the use of plants following a chance visit as a 14 year-old schoolboy to London's Imperial Institute, alas no more, having long been replaced by the College of Science and Technology. At that age I was not impressed by the serried rows of bottles on display in the Museums at the Royal Botanic Gardens, Kew, just across the river from where I went to school. A regular Saturday visitor to the Imperial Institute, I became hooked by the marvellous displays of dioramas and artefacts and resolved to become a tropical agriculturist. As I have tried to explain in Wickens (1990), economic botany in the United Kingdom was then equated with tropical agriculture and the Empire. Consequently I read agriculture and botany at Aberystwyth, the latter a subject that my grammar school education had failed to provide. In those halcyon years DNA, computers, rock-and-roll and top of the pops were still unknown

However, by 1952, when I first arrived in Africa, it was a generation too late for 'economic botany' to make a meaningful contribution to agriculture as the colonies were already in the process of being handed back to their indigenous inhabitants. Nevertheless, I learnt much through observation and inquiry as I practised my various trades in agriculture, soil and water conservation, ecology, land use, etc. in what were then known as Northern Nigeria, Northern and Southern Rhodesia and the Sudan. Living in a country teaches one more about the significance of plants in the lives of people than it is possible to learn through expeditions. An allergy to tropical grasses hastened my unwilling return to my home country where I was fortunate enough to obtain employment as a taxonomist at the Royal Botanic Gardens, Kew and later to be placed in charge of the Survey of Economic Plants of Arid and Semi-Arid Tropics, thereby leading to my inheritance and belated appreciation of those much despised bottles of my youth! The what, how and why some plants are used continues to fascinate me in my retirement. I can only hope that others will be equally fascinated.

In planning this book I have tried to take account of what I now know and, with hindsight, what I should have known. My academic education was based on broad principles, while in the latter half of the 20th century education has tended to be more job specific. Economic botany too has changed from and understanding of the biology, culture and utilisation of plants and plant products to a more detailed understanding of the chemistry and genetics involved. For example, during the past 50 years there have been enormous advances in biotechnology and in the search and development of new phytochemicals for the pharmaceutical and chemical industries. Tremendous advances have been made in plant breeding following with the develpment of genetic engineering, advances that in my opinion have too quickly used for profit before fully investigating their possible impart on the environment. The need for certain past usages may have changed to meet the requirements of national economies and developments and new uses found and developed for old crops, such as the growing of flax in the UK with for linseed oil instead of fibre.

This book is not intended as a source of detailed information on any one topic. Instead I have attempted to briefly discuss the major disciplines and topics relating to economic botany and to demonstrate some of the general principles and terminology encountered in the development of major plant products, procedures which, hopefully, may be usefully adapted for improving the utilisation of other economic plants. I have also tried to make each chapter as self-contained as possible and as a result some repetition has been necessary.

My own experiences have taught me that taxonomy, the proper identification and naming of a plant, is the indisputable foundation stone of economic botany, that economic botany is a multidisciplinary subject impinging upon a number of other disciplines and that a holistic approach must be followed. Also, it is equally important to understand the full potential of a plant as well as the processes involved in producing the end product.

Finally, economic botany, as I understand it, is an honourable profession concerned with the use of plants by the people for the people. Unfortunately many multinationals now misinterpret this by placing emphasis on a monetary interpretation of 'economic', often regardless of the needs of the peoples in the developing countries.

Gerald E. Wickens
Aylsham, Norfolk, UK

Chapter 1

Economic Botany

Although impossible to prove, it is reasonable to assume that the use of plants by man must date back to the origins of mankind. The earliest known record is believed to be the collection of stones (endocarps) from the fruit of *Celtis australis* (hackberry) plus clear evidence of the use of fire associated with the remains of Peking man (*Homo erectus*) from the Middle Pleistocene deposits at Zhoukoudian (Choukoutein), China (Day, 1969; Renfrew, 1973). Monod (1970) noted a similar association of endocarps of *C. integrifolia* with Neolithic man from the Ahaggars in central Sahara. The study of such evidence is sometimes referred to as *archaeo-ethnobotany*, and more often by the less clumsy *palaeo-ethnobotany*.

Neither should the uses of plants by earlier civilisations be regarded as being of little importance today. For example, the use of mouldy bread by the Ancient Egyptians predated Alexander Flemming's discovery in 1928 of *Penicllium notatum* by some 3000 years. Perhaps if more attention had been paid to the scrolls of the Pharaohs, or to a similar use of mouldy bread in medieval Europe for dressing wounds, the value of penicillin would have been identified much earlier (Böcher, 1963). Similarly, the 1st century AD herbal lore of Dioscorides, to which most of the European herbals owe some contribution, was investigated by Riddle (1985) and was found to be meaningful in terms of present-day pharmaceutical knowledge. Moreover the early Chinese herbals, such as the Ehr Ya of 3000 BC, Shu Ching of 1000 BC, and Pen Tschao Ko Kua of 1295 AD, are now undergoing a closer scrutiny by the Chinese, with a number of international pharmaceutical organisations showing a keen interest in their findings (Barrau, 1989; Wickens, 1990).

The early publications on the use of plants by man referred mainly to their medicinal properties, and their lore enshrined in the herbals (see Chapter 16). Apart from medicinal lore, other plant uses were not necessarily included. Indeed, many of the early Greek and Roman authors did not consider such non-medical information respectable. Much later, the scope of the herbal became, for a brief period, almost

1

encyclopaedic in its coverage, as shown by the title page (Fig. 1) from Green (1816, 1824).

Even 17th and 18th century scholars persisted in their contempt for the non-medical usage of plants. The German botanist Rivinus (1690) cited by Barrau (1971) went so far as to declare that botany should be restricted to the study of plants and not their properties. Even Linnaeus (1751), possibly ill-at-ease with the systematics of plant domesticates, questioned whether domestic plants were worthy of study. Yet regardless of such reservations Linnaeus included observations on plant uses in his own writings, e.g. *Flora lapponica* (1737), *Flora suecica* (1745), and those of his pupils, e.g. *Flora economica* (1748), a dissertation attributed to Elias Aspelin on the uses of Swedish plants, and *Plantae esculentae* (1752), a dissertation by J. Hiorth on food plants.

The posthumous publication from 1741-1755 of the seven volumes of *Herbarium amboinense* by the Dutch botanist Georg Rumphius (1627-1702) on plant usage in Ambon, Indonesia provided the earliest source of information for that part of south-east Asia. An important contribution on extra-tropical plants of potential industrial use was later made by the German-born Government Botanist for the Colony of Victoria, Sir Ferdinand von Mueller (1885).

In 1819 the Swiss botanist Augustin de Candolle used the term **botanique appliquée** for the study of the man's use of plants, yet the subject still failed to gain the recognition it deserved. It was not until the mid-19th century that economic botany finally became recognised as a subject in its own right, with Thomas Archer (1853, 1865), Alphonse de Candolle (1882) with his famous history of cultivated plants, and George Boulger (1889), being recognised among the fathers of modern economic botany. The standard work for much of the 20th century was that of Albert Hill (1937, 1952), the more recent works reflecting the increasing importance of phytochemicals (Wickens, 1990).

1.   ECONOMIC BOTANY DEFINED

What is economic botany? Kalkman (1989) succinctly defines it as *"Within botany it is that field of knowledge, study and research that is concerned with plants used by man."* My own more detailed definition is a revision of that first proposed by Wickens (1990): *"Economic botany is the study of plants, fungi, algae and bacteria that directly or indirectly, positively or adversely affect man, his livestock, and the maintenance of the environment. The effects may be domestic, commercial, environmental, or purely aesthetic; their use may belong to the past, the present or the future."*

THE

UNIVERSAL HERBAL;

OR,

BOTANICAL, MEDICAL, AND AGRICULTURAL

DICTIONARY

CONTAINING AN ACCOUNT OF

All the known Plants in the World

ARRANGED ACCORDING TO THE LINNEAN SYSTEM.

SPECIFYING THE

USES TO WHICH THEY ARE OR MAY BE APPLIED, WHETHER AS FOOD, AS MEDICINE, OR IN
THE ARTS AND MANUFACTURES

WITH THE BEST

METHODS OF PROPAGATION

AND THE

MOST RECENT AGRICULTURAL IMPROVEMENTS

Collected from indisputable Authorities

ADAPTED FOR THE USE OF

THE FARMER-THE GARDENER-THE HUSBANDMAN-THE BOTANIST-THE FLORIST-
AND COUNTRY HOUSEKEEPERS IN GENERAL

BY THOMAS GREEN.

THE SECOND EDITION, REVISED AND IMPROVED

VOL. 1

LONDON:

PRINTED AT THE CAXTON PRESS, BY HENRY FISHER,

Printer in Ordinary to His Majesty

PUBLISHED AT 38, NEWGATE-STREET, AND SOLD BY ALL BOOKSELLERS

FIGURE 1. A reconstruction of the title page from Green (1816) showing the wide range of topics to be found in his herbal

The word 'economic' is used in the sense of utilitarian rather than of monetary gain, although the latter should not be ignored. It is perhaps of interest that Stearn (1976) noted that the word 'botany' had utilitarian origins, evolving from the Ancient Greek word βοτανη, meaning fodder or grazing for cattle, which was later extended to herbs used in medicine. The eventual evolution of botany from herbalism is briefly discussed in Chapter 16.

The direct effects refer to the use for food, feed, fibre, fuel, medicine, etc. while the indirect effects can, for example, reflect otherwise 'useless' plants that harbour beneficial pollinators, while the adverse effects recognise the non-beneficial role of weeds, pathogenic fungi, drugs, etc.

The above definition of economic botany not only embraces the vascular and non-vascular plants but also, according to the Five Kingdoms classification (see Chapter 2), non-plants such as the fungi, algae and bacteria. The role of the non-plants are probably best known to the layman by their contribution to food and drink (edible mushrooms, fermentation for beers, spirits and wines), N-fixation, pathogenic fungi, and alginates. In practice their economic contribution is far, far greater, and in order to emphasise this fact Chapters 18 and 19 are devoted to their role.

Economic botany is a multidisciplinary study that involves not only the purely botanical disciplines of taxonomy, ecology, physiology, cytology, biochemistry, pathology, etc. but to some extent also those aspects of agriculture, forestry and horticulture concerned with plant breeding, propagation, cultivation, harvesting, manufacture and the economics of production and marketing (Shery, 1972; Wickens, 1990). Other disciplines where plant life impinges on man's survival and well-being includes archaeology and palaeoarchaeology, anthropology, sociology, economic history, economic geography, conservation, etc. Even a basic knowledge of etymology of scientific and vernacular names and plant products can reveal important information on former usage, development, product definition and past distribution of a species, as pioneered by Hehn (1870), cited by Harris (1998), although a more holistic approach is generally advisable.

Thus, the vernacular name *bu hobab*, provided by the Venetian Prospero Alpino (1592) for the fruit of *Adansonia digitata* (baobab) found in the Cairo souk, was probably invented by the Cairo merchants since the baobab does not occur in Egypt, although it does occur in neighbouring Sudan, where the Arabic name for the tree is *tebeldi*, and the fruit *gongoleis*. The fruit pulp, *lobb*, however, is used medicinally, and the late Egyptian botanist Mohammed Drar has argued that baobab could be a corruption of *lobb*, which could well have been marketed in Cairo. Others have considered *bu hobab* to be a corruption of *bu hibhab* - 'the fruit with many seeds', although there is no evidence for this name having ever been applied to the fruit of *Adansonia digitata*. Nonetheless I find the argument in favour of *bu hibhad* convincing, especially since trees planted in Egypt are known as *habhab*. It is concluded that common vernacular name of baobab is a European corruption. The etymology of vernacular names is certainly not always straightforward (Täckholm, 1970; Wickens, 1982).

Regardless of a person's specialist discipline, the complete economic botanist should always try to follow a holistic approach to economic botany. Nowadays this needs to include an awareness of the ethics and morality of plant patents and plant biotechnology, as well as any possible social impact from the source of supply through to the end-user. Ultimately it is chance, history, economics and politics that finally determine the development of plants and their products.

Taxonomy is the foundation stone on which economic botany is and must be based. Accurate identification and correct naming are absolutely vital. Mistakes in the identification of plants eaten for food or used for medicine can obviously have drastic consequences. Sometimes errors in identification can have quite surprising repercussions. The failure to recognise the existence of two sympatric species of African cedars on Mount Mlanje, Malawi, i.e. the endemic *Widdringtonia whytei* and the more widespread *W. nodiflora*, resulted in plantation mismanagement and the economically desirable *W. whytei* becoming in danger of extinction (Pauw and Linder, 1997).

Economic botany charts the progress on the use of a plant in the wild, from its casual acceptance to its eventual role in the domestic, national and international economies. For the majority of species this implies domestication, i.e. selection and cultivation. There are exceptions, such as *Salicornia europaea* (sea samphire), *Bertholletia excelsa* (Brazil nut), *Laminaria* spp. (kelp) and many rangeland forage plants, that are mainly harvested from the wild, although some form of conservation management is implied for their sustainable production. Some ancient cultivated plants are of hybrid origin and are unknown in the wild, such as *Psophocarpus tetragonolobus* (winged bean) and *Zea mays* (maize) as well as some *Dioscorea* spp. (yams), and it is only since the development of genetics that plant breeders have been able to discover their progenitors. Of course not all potentially useful plants make the grade to the international markets. *Aspalanthus linearis* (rooibos tea) from South Africa is one of those rare examples of the development from a wild species to a cultivar and the international health markets during the 20th century, while a number of other plants with interesting phytochemicals, such as *Vernonia galamensis*, could well make the grade early in the 21st century.

All the major crops of today have undergone some form of domestication by early civilisations in various parts of the world. Indeed, over a century ago Candolle (1882) wrote *"Men have not discovered and cultivated within the last two thousand years a single species that can rival maize, rice, sweet potato, breadfruit, date, the cereals, millets, sorghums, the banana, the soya."* As Heiser (1986) points out *"The situation has changed little, if at all, in the last 100 yr. ..... The so-called 'new crops', amaranth and winged bean, that have received considerable publicity in recent years are, in reality, both fairly old domesticated plants."* Coles (1970) even goes as far as to state *"Every food plant of major importance to mankind was grown in the Neolithic stage of culture."*

Eight major world **centres of diversity** of cultivated plants, together with their subsidiary centres, were recognised by Vavilov (1926, 1951) and were considered to represent reservoirs of genetic diversity. He presumed that they also corresponded to the **centres or origin** of the various cultivated plants. It is now recognised that such centres of diversity do not necessarily correspond to centres of origin. In recent years more critical studies have demonstrated that the domestication of the earliest crops, *ca.* 9000-10,000 years ago. were located in four centres (Table 1), the home of their wild relatives. Further domestication then took place as these crops and their cultivation spread into **secondary centres** or **regions**, which did not contain their wild relatives. In these secondary centres the crops were subjected to different environmental and cultural selection pressures.

Table 1. Areas of origin of cultivated plant diversity of staple food crops (original table prepared by Susannah Brown and reproduced from Hoyt (1988) by kind permission of E. Hoyt

Nuclear centres of agricultural origins (Hawkes, 1983, 1991)	Regions of cultivated plant diversity (Vavilov, 1926, 1951)	Rice	Millet	Wheat	Barley	Maize	Squash	Tomato	Potato	Phaseolus

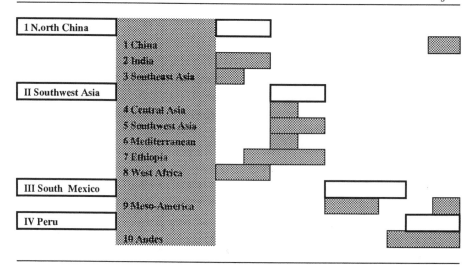

Further introductions from different areas within the original distribution would then allow hybridisation between previously isolated segments of the original gene

pool, or bring the introduced crop into fresh contact with wild relatives from which it had been isolated. While the secondary regions will not necessarily contain any unknown character states from the original centres of diversity, they will contain reshuffled character states and their mutations from within the secondary centres, and these can be a significant source of genetic material for the plant breeder (Hawkes, 1983; Pickersgill, 1998).

According to Gentry (1972) the requirements that need to be met in order that a wild plant can be brought into cultivation are: (1) The plant must be capable of yielding the required product in relative abundance; (2) It must be responsive to the artificial environment created by man; (3) The plant should exhibit sufficient genetic variation for the intensive selection of varieties; and (4) It must be economic to cultivate.

While domestication generally implies development as an agricultural, horticultural or forestry crop, the boundaries between the three disciplines and economic botany are ill-defined, the less advanced forms of agriculture, for example, being more interesting to the economic botanist than modern, high tech agriculture. Indeed, the older agricultural textbooks often included valuable sources of information on what are now regarded as non-agricultural economic plants, e.g. van Gorkon (1880, 1881) for tropical agriculture (cited by Kalkman, 1989), and Low (1847) for UK agriculture. The latter even includes brief notes on rice, opium, resins, herbal teas and coffees!

There are also practical difficulties in defining the boundaries between agriculture, horticulture and forestry. For example orchard crops may be, according to national interests, included under agriculture, horticulture or forestry. In some countries the planting of amenity trees and shrubs are even considered to be a function of forestry rather than horticultural departments. Similarly, range management can be under the control of agricultural, forestry, or even veterinary departments (Wickens, 1990).

The English use of the word '*agriculture*' probably originated from France during the 17th century from the Latin *agricultura*, originally *agri cultura*, the cultivation of land (Long, 1994). Agriculture is presently defined by *The Oxford English Dictionary* (1971) as '*the science and art of cultivating the soil, including the allied pursuits of gathering in crops and rearing livestock.*' **Horticulture** is the discipline concerned with ornamental plants, and non-agricultural crops. In the narrow sense it is concerned with the scientific cultivation of fruit, vegetables, flowers, trees and shrubs. It is also used to describe the commercial production of such crops on specialised holdings, and less commonly as part of normal agriculture (Dalal-Clayton, 1981). Although horticulture, like agriculture, bears Latin roots it is, contrary to expectations, post-Roman in origin. The first combination of the Latin *hortus* (garden) and *cultura* (cultivation) as *horticultura* is attributed to Lauremberg (1631), while the first English use of 'horticulture' is to be found in Holland (1678), where it is defined as the '*tillage, dressing, or improvement of Gardens as Agriculture of* *other Grounds'* (Janick, 1986; Simpson and Weiner, 1989). Lay people sometimes have difficulty in distinguishing between the role of horticulturists and botanists, especially among the staff of botanic gardens. My somewhat flippant response is that the former are taught some botany while the latter are not taught horticulture!

There has been a welcomed change in the role of forestry in recent years from the growing of trees for wood and wood products to the management of land as a natural resource for the additional provision of non-wood forest products (NWFP). These refer to plants and plant products used for food, forage, pharmaceuticals, plant exudates, etc. and for wild animals, birds, fishes, reptiles and insects as sources of food, hides, honey, etc. It also includes forestry services for the management of the range for domestic livestock and wild animals, and as parks and reserves for flora and fauna conservation, recreation, hunting, scenic and historic sites, soil and water conservation, etc. This represents a return to the earlier meaning of a forest as unfenced woodland or open, mainly treeless areas reserved for hunting, such as the royal forests or game reserves of Charlemange (*ca.* 742-814 AD). The word is probably from the Latin *foris,* for outside, outdoors (Dalal-Clayton, 1981; Wickens, 1991; Long, 1994).

2.   ETHNOBOTANY

There have been several attempts at distinguishing between plants used by the so-called 'developed' and 'primitive' societies', the distinction between the two societies being biased according to 'the eye of the beholder'. This viewpoint is made even more difficult by the clinal nature of the differences between societies, an issue which will be further discussed when defining ethnobotany. The French botanist Alphonse Rochebrune (1876), for example, used the term '*ethnographie botanique*' for the study of plants of human historical interest, especially those from archaeological sites (Barrau, 1972), apparently with the implication that if it is archaeological it must be primitive!

Despite the interest shown by European botanists in the uses of plants by past and present-day peoples of the Old World, the term ethnobotany owes its origins to the exploration and colonisation of the New World, which undoubtedly stimulated an interest in its rich flora and plant uses by what the colonising Europeans regarded as 'primitive societies'. It was the American Stephan Powers (1875) who first attempted to define what we now refer to as ethnobotany by the term '*aboriginal botany*' to describe the study of plants used by the Neeshenam of the Bear River, California for food, medicine, fabrics, ornaments, etc. But it was in a public address in 1895 by the botanist Dr John Harshberger to the University of Pennsylvania's Archaeological Association that '*ethno-botany*' was first used to describe "*the study of plants used by primitive and aboriginal people.*" The first paragraph of a newspaper report on the speech (Anonymous, 1895) is reproduced below.

"Some new and interesting ideas were presented by Dr John Harshberger in his last lecture on "Ethno-Botany" before the University Archaeological Association at the University. Ethno-botany deals with the plants used by primitive and aboriginal people, such as the Indians of America, and has for its object the study of the vegetable products employed for various purposes, as fibre and food. The main difficulty presented is that of the identification of the materials. Ethno-botany is of service in elucidating the culture position of primitive people. We can divide the races of mankind into pastoral, agricultural, semi-civilised, and civilised. The American Indian never occupied the pastoral stage, never had herds of domesticated animals, but emerged directly from the hunting into the agricultural stage."

The full text of the address is given in Harshberger (1896), where he defined the purposes of ethnobotany as: (1) An aid to elucidating the cultural position of 'tribes' according to their use of plants and plant products; (2) Evidence of the past distribution of useful plants; (3) Evidence of the ancient trade routes followed in the exchange of plant products; and (4) Of use in suggesting *"new uses of plants of which we were in ignorance."* The address provides the distinct and now unacceptable impression that ethnobotany applies to non-Caucasian Americans; the folk lore of the Caucasian Americans is not mentioned.

The term 'ethnobotany' was immediately accepted and widely used, replacing the earlier terms of 'aboriginal botany' and 'ethnographie botanique'. Its use has now been extended beyond the confines of 'aboriginal and primitive people' to include the use of plants by both tribal and atribal communities without any implication of primitive or developed societies (Ford, 1978; Wickens, 1990).

In a thoughtful discussion Jones (1941) notes the increasing broadening of Harshberger's original concept, and redefined ethnobotany as " ... *the study of the interactions of primitive man and plants ...."*, the term 'primitive' being employed in the sense of non-literate or aboriginal peoples in order to restrict its use to the less complex societies. He also commented on the impingement of ethnobotany on the established disciplines of economic botany, ecology, conservation, and agriculture, and its marginal position between plant sciences and anthropology. The role of ethnobotany was also defined as correlating the ecological impact of primitive man on the environment, archaeological plant data on the early use and distribution of plant species, and the anthropological information on the adaptation of primitive man to the plant environment, the usage of plants, and their role in the local economy.

Ford (1978) defined ethnobotany *"the study of direct interrelations between humans and plants."* He also recognised that most of the societies studied by earlier economic botanists are now part of a global economic system, also of the continuing need to redefine the term in order to meet contemporary conditions. Heiser (1985) followed Ford (1979) in defining ethnobotany as the study of plants in relation to people, and considered economic botany as dealing with plants and modern society. Heiser further speculated that since ethnobotany embraces both fields, economic botany could perhaps be considered as a subdivision of ethnobotany with monetary implications! Prance (1991) emphasised the multidisciplinary nature of contemporary ethnobotany, and the need to document tribal knowledge before the information is lost. Dr Cath Cotton (1996) in her excellent book on ethnobotany follows Jones (1941) in considering ethnobotany "*...to encompass all studies which concern the mutual relationships between plants and traditional peoples.*"

Because of its New World origins the definition of ethnobotany is biased towards a New World concept of primitiveness and fails to recognise the nations of the Old World with their long history of trade, cultural development, and periods of colonial and ethnic rule. While the idea of primitive or traditional peoples *sensu* Jones (1941) utilising wild plants and their products is readily understood, there is an inevitable gradual transition between such usage and the introduction of such people to modern agricultural crops and the industrial products of the more mixed rural and urban populations of modern societies. For example, the primitive agricultural economy of the Kayo Indians of the Amazonian rain forest involves not only the cultivation of several indigenous domesticated and semi-domesticated crops, but also the introduction of useful plants from elsewhere in the rain forest. Such is the high standard of management of their swidden culture that there appears to be no adverse effect on the forest environment (Posey, 1984). This example, while well within the New World concept of primitive ethnobotany, clearly demonstrates an underlying sophisticated approach that is far from primitive.

Elsewhere colonial and post-colonialism has brought about western 'civilisation' to many primitive or traditional peoples in both the Old and New Worlds. Thus, influenced by population expansion, deforestation, land privatisation, agricultural development, etc. Brokensha and Riley (1986) and Riley and Brokensha (1988) have documented the changes that have taken place since 1970 in Embu District, Kenya among the traditional Mbeere culture. The social organisation has changed dramatically, communal life has given way to individualism, traditional subsistence economy is being orientated towards a capitalist society, and the gap between the rich and the poor is widening. It is a process that is a microcosm of the irreversible changes that are now taking place elsewhere in the world. Ethnobotany is giving way under 'civilisation' to economic botany.

Historically the ethnobotany of the individual tribes that formerly occupied the British Isles have became similarly blurred as detribalisation followed in the wake of Roman, Saxon, and Norman, invaders, although traces of early ethnobotany can still be found in folklore at county or regional levels. Today urban areas, such as London, are becoming increasingly multiracial and, while the various ethnic flavours may still be recognisable, they too are becoming blurred as traditional ethnic customs adapt to changing economic conditions. The various ethnic food dishes, for example, become modified by alternative food sources. This last example also serves to emphasise the urgent need to record plant lore now, before it is forgotten by the increasingly urbanised generations.

With hindsight it is unfortunate that Harshberger (1896) failed to consider the impact of 'civilisation' on 'primitive societies' since, given time, the economic usage of such tribal societies will inevitably mingle with the national economy.

Ethnobotany forms one end of a continuum involving the development of plant resources, through post-ethnobotany to the allied disciplines of agriculture, forestry and horticulture, albeit, as discussed in Wickens (1990) there is no clear cut distinction between these disciplines.

In this development scheme there has to be a point where ethnobotany as we understand it becomes unrecognisable. As a consequence I would like to redefine ethnobotany as: *"To encompass all studies which concern the mutual relationships between traditional peoples and the past and present use of indigenous plants and primitive cultivars, the latter not necessarily indigenous."* This definition also embraces that of Nabhan (1985a) for native crops as *"Domesticated plants cultivated prehistorically or protohistorically within a region by its indigenous cultures since prehistoric or protohistoric times. Although they need not be endemic to the region of concern, their landraces should have been grown long enough in the region to exhibit morphological or physiological adaptations to the soils and climates found there."* It should be noted that the definition of ethnobotany emphasises the past and present usage of plants, but not future uses, which are considered to belong to the realm of economic botany.

Since ethnobotany is an identifiable stage within economic botany, its definition raises the question as to whether there should be additional terminology and definitions for the post-ethnobotanical use of plants for modern domestic and industrial purposes. My personal opinion is that we have managed very well without any, and because of the multidisciplinary nature of economic botany *sensu lato*, it would create more problems than it would solve.

3.   THE ROLE OF PLANTS AND PLANT PRODUCTS

Plants may be used for a variety of purposes, including food, fuel, fodder, medicines, timber, fibre, phytochemicals, shelter, shade, hedges, green manure, N-fixation, soil stabilisation, etc. as well as providing contributions to the plant breeder's gene pool. Without plants life on earth would not be possible. Plants provide directly, or indirectly through a carnivorous food chain, the necessaries for human and animal survival, including the protection of the environment against erosion and atmospheric imbalance. Indeed, it is doubtful if there is any plant that does not contribute in some way, no matter how small or how indirectly, to the pool of economic plant resources. Weeds, for example, may suppress crop growth, hinder cultivation, harbour pests and diseases, but they can also contribute to soil fertility, offer protection against erosion, possibly harbour desirable pollinators, contain useful genes for the plant breeder, or even provide some food and medicines. Even the more unpalatable and spiny of weeds could play an important role in stabilising overgrazed lands from which livestock should be permanently excluded. An example of the beneficial use of a weed is to be found among the Mexican homesteaders, who have allowed the composite *Melampodium divaricatum* to grow around tomatoes because it provides protection against the wind and maintains humidity (Alvarez-Buylla Roces *et al.*, 1989). The philosophy of the economic botanist must always be to seek the best in all plants, no matter how obnoxious or insignificant that plant might appear to be. Also, when a usage becomes redundant, an alternative role should be sought.

The main factor affecting the utilisation of plants is usually economic, but not necessarily in monetary terms; barter and availability are also involved. Availability in the sense of season, abundance and harvestability, can be priced by both monetary and bartering societies and in monetary societies can also be considered in terms of time and its most economical employment to suit the greatest need. Factors affecting the successful introduction and marketing of new crops are further discussed in Chapter 7.

Plant products are required to be easily grown, inexpensive in both cost and labour, sufficiently abundant for the required needs, simple to prepare or manufacture, readily available (wars, political boycotts, etc. permitting) and superior in quality to other alternative products. To some extent there is also an element of luck as to which is chosen from two or more plants with equal attributes. Plants and their products that are later rejected are not necessarily a failure; technological advances, biochemical requirements, fashion, economic and political changes, etc. may result in either reinstatement or alternative uses being found. Certain usage may cease because their synthetic alternatives are more uniform and cheaper to produce, e.g. synthetic dyes and fibres, and many pharmaceutical products. Others have failed because of cheaper and more productive plant alternatives. *Hevea brasiliensis* (Pará rubber), for example, has now replaced *Funtumia elastica* (Lagos silk rubber), *Landolphia* spp. (landolphia rubber), *Manihot glaziovii* (Ceará rubber), etc. in the world markets. Yet *Parthenium argentatum* (guayule rubber), although twice as expensive as Pará rubber, continues to be grown because it is the only alternative to Pará rubber suitable for use as arresters on aircraft carriers, tank track pads, rebuilding military aircraft tyres, etc. It has a strategic value (Whitworth and Whitehead, 1991).

4.   REGIONAL ATTITUDES

The interest shown in economic botany varies from country to country and continent to continent, and is largely affected by socio-historical factors. European settlers in North America, Australia, South and East Africa soon found there were areas where their introduced traditional crops were unsuitable and new alternatives had to be found. In Europe many of the agricultural and forestry crops and their husbandry have been well established for over 2000 years, while other crops, such as the potato and tomato were introduced from the New World in the middle of the 16th century. Until recently there has been little incentive to introduce further new crops. The changing global economy and the food surpluses within EU countries has now stimulated an interest in seeking new alternative crops, especially oilseed crops.

In the UK economic botany has been mainly associated with tropical plantation crops. This unfortunate interpretation owes much of its origins to the 19th century when the Royal Botanic Gardens, Kew had the responsibility for advising the Colonial Office on suitable economic crops for growing in the colonies, albeit largely by trial and error. As a result of which economic botany came to be regarded as being synonymous with tropical agricultural botany. As a consequence, economic botany *per se* was not made a major academic subject in the UK, although a series of lectures might have been included in the agricultural botany curriculum. The quest for alternative crops for the EU has stimulated a revival of interest in economic botany at Kew and other institutions, and even an MSc in Ethnobotany at the University of Kent, Canterbury.

In the past, those countries with colonial interests had sought commercial export crops from their colonies, largely for marketing within their colonial empires. Such plantation crops paid for the opening-up and development of the individual colony, although local crops and ethnobotanical studies tended to be neglected. On acquiring independence, the former colonial countries have had to compete against each other for the world's markets. Understandably many have often preferred to ignore the expertise of their former masters. During the first few years of self-determination, especially among many of the former African colonies, far too much reliance was placed on unsuitable European technologies and life style including, for example, the use of synthetic pesticides and a preference for wheat instead of traditional cereals. The following global economic depression has now meant that these countries have had to make greater use of their own natural resources and rely less on expensive imports. They have, of necessity, developed an increased interest in the utilisation of their own ethnobotanical resources. The classical example of this trend has been the 'Green Revolution' in India, described by Leaf (1983).

The US plant collectors (economic botanists) were, and still are, actively engaged in seeking new crops from both home and abroad. From the arid south-west and Mexico the ethnobotanists have introduced *Cucurbita foetidissima* (buffalo gourd), *Parthenium argentatum* (guayule), *Phaseolus acutifolius* (tepary bean) and *Simmondsia chinensis* (jojoba). From China came *Glycine max* (soya), and from Australia *Atriplex nummularia* (old man saltbush), etc. Economic botany can be a major academic subject in the USA, although some courses appear to offer little distinction between economic botany and agricultural botany (R.E. Schultes, pers. comm. 1986). Neither is it unusual for a plant geneticist to regard himself/herself as an economic botanist rather than a plant geneticist.

Despite its rich flora of more than 20,000 species, Australia has singularly failed to produce any new food crops apart from *Macadamia* spp. (macadamia nut), and that was domesticated following an introduction to Hawaii! Currently efforts are being made to domesticate *Santalum acuminatum* (quandong) for its edible fruit (Rivett *et al.*, 1989). The reason for this failure appears to be that the culture gap between the settlers and the indigenous hunting and gathering Aborigines is so wide that the settlers failed to appreciate that there could be any edible plants used by the Aborigines worthy of domestication. However, recent ethnobotanical studies, such as those by Brand and Cherikoff (1985), have demonstrated that many of the indigenous food plants are nutritionally superior to European foods. Also, unlike the other continents, Australia's flora has evolved in the absence of major ruminant herbivores, consequently its rangelands are ill-adapted to grazing by introduced sheep and cattle. The search for new fodder plants, especially from tropical America (*Macroptilium* spp. and *Stylosanthes* spp.) and Africa (*Cenchrus ciliaris*, etc.) have become increasingly important. On the plus side, Australia's contribution to forestry (*Acacia* spp. and *Eucalyptus* spp,) has been enormous.

In South Africa there has been marked differences between European and traditional agriculture. The former was largely based on Mediterranean and subtropical crops, such as *Citrus* spp. (oranges, etc.), *Nicotiana tabacum* (tobacco), *Saccharum officinarum* (sugar cane), and *Vitis vinifera* (grapes). The traditional agriculture was based on indigenous crops, including *Citrullus lanatus* (watermelon), *Pennisetum glaucum* (pearl millet) and *Sorghum bicolor* (sorghum). The introduced *Arachis hypogea* (groundnuts), *Gossypium* spp.(cotton) and *Zea mays* (maize, known in southern and East Africa as mealies) are grown by both societies. Although economic botany may not be sufficiently well recognised as an academic subject, government economic botanists have long been actively engaged in investigating the local flora and its potential for development (Arnold *et al.*, 1985).

The pattern of European agriculture in East Africa resembles somewhat that of South Africa, plus such introduced crops as *Agave sisalana* (sisal) from Central America, *Camellia sinensis* (tea) from China, *Coffea arabica* (coffee) from Ethiopia, and *Tanacetum cinerariifolium* (pyrethrum) from the Balkan Peninsula. By far the most active period of economic botany research on East Africa was during the pre-World War I period when Tanzania, then known as German East Africa, and Germany, like the other colonial powers, was actively engaged in the search for growing suitable indigenous or introduced export crops. These included *Acacia* spp. (tannins), Ceará rubber from South America, Lagos rubber from West Africa and, according to Anonymous (1906), guayule rubber from Central America. There is now the stimulus of a general economic depression for the East African countries to make their own decisions on how to make better use of their local plant resource.

5.   LIMITATIONS

The limitations imposed on economic botany are largely socio-economic rather than biological. New crops and crop products, from both cultivated and wild species, take time to be accepted by the farmers or gatherers as well as by the consumers. Their husbandry and labour management must fit into the existing cropping systems. The crop must also be economically competitive and free from handling and marketing problems. The term 'crops' is used here to include both cultivated and wild species since, for example, virtually all Brazil nuts are harvested from wild trees.

Economic requirements are for cheap, mass production, i.e. monoculture. It is more profitable to have a few high yielding crops than produce a wider range, each requiring different handling, packaging, marketing techniques, etc. For similar reasons, the international seed firms have profited by reducing the number of cultivars available to the grower for any particular crop. Even more serious is the link between some international seed firms and those dealing in agrochemicals, where the former are geared for the products of the latter.

It is only fairly recently that attention has been drawn to the dangers of relying on a relatively small gene pool to meet, for example, the world's food requirements. Among the several examples cited by Plucknett *et al.* (1987) is the Irish potato famine of 1846. The Irish potato had been derived from a few clones introduced from England which, in turn, were descended from two samples brought from South America to Spain in 1570 and England some 20 years later. The limited gene pool and the susceptibility of the Irish potato to *Phytophthora infestans* (late blight fungus) resulted in total devastation. By contrast, the mixture of modest-yielding landraces of the potato and its 8000 years of domestication in its native Andes offers some protection against such a devastating disease.

For some species the main biological problem is due to poor matching of the seed source and environment. Foresters have long been aware of the importance of matching the provenance or origin of the seed within its natural distribution to the environment of its intended introduction. Nevertheless, where possible more attention should be given to improving indigenous crop species rather relying on introduced species and involving a limited gene pool. The C_4 sorghum, for example, is already physiologically better adapted to a tropical environment than C_3 wheat. Its main disadvantage is that it is under-researched and, unfortunately, the more basic initial research requirements appear less intellectually satisfying than using more advanced techniques, such as genetic engineering for wheat. Yet the input required to find and adapt a wheat cultivar could have just as profitably, or possibly more so, have been spent in improving the sorghum.

In the drier tropics the vagaries of the climate, selection and breeding should be for acceptable yields of rain-fed crops over a wide range of climatic conditions and not for high yields within a limited climatic range. There is even a case for a variable cultivar that will ensure some production even in the driest of seasons, rather than a uniform crop which could fail entirely. The western-trained plant breeder should take note of the use of such variable cultivars by local farmers in, for example, northern Sudan.

However, as Arnon (1992) has pointed out, preconceived ideas based on homoclimates can be very misleading. Israel, for example, has obtained excellent results with staple crops from a wide range of countries and continents. In another example, Arnon citing Staples (1981) provides the example of how a number of lines of tomatoes especially developed for cool short season areas were also found to provide some of the best heat tolerant lines. The reason for this apparent paradox being the basic physiological stress mechanisms involved in stress tolerance where the greater thermostability of certain enzymes may function at both high and low temperatures.

Another limitation, which has already been mentioned, is the possible synthesis of a natural product and the advantages of a cheaper and more uniform product. This has certainly been done for many pharmaceutical products, although it is interesting to note that there is now a small but quite strong trend back to natural products.

Finally, there is the limitation of information and information transfer, especially among the developing countries where there are both language barriers and a shortage of literature. Individual countries must cooperate with their neighbours, since it is impossible to obtain a true assessment of the potential value of any economic plant until all its uses are known throughout its distribution range and where similar uses of the plant, especially by widely separated societies, can be regarded as significant.

Chapter 2

Plant Collecting, Taxonomy and Nomenclature

The purpose of this chapter is to provide some basic information on those aspects of plant collecting, taxonomy and nomenclature that are relevant to the applied scientist. It is not intended to provide a comprehensive treatment of these subjects; although it is hoped that it will make the applied scientist better informed about what to them may appear to be a source of irritation to their immediate research projects.

1.   PLANT COLLECTING

Plants are collected for identification and preservation as a permanent record of what has been found, where and, if known, how used. Duplicates of plants used for research purposes, including biochemical research, germ plasm collections, etc. should also be preserved as *voucher specimens*, providing an essential record against misidentification and a means of correcting identification should any future taxonomic revisions result in a change in taxonomic status as well as a means of checking identities when faced with conflicting research results. In the past a great deal of expensive and potentially valuable research has been invalidated because later taxonomic research has rendered the original identification suspect and the absence of a voucher specimen has made any redetermination impossible. The existence and location of such voucher specimens should always be cited in the original publication; fortunately, most authors appear to be aware of the desirability of such citations.

There are three phases of plant collecting for economic botanists. The first is what might be termed a reconnaissance ethnobotanical survey of an area to determine what plants are there and how they are being used. Most collections are of this nature and their reports constitute the bulk of the ethnobotanical literature. Should more intensive studies be required to investigate possible utilisation, a substantial investment would be required. Even the cost of a toxicological investigation of a potential food additive to meet EU requirements could be a minimum of £50,000 per substance, with complete clearance costing a minimum of £250,000 (Anderson, 1985).

The second phase is the collection of bulk material from different localities for comparative analysis before proceeding to the third phase, the collecting of living material for propagation. For cultivated material, where the background information is already known, the earlier phases may be omitted. Thus, the forester collecting propagating material for provenance testing, or a plant breeder introducing new germ plasm, may have already have found from an examination of available herbarium material, etc. that the species or variant thereof is present in a certain area.

The collection of good and well prepared herbarium specimens is a prerequisite for proper identification. This usually means flowering and/or fruiting material; the vegetative state is rarely a reliable means of identification. As has already been mentioned, herbarium specimens act as a voucher specimen to support the identity of plant material that has been discussed, analysed or propagated. Photographs alone are generally not acceptable for identification purposes because they rarely show the detail required for identification. They can be, however, extremely valuable for illustrating habitat, habit, colour, etc.

An examination of herbarium specimens from the country concerned will provide some indication of the months when flowering and fruiting may be expected. Even when aware of such timings the vagaries of the climate, especially in arid and semi-arid regions, can upset even the most careful planning. Unfortunately most plant collectors have to keep to a timetable and can rarely change their itineraries because of a failure of the rains! If seed collecting was the objective, it may be possible to collect cuttings instead, using soil polymers to ensure their survival, a technique first demonstrated by Aronson *et al.* (1990).

When in doubt as to what plant organs are required for identification do consult a taxonomist. For example, *Dioscorea* spp. (yams), *Ficus* spp. (figs) and palms, require specialist advice on what and how to collect. Certain grasses too can only be distinguished if it is known whether they are annual or perennial, hence the necessity of collecting rooting specimens, while bulbous species, such as *Allium* spp. (wild onions) and *Crocus* spp. (crocus), may require the tunic of the bulb or corm for proper identification. The collector, however, is warned that there are national and international regulations for the protection of endangered species and their collection may require special dispensation in advance (see Chapter 3). There are also quarantine regulations which may prohibit the importation of certain species and/or their organs, e.g. potato tubers into the UK. Plants growing in soil are also prohibited by many countries because of the possible presence of eelworm.

The various plant collecting techniques are described in Lawrence (1951), Davis and Heywood (1963), Hawkes (1980), Wormersley (1981), Bridson and Forman (1992), amongst others. The method used will largely depend on the climatic conditions in the country concerned. What ever the method it is essential that all specimens are properly labelled, stating the collector's name and serial number of the specimen; jeweller's tags are ideal for this purpose. Plants pressed between sheets of ordinary newspaper will dry rapidly in the sun in arid and semi-arid regions (airmail editions are unsuitable as drying paper but ideal for packing dried specimens). In the humid tropics special drying paper and the use of corrugates and spirit lamps, or preservation in alcohol may be necessary. Where possible specimens should be dried as rapidly as possible in order to preserve colour and prevent fungal growth. Excessive drying so that the specimens become brittle must be avoided. Sufficient pressure should be applied to ensure that pressed specimens remain flat, but are not crushed during the drying.

The collector's field notes, which will eventually by incorporated with the specimens, should contain the following information: (1) Collector's name and serial number of the specimen. A straight serial number, avoiding complicated codes involving country, place of collection, etc. is preferred. Do not start the serial sequence afresh each year or for a change in location as this has been known to cause unexpected complications in the herbarium when the same species has been recollected elsewhere under the same collector's number; (2) If known the full botanical name or generic name of plant should be given; (3) Vernacular name and dialect, if known. The collector should take care that a true vernacular name is given and not the local word(s) for 'weed', 'grass', etc. or even 'I don't know'; (4) Locality, i.e. country, province, nearest town or village - a grid reference is always helpful for remote areas and is essential in some countries where several towns and villages bear the same name; (5) Altitude, preferably in metres at the collecting site; if known, the altitudinal range can be added as an observation under habitat; (6) Habitat, including aspect, type of terrain, soil and ecological community, e.g. *Acacia senegal* savanna woodland on sandy soils; (7) Botanical observations, including habit, whether annual, biennial or perennial, height, tree or shrub, branching of trunk, evergreen or deciduous, flower colour (colours can change on drying), number of floral parts, etc.; (8) Environmental observations, pollinators, etc. if known; (9) Uses, if used for food, how is it prepared, etc. If medicinal, how is it prepared, quantities used, nature of disease, etc. If grazed, by what animals, season, preference, availability? (10) Date of collection - day, month and year; (11) A list of additional supporting samples collected relating to the herbarium specimen, e.g. bark and/or wood sample, spirit material, fruit, seed, live material, artefacts and photographs; and (12) Location of duplicate herbarium specimens using the standard abbreviations for herbaria as given in the *Index Herbariorum*, part 1 (Holgren *et al.*, 1981). There is a tendency, of which I too have been guilty, of writing the minimum of information on the herbarium labels and presenting the full information in a publication. Unfortunately, it is almost impossible for most herbaria to later insert the literature reference on the herbarium label once the specimen has been laid away in the herbarium.

Although not always obligatory, it is always a good policy to donate one set of a plant collection to either the local or national herbarium of the country in which the collection was made. This is especially helpful where developing countries are trying to build up their own collections and expertise. In addition it is often advisable to donate a further set to one of the major herbaria, especially if the collection is from a remote and little known area in which that particular herbarium is interested.

1.1   Literature Studies

Time spent on pre-field studies in a herbarium and library is never wasted and can play an essential role when planning expeditions. Floras, check-lists, ecological reports, etc. will give some indication of what plants are likely to be present in a particular area. The plants can then be checked for recognition purposes against herbarium specimens. See Frodin (1984) for a guide to the world's major floras. The reports of early botanical explorers from between *ca.* 1850 and 1940 should not be neglected because they were often able to provide greater detail about plants, their environment and uses than is now acceptable by modern publishers. Their slower pace of travel was also an asset when making observations and collecting information. Unfortunately basic information on plant uses are far too often repeated in other publications, often without any acknowledgement of the source, the repetitions often giving credence to wrong or misleading information.

Two categories of publications are recognised by Bhat (1997): (1) *Primary publications* providing reports in books (e.g. floras) and journals on current work or reviewing and analysing recent advances in knowledge. Many major journals now provide such information in electronic form; and (2) *Secondary publications* present the summarised, 'digested', or otherwise processed information obtained from primary sources. A much higher proportion of secondary than primary publications are available in electronic form. Where possible the primary references, if known, should always be consulted since a great deal of relevant information is often omitted or even inaccurately summarised in the secondary sources.

1.1.1   Databases

Bhat (1997) also recognises two categories of databases (1) *Secondary Electronic Databases*, e.g. bibliographic databases which provide summaries of information contained in primary publications, to which the user is directed for further information; and (2) *Tertiary Electronic Databases*, e.g. electronic compendia, encyclopaedias and other reference works, and provide detailed information on a range of specific topics; they may or may not indicate literature sources.

Although computerised tertiary databases provide a ready access to a wide range of information, such information is invariably selective and condensed. Nevertheless, tertiary databases can be an extremely valuable tool, provided a meaningful system of descriptors are used and are backed by references to the information source. Unqualified and crude descriptors such as 'used for medicine', are rarely helpful, apart from identifying the species concerned, and can even be misleading when used out of context.

Examples from among the several databases operated by the Royal Botanic Gardens, Kew include the Economic Botany Bibliographic Database and the Survey of Economic Plants for Arid and Sem-Arid Lands (SEPASAL) database. The latter embraces known economic uses of plants together data on their taxonomy, distribution, environmental parameters, ecology, phytochemistry, ecophysiology, cultivation, literature references, etc., thereby combining information gleaned from the herbarium specimens and literature, for use by interested parties. The standardised descriptors for the uses of plants, in their cultural context, are listed by Cook (1995). When the database is used in conjunction with Kew's Plant Anatomy Database provided by the Jodrell Laboratory, it is even possible to suggest alternative species for replacing endangered trees.

It is also planned to integrate data from Brazil's Plantas do Nordeste programme with the SEPASAL database for the joint Kew/Brazil contribution to the sustainable development of the local resources in the semi-arid north-east Brazil. Other Kew collaborative examples include a joint project with the Poisons Unit, Guy's & St. Thomas' Hospital Trust to provide a CD-ROM entitled *Poisonous Plants in Britain and Ireland* and available from HMSO, for the identification of both native and cultivated poisonous plants.

2.   THE USE OF TAXONOMY

Taxonomy is the study of the principles and practices of classification, in particular the study and description of variations in living organisms and the subsequent identification, description and classification, for the compilation of local and regional floras. Taxonomy is often used more loosely to include *systematics*, which is the scientific study and description of the variation in living organisms and the relationships that exists between them (Tootill, 1984).

For such applied plant sciences as economic botany, plant breeding, ecology, agriculture, etc. some knowledge of plant taxonomy is essential. In fact plant taxonomy is the basic tool for all disciplines concerned with living organisms. It is the function of taxonomists to identify plants accurately and to provide the correct name that identifies a taxon from all other taxa. Taxonomy also expresses affinities and relationships at all hierarchical levels, an essential knowledge for the plant breeder and disciplines concerned with phytochemicals, etc.

The practical application of taxonomic relationships and the subsequent reassessment of taxonomic affinities is well illustrated by the discovery in what was formerly considered as the monotypic leguminous tree *Castanospermum australe* (Moreton Bay chestnut) from the coastal forests and beaches of eastern Australia of the indolizidine alkaloid *castanospermine*, which has a possible application in the treatment of AIDS.

The screening of what was presumed to be the closely related species *Alexa leopetala* from South America provided a further source of castanospermine. It has also been detected by gas chromatography in seven other species (Anonymous, 1989). Although it has been suggested that the genus *Alexa* should now be included in *Castanospermum*, the necessary combinations have not been made. Similarly, a study of the volatile oils from the leaves of *Cupressus dupreziana* and *C. sempervirens* by Pauly *et al.* (1983) showed them to be almost identical and led to the relationship of former Saharan montane endemic being regarded by some as var. *dupreziana* of the Mediterranean *C. sempervirens*, which in turn had to be known as var. *sempervirens*.

The taxonomist usually bases relationships on observable macroscopic and microscopic characters, i.e. phenetic relationships. Unqualified, the words affinity and relationship suggest some form of genetic or evolutionary relationship. To the taxonomist the terms have a number of meanings: (1) *Phenetic relationship*, the overall visual similarity between two or more taxa; (2) *Phylogenetic relationship* or *evolutionary kinship* which, because it is more difficult to recognise, is less commonly used; (3) *Genetic relationship* expressed in terms of genome similarity; (4) *Cytogenetic relationship* expressed as the ability to cross; and (5) *Chemical relationship* expressed in terms of comparative biochemical data. Other categories may also be recognised. Unfortunately for the applied scientist there is often far too little supporting data accompanying phenetic relationships at species level, or where it is available, still needs to be incorporated into the taxonomic system. However, recent advances in molecular and biochemical data, etc. has led to a much better understanding of angiosperm affinities at family level (Davis and Heywood, 1963; Nandi *et al.*, 1998).

2.1   The Role of the Herbarium

The main function of a herbarium is to store preserved plant material indefinitely, preferably in a systematic manner, for future identification and research. For ease of retrieval, specimens should be laid away on the herbarium shelves in a systematic manner based on families, genus and species according to recognised monographs. In small herbaria the families may sometimes be in alphabetical sequence, although such a simple system has the major disadvantage of separating related families, a disadvantage that increases as the herbarium expands its holdings. Specialised herbaria dealing with a single family, e.g. Gramineae, will be similarly disadvantaged unless the subsequent alphabetical arrangement is based on subfamilies and tribes.

Most herbaria do follow a recognised classification system which, once introduced, is extremely difficult to change. Consequently the choice of classification is important, bearing in mind the obvious difficulties in presenting a 3-dimensional evolutionary tree as a linear system. The national herbaria of the British Commonwealth usually follow the *Genera plantarum* of Bentham and Hooker (1862-83), others may follow the Englerian system as set forth in the *Genera*

Siphonogamarum (Dalla Torre and Harms, 1900-1907) or its revision, *Syllabus der Pflanzenfamilien* (Engler and Diels, 1936), while in the United States the classification of Hutchinson (1926, 1934), *The Families of Flowering Plants*, has gained considerable popularity. Yet other systems are noted in Lawrence (1951) and Brummitt (1992), while Mabberley (1997) follows the more recent system of Kubitzki (1990-).

Herbaria also perform a second function, that of taxonomic research and its application through floras and other means. However, many herbaria do themselves a grave disservice by not adequately advertising how their collections and research can be beneficially used by other disciplines, including economic botany. The Royal Botanic gardens, Kew, for example is currently collaborating with agencies in China to develop an Authentication Centre for International Use of Traditional Chinese Herbal Medicine to provide an authentication service of Chinese herbal materia medica in order to promote the use of accurate and high quality herbs.

It cannot be emphasised too strongly that herbaria are a valuable essential resource for plant use and development which should be more widely used by all disciplines. Indeed, it has always been a surprise to me how little plant breeders make use of Kew's herbarium resources.

2.2   Plant Identification

Although the smaller herbaria are usually quite capable of rapidly naming a collection, many rarely have the extensive bibliographic resources necessary for keeping up-to-date with the current taxonomic literature, particularly nomenclatural changes. Cross-checking identification with a major herbarium is often advisable, especially where a specialised collection are involved, e.g. grasses and legumes. Although confusing, provided a plant has been correctly identified, it is not too much of a disgrace if an out-of-date name is applied instead of the currently accepted name. While the updating of synonyms from the literature is usually not too difficult, it can be very confusing should both the accepted name and its synonym occur in different papers within an edited volume or journal. The misapplication of a name, i.e. a wrong identification, however, can be disastrous. It is again stressed that the provision of a voucher specimen is essential if any error is to be corrected later.

The identification of an unknown plant involves determining whether it is within the currently accepted range of variation of an already known species or whether it is new to science; this can be a very time-consuming task. Recent systematic treatments in monographs and regional or local floras should ensure, in most cases, correct determination and naming. For new species it is customary herbarium practice to determine the correct name according to the *International Code of Botanical Nomenclature* (see Section 3.1 below).

The first procedure is always to determine the family, then through the family to genus and species. Attempts by an amateur to complete the final determination by matching with already named herbarium specimens without reference to flora keys is not recommended, especially in genera where a large number of species are involved. Even for the larger herbaria complete representation of the world's flora has yet to be achieved. Furthermore, maintaining a collection that is fully up-to-date with the current literature, nomenclatural changes, etc. is an impossible task. Consequently, unless it is known from the determinavit slips (identification labels attached to herbarium sheets) that the herbarium collection has been recently revised, care must be taken in matching specimens - there is always the possibility that wrongly identified specimens may have been included. When in doubt always consult the taxonomist concerned.

2.3 The Use of Collections

The major herbaria are mainly national institutions, usually containing an adequate representation of the world's flora, and often with regional specialisation. Other herbaria may be attached to museums, universities or research institutions, where the current research programmes may limit the interests of the herbarium but may, nevertheless, contain important local or specialised collections. A register of the world's public herbaria, their holdings and specialisation is maintained in the *Index Herbariorum*, part 1, compiled by Holgren *et al.* (1981) and updated as necessary.

Apart from naming, what further information may be obtained from a herbarium collection that is likely to be of interest to the economic botanist in his work, especially when planning collecting expeditions? If properly labelled by the collector, herbarium sheets can provide the following useful information: (1) The range of variation within a species. This may be expressed in a number of ways, e.g. height, size of fruit, number of seeds, etc. It is possible that the variations may form geographical patterns that can be correlated with climatic or ecological factors (White, 1962). If the species has been recently revised it may well be that such geographical patterns have already been given taxonomic rank at subspecies or variety level, e.g. the interspecific variation in the potential oilseed plant *Vernonia galamensis* by Gilbert (1986). In other cases there may be minor, sporadic or even isolated changes involving one character, such as flower colour or leaf shape, taxonomic characters that might otherwise not deserve taxonomic recognition above the rank of forma but perhaps of interest to horticulturists. Other taxonomic characters, such as the ability to retain seeds in plants that normally shed their seeds irregularly, could be of particular value to a plant breeder interested in the ease of harvesting from a seed crop; (2) The range of variation between closely related species or other possible breeding material. The plant breeder, in his search for characters to improve existing cultivated species, may well save much unnecessary fieldwork by first establishing from herbarium specimens what variation is likely to be found and where; (3) The distribution of a species. The major herbaria will usually provide adequate evidence of a species, country by country. Some, with specialised collections of a geographical region, may provide even greater detail of the distribution within a country. Even a major herbarium, such as the Royal Botanic Gardens, Kew, with large collections from East Africa, has to supplement the information required for compiling the *Flora of Tropical East Africa* by examining additional material from the collections held at the Natural History Museum in London and the East African Herbarium in Nairobi, Kenya, etc.

Unfortunately there is often an understandable tendency for collectors to under-collect or ignore well-known conspicuous or dominant species with which they are familiar, so that the full extent of the distribution and variation within such species are not apparent from the herbarium collections. The extremely poor holdings of the widely distributed, conspicuous and well-known baobab, *Adansonia digitata*, from tropical Africa is a good example of this problem (Lucas, 1971; Wickens, 1982). The mapping of plant distributions is discussed in some detail by Stearn (1951) and summarised by Davis and Heywood (1963). The *Index Holmensis* (Tralau, 1969-81) provides a useful but as yet incomplete index to plant distribution maps. Such maps can provide a wealth of information for the collector, especially when correlated with other information, such as phenological data and morphological information. The use of distribution overlays and climatic maps is often the simplest method of obtaining information on the climatic range of a species, information that is often vital when matching the climate for the provenance of a seed source with the area where it is to be introduced, a practice more widely used by foresters than by agriculturists and may well account for why many potentially useful introductions fail; (4) Phenological information, i.e. the seasonal or periodic behaviour of plants. An examination of herbarium sheets and collecting dates can provide useful information regarding times of flowering, fruiting, new leaf and leaf fall. The phenology of *Faidherbia albida*, syn. *Acacia albida* in East Africa, for example, did not make sense until it was demonstrated by climatic overlays that a bimodal climate was involved (Wickens, 1969). While it is usually obvious whether a tree is deciduous or evergreen, it requires a more careful study to decide whether a species is only briefly deciduous. Regrettably collectors rarely note such information and even more regrettable the information is often omitted from the published floras. Such information is of particular importance when dealing with browse species where there is a need to know when the browse is available. Again, collectors rarely note the seasonal availability of browse. Plant conservation interests now often make it illegal to uproot many herbs and grasses and collectors should be aware of the penalties for so doing. Fortunately most existing collections will be able to provide at least some rooting specimens, from which it should be possible to decide whether a plant is annual or perennial, while the length of the growing season may suggest whether an annual is in fact an ephemeral; (5) Habitat, good herbarium sheet labels should provide information on plant communities, topography, soils and altitude. Such information can be quite meaningful once it has been converted to a standard format; (6) Climate is rarely mentioned yet it is essential to those concerned with plant introduction and cultivation. As already stated, this information can be painstakingly obtained by comparing an overlay of the plant's distribution with a good climatic map, such as that of Walter and Lieth (1960-1967). The main problem, especially where accuracy is concerned, is to obtain a grid base map for the plant distribution at the same scale and projection as the climatic map. The use of computer graphics promises to be of considerable assistance where large quantities of data have to be processed; (7) Unfortunately, a great deal of the information on the use of plants appears to be anecdotal, relying heavily on verbal and literature sources, supported by relatively few *voucher specimens* in the herbaria authenticating the identity of the plants. In practice collectors often use the herbaria (and floras) for identification purposes, unwittingly withholding any information on usage for future publications, unaware of the value of including such information with the herbarium specimen. Investigations in a number of the world's herbaria, e.g. Harvard (Altschul, 1973, 1977), New York (Reis and Lipp, 1982), Pretoria (Morris and Manders, 1981) and Florence (Innamorati, 1973) have shown that less than 2% of the herbarium sheets contain any information, even for plants with well-known uses. While it is possible without a computerised cataloguing system to trace a voucher specimen cited in the literature, it is rarely practical to annotate specimens that have already been laid away in the herbarium with any subsequent literature reference. The value of the herbarium specimen as a voucher specimen is consequently lost. Even when available, much of the information is vague. Remarks such as 'grazed by stock' require further elaboration since the beast in question has not been identified; the presence of possibly less palatable species, seasonal availability, etc. have also to be considered when evaluating the grazing potential of the plant. Other vague remarks include 'used as medicine' (how and for treating what?), 'used for stomach complaints' (no clinical details given), 'eaten as food' (part used, preparation, etc.) and even 'widely used'!

3.   BASIC PLANT NOMENCLATURE

For the purpose of all communications it is essential that communicants should have a common language that can be understood without ambiguity. In the biological sciences this is extremely important, especially when, for example, communicating internationally on dangerous and potentially dangerous substances. Medicinal and toxic plants fall into this category and are probably the concern of many readers of this book. Scientific names are unambiguous, vernacular names are not.

The remarkable confusion that exists in the use of local vernacular names is not confined to the developing countries. Thus, according to Grigson (1958), in the British Isles there are 12 English vernacular names for *Atropa belladonna* (deadly nightshade) and 24 for *Solanum dulcamara* (woody nightshade), while *Taraxacum officinale* (common dandelion) has 52! Even more disconcerting, 'bachelor's button' and 'bird's eye' each refer to no less than 16 different species, of which three species,

Geranium robertianum, Silene dioica, syn. *Melandrium rubrum* and *Stellaria holostea,* are known by both vernacular names. The problem has been aptly summarised by the antiquarian Singer (1927) who wrote: *"Scholars who have sought to identify plant-names have far too often treated the semantics of these words as though they were stable. As in modern England, so in Anglo-Saxon England and in classical antiquity, the same species of plant is and was often known by different names in different regions, and different plants in different regions by the same name. Moreover, in any region the significance of plant-names changes with time. How labile then must have been the significance of the plant-names in the absence of any accurate standard, over a series of medieval centuries that had no idea of scientific standards. "*

To add to the confusion other nations (the United Nations recognises 192 nations) have similar problems with their vernacular names. This is compounded by the fact that there are, at a conservative estimate and depending on the criteria used by linguists to distinguish between languages and dialects, some 5000 and possibly 10,000 languages being spoken in the world today. To add further botanical confusion, some 'primitive' societies have separate names based on different phenologically stages in the growth of a useful species, yet group other plants, such as 'sedges', without any taxonomic distinction. There are also considerable problems in transcribing an unwritten language into Roman script, depending on acuteness of hearing, knowledge of the language and its dialects, plus the variations in spelling attributable to the translator's mother tongue. The translation of a non-Roman script into Roman is equally beset with problems, as witnessed by the extreme variations in spelling of the English, French and Italian versions of vernacular names from the Horn of Africa. Although phonetic spelling overcomes these difficulties it is too specialised and impractical for use in, for example, a popular or even national flora. There is clearly a major communication problem in trying to recognise species, especially cosmopolitan species, by their vernacular names, especially when the language used and district is not indicated.

Even within the English language there is the possibility for misunderstanding in the use of such a common name as 'corn'. In North America 'corn' refers specifically to maize (known as 'mealies' in South Africa), while in the UK it is often used as a general term for all cereal crops, although in Ireland, Scotland and Wales its use may be restricted to oats, whilst in England it is usually applied to wheat, or, in the eastern counties, barley (Dalal-Clayton, 1981).

Timber trade names too can be equally confusing, for example, the name 'mahogany' without a geographical modifier, can refer to African species of *Khaya* as well as the botanically unrelated *Shorea* from the Philippines. Even referring to African mahogany is not very helpful since the name is used for four species of *Khaya* (Chudnoff, 1980). The use of provenance names in addition to trade names can add to the confusion. For example, the fibre African bass, which is obtained from

Raphia hookeri, is also known under the provenance names of Sherbro, Sulima and Calabar bass.

The use of vernacular or trade names can also have unexpected consequences. For example, *Citrus reticulata* (clementine, mandarin, satsuma, tangerine) was first imported into the UK from Tangier as 'tangerine' (Mabberley, 1997) and until HM Customs were made aware that only one fruit was involved they claimed different import dues for mandarin and tangerine! A number of other examples of multiple common and vernacular names are to be found in the Subject Index.

3.1   International Code of Botanical Nomenclature

Communication within a country by using common names is clearly difficult; the confusion is even greater internationally. In order to avoid any ambiguity regarding identity, all plants have been classified according to the binomial system developed by Carl Linnaeus (1753) in his *Species Plantarum.* The plant is given a binomial name consisting of a generic name and specific epithet and described in Latin according to the *International Code of Botanical Nomenclature (ICBN)* (Greutner *et al.*, 1994); the Code being regularly updated at International Botanical Congresses. The principles of the Code are discussed by Davis and Heywood (1963), while a useful introduction for beginners has been provided by Jeffrey (1982). The Animal Kingdom, or Animalia, is similarly classified according to a somewhat similar Code, although recently there have been a move towards unifying the nomenclature all biological organisms under a common code, the BioCode (Greuter *et al.*, 1996), but until or if that occurs, the existing ICBN remains in force.

However, while taxonomists must publish names according to the Code, there is not always universal agreement about what constitutes a *taxon* (a taxonomic group at any rank within the hierarchy). There are the 'lumpers' who usually like to unite 'related' species at subspecies level, and the 'splitters' who appear to consider what the 'moderates' would consider at variety level as being species. Then there are those names which, according to taxonomic research, should no longer be used and are known as *synonyms*; however, they may still be used by those who are unaware for some reason or another that they are no longer an accepted name. Researchers into the plant literature must be constantly aware that a great deal of additional information about a particular species is often to be found by also searching under its known synonyms. Tracing accurate information on species that have been split is much more difficult. Name changes may irritate the uninitiated, but when following sound taxonomic investigation they provide a clearer appreciation of biosystematic affinities. Table 2 well illustrates the various generic and family placements assigned to what is now known as *Simmondsia chinensis*, the sole member of the monotypic family Simmondsiaceae.

3.2   The Classification of Plants

Systematics is the scientific study and description of the variation in living organisms and the relationships between them, with the emphasis on their evolution. It includes a number of specialised fields, such as chemosystematics, which are valid in their own right. *Taxonomy* is the study of the principles, practices, procedures and rules of classification and as such can be included in the wider term of systematics although sometimes the terms are loosely regarded as synonymous.

Cladistics is a discipline within systematics which aims at reconstructing the genealogical descent of organisms by means of objective and repeatable analyses, and from this pattern to propose a falsifiable hypothesis of natural classification or phylogeny. In the past taxonomic relationships have relied on morphological considerations, especially floral characters, that can be readily observed by the eye or light microscope. More recently support for or against taxonomic decisions based on purely visual characters has been obtained from biochemical, anatomical and cytological observations and it is the lack of such support that had created some anomalies which traditional taxonomy was unable to resolve. During the past decade DNA sequence analyses have revealed many surprising and unexpected relationships that conflicted with traditional morphological groupings as well as confirming others.

TABLE 2. The nomenclatural history of *Simmondsia chinensis*

Taxon	Family	Reference
Buxus chinensis Link	Buxaceae	Link, 1822, p. 386
Simmondsia californica Nutt.	Garryaceae	Nuttall, 1844, p. 400
Brocchia dichotoma Mauri ex Ten.	Euphorbiaceae	Tenore, 1845, p.80
Simmondsia californica Nutt.	Euphorbiaceae	Lindley, 1846, p. 281
Buxus californica Hort. ex Baill. *nomen*	Buxaceae	Baillon, 1859, p. 66
Simmondsia pabulosa Kellogg	not given	Kellogg, 1859, p. 21
Simmondsia california Nutt. syn. *Brocchia dichotoma* Mauri ex Ten. *Simmondsia pabulosa* Kellogg	Buxaceae Tribe Simmond-sieae	Müller [Argaviensis], 1864, p. 9, 22
Simmondsia	Simmondsiacées *nom. inval.*	Tieghem, 1898, p.103
Simmondsia chrysophylla Hort. ex Gentil	*nomen*	Gentil, 1907, p. 177
Simmondsia chinensis (Link) C.K.Schneid. syn. *Buxus chinensis* Link *Simmondsia californica* Nutt. *Brocchia dichotoma* Mauri ex Ten.	Simmondsiacées *nom. inval.*	Schneider, 1907, p. 141
Simmondsia chinensis (Link) C.K.Schneid.	Simmondsiaceae	Reveal and Hoogland, 1900, p. 206

Molecular taxonomy combined with classical taxonomy and including biochemical, anatomical and palynological evidence have provided new insights into taxonomic relationships. While much of the classical classifications have been confirmed, there are some surprising new results which, on closer inspection, now appear logical. This is the case with the Simmondsiaceae, whose earlier affinities with the Buxaceae and Euphorbiaceae have always appeared unsatisfactory. Although the Simmondsiaceae remains a monotypic family, its affinities are now seen to be with the Caryophyllidae, close to the Caryophyllales and widely separated from *Euphorbia*, whose affinities are with the Malpighiales, while the Buxaceae is now placed in the Eudicots, a small group of uncertain lineage (Nandi *et al.*, 1998; Chase, 1999).

In earlier classifications the Plant Kingdom was divided into the Spermatophyta (seed plants), Pteridophyta (ferns and fern allies), Bryophyta (mosses and liverworts), Mycophyta (fungi and lichens), Phycophyta (algae), Cyanophyta (blue-green algae), which are included by some systematists with the algae, and the Bacteriophyta (bacteria). Such classifications are intended to show in a two dimension format based largely on observable sexual and morphological characters what is essentially a three dimensional evolutionary tree from which many of its ramifications are now missing. Inevitably there differences of opinion about relationships at all levels of the hierarchy depending on the interpretation of the evolutionary criteria used. *"That which is a species to one taxonomist may be a subspecies to another, and that which is a family to one may be an order to another"* (Mason, 1950, cited by Davis and Heywood, 1963). Fortunately, a consensus of taxonomic opinion is generally able to decide what is acceptable and what is unacceptable.

The most recent classification (Table 3) involves all living organisms, recognising two Super Kingdoms, the Eukarya, which are basically nucleate organisms, and Prokarya, non-nucleate organisms. At the next level are the Kingdoms. Within the Eukarya these are the Animalia, Plantae, Fungi, and Protoctista (algae, etc.). The sole Kingdom representing the Prokarya is the Bacteria, containing amongst others, such important nitrogen-fixing genera as *Anabaena, Nostoc*, etc. formerly belonging to the Cyanophyta. It should be noted that viruses, which are only capable of replication by modifying the genetic machinery of the living host cells and have no life outside the host cells, are not included in this classification (Margulis and Schwartz, 1998; Bailey, 1999).

At the new level of Kingdom most organisms other than Animalia can be regarded as economic 'plants'. However, while recognising the Five Kingdoms classification and that the 'algae' as a group no longer has any taxonomic significance, it is sometimes convenient for the purposes of this book to refer to certain organisms as 'algae' and 'non-algae'. The Kingdoms are further subdivided into Subkingdoms, Phyla, Classes, etc. in what is basically a hierarchical classification, intended as far as possible to be a natural classification firmly based on a correlation of characters and discontinuity of variation.

TABLE 3. The Five Kingdoms classification of living organisms. The Protoctista are not fully represented here (Hawskworth *et al.*, 1995; Margulis and Schwartz, 1998; Bailey, 1999)

Super Kingdom	Kingdom	Sub Kingdom	Phyla	Class
Eukarya	Animalia			
	Plantae	Thallophyta	Anthophyta	Monocotyledonae
			(*Angiospermophyta*)	Dicotyledonae
			Gnetophyta	Gnetopsida
			Coniferophyta	
			Ginkgophyta	Ginkgoopsida
			Cycadophyta	Cycadopsida
			Filicinophyta	Leptosporangiatae
			(*Pteridophyta*)	Eusporangiatae
			Sphenophyta	
			(*Equisetophyta*)	
			Lycophyta	
			(*Lycopodophyta*)	
			Psilophyta (whisk ferns)	
			Anthocerophyta	Anthocerotae
			(hornworts)	
			Hepatophyta	(*Hepatiaceae*)
			Bryophyta	(*Musci*)
	Fungi		Ascomycota	
	see Hawksworth et al. (1995		Basidiomycota	
			Zygomycota	
	Protoctista		Acrasiomycota	
			(cellular slime moulds)	
			Chlorophyta (green algae)	
			Chrysomonada	Chrysophyceae
			(golden-brown algae)	
			Chytridiomycota	
			(fungal-like organisms)	
			Cryptomonada	
			(*Cryptophyta)*	
			Diatoms	
			Dinomastigota	
			(*Dinoflagellata*)	
			Euglenida (euglenas)	
			Eustigmatophyta (eustigs)	

Super Kingdom	Kingdom	Sub Kingdom	Phyla	Class
Eukarya	Protoctista		Gamophyta (conjugating green algae)	
			Haptomonada (*Haptophyta*)	
			Labrinthulomycota (slime nets)	
			Myxomycota (plasmodial slime moulds)	Myxomycetes Protosteliomycetes
			Oomycota	
			Phaeophyta (brown algae)	
			Plasmodiophora (*Plasmodiophoromycota)*	
			Rhodophyta (red algae)	
			Xanthophyta (yellow-green algae)	
Prokarya	Bacteria	Archaea (Archaebacteria)	Euryarchaeota (methanogenic, halophilic bacteria) Crenarchaeota (thermoacidophilic bacteria)	
		Eubacteria	Actinobactera (Gram +ve bacteria) Cyanobacteria (Gram -ve bacteria)	

It should be noted that all the available rankings are not always necessary within a particular classification since most groups do not lend themselves to rankings at all hierarchical levels. It is mainly at the classification level of family and below, i.e. genus, species, subspecies, variety and forma, that are of particular interest to the applied scientists.

Classification involves assembling plants into taxonomic groups based on their relationships, the latter by analysis and synthesis, i.e. whether it is more important to emphasise resemblance or differences between plant characters, with tradition often influencing variations in ranking from group to group. The species *Crotalaria juncea* (sunn hemp), an important fibre crop and green manure, is taken as an example of how the hierarchical levels of plant classification within the plant kingdom function, and is adapted from Davis and Heywood (1963), Polhill (1968, 1981) and Bailey (1999). It should be noted that not all of the available hierarchical levels are necessarily utilised.

Kingdom: *Plantae*
 Subkingdom: *Thallophyta*
  Phyla: *Anthophyta* (formerly *Angiospermophyta*)
   Class: *Dicotyledonae*
    Order: *Leguminales* (= *Fabales*)
     Suborder:
      Family: *Leguminosae* (= *Fabaceae*)
       Subfamily: *Papilionoideae* (= *Faboideae*)
        Tribe: *Crotalarieae*
         Subtribe:
          Genus: *Crotalaria*
           Subgenus:
            Section: *Calycinae*
             Subsection:
              Series:
               Subseries:
                Species: *Crotalaria juncea*
                 Subspecies:
                  Variety:
                   Subvariety:
                    Form:

The ***species*** is, for most purposes, widely accepted as the basic unit of classification, yet no other taxa are as controversial and difficult to define. The ***taxon***, plural ***taxa***, is any taxonomic group regarded as the smallest natural population permanently separated from each other by a distinct discontinuity in the series of biotypes, a ***biotype*** being a collection of individuals that are genotypically essentially the same (Davis and Heywood, 1963). The distinctiveness of the discontinuity can vary greatly from species to species. In some instances it is a historic concept of what constitutes a species within a genus, a concept that can vary from family to family, and from taxonomist to taxonomist!

Usually the taxonomist describing a species intends the specific epithet to be meaningful. Thus '*edulis*' and '*officinalis*' respectively indicate edible and medicinal plants. Nevertheless, the literal meaning of an epithet should be accepted with caution. For example, the species *Buxus chinensis*, now known as *Simmondsia chinensis* (jojoba) from the south-western USA and north-eastern Mexico was given the epithet '*chinensis*' because the botanist Johann Link (1821-1822), unfortunately muddled his collections from California and China, and mistakenly believed he was dealing with a plant collected from China. The generic name honours the contemporary English botanist and explorer Thomas William Simmonds. Later Thomas Nuttall (1844) described what he thought was a new species from California as *Simmondsia californica*. This was closely followed by a third species, *Brocchia dichotoma* by Tenore (1845), named in honour the Italian botanist Giovanni Battista

Brocchi. The three names represent a single species and, according to the ICBN, the earliest published specific epithet has priority over the others. The later names are therefore reduced to synonymy. Tieghem (1898) recognised *Simmondsia* as belonging to a separate family but his proposed family name of Simmondsiacées is invalid because Art. 18.4 of the *ICBN* requires a Latin termination (Table 2). There is an obvious moral from this muddled nomenclature in that collections should be clearly and fully labelled. Further confusion was also caused by differences in opinion regarding the generic and family relationships.

Hybrid species are generally indicated by the use of the multiplication sign, e.g. *Agrostis stolonifera* × *Polypogon monspeliensis*, or by the prefix *notho-* to the term denoting the rank of the taxon. While it is usually preferable to place the binomials or epithets of the hybrid formula in alphabetical order, as shown above, the direction of the cross can also be indicated by including the appropriate sexual symbols in the formula, or by placing the female parent first. If a non-alphabetical sequence is used, its basis should be clearly indicated.

Exceptions may be made where long usage has made any change of name undesirable. For example, although of hybrid origin and not found in the wild *Triticum aestivum* (wheat) is treated as a species even though its genomes have been shown to be composed of those from *Triticum monococcum, Aegilops speltoides* and *A. squarrosa*.

The *subspecies* are assumed to have a common origin and are nowadays largely regarded as forming populations within a species showing more or less distinct regional differentiation and morphology. The populations of different subspecies may be isolated from each other by area or altitude, ecologically, temporally or by pollination mechanisms, such as differences in exposure of anthers and stigma requiring different pollinating agents. It is axiomatic that there must be two or more subspecies recognised within a species, one of which will be based on the description of the parent species and will consequently bear the same epithet as the parent. The other subspecies, irrespective of whether new to science or as a result of combining other taxa with the parent species, will be named in accordance with the latest edition of the *ICBN*.

Although subspecific rank is intended to indicate that a relationship exists, it does not and cannot indicate evolutionary trends within the species since the Code gives precedence to the oldest validly published name as being the correct name for the species; exemptions from this rule will be found within the Code but will not be considered in this introduction to botanical nomenclature. For example, the cultivated olive, *Olea europaea* subsp. *europaea* has clearly been derived from the wild olive, *O. europaea* subsp. *cuspidata*, a subspecies based on *O. cuspidata*. The latter name was first published by George Don in 1837, while *Olea europaea* was published by Linnaeus (1753) and therefore takes precedence (Green and Wickens, 1989). The hypothetical and evolutionary correct '*O. cuspidata* subsp. *europaea*' would therefore, be considered an illegitimate combination! It should also be noted that the full author citation for *O. cuspidata* is Wall. ex G.Don, which indicates that the specific epithet *cuspidata* was first given by Wallich but without a published a description; *O. cuspidata* Wall. was, therefore, a *nomen nudum* with no taxonomic standing until validated by the description of George Don (1837). Note that the abbreviation of Wallich to Wall. is that accepted by Brummitt and Powell (1992) in their rationalisation of author names and their abbreviations.

The *variety*, like the subspecies, has been variously used in the past. Today it is widely used for local facies of a species that are morphologically distinct and occupying a limited geographical area; the emphasis is very much on the small scale, more localised distribution, compared with the large-scale regional basis of the subspecies. The term is also used where the precise nature of the variation is not yet properly understood (Davis and Heywood, 1963).

Botanical varieties must not be confused with commercial agricultural, horticultural or silvicultural varieties produced by plant breeders. Such cultivated varieties or cultivars can be morphologically indistinguishable from botanical taxa but recognisable to the breeder by such non-taxonomic characters as length of growing season, disease resistance, etc. and are dealt with in Section 3.2. as they are named in accordance with the *International Code of Nomenclature for Cultivated Plants (ICNCP)*. However, nothing precludes the use of cultivated plant names published in accordance with the requirements of the ICBN.

The *form* is the lowest formal rank recognised by the ICBN and is generally regarded as being distinguished by a single or few linked characters, with a sporadic and not distinct distribution, often as single character variations that occur within a population, e.g. flower colour or leaf shape. Such characters are often of considerable horticultural interest. However, since single characters are often chosen for forma recognition, it is possible to have the absurd situation where two or more characters, e.g. flower colour and leaf shape, both occurring on the same plant and hence providing the possibility of referring the same plant to two or more forma (Davis and Heywood, 1963).

The *genus* is generally regarded as a natural assemblage of more or less related species and that genera, in turn, can be separated from other genera within the family. The overriding consideration is whether it can be naturally delimited. At times the generic limits are poorly defined, especially in large and natural families.

There are two kinds of *family* recognised, definable or natural families, such as the Cruciferae, whose constituent genera are clearly related, and indefinable families, such as the Rosaceae wherein there is a considerable diversity of floral and vegetative characters delimiting the genera. With few exceptions a family name is based on a generic name within that family, e.g. Ranunculaceae - genus *Ranunculus*. The exceptions, which are permitted under the *ICBN*, allows the use of alternative names based on the earliest generic names. The accepted alternatives sanctioned by long usage are: Compositae or Asteraceae, Cruciferae or Brassicaceae, Gramineae or

Poaceae, Guttiferae or Clusiaceae, Labiatae or Lamiaceae, Leguminosae or Fabaceae, Palmae or Arecaceae, and Umbelliferae or Apiaceae.

The names of fungi and lichens are also governed by the Botanical Code. However, provision has been made within the fungi for the separate naming of perfect and imperfect states within the Ascomycota, formerly Ascomycetes, and Basidio-mycota, formerly Basidiomycetes. While the imperfect (asexual, conidial) state or *anamorph* is often well known, the perfect (sexual, ascal or basidial) state or *teleomorph* may be unknown or rarely seen and it is, therefore, convenient to provide separate names. While the name of the anamorph may be used with reference to the imperfect state alone, the teleomorph name must be used (except for lichen-forming fungi) when referring to both the perfect and imperfect states or **holomorph.** For lichens their names refer to their fungal component. Where applicable such names are also applicable to the lichen fungi in their non-lichenised condition and compete with ordinary fungal names according to the Botanical Code. The lichen algae must bear independent names (Jeffrey, 1989).

3.3 Taxonomic Treatment and Literature References

The taxonomic treatment of plants through their hierarchical classification is unfortunately very uneven. Some families maybe neglected because they are too obscure, unimportant or uninteresting, while others may be popular and overworked. Still others are so large or so difficult that they require a lifetime's devotion. Furthermore, taxonomist are human and vary in their ability, standards and training, as well as being influenced by the availability of comprehensive herbarium collections and botanical reference libraries. As McNeil (1996) points out, it must also be remembered, especially by taxonomists, that *"Biological nomenclature is not an end in itself. It is not even part of scientific endeavour; it is a regulatory system that seeks to serve the needs of science".*

For a plant's name to be complete it should be followed by the name of the author who first published the name. The author's name is used by taxonomists when tracing the original description and date of publication, part of the research process for ensuring that the accepted name is used in a revision or flora. Authorship is not just honorific, it also serves an essential purpose in distinguishing between different taxa that have been given the same name when described by different authors. For example, *Suaeda fruticosa* Forssk. ex J.F.Gmel. must not be confused with *S. fruticosa* (L.) Dumort, the latter being a synonym of *S. vera* Forssk. ex J.F. Gmel. Without the author citation such names used in the literature would be of uncertain application. For ease of reading, the authors are omitted from this present text, but are included in the taxonomic index, which is a perfectly satisfactory system for non-taxonomic publications.

All botanical names, or at least all those that have been traced, with their authors and relevant publication, are published by the Royal Botanic Gardens, Kew in the

Index Kewensis (*I.K.*) compiled by Jackson (1893-95) and subsequent quinquennial supplements. These are now available from Oxford University Press on CD-Rom. The *I.K.* is the internationally accepted register for all published names, both acceptable and unacceptable. Until the 1981 Supplement only names down to the rank of species were published, since then subspecies, varieties, etc. have also been included, making the tracing of their origins a simple task. Because some of the early taxonomic concepts were somewhat arbitrary the original Index and the early Supplements should be used with care and studied in conjunction with the present Code.

A useful dictionary of the higher plants (angiosperms, gymnosperms and ferns) has been compiled by Mabberley (1997) and contains the currently accepted generic and family names as well as the more commonly used vernacular names, together with notes on their uses, and is based on a more exhaustive publication for the taxonomist by Willis (1973). The generic names currently accepted by the Royal Botanic Gardens, Kew, and their synonyms are given by Brummitt (1992), while the authors of botanical names and their accepted abbreviations are to be found in Brummitt and Powell (1992); both works are followed in this publication.

3.4   Name Changes

Many non-taxonomists are both confused and irritated by the constant changes in botanical names. As already mentioned, all botanical names have to meet the strict requirements of the *ICBN*, a code that has been specifically designed so that professional and amateur taxonomists from all nations, irrespective of language, will be able to understand the procedures for the scientific naming of any plant in the world. Anomalies in the Code are rectified by regular revisions, the latest revision being edited by Greuter *et al.* (1994).

Name changes are the result of painstaking taxonomic research. With some 265,000 flowering plants and approximately 950,000 names listed in the *Index Kewensis*, it is inevitable that some errors will have been made. Some species may, due to inaccurate research, have been placed in the wrong genus, or worse still, in the wrong family! In some cases later research may have shown a taxon to be separable into two taxa. In other cases the same taxon has unwittingly been described more than once, either through ignorance based on poor communications, or because of insufficient material. In the latter case the full range of variation within a taxon had been imperfectly understood and different names had been given to what has eventually proved to be a continuum.

Other names may have been misapplied to the wrong taxon. Thus the East African leguminous shrub *Colutea abyssinica* had been wrong attributed by Baker (1929) to *Colutea istria* from the Middle East. Consequently the synonymy for *C. abyssinica* will now show the misidentification as *C. istria sensu* Baker f. In other examples two taxa have been combined because they were considered to be identical, such as the case with the Indian species *Balanites roxburghii*, which was considered synonymous with the African *B. aegyptiaca*. More critical examination involving a wider range of characters has now shown the two species to be distinct (M. Sands, pers. comm. 1989).

Offences against the Botanical Code are also reasons for rejecting a name. However, it is not the purpose of this chapter to enter further into the highly specialised subject of botanical nomenclature. It is sufficient for the non-taxonomic reader to understand that there are always sound reasons for name changes and that they must be accepted with good grace.

4.   NOMENCLATURE OF CULTIVATED PLANTS

According to Jeffrey (1968) cultivated plants differ from wild species by the following criteria: (1) They rarely have a population structure; (2) They neither occur in natural areas nor natural habitats; (3) They grow in simple, often monospecific communities; (4) Their growth is often more or less modified by man; (5) Their intensity and range of variation tends to be greater than for the parent wild species; (6) They are not products of natural selection so that many varieties of cultivated plants would be eliminated in the wild; (7) Varieties are often undergoing rapid genetic change; (8) Cultivation may not always bring about a breakdown in *allopatry*, i.e. taxa having different areas of distribution, but can cause genetic isolating mechanisms that exist in nature to be either broken down or circumvented; (9) Hybridisation and introgression are further sources of variation; (10) As a result of artificial propagation, which is often apomitic, those varieties which, because of their inability to breed would have no systematic importance in wild plants, may come to contribute largely to the numbers and variations of a cultivated crop; and finally (11) Similar selection pressures operating on genetically similar but distinct lines, may evoke similar responses in those lines giving rise to parallel varieties, the homologous series of Vavilov, a phenomenon by no means confined to cultivated plants, but often exhibited by them to a marked degree. Such differences between cultivars and their parent wild species have long been recognised by the use of fancy names for many plant domesticates, especially food plants and ornamentals.

Because cultivated plants are essential to cultivation it is important that a precise, stable and internationally accepted system should be available for their naming, and to provide uniformity, accuracy, and fixity in the naming of agricultural, horticultural and silvicultural cultivars, which are normally given fancy names, such as the apple 'Cox's Orange Pippin', barley 'Proctor', and *Juglans regia* 'King'. The common names of genera and species such as beech for *Fagus*, potato for *Solanum tuberosum*, hollyhock for *Althaea rosea*, Jerusalem artichoke for *Helianthus tuberosus*, and rye for *Secale* are not recognised by the *International Code for Cultivated Plants* (Brickell, 1980).

4.1   The International Code of Nomenclature for Cultivated Plants

In order to rationalise the use of cultivar names the *International Code of Nomenclature for Cultivated Plants* (*ICNCP*) has been developed in parallel with the *International Code for Botanical Nomenclature* (*ICBN*), both dealing with the scientific names of plants. Essentially the *ICBN* deals with the formation and use of latinised plant names, whereas the *ICNCP*, with the exception of graft-chimaeras, deals with the epithet, the final, man-made part, thereby providing a precise, stable and internationally acceptable system. The *ICNCP* follows the *ICBN* at least to the level of genus and even to specific level and below provided they are identifiable with botanical taxa in those ranks. In addition, cultivated plants may be assigned to one or more cultivated plant taxa. The *ICNCP* is also regularly updated, the latest being edited by Trehane *et al.* (1995)

A *cultivar* is defined as an assemblage of cultivated plants which are clearly distinguished by any character, e.g. morphological, physiological, cytological, chemical, or others, and which, when reproduced either severally or individually, retains its distinguishing characters. It is the basic taxon of the cultivated plant. The term cultivar does not equate with the botanical rank of variety or forma, neither can the words 'variety' and 'forma', or their equivalent in other languages, be used as a synonym for cultivar as defined by the *ICNCP*. However, in certain national and international legislation, or other legal construction, the word 'variety is a statutory or otherwise legal term used to denote a proven variant and thus equates to the *ICNCP* definition of cultivar. The abbreviations cv. and var. are not acceptable.

4.2   The Naming of Cultivated Plants

The *ICNCP* deals with the definition, formation and use of three taxonomic groups of cultivated plants: the cultivar, cultivar group, and the graft-chimaera, and includes both sexually reproduced and clonal material. A *clone* is a genetically uniform assemblage of individuals (which may be chimaeral in nature) obtained originally from a single individual by asexual propagation, i.e. by cuttings, divisions, grafts, or obligate apomixis.

For cultivars a *cultivar name* can be given to the following categories: (1) *Clones* that are asexually propagated from any part of a plant, e.g. *Asparagus officinalis* 'Cadet', *Solanum tuberosum* 'Wilja'; (2) A *tipophysic clone*, which must be asexually derived from a particular part of a plant, e.g. *Abies amabilis* 'Spreading Star' cloned from lateral growth of the parent plant; (3) *Cyclophysic clones* derived from a particular phase of the plant's growth cycle, e.g. *Chamaecyparis lawsonsiana* 'Ellswoodii' obtained from juvenile cutting material, and *Hedera helix* 'Arborescens' obtained from adult material; (4) Clones derived from aberrant growth, e.g. *Picea abies* 'Pygmaea', a dwarf plant derived from sprigs of a witches' broom; (5) *Graft-chimaeras*, which are composed of tissues from two or more different plants in intimate association and originating from grafting; they are not sexual hybrids, e.g. +*Crataegomespilus* 'Dardarii' combine the tissues of *Crataegus monogyna* and *Mespilus germanica*; (6) Assemblages of individuals grown from seed derived from uncontrolled pollination may be given cultivar names if they can be consistently distinguished by one or more characters, even though such individuals may not necessarily be genetically uniform, e.g. a line resulting from repeated self-fertilisation or inbreeding, e.g. *Zea mays* 'Wisconson 153', or a multiple line made up of several related lines, e.g. *Trifolium repens* 'Star', and F_1 hybrids, the result of a repeatable single cross between two pure bred lines, e.g. *Brassica oleracea* 'King Arthur'; (7) An assembly of plants grown from seed which has been repeatedly collected from a particular provenance and are clearly distinguishable by one or more characters, i.e. a **topovarient**, e.g. *Eucalyptus camaldulensis* 'Lake Albacutya' from the environs of Lake Albacutya, Victoria; (8) An assemblage of genetically modified plants showing new characteristics following pollination by alien genetic material, e.g. *Phlox drummondii* 'Sternenzaubers' with a number of different colour forms, all of which are characterised by the same star-like shape of the corolla, although in practice such man-made cultivars are marketed under trade-mark names.

The **cultivar group** consists of an assemblage of two or more similar named cultivars within a genus, species, nothogenus (hybrid genus) nothospecies (hybrid species) or other denomination class, e.g. cultivars of *Festuca rubra* are represented by three cultivar groups, a Hexaploid, Non-creeping Group, a Hexaploid Creeping Group, and an Octoploid Creeping Group, each with a distinct set of attributes. Cultivars may also belong to one or more cultivar groups if they serve a practical purpose, e.g. *Solanum tuberosum* 'Desiree' (Maincrop Group) or *S. tuberosum* Desiree (Red-skinned Group).

For **graft-chimaeras** where the rootstock is a separate cultivar, the plants grafted onto the rootstock are named according to the scion. Thus, *Malus* × *domestica* 'James Grieve' grafted onto a *M.*× *domestica* 'M9' rootstock retains the 'James Grieve' epithet despite the dwarfing effect of the rootstock.

Hybrids are named in alphabetical sequence, or where the female parent is known, it is placed first. The male and female symbols may be added if desired (Brickell, 1980).

The registration for cultivar names is of the greatest importance for nomenclatural stability and International Registration Authorities are charged with the registration of those cultivars and cultivar groups that are part of that particular authority's remit, and for their publication. For example, the Royal Horticultural Society acts as the international registration authority for the cultivar names of *Lilium*, and within the UK the Plant Variety Rights Office is responsible for the national registration of field crops (Jeffrey, 1989; MAFF, 1998b).

A cultivar name must be freely available for use by any person to denote the cultivar whose name it is. A **trade designation** is used in place of the accepted cultivar epithet when the accepted epithet is considered unsuitable for marketing purposes. They may be cited together, or in juxtaposition to the trade designation, e.g. *Alstroemeria* 'Stakrist' KRISTINA. The names may only be used in conformity with trademark laws. A cultivar name cannot, in general, be regarded as a trademark (see Chapter 6 for further discussion on trademark names) (Brickell, 1980).

The generic names recognised are: (1) Botanical genera, e.g. *Lilium, Pinus, Triticum*, and common names used in the generic sense, e.g. lily, pine, wheat; (2) Botanical names of intergeneric hybrids, e.g. ×*Triticosecale, Sorbopyrus*; and (3) Botanical and common names of intergeneric graft-chimaeras, e.g. the graft hybrid between *Crataeguus* and *Mespilus germanica* (Bronvaux medlars) known as +*Crataegomespilus*.

Names recognised at species level include: (1) Botanical and common names of species, e.g. *Lilium candidum* (Madonna Lily); (2) Botanical and common names of interspecific hybrids, e.g. *Tilia* × *europaea*, syn. *T.* × *vulgaris* (common lime); (3) Botanical and common names of interspecific graft-chimaeras, e.g. *Syringa* + *correlata*; (4) Botanical and common names of particular interspecific combinations of an intergeneric hybrid, e.g. ×*Citroncirus webberi* (citrange) and ×*Cupressocyparis leylandii* (Leyland cypress).

The Leyland cypress has an interesting history. It resulted from the accidental crossing of *Chamaecyparis nootkatensis* (Nootka cypress) with pollen from a nearby *Cupressus macrocarpa* (Monterey cypress) on the Leighton Estate, Welshpool in 1888, and later named after C.J. Leyland, the British botanist and brother-in-law of the estate owner. The six seedlings were transferred to Haggerston Hall, Northumberland in 1892-3, five of which still exist, and are the parents of a number of cultivars by vegetative propagation of ×*Cupressocyparis leylandii*, clones 1 to 5. It is now the most planted (and hated) garden hedge and shelterbelt tree in the UK (Bean, 1970).

Several categories of cultivars are recognised according to their mode of reproduction: (1) One clone or several closely similar clones, e.g. *Rubus nitidoides* 'Merton Early' and the potato 'Bintje'; (2) One or more similar lines of normally self-fertilising individuals or inbred lines of normally cross-fertilising individuals, e.g. *Triticum aestivum* 'Marquis' and *Zea mays* 'Wisconsin 153A'; (3) Cross-fertilising individuals which may show genetic differences but having one or more characters by which it can be differentiated from other cultivars, e.g. *Lolium perenne* 'Scotia' and *Medicago sativa* 'Ranger', the breeder seed of the latter was obtained from crossing five seed-produced lines, each maintained under isolation; (4) An assemblage of individuals reconstituted on each occasion by crossing. These include: (i) single-crosses, e.g. *Sorghum* 'Texas 610'; (ii) double-crosses, e.g. maize 'US 13', a double-cross involving four inbred lines, three-way-crosses and top-crosses, and (iv) intervarietal (intercultivar) hybrids, e.g. 'H-611', an intervarietal maize hybrid of 'Kenya Flat White' and 'Ecuador 573'; (5) One clone or several similar clones which have a habit of growth that is clearly indistinguishable from the normal habit and is retained by appropriate methods of propagation, e.g. *Chamaecyparis pisifera*

'Squarrosa Intermedia' which is a juvenile form, *Sequoia sempervirens* 'Prostrata' which is prostrate form and *Picea abies* 'Pygmaea', a witches broom form; and (6) Somatic hybrids obtained by a parasexual process, such as protoplast fusion, and are expressed by using the multiplication sign within parenthesis, e.g. *Nicotiana glauca* (×) *Nicotiana langsdorfii*.

In a number of cultivated and mainly agricultural crops, intensive breeding has taken place, and with it a need to group into categories those cultivars with similar characteristics and relationships. Since the *ICNCP* was not designed to cover individual crops and where a hierarchy of categories has been applied, their use is not governed by the Code. These hierarchies of categories refer to landraces, cultigens, convarieties, etc. As such they are intermediate categories developed or invented by breeders to fit their particular purposes and are sometimes used in a hierarchical fashion between the ranks of species or species hybrid and the cultivar (C.D.Brickell, pers. comm. 1989).

Notable problems in the application of the *ICNCP* occur with complex polyploid cultivars, such as in the classification of the cultivars of *Musa* spp. (banana) and *Gossypium* spp. (cotton). This involves the abandonment of formal nomenclature according to the *ICBN* and adopting a genome nomenclature in which the clone is referred to the genus in accordance with the *ICNCP* together with its appropriate genome group. For example, *Musa* (AAA Group) 'Gros Michel' is a triploid cultivar of *M. acuminata* from Jamaica, while clones of the Cavendish Subgroup of the AAA Group include "Dwarf Cavendish" bananas from the Canary Islands. For most practical purposes the cultivar name *Musa* 'Gros Michel' can be used and is perfectly acceptable under the *ICNCP* (Simmonds and Shepherd, 1955; Simmonds, 1959, 1962; Purseglove, 1985; Stover and Simmonds, 1987).

A similar arrangement is made for cotton, *Gossypium* spp. For example, in pre-history the AA genomes of the Old World diploid species *G. arboreum* (tree cotton) and *G. herbaceum* crossed with New World DD genome species to produce by 1400 BC the tetraploid *G. hirsutum* (upland cotton), a short staple cotton cultivated in Mexico, and *G. vitifolium*, syn. *G. barbadense* (Sea Island cotton) with long, silky fibres, which was cultivated in Peru by 1600 BC (Hutchinson *et al.*, 1947; Saunders, 1961; Fryxell, 1965; Brücher, 1989).

Chapter 3

Environmental Considerations

The purpose of this chapter is to create an awareness of some of the many problems associated with plant introduction and management in relation to the environment. It is by no means an exhaustive treatment since case studies could well run to several volumes of text.

1.   THE ENVIRONMENT

The *environment* is defined as the physical and chemical surroundings of an organism, i.e. the conditions in which an organism lives. This is a function of the climate, topography, soils and nutrition and should not to be confused with the *habitat* which is the normal locality occupied by an organism or group of organisms (Tootill, 1984). In the wild individual plants have to compete for space, light, moisture, nutrients, etc., conditions that may bear little or no resemblance to those found in cultivation. Cultivation produces an unnatural environment, providing freedom from competition, pests and diseases, control of soil nutrients and, when irrigated, control of soil moisture. It is the role of the agriculturist, horticulturist and forester to manage the environment and, through selection, breeding and crop management, to manipulate the growth of the plants in order to produce the desired and sustainable effects without undue damage to the natural environment.

An *ecosystem* is defined as a community of interrelated living organisms, i.e. *biotic factors*, and their environment, i.e. *abiotic factors*, and may be large or small, simple or complex in structure. The two major abiotic factors that determine an ecosystem are climate and soil, while the nutrient cycling and energy flow activities of the two biotic factors, the flora and fauna, are regarded as secondary. The nutrient cycle is initiated by organic compounds formed by the plants (*the producers*) during photosynthesis which, either directly through herbivory or indirectly through carnivory, will provide food for the animals (*the consumers*). Ultimately plant and

43

animal remains are broken down by micro-organisms and small invertebrates (*the decomposers*), releasing mineralised nutrients to be taken up by the plants and thereby completing the cycle. A parallel energy flow also occurs where the radiant energy from the sunlight is used in the photosynthetic processes and converted into chemical energy, which is then utilised by both the consumers and decomposers.

The mapping of ecosystems poses a number of problems. Because the abiotic and biotic factors interact, no one factor alone should be used for constructing ecological subdivisions. The fauna are too mobile, migratory animals and birds often traversing several zoogeographical regions. The flora is static, and are believed by many to reflect prevailing ecological conditions. However, Walter and Breckle (1985) consider this to be a false assumption since plant formations are the result of plant migrations from past climatic changes, and plant life-forms can be ecological or even taxonomic in origin. Such considerations have led them to regarding the floristic concept of plant formations to be artificial and oversimplified, and consequently of only secondary importance in the defining of an ecosystem. They also consider the use of soil types even less satisfactory for major subdivisions because soil profiles cannot be seen and can only be examined by random sampling. Soils are also historical in origin, influenced by the geology of the substrate and palaeoclimatic weathering. Moreover, the soil and vegetation form an inseparable whole, where the soil supplies the water, nutrients and matrix to support both the plant and a rich soil fauna and flora. Only the climate is considered free from any historical influences since it is determined by present global circulation controlled by the changes in the distribution of the continents and oceans. Although there have been past climatic changes, there can be no fossil climates. The climate remains as the sole primary factor influencing the soil, vegetation and, to a lesser extent, the fauna, and it is only at the microclimatic level that the environment is affected by the soils and vegetation. The macroclimate is consequently regarded as particularly suitable for the subdivision of the geobiosphere and the delimitation of ecosystems.

2.   THE CLIMATE

A number of attempts have been made to define the world's climates, many used formulae, such as that of Thornthwaite (1948), based on precipitation in relation to temperature and/or evapotranspiration. Examples include that of Köppen and Geiger (1936) who considered the vegetation as the best means of expressing climate boundaries and failed to take into account the influence of geology, soils and human interference on some vegetation types. Gaussen (1955) developed a classification based on the number of dry months in a year, while Troll (1965) emphasised the duration of dry and humid months; the latter being defined as a month in which the mean monthly rainfall exceeds the potential evapotranspiration. For other examples see Le Houérou *et al.* (1993).

From a practical point of view climatic indices *per se* do not provide any information about seasonal variations, such as the distribution and variability of rainfall, maximum and minimum temperatures and length of the dry season, all of which affect plant growth. Neither do they provide information on the redistribution of rainwater within the ecosystem resulting from variations in relief and soil textures, and their effect on plant water availability (Walter, 1963; Wickens, 1992).

A more graphic representation of the climate as a whole, showing its seasonal course, may be obtained by the use of ecological climatic diagrams, based on the ombro-thermal curves of Gaussen (1955), such as those of Walter & Lieth (1960-1967) based on a system originally developed by Bagnouls and Gaussen (1957). For any particular climatological station climatic diagrams provide information on: (1) The annual pattern of temperature and rainfall; (2) The wet and dry seasons characteristics and their intensity; and (3) The occurrence or otherwise of a cold season, including the months in which early and late frosts are recorded.

There can be no 'normal climate' based on mean values since every year is different. For plants, periods that are either too cold or too warm, too dry or too humid, are more important than the average conditions, the frequency of such extremes being especially important for agriculture and forestry. By constructing climatic diagrams using values for single years taken over a period of at least 20 years instead of mean values, it is possible to show how often such extremes occur (Walter, 1963).

Using such climatic diagrams, Walter and Breckle (1985) recognised nine major low-altitude climatic zones and nine major ecological zones of the world. Their extremely close correspondence led to the recognition of nine zonobiomes (ZB):

ZB I   Equatorial zonobiome, with a diurnal climate (perhumid)
ZB II   Tropical zonobiome with summer rainfall (humid-arid)
ZB III   Subtropical arid biome (desert climate)
· ZB IV Zonobiome with winter rain and summer drought (Mediterranean arid-humid)
ZB V   Warm-temperate (oceanic) zonobiome
ZB VI   Typical temperate zonobiome with short frost period (nemoral)
ZB VII   Arid-temperate zonobiome with cold winters (continental)
ZB VIII Cold-temperate zonobiome with cool summers (boreal)
ZB IX   Arctic and Antarctic zonobiome

Intermediate zones between zonobiomes with a smooth transition between the two are termed 'zonoecotones' (ZE) and designated by the roman numerals pertaining to the neighbouring zonobiomes, e.g. ZE III-IV, thus avoiding the drawing of artificially sharp boundaries.

However, it should be borne in mind that the climates are no longer static. As a result of global warming major climatic changes are now taking place with warming in the high latitudes and a poleward advance of monsoon rainfall (Parry, 1990; Le

Houérou, 1996), although it may well into the middle of the this 21st century before any significant global change in the ecosystem boundaries become apparent.

2.1   Rainfall

Matching climates based on total rainfall and temperature would appear to provide a reasonable assessment for plant introduction and indeed has often been very successful. However, total rainfall can be misleading. For example, Jerusalem with 600 mm rainfall and a dry season of 6-7 months is classified as semi-arid. while London with 620 mm rainfall regularly distributed through the year has a humid climate (Arnon, 1992), i.e. it is rainfall distribution rather than total rainfall that is important. This is well demonstrated by Ethiopian germ plasm of the potential oilseed crop *Vernonia galamensis* which, in its natural semi-arid habitat, produced a few seed heads with an even seed maturation. Introduction into an area with higher and prolonged rainfall resulted in the production of secondary seedheads and an irregular seed maturation. However, with sufficient initial rainfall to establish good stands but insufficient to produce secondary seedheads followed by nil or very little rainfall, an even seed maturation is ensured. Too high a rainfall resulted in crop failure, apparently due to excessive soil moisture (Perdue *et al.*, 1986). In general crop yield levels are determined by the amount of rainfall above the minimum required for crop maturity, while rainfall in excess of optimum requirements can reduce crop yields from waterlogging (Arnon, 1992). For a discussion on water stress see Chapter 5.

The *relative humidity* (*RH*) and *saturation deficit* (*SD*) should also be considered in addition to rainfall, especially in the drier regions. For example, in the Sahel it was found that certain introduced species were unsuccessful except in the more humid maritime zone, yet they grew satisfactorily elsewhere in Africa under similar rainfall regimes (Le Houérou *et al.*, 1993).

2.2   Temperature

High and low temperature stress can limit plant distribution, especially frost, where the stage of development, degree and length of exposure can be critical. For example, adult plants of *Simmondsia chinensis* tolerate high daily temperatures ranging through 30-40°C during the morning, as well as high daily extremes of 43-52°C shade temperatures or temperatures as low as -9°C without serious damage. The flowers, however, can be destroyed by a late frost, while the seedlings are sensitive to light frosts of -3 to -4°C (Gentry, 1958). For further discussion on temperature stress see Chapter 5.

3.   THE SOILS

Soil characteristics in any one place are the result of the combined influence over long periods of time of climate and living matter acting upon the parent rock material, as conditioned by relief and also the effects of the cultural environment and man's use of the soil. Rainfall, temperature and wind are the major erosion agents acting on the parent rock and transporting the soil materials. Peat soils differ in being organic in origin, although they may also contain wind- and water-borne mineral material.

Erosion is both constructive and destructive, depositing unconsolidated parent mineral materials. The raw soils are eventually stabilised by vegetation to form a natural landscape which, in turn, can be destabilised by later erosion. *Natural erosion* processes can be gradual or catastrophic and are caused by geological uplift, volcanic explosions, climatic changes, and natural wear and tear caused by the ever changing vegetative cover, rainfall, etc. *Accelerated erosion* results from the disturbance of the natural landscape by man and his livestock through cultivation, deforestation and excessive grazing, thereby exposing the soil to the wind and the rain (Soil Survey Staff, 1951). Fortunately land and livestock management techniques, including the use of plants, can be used to reduce accelerated erosion to within acceptable limits.

Soils are classified according to their natural relationships based on their major characteristics and production potential, thus providing planners with essential information regarding their potential restraints and benefits, as well as their response to management practices. Prior to the 1970's soil classifications were developed for use by individual countries or groups of countries. Even attempts at a global classification, such as the pioneer Russian soil classification (chernozems, serozems, podzols, etc.) and the old American system (Soil Survey Staff, 1937) were biased towards national interests and experience, although the latter system drew some of its classification from the Russian. In view of the large number of classifications and the confused application of the nomenclature, a number of new systems have now been developed. The most widely recognised systems being the American Comprehensive Soil Classification System (Soil Survey Staff, 1975) and the FAO/UNESCO Soil Map of the World (FAO-UNESCO, 1985). See Arnon (1992) for further discussion.

The soils provide a rooting medium in which plants (apart from free-floating aquatic and parasitic plants) find support, obtain moisture and nutrients. The depth of the rooting system varies according to species and can be restricted by such factors as soil depth, the ability of the roots to penetrate the soil, height of the water-table, etc. Some plants, known as *phreatophytes*, have root systems that rely on the water-table for water during the dry season. Others are shallow rooted and take advantage of the slightest shower, while others, by timing their development early in the growing season, can utilise the available moisture without competition from later developers. The available soil moisture is that which can be absorbed by the plant from the smaller pore spaces in the soil (*capillary water*) and is reflected in the soil texture.

Clay soils, for example, have a low soil water availability due to some of the water being bound by the clay colloids (*hygroscopic water*), while sandy soils, due to their large pore spaces, is unable to retain capillary water against gravity and the resulting *gravitational water* is of a transitory nature and drains away (see also Chapter 5).

3.1   Soil Nutrients

Soil plant nutrients are provided from the weathering and transportation of minerals from the parent material and the recycling of organic material, although all soil minerals are beneficial to the plant. High levels of salt and alkali can only be tolerated by plants adapted to such conditions, i.e. *halophytes* and *gypsophytes*. See Chapter 5 for further information on halophytes. High levels of aluminium and manganese, for example, can also be toxic. Strains of plants that develop a tolerance to mineral toxicity can be used in the primary colonisation of mine waste and other contaminated soils; see also Chapter 20.

3.2   Soil Micro-Organisms

The symbiotic relationship between micro-organisms and seed plants in nitrogen fixation is of particular importance in agriculture. It contributes directly to an increase in productivity of both forage and pulse legumes. The productivity is relative since, provided the mineral status is satisfactory, the results will be influenced by the strain of the appropriate inoculant. A less direct contribution may be made to the growth of rice by the aquatic fern *Azolla*.

While our knowledge of the rhizobia associated with the cultivated legumes is fairly well documented, that of the family as a whole is still very poor. While N-fixing bacteria are usually associated with the Leguminosae (mainly within subfamilies Papilionoideae and Mimosoideae); they are also found among members of other angiosperm families. Research so far has revealed microsymbionts, either as associated or free-living bacteria or cyanobacteria, to be present in members of the Betulaceae, Casuarinaceae, Datiscaceae, Elaeagnaceae, Gramineae, Gunneraceae, Myricaceae, Rhamnaceae and Rosaceae. Cyanobacteria are also known in the Cycadaceae, Boweniaceae and Zamiaceae of the gymnosperms (Sprent and Sprent, 1990). See Chapter 5 for further discussion.

4.   BIOLOGICAL CONSIDERATIONS FOR PLANT INTRODUCTIONS

No plant survives in isolation and it is important to ensure that when plants are introduced into another ecosystem that: (1) They are compatible with the environment, e.g. provenance, photoperiodism, dioecy, etc.; (2) The necessary supporting biota are available for their survival, e.g. pollinators, micro-organisms, etc.; (3) That they are not susceptible to alien pests and diseases; and (4) In the absence of any biological control associated with their native habitat they do not become invasive.

4.1   Provenance

The genetic constitution of a species may be expected to exhibit intra-specific variation within its range of distribution as a genetic adaptation to a range of environmental conditions, and for the variation to be distributed in such a way that neighbouring plants tend to resemble one another. For example, much of the variation in forest trees is clinal, and Davis and Heywood (1963) quote the suggestion by Böcher (1960) that the clear-cut ecotypes described in many wind-pollinated outbreeding tree species are probably parts of a cline where their apparent distinctness is due to randomly selected segments from different localities. The expression may be physiological or, for trees, refer to timber properties, characters that do not lend themselves to taxonomic recognition. In other examples adaptation may be expressed morphologically and may thus be taxonomically recognisable as subspecies or varieties within which genotypic variation may again be expressed. Such expressions are reflected in the adaptability of a species when introduced. It was following, for example, the variability expressed in provenance trials of *Vernonia galamensis* that a taxonomic revision was deemed necessary. An independent revision by Gilbert (1986) was indeed found to conform to these provenances.

Such provenance expression is of particular importance to foresters, where the suitability of an introduced species from a particular seed batch is subject of a long-term investment. Foresters refer to the place of origin of the seed, and hence its genotypic constitution, as its **provenance**. For example, *Pinus halepensis* is a native of the Mediterranean region, distributed from Spain to Greece, and the Cilician plain of Turkey and is widely cultivated elsewhere, while the closely related *P. brutia* is a native of the eastern Mediterranean region, from southern Italy to western Caucasus and western Syria. Both species were among the most successful for growing on terra rossa and redzina soils. Provenance trials in Israel have shown that the most promising seed sources for *P. halepensis* were from the lower altitudes of Greece and other Israel introductions, while the preferred seed sources for *P. brutia* were from the lower elevations on the Mediterranean coast of Turkey (Tutin *et al.*, 1964; Davis, 1965; Weinstein, 1991).

Similarly, agriculturists and horticulturists recognise the variations within a taxon as **races**, with the selection of suitable races for cultivation in particular areas and under particular conditions. Disease resistance, frost hardiness, drought resistance, early ripening, seed retention, quality and quantity of the oil content, etc. are all examples of characters that cannot be taxonomically recognised, whereas flower colour, if relevant, could warrant taxonomic recognition.

4.2   Photoperiodism

In many plants the relative length of light and dark periods (*photoperiodism*) exert a decisive effect on plant behaviour and may even determine if and when the plant passes from the vegetative to reproductive phase. Day length is largely a function of latitude and varies according to the season. At the time of the equinoxes, the two times in the year when the sun crosses the celestial equator, day length is practically the same at all latitudes. In early summer day lengths between latitudes 10° and 3° are 13-14 hours; between 30° and 50° they are 14-16 hours, and at 60° approximately 19 hours. The actual length of the light period is more important than the intensity of the light. See Chapter 5 for further details.

Moving a plant to another latitude so that it is out of phase with its day length can cause the plant to remain at a certain stage of development or even die because it is behaving in an unnatural manner, such as attempting vegetative growth during winter or too late in the spring. Day length is certainly an important factor for a forester when considering provenance selection from within a wide-ranging species which may have developed well-marked photoperiodic races or ecotypes (Fitter and Hay, 1985).

4.3   Dioecy

Dioecy is where the male and female reproductive organs are separated on different individuals, making cross fertilisation obligatory and thus preventing isolated individuals from reproducing. Collectors introducing breeding populations beware! The wind-pollinated *Simmondsia chinensis* (jojoba) is such an example. The male and female plants grew in such close cohabitation that it took a number of years before the fact that there were male and female plants was recognised. Strangely, in California the ratio of male to female plants is virtually equal, while in Arizona the ratio is more than four to one. In cultivation one male to six female appears to be a satisfactory ratio. Similarly, for the date palm, *Phoenix dactylifera*, one male to 50 or more females is a satisfactory ratio and pollination may even be assisted by placing cut portions of the male inflorescence in the receptive female inflorescence (Brooks, 1978; Bloomfield, 1985; Purseglove, 1985).

The purpose of this chapter is to create an awareness of some of the many problems associated with plant introduction and management in relation to the environment. It is by no means an exhaustive treatment since case studies could well run to several volumes of text.

4.4   Pollination

The process of pollination in seed plants involves the transfer of pollen from the male reproductive organs to the female, and is usually effected by intermediary agents such as insects (*entomophily*), birds (*ornithophily*), bats and small marsupials (*zoophily*), (wind (*anemophily*), or water (*hydrophily*); self pollination, where the anthers and stigmas mature simultaneously within the same flower (*homogamy*) may also occur. An extreme example of homogamy occurs in cleistogamous flowers (which never open), with the pollen germinating on the stigma within the closed perianth; in some examples the flowers remain closed until after pollination has occurred.

In the case of entomophily, flowers may be pollinated by a number of different insects; such flowers are termed *allophilic*. Other species can only be pollinated by one specific agent, e.g. *Ficus* spp. (figs) and are termed *euphilic*. Long-tongued moths and butterflies are able to pollinate flowers with long corolla tubes, such as the moth-pollinated Madagascan epiphytic orchid *Angraecum sesquipedale* with its 45 cm long spur!

Bird pollination is chiefly by members of the Trochilidae (humming birds) and Coerebidae (honey-creepers) of the Americas, the Zosteropidae (spectacle birds), Nectariniidae (nectar and sun birds) and Dicaeidae (flower-peckers) of Africa and Asia, the Meliphagidae (honey-eaters) and Trichoglossidae (brush-tongued parrots) of Australia, and the Drepanididae (honey-creepers) of Hawaii. Ornithophily occurs in a number of the Leguminosae subfamily Caesalpinioideae, including species of *Amherstia*, *Bauhinia*, *Brownea*, *Caesalpinia* and *Delonix*, and in subfamily Papilionoideae, *Erythrina* (coral tree) and many others. Other examples of bird pollination include *Bombax* spp. (silk-cotton trees) of the Bombacaceae, and *Hibiscus* spp. of the Malvaceae.

Bat pollination is reported in a number of families, including the Bignoniaceae for *Cresentia cujete* (calabash tree), *Kigelia africana* (sausage tree) and *Oroxylum indicum* (midnight-horror), the Bombacaceae for *Adansonia digitata* (baobab), *Ceiba* spp. and *Durio* spp., the Leguminosae subfamily Caesalpinioideae for *Bauhinia megalandra* and *Eperua falcata*, and subfamily Mimosoideae for some Asiatic members of *Parkia* - the American species are probably pollinated by birds, as are also some night-flowering *Cereus* of the Cactaceae.

Our knowledge of pollinating agents is still largely anecdotal and imperfect. For example, bat pollination has been reported for *Adansonia digitata* and, significantly, the distribution of the tree coincides with the distribution of the bat pollinators *Eidolon helvum*, *Rousettus aegypticus* and *Epomorphus wahlbergii*. In East Africa *Galago crassicaudatus* (bush baby) is also believed to aid pollination, while the strong carrion smell of the baobab flowers is also attractive to *Chrysomyia marginalis* (bluebottle), three nocturnal moths, *Heliothis armigera* (American bollworm), *Diparopis castanea* (red bollworm), and *Earias biplaga* (spring bollworm), in addition to a number of Hymenopterous insects; there is also a suggestion that the lightly held pollen could also be wind dispersed (Wickens, 1982). It is clear from this one example that pollination in supposedly zoogamous flowers can involve a number of vectors.

When plants are introduced outside their natural habit pollination there may be a problem for zoophilous plants, especially where euphilic species are concerned. The figs are an often cited example of a euphilic species, undergoing a complicated process of cross-pollination by minute fig wasps, *Blastophaga* spp., with each species of wasp being specific to either a given species or a few related fig species. As a further complication, *F. carica* (cultivated, common or Adriatic fig) has no staminate flowers and the seedless fruit develop parthenocarpically, while the widely cultivated Smyrna fig, although lacking staminate flowers, requires pollination for fruit development. Pollination is achieved by caprification, the suspension on the branches of the Smyrna fig of inflorescences of 'wild' caprifigs, from which the specific fig wasp *B. psenes* are about to emerge. When introducing figs into a new area the cultivars of the Adriatic fig should be used unless the caprifig and its associated fig wasp can be introduced with the Smyrna fig (Hill, 1954; Purseglove, 1987).

A somewhat similar problem occurs with the oil palm, *Elaeis guineensis*, where the monoecious palm bears either male or female inflorescences, rarely male and female, or even more rarely hermaphrodite flowers, with individual palms passing through alternating phases of male and female inflorescences, thus making cross pollination obligatory. In its native Africa, the weevil *Elaeidobius kamerunicus* is the pollinating agent, while on older trees in Malaysia it is the less efficient *Thrips hawaiiensis*, with the younger trees receiving manually assisted pollination. The introduction of the weevil from West Africa to Malaysia in 1981 has eliminated the need for hand-pollination and increased yields by 20%, while in Sabah, where the thrip does not occur, by 53% (Syed 1979 cited by Webster & Watson 1988).

Humidity may affect pollination in, for example, the date palm, *Phoenix dactylifera*. If the humidity is too high the flowers will not open and pollination cannot take place and the palms are sterile. Such environmental factors as sunshine, high temperatures, low humidity, low rainfall and adequate soil moisture are prerequisites for good fruit production. A further problem is **metaxenia**, where the tissues outside the embryo sac are influenced by the pollen source, resulting in differences in fruit size, quality and time of ripening.

Cross-pollination, both between species and infraspecific taxa is always a danger when plants that hitherto had been spatially separated are brought close together and can create serious problems in forestry, agriculture and horticulture for plant breeders, conservationists, etc. (Corner, 1964; Faegri and Iversen, 1964; Harder *et al.*, 1965; Purseglove, 1985, 1987; Dransfield, 1986).

4.5   Pests, Diseases, and Invasiveness

In general a species within its natural habitat can be expected to have achieved a balance between survival and succumbing to pests, diseases and competition with other species. Outside its natural habitat it can be either subjected to or free from similar or different pests, diseases and competition. For example, in India damage by bruchid beetles to the seeds of *Acacia nilotica* subsp. *indica* ensures a low regeneration. When the tree was introduced into Queensland for browse, fodder and shade, the absence of bruchids led to thicket formation and in 1959 for the species to be declared an obnoxious weed. *Bruchidius sahlbergi* had to be introduced from India to help control viable seed production (Hall, 1964).

Colonisation and the importation seed of European crops has resulted in many European weeds being introduced to many parts of the world. For example, Salisbury (1961) notes that 70 British weed species were recorded in Tasmania a century after the first British settlements, while according to Crompton *et al.* (1988) 50% of Canada's weeds are of Eurasian origin and only 3% from elsewhere in the New World.

The horticulturists must also take their share of the blame for introducing invasive weeds. The introductions of *Fallopia japonica* (Japanese knotweed) into the UK during the first half of the 19th century proved to be extremely aggressive, and almost impossible to eradicate; it is now officially regarded as Britain's most pernicious weed (Maby, 1996). Most of our greenhouse pests too are not native, including *Trialeurodes vaporariorum* (whitefly). Many of our garden pests have also been imported on plants from abroad, e.g. *Dasyneura gleditsiae* (Gleditsia gall midge), *Contarina quinquenotata* (Hemerocallis gall midge), *Liloceris lilii* (lily beetle), *Liriomyza huidobrensis* (South American leaf miner), *Macrosiphum albifrons* (lupin aphid), etc.

The risk of importing pests and diseases are generally regulated by individual countries with varying degrees of success - the importation of only soil-free root stocks being a common regulation. The dangers of foreign soil importation is demonstrated by the recent introduction into Great Britain of the Australian and New Zealand flatworms, and were probably brought into the country in the rootballs of imported plants. They have had a devastating effect on the indigenous earthworm population in certain areas of the country and with the strong probability that they will eventually spread throughout the country. While rigid inspection for *Leptinotarsa decemlineata* (Colorado potato beetle), a native of Central America and now a major pest of potatoes (also of aubergines, peppers and tomatoes) in Europe has so far prevented its entry into the UK. However, inspection failed to prevent the entry of Dutch elm wilt, the preferred name but more commonly known as Dutch elm disease, caused by an aggressive strain of *Ceratocystis ulmi*. The fungus was imported on elm logs from Canada to England in the mid 1960s, the resulting pandemic killing *ca.* 10.6 million elms (Holliday, 1990).

Viruses are also a problem. Some 400 plant viruses are known, many of which are transmitted by aphids, such as the potato virus, which is capable of reducing crop yields by 40%. Others may be transmitted by infected seed or pollen. Since viruses are generally not present in meristematic tissue it is possible in many cases to obtain virus-free stocks by tissue culture of meristem explants (see Chapter 6). Some virus infections are symptomless, apart from reductions in yield, others manifest themselves as mosaics, leaf spots, and deformities. The earliest known record of virus symptoms being the strikingly regular colour breaking of *Eupatorium japonicum* caused by the Eupatorium yellow vein virus and recorded in a poem by a Japanese empress in 752. Similarly, the colour breaking on the tepals of tulip cultivars painted by 17th century Dutch artists can be attributed to the tulip breaking virus (Tootill, 1984; Holliday, 1990).

5.   WHY INTRODUCE?

Plants are introduced mainly to supply a need in the economy of a nation's agriculture, horticulture or forestry. This need may for new products or new species, races or provenances to improve the health and productivity of existing species. This is often accompanied with belief that because the introduction has a proven record elsewhere it must be desirable; the potential of indigenous species for the same or similar purpose is often overlooked. A plant with a proven record, probably after having undergone some selection and breeding for increased productivity elsewhere, has the undoubted advantage of having a known research input and cultural record. This clearly gives the introduced species an apparent research advantage for the recipient/breeder/grower over local indigenous species whose potential, although an unknown quantity, could be even greater. Thus, for mainly historical reasons the major cereals, wheat, barley, maize, and rice, have been intensively researched and improved and the minor cereals under-researched. It is the writer's belief that if even a small proportion of the investment that had been made in the major cereals could be spent in the developing countries on the minor cereals, they too could become 'major' cereals. Also for historical reasons and prejudices, Australian settlers have, because of the large cultural gap, ignored the food resources of the Aborigines for generations and it is only recently that research has shown many of their foods to be nutritionally better than those forming the traditional Caucasian diet (Brand and Cherikoff, 1985). Similar findings have been noted for southern Africa (Arnold *et al.*, 1985) and south-western USA (Nabhan, 1985b).

5.1   Agronomic Considerations

New plants intended for cultivation are usually imported as seed since they are easier to transport and do not have such severe phytosanitary restrictions as those required for organs such as bulbs, corms, tubers, rhizomes, rootstocks and cuttings. New introductions may well require the adapting of existing, or the introduction of new, husbandry techniques, including pollination, harvesting and post harvest practices, as well as fungal and insect pest control. There are also logistic problems. For example, how does the new crop fit into existing cropping systems? Are there peak labour and equipment requirements that would interfere with existing management systems? Labour requirements have to organised. How many workers can be fully employed throughout the year and when and how high are the peak labour requirements? Is part-time labour readily available as and when required? Finally, is it economic?

5.2   Market Forces

The subject is discussed in greater detail in Chapter 7 and is only briefly considered here. For profitability there must be a market demand where the buying price adequately compensates for the cost of production. The buyers require standards for both quantity and quality. Irregular or unpredictable supplies are unsatisfactory. Even if these criteria can be met, the necessary combination of agronomic and market characteristics are necessary to stimulate any major development. There is also an element of luck in whether or not a crop is accepted. For example, *Metroxylon* spp. (sago palms), especially the spineless *M. sagu* and its prickly *'M. rumphii'* form, are natives of the freshwater swamps Southeast Asia and Melanesia, occupying a self-sustaining environment requiring neither fertilisers nor cultivation and weeding. Not only do they grow in swamps unsuitable for other crops, they can also compete with rice in terms of starch production, with yields of 5-25 t ha^{-1} dry weight of starch after harvesting at 8-15 years. Although used locally for food sago is also a potential industrial crop, yet many thousand of hectares of wild sago palm in Indonesia remain unharvested (Syed,1979), cited by Webster and Watson (1988).

6.   LEGISLATION

It is clear from the dangers associated with plant introduction outlined above, especially with regard to pests, diseases, invasiveness, genetic contamination, etc. that the indiscriminate movement of plants around the world could create serious problems and must be controlled. It is essential therefore, that plant collectors should be familiar with the export and import regulations regarding prohibited plants, live plants, seeds, fruits, etc. and also the permitted growing medium/soil in which living plants can be transported (see Chapter 7 for further details regarding the UK).

Endangered species, i.e. rare species whose survival is endangered, are subject to the International Convention on Trade in Endangered Species (CITES) under which it is illegal to collect wild living material covered by the convention, although trade in cultivated material is permitted. Trade in such species refers not only to the commercial exchange between nations but also the collection of plants by private individuals and research organisations. In addition, individual nations may legislate to control the genetic material of their own economic and potentially economic plants by banning the collection of such plants.

Chapter 4

Plant Conservation

The world's human population has reached 4.8 billion and is expected to double before it stabilises (Plucknett *et al.*, 1987). At the same time the environment is being destroyed at an unacceptable rate. For example, 11.3 million ha out of a global total of more than 4,000 million ha of forest are currently being cleared annually, 45 % of which is attributed to shifting cultivation and long fallow agriculture. Many countries, especially in Africa, will be unable to meet their domestic fuel requirements. It has also been estimated that by the year 2000 no less than 64 countries - 29 of them in Africa - will be unable to feed themselves from their own natural resources (Committee on Forest Development in the Tropics, 1985); plant resources other than food are similarly affected. The time has passed but the threat still remains. In view of the global increase in population and the concurrent degradation of the environment, increasing productivity without further degrading the environment is an obvious necessity. While it is widely recognised that social and economic pressures are often the cause of environmental degradation, it is unfortunate that land degradation, depletion of renewable resources and population increases, in turn, further increase the social and economic pressures.

The effects of desertification brought about by over-grazing, over-cultivation and deforestation may be responsible for climatic as well as environmental deterioration. The increased albedo effect is believed to suppress convective cloud and rainfall in arid regions, thereby aggravating the consequences of desertification (Grove, 1985), although other authorities disagree that the albedo is responsible. The possible effects of global warming on the world's agriculture suggest that there could be a small net decrease in the global food production-capability in some crops. Certain countries that are currently net exporters of grain could have their production potential reduced, thereby affecting global food prices. Fortunately agriculture has the ability to adjust, within given economic and technological constraints, to a limited rate of climatic change (Parry, 1990).

An early example of this climate change are the recent crop failures in the Sahel arising from the combined effects of climatic drought aggravated by desertification. The result was dramatic in its swiftness and intensity and has doubtless led to the loss of a number of cultivars adapted to that environment. The predicted results of global warming, although as yet impossible to confirm, increase the urgency of the necessity to conserve for the future, especially from the arid and semi-arid regions where the effects could be as dramatic as those for the Sahel.

As already briefly discussed in Chapter 1, plants are widely used for a variety of purposes, although relatively few are of global commercial importance for food, timber, fibre, medicines, etc. There are, for example, some 12,500 species of edible plants listed by Kunkle (1984), yet 90% of the world's food is obtained from 103 species (Prescott-Allen and Prescott-Allen, 1990). This makes the world's food supply extremely vulnerable to possible catastrophic events and economic ransom. The economics of increasing monoculture, and the standardising of cultivars by international, commercial and governmental organisations will, by limiting genetic variability, increase both the probability and extent of any future catastrophic events.

1.   WHY CONSERVE?

A species lost is lost forever together with all it attributes. This may not appear to be serious but no species survives alone, each is part of an intricate mesh involving producers, consumers and decomposers. Those that are entirely dependant on the lost species will themselves become extinct, others will be weakened and, as nature abhors a vacuum, others will be strengthened. The environment will be changed. At our present state of knowledge we neither know all the organisms involved nor their role. We destroy at our peril.

It is essential for our future survival that the genetic resources of wild, cultivated and potential crop species together with their wild relatives should be conserved so that they may be available for both present and future breeding programmes. They are required to ensure the sustainability and adaptation of crops *sensu lato* to the changing environmental and socio-economic needs of both the present and future human and livestock populations. In addition, there is no way of knowing what useful plant attributes awaits discovery. There are, for example, more than 10,000 different chemical compounds that have been isolated in recent years from plant tissues. The screening and developing of other little-known or yet to be identified species is likely to result in the future discovery of numerous other products and uses. Improved genetic knowledge and new technologies, such as genetic engineering, may permit the utilisation of genetic material from an increasing range of species. Species that were previously regarded of no economic importance are now being used to develop resistance to pests, diseases, and physiological stresses. While the attitudes and requirements of future generations cannot be reasonably predicted, this present generation has the responsibility of keeping all options open and prevent to the best of our ability the depletion or destruction of natural areas and their genetic diversity (Hoyt, 1988; FAO, 1989).

The need to conserve plant genetic material may be summarised as follows: (1) To widen the existing genetic base of cultivars in order to increase their productivity by increasing both their yield and environmental acceptability as well as making them less vulnerable to pests and diseases; (2) To make better use of wild and little-known species, either for the improvement of existing crops or to develop as new crops; and (3) To preserve crop relatives whose genetic attributes may or may not be known at present, but may or may not have a potential for crop improvement.

There are a number of examples where genetic uniformity has made a crop vulnerable to pandemics of pests and diseases. In the Irish potato famine of 1845-7, an estimated 1 million people died of starvation and disease and a further 1.5 million were forced to emigrate. The Irish potato crop was based on a small number of introductions and was consequently genetically highly uniform, so that when given the appropriate conditions the entire crop could be susceptible. Suitable humid conditions occurred during 1845-7 and the entire crop was destroyed by *Phytophthora infestans* (late potato blight); fortunately more resistant cultivars are now available.

Similarly, in India, the Bengal famine of 1943 was caused by *Cochliobolus miyabeanus* brown spot disease which, aggravated by a typhoon, destroyed the rice crop. Highly resistant commercial rice cultivars are now available. Again, in the early 1970s, grassy stunt virus transmitted by *Nilaparvata* spp. destroyed more than 116,000 ha of rice in India, Sri Lanka, Vietnam, Indonesia and the Philippines. A wild relative, *Oryza nivara* from Madhya Pradesh, which is an annual form of the perennial *O. rufipogon*, now confers resistance to *O. sativa* for both grassy stunt virus and *Pyricularia oryzae* (blast) (Salaman, 1949; Sharma and Shastri, 1965; Prescott-Allen and Prescott-Allen, 1983; Hoyt, 1988).

There are a number of reasons for the need to diversify and make use of hitherto little or under-exploited wild species or minor crops, especially in the developing countries. The economics of importing a particular crop or crop product may be prohibitive so that alternative substitutes have to be found and grown by the country concerned. Thus, the high cost of synthetic pesticides has considerably increased interest in natural pesticides among the developing countries. The 'Green Revolution' in India well demonstrated how a nation could utilise its own natural resources and reduce reliance on imports, although the beneficial results have been somewhat negated in recent years by an excessive use of agrochemicals and the monoculture of export crops to the detriment of biodiversity and the environment. Another reason is the strategic necessity to reduce reliance on a particular country or region. The search during the past half century for alternatives to the Pará rubber grown in the Far East is one such example.

Increasing populations have led to attempts to develop less favourable soils and climates, with particular attention regarding suitable new crops for arid and saline soils. Over-production, resulting 'grain mountains' in Europe and North America, has created yet another reason why farmers need to seek alternative new crops, such as those suggested for the UK by Dover (1985), Carruthers (1986) and MAFF (1994a).

It is an impossible task to predict what plants, known and unknown, may be required for the future. Farnsworth and Soejarto (1985), for example, have attempted to quantify the loss to medical science arising from the destruction of North American plants by the end of the 20th century. They have calculated that out of the 250,000 flowering plants in the world, about 10% will be extinct by the end of the 20th century. By inference, some 2067 North American species may, at the current rate of extinction, also be expected to be lost. It has also been estimated that for every 125 species subjected to thorough pharmacological examination, one will eventually become important as a drug source. Thus, they have projected that 16 potentially useful drug plants (2067÷125) will become extinct, which they estimated (in 1980 dollars) would be worth $3.348 billion in the year 2000 AD. To understand the importance of this calculation it should be appreciated that, unlike scientists in many other countries, American scientists have paid little attention to the medicinal properties of the North American flora. Hence, if all the 2067 species expected to become extinct had been thoroughly investigated, it is probable that 16 could have produced drugs.

It is obviously impossible to conserve sufficient breeding stock to maintain the genetic diversity of all of the world's plant species, although efforts are being made to safeguard many of the endangered species recognised by IUCN and WWF through *in situ* and *ex situ conservation*.

2.   *IN SITU* AND *EX SITU* CONSERVATION

The *in situ* and *ex situ* approaches to conservation complement each other and the distinction drawn between them has often been over-emphasised. Unfortunately it is unrealistic to expect that wild relatives in their natural environment can be perpetually conserved *in situ*, since nature reserves, national parks and other protected areas are not always inviolate - they remain vulnerable to loss or destruction and to plant succession. There is also the possibility of political barriers to free access to the plants. Likewise, *ex situ* conservation, irrespective of whether in seed banks, tissue cultures or field genebanks, is vulnerable to human failings, natural disasters and technical problems such as power cuts, fires and floods. Furthermore, *ex situ* conservation disrupts the processes of evolution found in wild populations (Hoyt, 1988).

Ex situ species in field genebanks or botanic gardens may also be exposed to pests and diseases for which they have no protection. The necessary biota for pollination, seed dispersal, root nodulation, etc. may also be unavailable and, if recognised in time, improvised. Yet another drawback is that for some species hybridisation may occur because they have been brought into the close proximity of species from which they had been previously been spatially isolated. Again, free access to the plants may be denied by the host country. Despite the disadvantages in both *in situ* and *ex situ* conservation, both approaches are necessary, in varying mixes for different species.

3.   SEED BANKS

Seeds offer the most convenient and natural plant organ for storage. Also, with few exceptions, every seed has a different genetic constitution, thereby enabling a wide range of genetic variability to be stored in a small sealed container. Most seeds can be dried to a low, *ca.* 5%, moisture content (on a wet weight basis) and stored at -20°C for possibly up to a century or more without loss of viability, although some genetic erosion may occur. Such seeds are termed *orthodox* and represent the majority of crops and their wild relatives. Seed banks are certainly the cheapest and most convenient storage method for conserving species with orthodox seeds.

However, there are two important categories of problem plants that cannot be conserved in this way. The first category are typically large and fleshy seeds which lack a natural dormancy mechanism and deteriorate rapidly, usually within a few days or weeks after collecting. Such seeds cannot have their moisture content reduced to 20% or below without injury, and are termed *recalcitrant*. They cannot be stored in a seedbank, but they may be kept for a few weeks or months if treated with a fungicide and placed in a polythene bag in moist sawdust or charcoal and kept aerated. These include a small number of major crops whose threatened wild relatives have an obvious and urgent need for *in situ* conservation. The second category are the vegetatively propagated crops. These are usually conserved vegetatively in the form of clones, either because of sterility or in order to conserve gene combinations that would be lost by conversion to seed (Prescott-Allen and Prescott-Allen, 1983; Smith, 1985; Hoyt, 1988; Withers, 1991). Such plants are dealt by *in vitro* conservation (see Section 4 below).

Three main types of seed banks for the storage of orthodox seeds are recognised: (1) *Working* or *short-term seed banks* intended as working collections for breeders and agronomists to use over one or two seasons. The seeds are kept at room temperature or, in hot, humid climates, in air-conditioned rooms. The key to their successful storage is to dry the seeds and maintain them at a low moisture content. This is frequently more important than temperature conditions. As a general rule, once the moisture content has been reduced, the life span of seeds can be expected to double for every 5°C that the temperature is lowered between 50 and 0°C; (2) *Medium-term storage seed banks* are where the dried seeds are kept close to freezing in glass or plastic bottles, or aluminium-foil packets. Such collections are readily accessible to the breeders and can be transferred to the working collections as and when required; and (3) *Long-term storage seed-banks* are not normally available for routine distribution or exchange. They are a safeguard against present and future loss of genetic resources. The dried seeds are kept in sealed bottles, vacuum-packed cans, or in laminated aluminium-foil envelopes at -10°C to -20°C, and may be expected to be preserved for a 100 years or more. The recommended viability for storing such accessions should be at least 90%. After storage the seeds need to be periodically checked to ensure that viability is being maintained. Should the viability fall below 85% the accession should be regenerated. This is in order to avoid the loss of rare genes and also because of the genetic changes that occur when viability declines (Plucknett *et al.*, 1987). It is because of the low humidity and ensured power supplies that the majority of the medium- and long-term seed banks are to be found in the developed countries of the temperate regions; there is no political implication involved.

Seed collecting is not a task that should be entrusted to the amateur or the inexperienced. Advice should always be sought from the appropriate seed bank before seed collectors start their collections. Proper documentation is always required by the seed bank regarding both the collecting site and the species and, where wild relatives are concerned, a herbarium voucher specimen will also be required in order to obtain and preserve evidence of a correct determination of identity. Good advance planning can save much time wastage in the field. The necessary information on where to go to find a species and its fruiting season can be readily obtained by consulting herbarium specimens (see Chapter 2). It must also be borne in mind that the vagaries of the climate, especially in the arid and semi-arid regions, can mean changes in the time of fruiting. Since it usually difficult to alter prearranged itineraries at short notice to cater for unexpectedly late or early fruiting of the desired species, collectors should always plan to collect alternative species or visit other localities. Taking cuttings of woody species is another possibility when seeds are not available. A suitable technique for transporting cuttings using soil polymers as the rooting medium is described by Aronson *et al.* (1990). The method has the advantage of placing fewer restraints on the season for collecting, but has the great disadvantage of providing limited genetic variability. *In vitro* techniques (see below) can be used for their multiplication and storage.

Where possible it is usually recommended that accessions should not be less than 5000 seeds. This is in order to permit adequate samples to be removed for testing germination. Collections of the same species from different collecting sites should never be mixed. Cloth or paper bags, which prevent the seeds sweating, should always be used for their collection and transportation; plastic bags should never be used. While in the field the seeds should always be kept as cool and dry as possible. The seeds should also be sent to the seed bank with the minimum of delay.

Many of the developing countries feel that their genetic resources are being taken for use by the developed countries and that they receive no benefit in return. It can also be argued that the developing countries cannot fully utilise their own genetic resources and that unless they are conserved by the developed countries they could be lost. The issue has been further complicated in recent years by questioning the morality and ethics of patenting genetic resources, especially those from the developing countries (see Chapter 6 for further discussion). Seed banks may also vary regarding granting access to the collections by nationals and non-nationals. Although those seedbanks recognised by the International Bureau of Plant Genetic Resources (IBPGR) will offer freedom of access to all responsible breeders and organisations, there are others who restrict access. As yet there is no satisfactory international solution to this problem.

4. *IN VITRO* CONSERVATION

The micropropagation of species through tissue culture for storage under 'slow growth' conditions is a useful method when there are too few seeds for satisfactory viability testing and storage in the seed bank, or where, in the absence of seeds, clonal propagation from bulbs, rhizomes and cuttings are the more usual methods of propagation. Such vegetatively propagated plants include a number of important staple foods, including root and tuber crops such as *Colocasia esculenta* (taro) and other aroids, *Dioscorea* spp. (yam), *Manihot esculenta* (cassava), *Musa* spp. (dessert and cooking bananas), *Solanum* spp. (potato) and several fruits, including citrus, pear, apple and *Prunus* spp. (Withers, 1991).

Tissue culture is being increasingly used for plants that set seed infrequently, in order to store organs collected from the wild when not in flower, and for recalcitrant species. Such recalcitrant species include members of the following economic genera: *Artocarpus* (breadfruit), *Camellia* (tea), *Castanea* (chestnuts), *Cinnamomum* (cinnamon), *Cocos* (coconut), *Hevea* (rubber), *Juglans* (walnuts), *Mangifera* (mango), *Myristica* (nutmeg), *Nephelium* (rambutan), *Persea* (avocado), *Quercus* (oak), *Swietenia* (mahogany) and *Theobroma* (cacao).

Although tissue culture is well suited to mass cloning of an individual plant, it is a time-consuming process since each species requires specially formulated techniques. There is also the complication with *in vitro* culture technology in that it can result in somatoclonal variation. Although this may be an advantage for breeding purposes, if the genetic characters are to be conserved, the release of such variation must kept to a minimum. Slow growth techniques for the medium-term storage of tissue cultures are now relatively well developed and gene banks using such techniques are now available (Woods, 1985; Hoyt, 1988; Williams, 1989;Withers, 1991)

Although not yet perfected the future solution to long-term *in vitro* conservation appears to be in '**cryopreservation**', i.e. the storage of frozen tissue at extremely low temperatures, such as liquid nitrogen at -196°C, where all biological processes would virtually cease and organs could then be preserved almost indefinitely, provided damage due to freezing and thawing could be overcome. Among other aspects still to be examined are the behaviour of different culture systems when exposed to such ultra low temperatures, crop-specific requirements and possible genetic erosion (Hoyt, 1988; Withers, 1991).

Chapter 5

Ecophysiology and Allied Disciplines

In order to manage both wild and cultivated plants for sustainable productivity it is necessary to have a basic understanding of the many factors involved in the interaction between the life cycle of a plant and the physical and biotic environments. It is the purpose of this chapter to provide those economic botanists *sensu lato* who are not specialists in physiology, anatomy, morphology and biochemistry, etc. with a very brief and elementary insight into some of the very many ways by which plants appear to react to and adapt to a particular environment.

The early botanists, such as Schimper (1898), Clements (1907), long ago recognised the interrelation between ecology and physiology, but it was Billings (1957) who first formalised the union with the terms *'physiological ecology'* and *'ecological physiology'*. Zoologists too recognised such an interaction within the animal kingdom, for which they coined the term *'eco-physiology'* and the term was first taken up by botanists in 1962 at UNESCO's Montpellier Symposium. It was here that Eckardt (1965, p. 9) first provided an acceptable definition for both botanists and zoologists. The following is taken from the English summary of his paper: *"Eco-physiology, in the acceptance of the term adopted in the proceedings of the Montpellier Symposium, deals with all relationships existing between living beings and their physical and biotic environment. It thus embraces not only the study of the adaptive structural and functional features which link the individual plant or animal to its specific environment, but also that of all forms of transfer and transformation of energy and mass connected with the dynamics of the ecosystem."*

The range of ecophysiological behaviour exhibited by species growing in the various global environments represented by the different climatic zones is so great that it is not possible to identify any one behaviour as typical for a zone. This observation finds agreement with the general observation by Walter and Breckle (1986) that the water supply to individual plants in humid and arid biomes does not differ greatly because there is a reduction in the total transpiring surface per square metre or hectare of ground surface which is in proportion to the decrease in rainfall.

65

1.   ENVIRONMENTAL RESTRAINTS

The survival of a plant within a community depends on achieving a balance between the abiotic constraints imposed by such environmental factors as rainfall and moisture availability, soils and fertility, etc. plus the competition and interactions between the biotic components by autotrophic green plants representing the *producers*, by mainly animals as the heterotrophic *consumers* of the vegetation and the micro-organisms and small invertebrates functioning as the *decomposers* (Walter and Breckle, 1985).

An extreme example of the interaction between plants and animals is from Mauritius where the killing for food and eventual extinction in the late 17th century of the flightless *Raphus cucullatus* (dodo) resulted in the failure of the seeds of the indigenous *Sideroxylon sessiliflorum* (tambalacoque) to germinate. This failure was attributed to the dodo eating the fruit and scarifying the seed. The problem was eventually overcome by force-feeding turkeys with the fruit. The introduced monkeys were also found able to scarify the seed (Temple, 1977; Owadally, 1979).

The major plant stresses are imposed by water, temperature, salt and heavy metals, against which survival depends upon physiological, anatomical and morphological adaptations and genetic make up. There is, however, often a problem in distinguishing between what are truly adaptive characters and those that are hereditary (Metcalfe, 1983). While it is convenient to consider these stress adaptations separately, it must be borne in mind that they are all interactive, functioning together in order to ensure survival in a stressful environment. For example, Parker (1968) considers water stress-resistance as being a characteristic that is the sum of the many parts of the plant, from whole organs to the fine structure of the cytoplasm, from the seed to the whole plant. Furthermore, the same stress adaptation may serve for different environmental factors. Thus, small evergreen leaves are to be found in plants growing on infertile soils and other unproductive environments, such as semi-arid areas, boreal and high-altitude forests, and bogs; areas where carbon investment in leaf is low (Fitter and Hay, 1989). Similarly, among the palms the wax covering on the leaves of *Ceroxylon* spp. act as an insulator against low temperatures, whereas in the case of *Copernicia prunifera*, syn. *C. cerifera* (carnauba wax palm) the wax has been shown to reduce water losses during the long dry season (see Section 2 below). Likewise, the storage rootstocks of some suffrutices whose aerial canopy die back annually may be an adaptation to drought and/or fire. A comprehensive account of the subject is given by Levitt (1980).

The physiological adaptations of plants to constraints are defined by Street and Öpik (1984) as: (1) *Stress resistance*, the ability to endure an externally applied stress; (2) *Stress avoidance*, the ability to prevent an externally applied stress from producing an equivalent internal stress in a plant; and (3) *Stress tolerance*, the ability to endure an internal stress induced by externally applied stress. In some species, by the application of simulated mild and/or gradually increasing stress, it is possible to develop *hardening* (*acclimation*) as a form of resistance to a particular stress.

1.1   Water Stress

The major constraint on plant life is undoubtedly water. In the past water stress has tended to be regarded as being mainly a characteristic of the low to medium rainfall regions. The meteorological anomalies in the Pacific due to the unusual El Niño oscillation has dramatically demonstrated how atypical periods of drought can also affect the rain forest, a factor that has been further emphasised by devastating forest fires.

Plants contain 70-90% water and they need water to stay alive, grow and develop; water is also needed to survive drought. Water absorption is by the plant's roots and is lost to the atmosphere through the leaves. It is the balance between absorption and respiration that determines the water status of the plant. Water serves to maintain turgor pressure in the plant cells and as a substrate in photosynthesis and hydrolytic reactions. Any changes in water supply, whether climatic or from irrigation, will have a profound effect upon the plant, affecting such key processes as nutrient uptake, carbohydrate and protein metabolism, and the translocation of ions and metabolites. Water is consequently intimately linked with plant growth and development (Slatyer, 1973).

Because of the complexity of plant/water relations, there is no single index of water supply in terms of available soil water, bulk air humidity or vapour pressure, that can be used to express the degree of water [deficit] stress suffered by a plant. Consequently it is usual to express water stress in terms of *plant tissue water potential*, *relative water content* or *water deficit* as a measure by which the plant's tissue water content has fallen below the maximum water content at full turgor, i.e. the optimum conditions for plant growth and function. Except in the most humid environments, plants are likely to suffer some degree of water stress on many occasions during their life cycle, the stress being accompanied by a reduction in photosynthesis and growth. Indeed, it has been estimated that losses in agricultural crops due to water stress exceed losses from all other sources combined (Langer and Hill, 1982).

During photosynthesis the uptake of CO_2 by the leaf is inevitably associated with water loss to the atmosphere through the stomata and a loss of turgor. Consequently the leaves of most plants are subjected to daily fluctuations of water stress and loss of turgor. Severe water stress will cause wilting of the leaf which, provided the plant roots are able to take up sufficient water from the soil during the night to restore cell turgidity, is known as *transient* or *incipient wilting*. The *permanent wilting point* is reached when the soil water content attains such a level on drying that all the soil water is firmly held by the soil and is unavailable to the plant roots, thereby leading to permanent wilting and eventually death. The permanent wilting point varies not only according to soil type but also to species. Because of its fine texture and high water retention by the clay colloids of a clay soil will hold more water at *field capacity*, i.e. the maximum water content that can be held by the soil against gravity. Clays also have a higher wilting point than coarser textured, freer draining soils (Street and Öpik, 1984; Fitter and Hay, 1989). This is in agreement with the simple formula for the Sudan where trees species requiring 3x mm of rainfall when growing on clay soils with neither run-off nor run-on water, will require 2x mm on sands (Smith, 1949). The Sudan, because of the more or less parallel alignment of the isohyets and the extensive, relatively uniform, sandy plains penetrated from south to north by the clay soils of the Nile valley, made such a relationship relatively easy to distinguish. In the field, by knowing two of the three variables, species, rainfall, and soils, I found I could predict the third with surprising accuracy.

Under conditions of acute water shortage some plants are able to avoid water stress by reducing their rate of water loss to very low levels by means of a range of morphological features and physiological mechanisms. The morphological features include thick cuticles, low surface/volume ratios, the possession of succulent water storage tissue, sunken stomata, rolling of the leaves, daytime closure of stomata, hairiness, and high reflectance. The evaporation of essential oils by aromatic species during the hot, dry summer months can also reduce water loss by creating a higher diffusion resistance layer around the leaves. Within this category of morphological adaptations there are many xerophytes, including succulent desert species such as, for example, the cacti. Despite such adaptations, the plants are still unable to withstand dehydration to any marked degree, so that water stress avoidance rather than water stress adaptation is obviously of supreme importance in arid environments (Street and Öpik, 1984). The ecophysiology of plants from the arid and semi-arid regions of the world are discussed in Wickens (1998).

Drought tolerance or *desiccation resistance*, is defined by Parker (1968) as the lowest relative humidity at which a plant can come to equilibrium without injury. This can not only vary from species to species but also between provenances and individuals within the same species. For example, accompanying the withdrawal of water from the cell there is probably some interference with enzymatic reactions and inhibition of other functions as the viscosity of the cell contents increases. There can also be changes in enzymatic reactions to favour one enzyme more than another, or shifts in pH values which could cause, for example, a tendency for starch to be changed to sugar; an increase in sugar certainly does occur in many plants subjected to drought. Plants may also adapt to drought stress by accumulating the 5-carbon imino acid *proline*. Such proline increase may account for as much as 30% of the free amino-acid pool. In legumes, proline is replaced by the aliphatic polyol *pinitol*. While it is possible that the special osmotic properties of proline enable it to contribute directly to plant water retention, it is also possible that the increase in proline or pinitol could be a symptom of some more fundamental adaptation to water stress (Harborne, 1988).

A number of desert plants of arid environments do not die when dehydrated, but enter into an anabiotic state of desiccation and revive after wetting. The African grasses *Sporobolus nervosus,* syn. *S. lampranthus,* and *S. staphianus,* for example, are able to revive after their water content has been reduced to 5-13% and, on rehydration, the leaves regreen. The hardy evergreen alpines *Haberlea rhodopensis* and *Ramonda myconi* (Pyrenean primrose) can also withstand dehydration down to 10% water content. The ability of such *resurrection plants* to revive after desiccation is attributed to the presence of the non-reducing disaccharide *trehalose*. Trehalose is also present in fungi, generally replacing sucrose in plants lacking chlorophyll and starch, as well as being the principal carbohydrate of insect haemolymph (Sharp 1990; Mabberley, 1997; Grant-Downton, 1998). Currently there is considerable interest in the possible use of trehalose for the preservation of blood plasma and inoculations as an alternative to freezing, a method which would be of inestimable value in the more remote regions of the world where refrigeration can be a problem. There is also the possibility of using genetic engineering to introduce trehalose into cereals so that the cereals could enter an anabiotic state during severe drought and recover completely when moisture conditions improve. Current research at the Quadrant Research Foundation, Cambridge is also considering its use in resurrecting dehydrated foods without denaturing.

1.2   Temperature Stress

The genetic and physiological understanding of resistance in plants to *heat stress* is still only imperfectly understood, especially in the absence of efficient and accurate laboratory tests to select the physiological traits for *heat tolerance*. Arnon (1992) cites the breeding of certain tomato cultivars, where 'Nacarlang' was developed for high-temperature conditions and was then found to set fruits better at temperatures too low for normal fruit set. Another cultivar, 'Cold Set', developed for the cool temperate, short-season regions, was found to be heat tolerant. Subsequently it was found that some of the best high-temperature tolerant lines had been developed in and for cool, short-season regions. The explanation of this apparent paradox would appear to be that the basic physiological mechanisms involved are concerned with stress tolerance rather than temperature tolerance *per se*. An additional mechanism could be the greater thermostability of certain enzymes serving a genotype at both high and low temperatures.

An added complication in characterising heat stress could be in the photosynthetic pathway followed, including the ability of some C_3 species to switch to the CAM pathway when stressed, although it would be premature to suggest any definite relationship (see Section 3.2). As Nobel (1996) has pointed out, CAM plants, such as members of the Agavaceae and Cactaceae, are very tolerant of high temperatures and generally not particularly tolerant of lw temperatures.

An added complication in characterising heat stress could be the photosynthetic pathway followed, including the ability of some C_3 species to switch to the CAM pathway when stressed, although it would be premature to suggest any definite relationship (see Section 3.2). As Nobel (1996) has pointed out, CAM plants, such as the agaves and cacti, are very tolerant of high temperatures and generally not particularly tolerant of low temperatures.

1.2.1   High Temperature Stress

In general the basic metabolism of living cells is heat sensitive because many of the molecules involved in the intermediary metabolic pathways are unstable 40°C, often with many of the macromolecules and enzymes becoming denatured and inactive with rising temperatures. Yet some forms of life have adapted to high temperatures, e.g. the Archaea (archaebacterium) organism *Pyrodictium*, which occurs in hot springs and exhibits optimal growth above 80°C and is even able to continue growth in temperatures up to 110°C.

Plants respond to heat shock due to above normal temperatures with the synthesis of *heat shock proteins* (*HSP*), accompanied by the suppression of normal protein synthesis. These processes appear to be responsible for the ability of the plant to withstand such increases in temperature. There is, for example, a strong correlation between the production of HSPs and the ability of some millet seedlings to tolerate rapid changes in temperature. Thus, there is now the possibility that once the responsible genes have been identified the temperature tolerance could be transferred to other crops (Harborne, 1988; Blundell, 1996).

Although heat injury is less of a problem than drought or frost damage, high temperatures can cause problems indirectly through water stress. Indeed, high temperatures are often a limiting factor for survival in the dry tropics, although it may be difficult to dissociate the effects of heat stress in such habitats from those of water stress. The lethal temperature, i.e. the *thermal death point* of a plant tissue, varies according to the period of exposure, with higher the temperature the shorter the period of endurance. The thermal death point for organs of temperate plants generally follow exposure for a few hours at temperatures between 45-55°C, while aquatic and shade plants may only survive temperatures up to 38-42°C. At the other extreme, desert plants, such as agaves and cacti, are able to survive air and tissue temperatures exceeding 55°C and have thermal death points at above 60-65°C. Fortunately, for much of the drier regions, the period during which a plant is subjected to potentially lethal temperatures is insufficient for the thermal death point to be reached, although the combination of high temperatures and drought can obviously spell disaster for seedling establishment. Dry seeds are able to withstand higher temperatures than other plant organs, an important survival strategy for survival during a hot dry summer lying on the soil surface (Street and Öpik, 1984; Nobel, 1996).

Soil surface temperatures are strongly influenced by the vegetation cover, temperatures increasing with soil exposure, and this has a direct effect on soil water evaporation which, in turn, also affects seed germination and seedling establishment. This can become an important factor in the recovery of desertified arid lands under conditions of water stress. For example, Cloudsley-Thompson (1977) records a daytime surface temperature of 84°C at Wadi Halfa in the Sudan, which would certainly limit plant establishment and survival. High soil temperatures are probably responsible for the poor to zero recovery of the woody vegetation of the Sahel following the dramatic drought of the late 1960s, even when livestock are excluded.

Some plants make use of the insulating effect of the soil for protection against high atmospheric temperatures by exposing the minimum of aerial parts to the atmosphere. The 'living stones' of south-western Africa, for example, are members of the geophytic, succulent genus *Lithops*, with the fleshy leaves so well camouflaged that they resemble stones embedded in the soil surface. The plants grow with most of the plant body buried in the upper soil surface, tolerating a maximum March soil surface temperatures of 43.7°C. Each year the plants produce one or more obconical bodies composed of a pair of fleshy leaves above a basal column. The leaf apices are flattened, often with transparent 'windows' through which the aqueous leaf tissues are visible. Variations in the clarity of the windows appear to be the sole thermal adaptation, enabling the plant to maintain a constant leaf temperature over a wide range of climatic conditions (Turner and Picker, 1993). Similarly, in the barren coastal desert of Chile there are the geophytic 'earth cacti' belonging to the Thelocephala group of *Neoporteria*. They bear small, brown aerial parts that are barely visible except when the plant is in flower or in times of extreme drought when the swollen subterranean storage tubers shrink, leaving a noticeable gap between the plant and the soil. The tubers also have thin root necks from which arise fine lateral roots radiating out at 1.5 cm below the surface, absorbing the condensed fog that provides the major or only source of water (Rauh, 1985; Hoffmann, 1989).

Plants growing in hot climates may show some adaptations for heat avoidance in order to keep their temperatures near or below air temperature. These may include highly reflective surfaces to reduce heat absorption, or making use of stomatal resistance for **transpirational cooling** to reduce shoot temperatures. Under favourable circumstances **stomatal cooling** will enable leaves to be cooled to 10°C below air temperature. However, such cooling varies from species to species and is greatly affected by environmental conditions as well as being dependent on an ample water supply for any significant cooling. Once leaf temperature has risen above that of air temperature, heat convection will result in heat loss, and this effect may be important for plants with small leaves (Street and Öpik, 1984).

1.2.2   Low Temperature Stress

Low temperature stress is experienced particularly by plants subjected to the long, winters of the cold temperate and polar regions of the world, where the temperatures reported for plant ecosystems can be as low as -60 to -70°C. In the polar latitudes arboreal growth becomes increasingly stunted towards the poles, and a *tree line* is formed where the summer temperatures are rarely sufficiently high for tree seeds to ripen; a similar transition also occurs in mountain zones (Metcalfe, 1983; Street and Öpik, 1984; Arnon, 1992; Archibold, 1995).

Even in the warm deserts seasonal low temperatures can limit plant growth. In both the Sonoran and Chihuahuan Deserts of North America the northern limit for various cacti species is the -1°C annual isotherm. It is the seasonal low temperature that mainly limits the cultivation of *Opuntia ficus-indica* (Indian or Barbary fig) in the 20 or more countries around the world where it is grown (Nobel, 1996). Similarly, in southern Africa the frost line marks the southernmost distribution of *Adansonia digitata* (baobab) (Wickens, 1982). Indeed, many warm climate species are unable to endure temperatures below 5°C, and sometimes not below 15°C. This is believed to be due to a gelling of the cellular membranes, and is known as *cold shock*, or *chilling injury*.

A specific mechanism for temperature regulation is absent in the majority of plants. Even the temperature within bulky organs is close to the outside air temperature, except in direct sunlight or on cold nights when both leaf and bud temperatures will often fall below ambient. However, some rather exceptional adaptations for keeping the internal temperature of plant organs above air temperature have been observed at high altitudes in the tropics, where the night temperatures throughout the year may fall to *ca.* -10°C. Their most distinctive feature being the production of giant rosettes, and the formation of diurnal *night buds*, where the freeze tolerant adult leaves close tightly over the central meristem. A copious supply of mucilaginous fluid in the buds also acts as a thermal buffer, enabling the meristem temperature to be maintained appreciably above the air temperature. Such examples include the giant *Dendrosenecio* and *Lobelia* of East Africa and species of *Espeletia* and *Puya* in the Andes.

The main resistance mechanism of plants to low temperature is *low temperature tolerance*. A form of freezing avoidance may be obtained for perennating organs such as buds, rhizomes and bulbs by their subsurface positioning, whereby they are maintained at near temperature equilibrium with their immediate environment. The effects of cold, however, depends not only on the lowest temperature reached but also on the rates of cooling and warming, the length of exposure to minimum temperatures, and subsequent treatment after warming. Not surprisingly, species in regions with cold winter climates are more resistant to cold than species from warm climates. The cold resistance of perennials of temperate and polar regions also undergo noticeable seasonal changes, with very much higher resistance in the winter months than in the summer. The increase in hardiness that occurs in the autumn is stimulated by falling temperatures and shortening day length, without which the plants would be unable to survive the rigours of winter. A greater resistance to frost injury is developed during severe winters than in mild winters. A loss of hardiness in the following spring occurs with the relative rapid rise in temperature. It is also possible to induce artificial hardening of a plant at any season of the year by chilling, a process known as *vernalisation*, although the resistance can be destroyed in mid-winter by warming (see also Section 6).

Freezing damage could be regarded as the result of mechanical disruption caused by ice crystals within the plant tissues. Since tissue dehydration occurs as water passes from the cells into the ice crystals, freezing damage could also be regarded as a form of dehydration damage caused by the denaturation of the cell proteins, especially those of the cell membranes. The freezing resistance of a tissue is a cryoscopic effect and is dependent upon the amount and state of the water in the plant cells.

During the period of frost hardening of perennials, there is a tendency for the total water content in the tissues to decrease and for the proportion of bound water to increase. This is due to an increased hydration capacity of the cell colloids. Despite this, many frost-hardy tissues still contain very appreciable amounts of freezable water. Freezing of the cell water may be avoided by the cryoscopic effect of the solutes present, an increase in soluble carbohydrates, mostly glucose, fructose and sucrose, being associated with *frost hardening*, although other sugars or polyhydric sugar alcohols may also be present. For *Robinia pseudoacacia* frost hardening is associated with an increases in the soluble protein content of the bark, while the high salt content of halophytes has resulted in freezing points as low as -14°C.

In the field *cold injury* is more frequent in spring when late frosts act on young, non-hardy tissues. Resistance also varies greatly between plant organs, the greatest resistance to cold injury is encountered in dry seeds, a factor that is utilised by seed banks for the long-term storage of seeds. Underground structures are much less resistant than aerial parts; they also have a more uniform resistance throughout the year because soil temperatures show less annual temperature variation and never fall as low as air temperatures (Street and Öpik, 1984; Harborne, 1988).

1.3   Salt Stress

According to Szabolcs (1993) nearly 10% of the world's land surface, i.e. *ca.* 954.8 ×10^{-6} ha, are affected by salinity. In addition approximately half of all the world's irrigation systems are subject to secondary salinisation, alkalisation, and water-logging, with *ca.* 10 ×10^{-6} ha of irrigated lands being taken out of cultivation annually due to salinity alone. Since 98% of terrestrial plants are non-halophytes, and where some plants, such as tomatoes, peas and beans are intolerant of salinity as low as 0.1% NaCl, salt stress is clearly an important global problem, especially for agriculture.

Stress due to high salinity may be caused by either water stress or by the toxic effects of the Na and/or Cl ions, the latter being considered the major stress factor. Salt accumulation, as well as producing toxic concentrations that stress plant growth, also reduces water availability to the plant by increasing the osmotic potential of the rhizosphere. In addition, high concentrations of Na ions in the soil can adversely affect the physical soil characteristics. For example, sodic soils, which have a high concentration of exchangeable sodium but a low concentration of total salts, can exhibit a poor soil structure and low water permeability, thereby further reducing water availability.

Because of their diversity halophytes are a potential source of new vegetable, forage and oilseed crops. Trials with sea water irrigation for biomass production have been encouraging, the most productive yielding 10-20 ton of biomass ha^{-1}, and with *Salicornia bigelovii* yielding 2 ton ha^{-1} of seeds containing 28% oil and 31% protein, comparable to soya bean in yield and quality (Harborne, 1988; Briggs, 1996; Glenn *et al.*, 1999).

Salinity is usually measured in terms of *specific electrical conductivity* (EC_e or EC_i) of saturation extracts at 25°C in decisiemens per metre (dS m^{-1}), where the criteria for salinity is EC_e or EC_i = 7.8 dS m^{-1}, if not considerably lower. For sea water EC_i = 50-80 dS m^{-1}, while underground brackish water typically varies from as low as 1.3 dS m^{-1} to as high as 10 or even 20 dS m^{-1}. Salt sensitive plants will show reduced growth with salinity levels as low as 2 dS m^{-1}, whereas salt tolerant plants will show no ill effects at levels as high as 10 dS m^{-1}. Most plants, however, will show signs of slight to moderate sensitivity at levels of 5-10 dS m^{-1} (Aronson, 1989; Briggs, 1996).

A saline habitat is defined as one where the NaCl concentration equals or exceeds 0.5%. Plants capable of growing normally in such concentrations are classed as *halophytes*. Street and Öpik (1984) recognise both *facultative* and *obligate halophytes*, the latter requiring at least 0.5% NaCl. The term, however, is subject to several interpretations depending on the author. Thus Sen and Mohammed (1993) regard *true halophytes* as plants such as *Salsola imbricata*, syn. *S. baryosma* and *Suaeda vera*, syn. *S. fruticosa*, that grow in more than 1.5% NaCl and resist extreme saline conditions. Species such as *Portulaca oleracea* and *Tamarix* spp. are examples of the many *facultative halophytes* capable of growing in either saline and nonsaline soils or in saline soils up to 0.5% NaCl. As the name suggests, *transitional halophytes* such as *Salvadora persica* and *Senna italica*, syn. *Cassia italica* grow in the transition zone between saline and nonsaline conditions, while the salt-sensitive *glycophytes* consist of those deep-rooted perennials that are able to tolerate low levels of salinity for only a short duration. Data on the maximum recorded salinity tolerance for individual species is given in Aronson (1989).

As well as having to satisfy their internal water requirements, halophytes also have to regulate the salt concentration. In order to absorb water from the saline soil solution with a low osmotic potential, it is also necessary for the osmotic potential of the cell sap in the vacuoles to be even lower. This is achieved in some plant by first absorbing salt and water until the osmotic forces of the soil water and cell sap are in equilibrium, after which no further salt can be taken up since it would then have a toxic effect upon the cells. Since the older roots are almost impermeable to salts they are able to absorb almost pure water from the saline soil water, and this under very high cohesive tension in the conducting vessels.

A limited resistance to salinity may be achieved in some species by avoiding the uptake of excessive amounts of salt. Although the salt is absorbed into the root system, very little salt is transported into the shoot, the exclusion mechanism being dependent on the properties of the plant's cellular membranes. Halophytes capable of growing in salinity levels of sea water, i.e. 3-5% NaCl, or even higher concentrations, have the ability to accumulate high levels of Na^+ and Cl^-, thus enabling them to maintain a low internal water potential and to avoid water stress deficits. For example, *Salicornia* can accumulate a 10% solution of NaCl in its tissues, while the perennial grass *Leptochloa fusca* (kalla grass) is able to sequest high levels of Na^+ and Cl^- in the leaf sheath, away from the blade.

The osmotic balance is provided by the synthesis of organic solutes, including amino-acids *sensu lato*, especially the imino-acid ***proline***, which is also involved in drought stress (Section 1.1). In general the concentration of proline in halophytes may reach ten-fold the level present in the free amino-acid pool of non-halophytes, although there are a few halophytes that do not exhibit enhanced proline levels. The precise mechanism by which proline functions in salt resistance is not known, but its presence is presumed to enable halophytes to withstand the high osmotic forces to which their cells would otherwise be subjected.

Other common halophyte characters are succulence, which is believed to be a means of diluting the salt within the plant with large volumes of water. Also the possession of excretory salt glands which enable many halophytes, especially non-succulent halophytes growing in saline habitats, to excrete excess salt and thereby prevent the accumulation of high and possibly toxic concentrations of salt in the shoot tissues. The excreted salt can often coats the leaf surface and even the ground below, the allelopathic effect of which can inhibit the growth of ground cover beneath the canopy (Pollak and Waisel, 1970; Street and Öpik, 1984; Walter and Breckle, 1985; Harborne, 1988).

Tolerance to salinity is genetically controlled, although the inheritance is relatively complex and varies with climate and other environmental conditions, as well as differences in susceptibility, especially among grass species and to a lesser extent among forage legumes. By understanding the metabolism of halophytes, how they respond to salinity and how they differ from non-halophytes, should help in the selection and/or development by conventional breeding or genetic engineering of crops with increased salt tolerance. *Lycopersicon cheesmanii* (Galapagos wild tomato), for example, can grow in sea water, albeit with reduced growth, while the cultivated tomato, *L. esculentum*, can barely survive in 50% sea water. The two species exhibit markedly different nitrogen metabolisms and toxicity tolerances. The wild tomato tolerates 200 mM Na^+ while 200 mM K^+ is toxic; the cultivated tomato exhibits the reverse reaction. Studies with the interspecific hybrids of the cultivated tomato revealed that the higher level of tolerance is a dominant genetic factor.

Selection for salt tolerance hybrids by backcrossing to a cultivated tomato resulted in tolerance to 70% sea water with associated fruit size, quality and yield increasing with each successive backcrossing (Arnon, 1992; Grattan and Grieve, 1993; Mayber and Lerner, 1993; Subbarao and Johansen, 1993). For further information on salt stress see Pesssarakli (1993).

2.   ADAPTATIONS TO THE ENVIRONMENT

There are considerable practical problems in distinguishing between behaviour that is truly adaptive and that which is hereditary. For example, the term *xerophyte* has traditionally been used to denote a plant restricted to a dry locality, while *xeromorphic* is used to denote plants which, from the appearance of their external morphology and internal histology, give the impression that they will only be found in dry localities. However, xeromorphic plants are not necessarily restricted to dry localities; in fact, the so-called xerophytic histological characters should more correctly be considered as xeromorphic. The *sclerophytes* also, while exhibiting many xeromorphic characters such as tough, leathery leaves, are characteristically trees and shrubs species of Mediterranean-type climates, belonging to such genera of the Mediterranean maquis as *Olea*, *Pistacia*, *Quercus*, *Styrax* and *Thymus*. The characters can be variously attributed to conditions of high light intensity, nitrogen deficiency or water deficit (Kummerow, 1973; Metcalfe, 1983).

In view of such problems, plus the diversity of environments, ranging from tropical rain forest to arctic tundra, and further complicated by the distribution of some species embracing both mesic and xeric environments, it is impossible to discuss environmental adaptations in any detail in this chapter. It is proposed, therefore, to limit the discussion to the more common stem, leaf and root adaptations to stress.

Some plants are able to avoid water stress by completing their life cycle before the onset of drought, i.e. the dry season. Included in this category are the *therophytes*, the ephemeral and annual plants which are susceptible to drought and can only survive the onset of drought as desiccation tolerant seeds. *Ephemerals* have a short life cycle that enables them to take advantage of the occasional favourable light rainfall and, under favourable conditions, produce more than one generation in a year. Should a favourable rainfall fail to occur, their seeds are capable of remaining dormant for a number of years. *Annuals* are less opportunistic, producing only one generation during the growing season, completing their life cycle during the rainy season and, like ephemerals, surviving the dry season as seed. Under water stress some perennials may also behave as annuals, completing their life cycle within the first rainy season. Such facultative behaviour may be found in the Namibian grasses *Centropodia glauca* and *Stipagrostis ciliata* (Jacobson, 1997).

Many ephemerals and annuals produce large quantities of seed, a survival strategy to ensure that at least some will survive animal and insect predation, the latter being a major source of loss from the *soil seed bank*. The various strategies by which seeds are intermittently dispersed and/or germinate in order to ensure their survival are discussed by Gutterman (1993). While such strategies are essential for plant survival under natural environmental conditions, they can present considerable problems for the cultivator.

2.1   Stem Adaptations

The function of a stem is to act as a pathway for the transport of water and nutrients between the organs and, if required, their storage. In addition the stem provides support and position for the leaves, flowers and fruits in order that they may function efficiently.

Those plants which have a high proportion of water storing parenchymatous tissue associated with a low surface to volume ratio are known as *succulents*. They have been defined by von Willert *et al.* (1992) as follows: "*A succulent (or succophyte) is a plant possessing at least one succulent tissue. A succulent tissue is a living tissue that, besides possible other tasks, serves and guarantees at least temporary storage of utilisable water, which makes the plant temporarily independent from external water supply when soil water conditions have deteriorated such that the root is no longer able to provide the necessary water from the soil.*"

For many non-halophytes succulence is a means by which plants are able to endure periodic droughts. Although succulence is generally considered to be a xeromorphic character associated with arid or saline environments, this is not necessarily true. Metcalfe (1983) cites the example of numerous cacti being successfully cultivated with their lower portions immersed in bottles of tap water and remaining perfectly healthy for a number of years. Stem succulence is relatively common and is not restricted to arid environments, the thick stemmed or *pachycaul* habit also occurs in more mesic environments, rarely in rain forests albeit sometimes dominating tropical montane communities (see Mabberley, 1973 and references therein). Columnar stem succulence is found in a number of the families, including the Bombacaceae, Cactaceae, Didiereaceae, Euphorbiaceae and Fouquieriaceae.

The ability of some desert plants, mainly in coastal and areas, to trap moisture from dew or fog is well documented. Adaptations by the plant to maximise the trapping of the dew or fog include divaricate branching patterns and rosette growth with upright leaves and spreading spines. The moisture from dew, fog or light rain is concentrated on the plant surface and falls to the ground as leaf drip and stem flow which, in the latter case, is then channelled along the root system into the soil This can significantly augment the water available to the plant (Glover and Gwynne, 1962; Glover *et al.*, 1962; Gindel, 1970; Rundel, 1982; Sundberg, 1985; Martinez-Meza and Whitford, 1996; Whitford *et al.*, 1997).

Various forms of stems may be adapted for food storage, especially starch, and/or water, and includes many species of economic importance for food, pharmaceuticals, ornamentals, etc. These include: **stems** - *Metroxylon sagu* (sago palm), *Saccharum officinarum* (sugar cane), **stem tubers** - *Helianthus tuberosus* (Jerusalem artichoke), *Oxalis tuberosa* (oca), *Solanum tuberosum* (potato), *Ullucus tuberosus* (ullucu), **rhizomes** - *Canna edulis* (Queensland arrowroot), *Dioscorea* spp. (yams), *Leontice leontopetalum* (lion's paw), **stolons** - *Trifolium repens* (white clover), **corms** - *Crocus* spp., *Gladiolus* spp., **swollen hypocotyle** - *Brassica rapa* Rapifera Group (turnip), *Cyclamen* spp., **rootstock** - *Astracantha microcephala*, syn. *Astragalus microcephalus*, (gum tragacanth). Also, according to Schimper (1898), the extremely succulent bark of the chenopod *Haloxylon aphyllum* is reputed to act as a water reservoir. Conversely, in extreme cases of drought some multistemmed members of the Chenopodiaceae, Compositae and Rosaceae are able to shed some of their limbs by producing **intercalary cork** between the stem segments, thus enabling the plant's water resources to be concentrated in only a section of the canopy (Gintzburg, 1963; Danin, 1983).

In functionally leafless species, where the leaves are small and vestigial, rudimentary or caducous, the process of photosynthesis is transferred to other organs, such as the stem. In such **apophyllous plants**, also known as **aphyllous**, **spartioid** or **switch-plants**, they often acquire a broom-like habit. In xeromorphic species the usually cylindrical stems may be longitudinally ridged or furrowed, thereby protecting the stomata in the furrows; the stems may also be winged, as in species of *Baccharis* and *Genista* (Metcalfe, 1983).

The ridge and furrow morphology, especially in stem succulents, may also function in a concertina-fashion as the stem water content increases and decreases. In other examples specialised, flattened and leaf-like, stem structures known as **cladodes** (**phylloclades**) perform the function of the leaf, from which they may be distinguished by the presence of surface buds. Such structures are represented by the 'leaves' of *Ruscus aculeatus* (butcher's broom), *Euphorbia enterophora* and the stem joints or pads of *Opuntia* spp. The cladodes should not be confused with **phyllodes**, which are modified petioles, some of the best-known examples of which are to be found among the Australian species of *Acacia* subgenus *Heterophylla* (see Section 2.2).

In the tropical rain forest trees tend to have relatively slender trunks in relation to their height, which is associated with their rapid growth to reach the upper canopy and compete with adjacent trees for light. As a consequence the trunks tend to be unbranched and bear a relatively small crown. To illustrate the withdrawal of competition for light Richards (1957) cites the South American leguminous tree *Eperua falcata* (wallaba) which, when planted in the Botanic Garden at Port-of-Spain, Trinidad, bore leafy branches to within a short distance from the ground. Similarly, forest trees bordering river banks will also branch nearly to the base of the trunk on the side exposed to the open water. The forester takes advantage of this trait by the close planting of trees to encourage trunk growth, and thinning later to avoid overcrowding and competition for nutrients and water. The ability for fallen or decayed trees to coppice readily is another characteristic although not necessarily confined to tropical forests since the fallen trunks of the African savanna tree *Adansonia digitata* (baobab) similarly develop coppice shoots along the length of the trunk.

Another forest adaptation is for *cauliflory*, the bearing of flowers on the secondarily thickened tissue of the trunk or branches, the flower buds developing from suppressed side shoots or, less commonly, from the phloem and forcing their way through the bark, e.g. *Theobroma cacao* (cacao). This is a trait that is most frequently found in the lower story and is possibly associated with pollination or fruit dispersal by bats (Richards, 1957; Tootill, 1984).

A group of woody plants of the lower rainfall regions known as *suffrutices* are woody at the base and bear only annual or short-lived aerial shoots, thereby keeping transpiration to a minimum. They are able to evade water stress during the dry season or periods of drought by having massive underground storage organs in the upper 2 m of the soil, with absorbing roots descending for 18 m to the water-table (Ferri, 1961; Parker, 1968). The stems of some suffrutices may even have a corky layer within the xylem, e.g. *Artemisia* spp. where it is believed that such layers occurring between the annual rings help to retain stem moisture.

2.2   Leaf Adaptations

Those woody perennials that retain their canopy throughout the year by continuously shedding and replacing a proportion of their leaves are referred to as *evergreen*. Many are from tropical regions not subjected to cold or dry seasons. Those of temperate and cold climates exhibit various adaptations against transpiration water losses. Thus, many coniferous species such as *Pinus*, bear needle-like leaves, others, like *Cupressus*. have scale-like leaves, thereby reducing their leaf surface area as a means of reducing water loss.

In the tropical rain forests young leaves tend to be brightly coloured red or crimson, limp and lacking in chlorophyll (also a striking feature of the *Brachystegia* savanna woodlands of tropical Africa). The colouring may possibly be a causal one connected with the nature of the climate and of no adaptive function, while the limpness is due to a delay in the development of the strengthening tissues. The mature leaves, regardless of taxonomic affinities, are predominantly large and sclerophyllous, typically oblong-lanceolate to elliptic in outline, margins generally entire, and often with a long and distinct acumen forming a 'drip-tip', often with a pulvinus at the junction of the petiole and the blade. The conspicuous exception to these generalisation are the *nanophylls* or *leptophylls* of members of the Leguminosae subfamily Mimosoideae. In contrast to the striking uniformity of the foliage in the upper story, a wide range of foliage is found in the undergrowth, where temperature and humidity are extraordinarily constant (Richards, 1957).

The majority of subtropical and temperate broad-leaved trees and shrubs are *deciduous*, shedding their leaves either at the end of the rainy season/autumn, or they are *drought deciduous*, shedding their leaves at the onset of drought; leaf abscission during periods of drought being a common means of reducing transpiration. Depending on the distribution of the rainfall some species, e.g. *Fouquieria splendens*, may even produce several sets of leaves each year. Although generally associated with trees and shrubs, deciduous leaves may also occur in some perennial herbs, e.g. *Anthyllis cytisoides* (Alados *et al.*, 1996) and the temperate, winter deciduous grass *Molinia caerulea* (purple moor grass).

Woody plants with tough, leathery and often evergreen leaves are known as *sclerophytes* and are characteristic of, but not restricted to, Mediterranean climatic regimes. Such heavily cutinised leaves generally have a low surface to volume ratio, which is regarded as an adaptation to high light intensity, nitrogen deficiency or water deficiency. They have the ability to endure water stress and continue to function, albeit at a slower rate, during the dry season. In general, the photosynthetic potential of their leaves tends to be substantially lower than that of deciduous plants, but this is compensated for by an increased leaf area and duration, and lower annual investment in replacement of leaf tissue. On the other hand, those evergreen trees and shrubs of the tropics which are not subject to dry seasons and periods of cold will have their shoots and leaves unreduced and continue to be physiologically active throughout the year. The leaves of sclerophytes also show a number of what might be considered xeromorphic features, such as thick outer walls to the epidermal cells and overlying cuticle, although the stomata are not always depressed (Meidner and Sheriff, 1976; Metcalfe, 1983; Walter and Breckle, 1985; Archibold, 1995).

A reduction in transpiration under dry conditions may also be obtained by a reduction in the actual leaf area. This reduction, as in the caesalps *Parkinsonia florida* and *P. microphylla*, is compensated for by the green bark carrying out as much as 50% of the annual photosynthetic output. A similar behaviour is also found in other families, including the Chenopodiaceae (*Haloxylon* spp.) and Euphorbiaceae (*Euphorbia* spp.). In other examples, such as the legumes *Caragana* spp. and *Parkinsonia aculeata*, the leaflets are shed, but the petioles persist, thereby reducing the transpiring surface by *ca*. 85%. A somewhat similar situation occurs with *Zygophyllum dumosum* (bushy caper bean) during the dry Mediterranean summer when the leaflets are shed and the cylindrical petioles, which act as water reservoirs and may persist for up to 3 years, become shrivelled and shrunken, inflating again within 2 days following the first effective winter rain (Danin, 1983).

Other desert and Mediterranean plants may bear *dimorphic leaves*, with relatively large leaves during the growing season and smaller leaves during the dry or dormant season, often with increased pubescence. Such seasonal dimorphism is characteristic of shrubs with mesophyll leaf dimorphy; it is not commonly found among the sclerophylls. Typical examples of dimorphic leaves may be found among the rocky heaths of the Mediterranean and Middle East, which abound in perennial species with superficial root systems that do not have access to deep soil water, e.g. *Thymus serpyllum* and *Salvia fruticosa*. Others may even replace their leaves with short lateral shoots known as **brachyblasts**, e.g. *Noaea mucronata*, *Salsola tetrandra* and *Sarcopoterium spinosum* or have stipules which persist after leaf fall and grow turgid under low soil moisture conditions (Westman, 1981, cited by Archibold, 1995; Danin, 1983).

Phyllodes are more or less flattened petioles that resemble and perform the function of a leaf. They are particularly prevalent among the Australian species of Compositae, Labiatae, Leguminosae, Myrtaceae, Pittosporaceae, Proteaceae, etc. Thus, in many 'evergreen' members of the mainly Australian *Acacia* subgenus *Heterophyllum* adapted to dry climates, e.g. *A. aneura* and *A. saligna*,. the photosynthetic function is taken over by phyllodes. Usually there is a complete loss of rachis and leaflets, although in some species the leaflets are present on juvenile plants or vestigial as in *A. victoriae*; stipular spines are usually absent. As a general rule, thin, cylindrical phyllodes predominate in arid areas, while in more mesic areas the phyllodes tend to be relatively broad and flat. Both forms, however, may occur in the same habitat or even on the same shrub in relation to their relative position to sun or shade (Tootill, 1984; Pedley, 1981; Walter and Breckle, 1986; Maslin, 1988).

The ability of leaves to roll or fold under water stress is a fairly common means of shielding the stomata against high transpiration losses, especially among the Gramineae. A similar protection may also be achieved when the stomata are in longitudinal grooves, as in *Nerium oleander* and *Nolina parryi*, where a lessening of leaf turgidity brings the lips of the grooves closer together (Went, 1971).

Many xerophytes have hairy leaves; these are generally to be found among members of the Caryophyllaceae, Chenopodiaceae, Compositae, Ericaceae and Solanaceae. In some xeromorphs the hairs are either air-filled or of a distinctive type. While it is widely believed that the indumentum can give a physiological advantage to xerophytes by shielding the stomata and thereby reducing transpiration loss, this is not always the case, since their removal has also been demonstrated to reduce transpiration. The situation is certainly not straightforward. *Seriphidium tridentatum*, for example, has both live hairs filled with protoplasm and dead, air-filled cells. The silky down of the living hairs on a freshly unfolded leaf promote transpiration by increasing the surface area, while a dense felt of dead hairs reduces transpiration and at the same time acts as a protective layer against direct sunlight. Furthermore, hairs may perform other functions, such as scattering incoming radiation or in breaking up soil-reflected radiation, and helping to keep the leaf cool by increasing the total radiating surface. For the halophytic members of the Chenopodiaceae the bursting of the salt-excreting, vesicular hairs deposits salt on the leaf surface, where their reflectance is believed to reduce absorption of solar radiation. An indumentum may also help in preventing insect attack (Harder *et al.*, 1965; Danin, 1983; Metcalfe, 1983; Archibold, 1995).

A waxy leaf surface is commonly regarded as an effective means for reducing transpiration water losses. But this is not necessarily so. In the carnauba wax palm it appears to be a response to low humidity since it is not formed if the humidity is too high, but for species of *Ceroxylon* palms, the wax acts as an insulator against lower temperatures (Corner, 1966; Andrade-Lima, 1981).

Leaves may also have a protective function, presumably against herbivores, through selective **aphylly** or **branch dimorphy**, such as occurs in the Saharan composite *Launaea arborescens*, where the outer 'sun branches' are aphyllous and it is the inner branches that bear leaves. Similarly, in the intricately branched Andean leguminous shrub *Adesmia spinosissima*, the protective spinose outer branches are likewise leafless, with the leaves and flowers confined to the inner branches. In other examples the leaf rachi, or spinescent stipules may persist after the lamina has fallen and serve to protect young growth in the following growing season, as in the legumes *Astragalus* spp., *Caragana* spp. and *Halimodendron halodendron*. This appears to be a relatively common characteristic of cold desert species. While in the African scandent shrub *Combretum aculeatum* the persistent petiole becomes indurated and performs a climbing function.

The adaptation of leaves for food storage, especially starch, sugar, and/or water by means of modified scale leaves or leaf bases of **bulbs** occurs in a number of species, including some that are a source of food, such as *Allium* spp. (onions, leeks, etc.) as well as numerous other species that are of horticultural interest. In addition to water storage, leaf succulence may also be a means of reducing the transpiration area. For example, *Agathophora alopecuroides* is able to dramatically reduce its transpiration surface by changing the shape of the leaves from cylindrical to globose (Danin, 1983).

2.3   Root Adaptations

Many of the trees in the evergreen tropical forests, irrespective of their taxonomic relationships, have a number of characteristic features that are usually not found in other environments, although whether they necessarily have any survival value is considered questionable by Richards (1957). These include wing-like expansions at the base of the trunks of many of the larger trees. Such buttresses are considered to provide support or anchorage to what are usually shallow rooted trees. Buttresses are quite common among the upper storey trees of the lowland rain forests, less prevalent in montane rain forests. *Fluting*, which involves the entire trunk, must not be confused with basal buttresses, which are modifications of the lateral roots and may take the form of stilt roots or plank buttresses, with intermediate forms also present.

Stilt roots are stout, woody, adventitious roots, arising from up to 1 m or more above ground level, where they may branch or sometimes anastomise, bending downwards to enter the ground and there giving rise to secondary or tertiary roots and rootlets. They are generally found among lower story species growing in poorly-drained forest soils, and are usually found among species that do not have horizontal, lateral roots. However, certain species that normally bear stilt roots when growing in poorly-drained or swampy soils may not develop stilt roots when growing on well-drained soils or even develop plank buttresses. Their development appears to be associated with such factors as flood levels or a continuously damp atmosphere near ground level rather than function for support. Stilt roots are also well represented in the mangrove and fresh-water swamps, the latter including *Taxodium distichum* (bald, south or swamp cypress) of the Florida everglades (Richards, 1957; Archibold, 1995).

Plank buttresses or *tabular roots* are more or less flat, triangular plates subtended between the trunk and a subsurface lateral root. They are particularly well represented in the lowland rain forests. Such buttresses can only occur in trees with a superficial root system, i.e. lacking a well-developed taproot and well-developed and generally, horizontal lateral roots. A striking feature of trees with well-developed plant buttresses is that the trunk, instead of tapering steadily upwards from ground level, is always more-or-less markedly tapered upwards and downwards from the top of the attachment of the buttress to the trunk. Because three buttresses are the minimum number necessary to provide mechanically efficient support, and because trees with one or two buttresses are comparatively rare, plank buttresses are regarded by some as being primarily an adaptation for support. Buttressing also appears to be more closely correlated with effective soil depth than with soil texture. A number of theories are discussed by Richards (1957), including that of Petch (1930) who postulated that due to the absence of a taproot, buttress growth was due to water movement in the trunk being conducted in the region above the superficial lateral roots. Later, anatomical work by Chalk and Akpalu (1963) cited by Archibold (1995) confirmed that species with well-developed buttresses could confine water movement to those parts of the trunk. Outside the rain forests buttress-like appendages can also develop on *Thuja plicata* (western red cedar) of the coniferous forests of the Canadian Pacific coast, and on *Ulmus americana* (white elm) growing in poorly aerated sites of south-eastern North America, and in Europe they may be found on *Populus nigra* 'Italica' (Richards, 1957; Archibold, 1995).

Other woody plants, known as *phreatophytes*, have deep roots tapping the water-table which, in evergreen species, permits ample transpiration throughout the year, e.g. *Conocarpus lancifolius* and *Larrea divaricata*. But not all phreatophytes are evergreen, some, including species of *Acacia* and *Prosopis*, are deciduous during the dry season. A distinction should be drawn between *obligate phreatophytes* which continue to tap the perennial water-table throughout the year and *facultative phreatophytes* whose deep roots are in contact with a deep but seasonally fluctuating ground water-table. The former favours a more mesophytic canopy and the latter a deciduous or more xerophytic foliage. It is only by the rigorous confirmation of water uptake by roots from the capillary fringe of a permanent water-table that a plant can be designated as being a true phreatophyte (Smith *et al.*, 1997). In tropical Africa the probably obligate phreatophytes *Acacia nilotica* subsp. *nilotica* and *Faidherbia albida* (possibly a facultative phreatophyte or even a provenance variant when growing away from riverine sites) are able to tolerate waterlogging for several months of the year. The respiration pathway of the root system under such anaerobic conditions has yet to be determined (see Section 4). The latter species also has the unusual habit throughout its entire distribution range of bearing leaves during the dry season and shedding them at the start of the rainy season, including areas with a bimodal rainfall. In South America *Acacia caven* behaves in a similar manner, but only for part of its distribution; the reason for such behaviour is not understood.

A more bizarre root system is that of *Adansonia digitata* (baobab) whose initially swollen, carrot-like taproot is remarkably short-lived, being replaced by a shallow and extensive lateral root system for stability, nutrients and water; hence its presence along the margins of dune depressions and absence in deep sands (Wickens, 1982).

Mucilaginous secretions from persistent root hairs and culms that cause sand particles to agglutinate and form a **rhizosheath** have been reported for a number of species growing on sandy soils, especially among the Gramineae. Danin (1996) found them on all the perennial grasses he investigated inhabiting desert dunes. They appear to be restricted to plants growing on sandy soils where they are believed to protect the roots from abrasion by sand grains when exposed by deflation. Rhizosheaths have also been recorded among other families of the monocotyledons, while the dicotyledons are less well represented, possibly because they have not been sought for.

Their function is still subject to speculation and requires further research, including why they are confined to plants growing on sandy soils and not heavier textures. It has been suggested by Oppenheimer (1960) that in the Chenopodiaceae they enable the root hairs to continue functioning during drought, while Volkens (1887) and Price, (1911) believe that the mucilage concentrated moisture around the sheath, thereby promoting water uptake in addition to providing protection against desiccation. Wullstein *et al.* (1979) and Wullstein (1991) certainly demonstrated the water content of rhizosheaths to be four times greater than that of control sand samples, with the rhizosheaths providing a nano-climate within which nitrogen-fixing bacteria were able to survive. Using acetylene reduction, Wullstein *et al.* (1979) found a relatively high rate of nitrogen fixation within the rhizosheaths of *Aristida purpurea, Elymus lanceolatus, Oryzopsis hymenoides* and *Stipa comata*, although the actual uptake of nitrogen by the host grasses was not demonstrated.

A somewhat analogous situation occurs in numerous South African grasses and in the herbs *Commelina* spp. and *Gerbera* spp., where the inner cortex of the root eventually become spongy due to the perforation of the disintegrating cell walls and form dead sleeves which readily absorb and store water. In the Mediterranean region the old and spongy roots of *Bromus rubens* (red brome grass) were similarly able to absorb 12% of their weight by capillary action on contact with water, while in the rhizomatous perennial grass *Lygeum spartum* (albardine) the strongly hygroscopic dead roots were able to absorb 100% of their original weight from moist air (Oppenheimer 1960).

The ability to produce **adventitious roots** on stems buried by shifting sands occurs in a number of psammophilous species belong to a number of families, among which are the Anacardiaceae, Asclepiadaceae, Chenopodiaceae, Compositae, Labiatae, Rhamnaceae, Rosaceae and Tamaricaceae. Other plants have the ability to trap sand, either within the plant canopy or on the leeward side, and to grow up through the accumulated sand, either by producing adventitious roots on the buried branches or by rapid growth to maintain their original rhizosphere. Such nebka-building plants are widely distributed both globally and taxonomically. I have also observed adventitious roots in the Sudan on the stems of *Leptadenia pyrotechica* overwhelmed by dune movements (Le Houérou 1986; Walter and Breckle 1986; Danin 1996).

Adaptations of roots for food and/or water storage by means of **swollen taproots** and **root tubers** are to be found in a number of species, many of which are of economic importance. They include: *Beta* spp., *Brassica* spp., *Ceiba acuminata*, *Cucurbita foetidissima*, *Daucus carota*, *Harpagophytum procumbens*, *Luetzelburgia auriculata*, *Manihot glaziovii*, *Marah gilensis*, *Proboscidea peruviana*, *Raphanus* spp., *Spondias tuberosa*, *Tylosema esculentum* and *Tragopogon porrifolius*.

2.4   Adaptations to Fire

Controlled burning can be a very useful tool for bush control and in range management, but when out of control the effects can be devastating. Fire can be a major hazard in those regions with a marked dry season, such as semi-arid grasslands, savannas, woodlands and heaths, where dry vegetable matter can accumulate during the dry season. It is no longer a hazard in, for example, the Sahel where prolonged drought and desertification have resulted in an extremely poor vegetative cover. While species rich in volatile oils, such as *Eucalyptus*, may be subject to spontaneous combustion. Fires are even possible in the rain forest, as is evident from the horrendous fires in Indonesia during 1998 following a freak dry period.

Sometimes the actual duration of grass fires can be surprisingly short, lasting for only 5 sec (Hodkinson and Griffin, 1982). In Mediterranean ecosystems Archibold (1995) cites references to temperatures of 700°C in Californian chaparral and reaching 850-1100°C in garrigue communities of the Mediterranean basin. Keeley and Keeley (1988) reported surface temperatures in the Californian chaparral can remain higher than 500°C for more than 5 min during some fires, but not exceed 250°C in others, while soil temperatures at 2.5 cm depth are mainly in the range of 50-200°C and often persisting for 30 min or longer.

Fire-adaptive traits that facilitate survival and/or reproduction have been reported in a number of species. In Australia *Xanthorrhoea* spp. (blackboy), with their highly inflammable skirts of thin dry leaves, survive fire because the burning **acaroid resin** from the leaf bases form a thick and almost fireproof residue protecting the phloem and growing point. Fire is even essential for flowering as blackboy will only flower following a fire. It has even been suggested that the absence of a cambium in tree ferns and tree-like monocotyledons could be a survival trait against frequent burning. In a number of the Palmae, the sheathing leaf bases, e.g. *Copernicia prunifera* (carnauba wax palm), also insulate the trunk from fire.

The insulating fibrous bark of *Sequoia sempervirens* (Californian redwood) can be up to an astonishing 1 m thick in mature trees, while the corky bark of mature *Thuja plicata* (western red cedar) can be up to 30 cm thick. The North American *Pinus contorta* (lodgepole pine) and *P. banksiana* (jack pine) also have a relative thick bark capable of surviving repeated scarring from surface fires, they also bear slow maturing cones that remain in the canopy for up to 25 years. In the latter species the resin bond sealing the cone scales melts at 60°C, releasing many of the heat resistant seeds. The seeds can even withstand temperatures of 150°C for 30-45 sec and 370°C for 10-15 sec, an important survival trait when combustion in coniferous forests occurs at 350°C.

Likewise, in Australia the seeds from most eucalypts are shed gradually over a period of 2 to 4 years. Fire greatly accelerates the process, often resulting in a massive seed fall following an intense fire. The suppression of ground cover and seed predators also affect germination and establishment. Most of the aerial 'seed-bank' retained in the capsules may be protected from fire by: (1) Insufficient fuel to produce crown fires; (2) Capsules situated beneath the leaves so that even when a crown fire develops the heat flux received by the capsule is small; and (3) Insulation of the seed within the capsule. However, when a high intensity crown fire does develop the crown is often totally consumed by the fire and it is unlikely that many seeds could survive. Provided there is a sufficient interval between the period fires for seed set, such fires may be required to sustain genetic diversity by facilitating seedling recruitment. The fruits of *Banksia ornata* (desert banksia) and other members of the Proteaceae also only open following a fire. Similarly, fire will cause the cones of the Australian conifer *Actinostrobus* to open and release the tightly contained seeds (Rotherham *et al.*, 1975; Hodgkinson and Griffin, 1982; Johnson, 1985; Walter and Breckle, 1985; Noble, 1982; Archibold, 1995).

Suffrutices are also able to avoid permanent damage from fire because their underground rootstocks are insulated by the soil. Such *geoxylic suffrutices* are to be found mainly in the savannas of southern Africa, with some examples in the campos cerrados of the Planalto of Central Brazil, e.g. *Attalea insignis,* and with very few in Asia. They are apparently absent in Australia, where their arboreal equivalent is represented by the lignotubers of the mallee species of *Eucalyptus.* The geoxylic suffrutices are sensitive to fire and die back to the base, even when only slightly singed. Following a severe fire all the subaerial parts are killed. Well before the onset of the rainy season the suffrutices send forth new shoots from a massive woody rootstock or *xylopodia*, often producing precocious flowers at the base of the partially developed shoots, thereby achieving a competitive advantage by completing their reproductive cycle before the start of the rains, unlike the associated grasses and herbs which do not develop their vegetative growth until the rainy season has begun (White, 1976; Walter and Breckle, 1985; Archibold, 1995).

The large, swollen and woody rootstocks known as *lignotubers* of the shrubby, multistemmed, mallee species of *Eucalyptus* from the Australian shrublands, are capable of producing numerous epicornic shoots known as *lignotillers*, which are protected from lethal temperatures by being underground. Their development is accentuated in soils of low fertility, especially where phosphorous is limiting, suggesting that they may have a wider adaptive function in some eucalypts than being a regenerative base following fire. In areas where the fires are infrequent some mallee species occur as 1-3 stemmed trees up to 10 m high and are known as *bull mallee*, while where fires are frequent they form 1-3 m high, multistemmed *whipstick mallee*. But when the bull mallee is burnt, it first regenerates as whipstick mallee before regaining the tree form, which suggests a fire-induced habit. The fire hazard is arises from the regular, but not necessarily annual, shedding of mallee bark during the main summer growing season (December to April), the remaining periderm is quite thin and offers little protection against fire. The shed bark, some of which may be caught up in the branches, also contributes to the fuel (Noble, 1982 Walter and Breckle, 1986). Similarly, in the swidden cultivation of Northeast Brazil *Orbignya phalerata* (babassu palm) survives clearing and burning, regenerating during the fallow from an apical meristem grows horizontally below the soil before emerging vertically (Hecht *et al.*, 1988).

The majority of perennial grass species are able to tolerate almost complete defoliation by fire (and grazing), although too frequent removal of aerial growth can be detrimental to the plant. The perennating buds occur just above the soil surface in tussock grasses, at the soil surface in sward-forming and stoloniferous grasses, and just below the surface in rhizomatous grasses, thereby ensuring survival when defoliated, the apical primordium being protected within a nest of concentric sheaths. The sharply pointed and hygroscopic seeds of *Heteropogon contortus* (bunched speargrass) and *Themeda triandra*, syn. *T. australis* (kangaroo grass) are able to survive because of their ability to screw themselves into the soil and thus obtain protection from lethal temperatures (Clayton and Renvoize, 1986; Archibold, 1995).

The high NaCl content (20% of leaf dry-weight) in the leaves of *Atriplex vesicaria* is believed to account for the low inflammability of the species, nevertheless, the species can be killed by fire, probably because there are no buds capable of growing into shoots present beneath the bark above or below ground. The abundant salty litter formed beneath the canopy of *Tamarix aphylla* (athel, or tamarisk) by the nightly drip of excreted salt 'tears' and the continuous shedding of twigs, kills any ground cover. In the absence of ground cover and the inability of the litter, because of its high salt content, to burn, makes the *Tamarix* a useful tree for growing as a fire break. Indirectly, the dense shade or the allelopathic effect of other species, such as *Artemisia* spp. in suppressing ground cover may also provide some protection against fire (National Academy of Sciences, 1980; Booth. and Wickens, 1988; Hodgkinson and Griffin, 1982).

For species with a low sprouting capability, reaching reproductive maturity within as short a period as possible has an obvious advantage. Annuals have such a capability and, in areas where fires are not a regular annual event, perennials too are capable of flowering and setting seed within a few years. The avoidance of fire damage, especially if accompanied by a heavy seed production plus a long-lived soil seed bank, will ensure successful seedling recruitment. Thick endocarp walls and tests also offer protection against lethal temperatures, both on the plant and in the soil. Conversely, fire may be necessary to crack the endocarp or testa before the seed can germinate, e.g. *Acacia aneura* (mulga) seeds, where the slow germination is due to the impermeability of the hard testa, and is enhanced following fire. Sometimes dormancy is broken when the litter covering the seeds is burnt off. In other cases germination occurs when the seeds are exposed to charred wood, with germination usually being delayed until the water-soluble oligosaccharide produced from heating hemicellulose has been leached from the char. the juvenile period during which seedlings and young saplings are susceptible to fire, may vary from 12 months for some species of *Acacia* to 6 or more years for some conifers. The former species are then virtually immune to fire while the latter may survive moderate fires (Hodgkinson and Griffin, 1982; Kruger and Bigalke, 1984; Archibold, 1995).

3.   PHOTOSYNTHESIS

Photosynthesis is defined by Bailey (1999) as the sequence of reactions performed by green plants, and a group of photosynthetic bacteria, i.e. containing the pigment chlorophyll, in which light energy from the sun is converted into chemical energy and used to synthesise carbohydrates and ultimately all the materials of the plants.

Three photosynthetic pathways are recognised: C_3 C_4 and CAM. These are briefly described below. The families with C4 or CAM pathways are shown in Table 4 and anaerobic respiration is briefly discussed in Section 4.

3.1 C_3 and C_4 Photosynthetic Pathways

The C_3 and C_4 pathways represent the two basic differences by which the internal products of photosynthesis are processed, albeit with considerable variations to be found within the two pathways, with the most widely investigated family being the Gramineae. The first product of CO_2 fixation by photosynthesis in most higher terrestrial plants is 3-phosphoglyceric acid and, since this is a three-carbon acid, the process is known as the *C_3 pathway*. Most plants found in the temperate regions are C_3 plants. The process is regarded as being relatively inefficient photosynthetically compared to the C_4 pathway, the C_3 plants generally having a lower CO_2 fixation rates and higher compensation points than C_4 plants. The ***compensation point*** being the minimum level at which photosynthetic uptake of CO_2 is exactly balanced by its respiratory release, i.e. where the rate of synthesis of organic matter is equal to breakdown by respiration.

TABLE 4. Families with C_4 and CAM species (Raynal, 1973; Edwards and Walker, 1983; Clayton and Renvoize, 1986)

Family	C_4	CAM	Family	C_4	CAM
DICOTYLEDONS			Saxifragaceae	+	
Acanthaceae	+		Vitaceae		+
Aizoaceae	+	+	Zygophyllaceae	+	
Amaranthaceae	+				
Asclepiadaceae		+	MONOCOTYLEDONS		
Boraginaceae	+		Agavaceae		+
Cactaceae		+	Bromeliaceae		+
Capparaceae	+		Butaceae		?
Caryophyllaceae	+	??	Cyperaceae		
Chenopodiaceae	+	+	Tribe Cypereae	+	
Compositae	+	+	Tribe Fimbristylideae	+	
Convolvulaceae		??	Gramineae		
Crassulaceae		+	subfam. Arundinoideae	+	
Cucurbitaceae		+	subfam. Chloridoideae	+	
Didiereaceae		+	subfam. Eragrostoideae	+	
Euphorbiaceae	+	+	subfam. Panicoideae	+	
Geraniaceae		?	Liliaceae		+
Labiatae		?	Orchidaceae		+
Nyctaginaceae	+				
Oxalidaceae		?	GYMNOSPERMS		
Piperaceae		?	Welwitschiaceae		+
Plantaginaceae		??			
Polygonaceae	+		PTERIDOPHYTES		
Portulacaceae	+	+	Polypodiaceae		+

Other species follow a modified carbon cycle where, in addition to the C_3 cycle, there is a second CO_2-fixing pathway, the Hatch-Slack cycle, where the first step in photosynthesis is oxaloacetic acid, i.e. a four-carbon decarboxylic acid and hence the designation of C_4 pathway. Most C_4 plants tend to occur in the drier and hot subtropical regions. Many of the world most important tropical grass genera are C_4 plants, e.g. *Pennisetum*, *Saccharum*, *Sorghum* and *Zea*, as are also many of the world's worst weeds. The C_4 pathway is extremely rare among arborescent species, and until recently only known from members of the Euphorbiaceae in the Hawaiian rain forest. It has since been found among some arborescent members of the Chenopodiaceae, e.g. *Haloxylon ammodendron, H. aphyllum* and *H. persicum* (Winter, 1981; Edwards and Walker, 1983; Noggle and Fritz, 1983; Tootill, 1984; Noble, 1991a; Hattersley and Watson, 1992).

C_4 plants possess the following characteristics: (1) Photosynthesis, unlike that in CAM plants, occurs simultaneously in the light; (2) Anatomical examination generally shows the presence of a Kranz structure (from the German Kranz, meaning wreath) with the cells arranged in concentric layers around the vascular system. Although an anatomical examination for the presence of Kranz cells is the quickest way of identifying C_4 plants, their absence does not necessarily mean that it is not a C_4 plant. As a further complication, the grass *Alloteropsis semialata* contains morphologically indistinguishable diploid and hexaploid races with either a C_3 non-Kranz or a C_4 Kranz variant leaf anatomy respectively; (3) Apart from some noticeable exceptions, C_4 plants have an optimum photosynthetic efficiency at *ca.* 10°C above that of C_3 plants; their photosynthetic efficiency is also higher; and (4) C_4 plants produce more glucose per unit leaf area than C_3 plants and consequently grow more quickly.

The experimental claims that C_4 plants are more efficient users of water than C_3 plants have largely proved to be erroneous. Because C_4 plants in their natural environment function at much higher temperatures than C_3 plants, their water demands are also considerably higher, although there are some exceptions, including some cultivars of maize, sorghum and pearl millet (Gibbs Russell, 1983; Le Houérou, 1984; Tootill 1984; Clayton and Renvoize, 1986; Noble 1991a; Hattersley and Watson, 1992; Mabberley, 1997).

3.2   Crassulacean Acid Metabolism

The third metabolic pathway for fixing atmospheric CO_2, known as the Crassulacean acid (CAM) pathway, was first discovered in the Crassulaceae. The CAM plants are very conservative in their water use due to the stomata remaining closed during the day to reduce transpirational water losses. The stomata open at night and the CO_2 assimilated is ultimately converted into malate and stored until daylight, when it is broken down and the released CO_2 is then fixed in the normal manner. The CAM pathway differs basically from the C_4 pathway in that all the three processes of initial fixation, decarboxylation and refixation occur in the same cells; the initial fixation occurs in the dark and the other two processes in the light, thereby conserving carbon and minimising respiration losses of CO_2 (Tootill, 1984).

The CAM pathway is an adaptation to water stress and water conservation and species following the CAM pathway are five times as numerous as C_4 species. The CAM plants also have a higher water-use efficiency than other pathways but a lower productivity. Consequently the short-stemmed members of the Cactaceae and many epiphytes from other families, are generally slow growing, although this is not necessarily true for all CAM species (Edwards and Walker, 1983; Noble, 1991b).

4.   ANAEROBIC RESPIRATION

The flooding of the roots and rhizomes of wetland plants imposes periods of oxygen deficiency due to the rapid microbial consumption of any existing oxygen and the slow diffusion of oxygen through the water. Anatomically the presence of root tissues with numerous, large, intercellular spaces known as *aerenchyma* is of major importance in facilitating root aeration in flooded soils and, in some cases, increasing buoyancy, e.g. the aquatic herb *Ludwigia repens*, syn. *Jussiaea repens* (Harder *et al.*, 1965). Other morphological adaptations include the development of superficial mats of adventitious roots, e.g. *Oryza sativa*, or the presence of the more vertical 'sinker roots' in trees, e.g. *Pinus contorta* (shore pine), and the presence of *pneumatophores* on species growing in mangrove communities and fresh water swamps.

Pneumatophores may take the form of knee roots, such as those of mangrove species of *Bruguiera*, the fresh water swamp species *Mitragyna ciliata* and the gymnosperm *Taxodium distichum* (swamp, southern or bald cypress), or as the negatively geotropic aerial roots of the mangroves *Avicennia, Ceriops, Laguncularia* and *Sonneratia* and fresh water species *Ploiarium alternifolium, Cratoxylum* and *Tristania*, also *Metroxylon sagu* (sago palm), a native of seasonally inundated and swampy habitats. The lenticels in the portion of pneumataphore above the water level act as a pump, supplying O_2 to the aerenchyma in the submerged portion; the aerenchyma also acts as an O_2 store when the water levels are high. The lenticels close during tidal flooding, thereby creating a negative balance within the aerenchyma due to the respiratory consumption of the stored air. This imbalance then enables air to be sucked in when the lenticels reopen at low tide. A secondary function, at least in mangrove swamps, is for the pneumatophores to elongate sufficiently in response to silt accretion to maintain the fine rootlets borne by the pneumatophore at a constant depth below the soil surface (Richards, 1957; Archibold, 1995; Schuiling and Jong, 1996).

Less subject to anaerobic conditions, the palm genera *Areca, Cocos, Elaeis, Oncosperma, Phoenix* and *Raphia* develop spiky, 3-6 mm long rootlets known as *pneumathodes* on both underground and aerial roots. In *Elaeis guineensis* (African oil palm) the pneumatothodes produced by the tertiary absorbing roots lie close to the surface. Consequently the ploughing oil palm plantations is not recommended since it would destroy many of the feeding and breathing rootlets. Although they are presumed to ventilate the underground roots, direct physiological evidence is lacking. Pheumathodes apparently do not occur on the swamp dwelling *Nypa fruticans* (nipa palm), instead large air cavities traverse the root cortex and connect with large air cavities in the buoyant leaf bases, the air being slowly pumped around by the rise and fall of the tide (Corner, 1966; Hartley, 1967).

In general, trees growing in waterlogged soils are shallow-rooted and prone to 'wind-throw'. There are also physiological adaptations involving catabolic pathways whereby energy is obtained from organic compounds in the absence of free oxygen. This involves a high turnover rate of many enzymes and the ability of the plant to turn gene expression on and off in response to environmental stress (Braendle and Crawford, 1987; Fitter and Hay, 1989; Crawford *et al.*, 1989; Monk *et al.*, 1989).

Some herbaceous species, e.g. *Iris pseudacorus* (flag iris) and *Juncus effusus* (soft rush), are capable of regularly adapting to seasonal anaerobic and aerobic conditions. Others, e.g. *Senecio vulgaris* (groundsel), have both tolerant and non-tolerant races to waterlogging. While some wetland plants are adapted to survive extensive periods of flooding, the terms *flooding resistance* and *anoxia*, i.e. resistance to an O_2 deficit as a result of flooding, are not synonymous. For example, while the roots of the flooding resistant *Glyceria maxima* (reed-grass) are highly sensitive to O_2 deficiency, they appear to be no less exacting in their O_2 requirements than the flood-intolerant *Pisum sativum*, both being anoxia intolerant.

Species that are resistant to flooding appear to use a range of survival mechanisms, including an ability of the roots to exclude toxins and avoid an oxygen deficit through internal transport within the plant tissues. In some rhizomatous marsh plants, for example, the hollow stems above the water line enable O_2 to readily diffuse down to the intercellular spaces of the rhizome and roots, thereby maintaining O_2 and aerobic respiration under what would be anaerobic conditions.

Tolerance to anoxia is the ability to sustain metabolism, at least at maintenance levels, through anaerobic pathways, an ability to be found in the truly anoxia tolerant *Echinocloa oryzoides*, syn. *E. oryzicola* - a paddy-rice mimic weed, *Schoenoplectus lacustris* (bulrush) and *Typha angustifolia* (lesser reedmace) (Harborne 1988; Drew and Stolzy, 1991).

5.   PHOTOPERIODISM

Photoperiodism is the physiological reaction of plants to day length, which is the climatic factor that has the greatest effect on species adaptation to a given latitude. The photoperiod response allows plants to time their vegetative and floral growth to suit the seasonal changes of the environment. Most species in the temperate regions together with their cultivars, exhibit their own preferential photoperiodic response for flower initiation, seed germination, bud-break, stem elongation, leaf-fall, etc. In the equatorial regions there is so little seasonal change in day length that photoperiodism has little effect (Evans, 1973, cited by Arnon, 1992; Street and Öpik, 1984; Fitter and Hay, 1989).

Long-day plants usually require more than 14 hours of light and either fail to flower or have flowering delayed if, depending on the species, the daily photoperiod falls below a certain critical day length. Such plants flower more readily the longer the daily illumination, with the duration being more important than the intensity of illumination. Long-day plants merely require a certain minimum daily photoperiod and not a rhythmical alternation between light and dark in order to flower. It is even possible to provoke flowering in long-day plants during a short-day period by interrupting the period of darkness.

Short-day plants require a day length of less than 10 hours before they will flower, i.e. a long period of darkness and normally a rhythmical alteration of light and darkness for flowering. The period of darkness must equal or exceed a certain critical period, with flowering promoted or at least facilitated by diminishing day length, the darkness having a maximum as well as a minimum length. It has also been demonstrated that the effect of a long night period can be nullified if interrupted by a brief period of light. Thus, the cultivated *Chrysanthemum* spp. for example, are short-day plants and will not flower in the temperate regions until autumn, when the period of dark exceeds 8 hours. Similarly, the potential oil-seed species from tropical Africa, *Vernonia galamensis*, is a short-day plant which, under greenhouse conditions in the USA, flowered in the spring and in the autumn (Harder *et al.*, 1965; Perdue *et al.*, 1989; Arnon, 1992).

The majority of typical tropical crops are short-day plants, while plants/crops of the higher latitudes, like those of the temperate species, are generally *long-day plants* and will usually require day lengths of 14 hours before they will flower. By selection and breeding it has even been possible to develop some short-day crops for the long-day temperate regions. Growers have also been able to manipulate day length to control the time when certain high value horticultural crops can be made ready for the market (Harder *et al.*, 1965; Arnon, 1992).

In addition to the well-defined long- and short-day plants there are a number of *indeterminate plants* requiring 12-14 hours of light daily. This may take the form of long days followed by short days (long/short-day plants) or the reverse sequence (short/long-day plants). There are also plants that are *day neutral* and show no response to photoperiodism. They are capable of flowering regardless of the amount of light they receive during the day and, as a consequence, the onset of flowering is controlled by factors other than day length. *Lycopersicon esculentum* (tomato) and *Nicotiana tabacum* (tobacco) are examples of such day-neutral plants, although the 'Mammoth' cultivars of tobacco will grow vegetatively under long-day conditions but will only flower when exposed to short days. Selection pressure to develop day length neutrality in crop species in order to widen their areas of production has been very successful in a number of major crops, including varieties of maize, a typical short-day crop, and now developed for regions with long-days (Fitter and Hay, 1985; Purseglove, 1987; Arnon, 1992).

The ***critical day length*** is the specific period of daylight required by a species in order to initiate flowering in long-day plants or inhibit flowering in short-day plants. In practice long-day plants will not flower if the length of the dark period exceeds a certain maximum and, conversely, short-day plants will not flower unless the dark period exceeds a certain minimum. Such periods are referred to as ***critical dark periods*** and must be continuous if they are to be effective; a period of uninterrupted darkness at night has a greater effect on plants than the duration of daytime light. The ***night-break effect*** is the phenomenon exhibited by many plants when a period of artificial light is administered during the night, regardless of whether for a brief period or of low intensity, thereby upsetting the plant's normal flowering rhythm; the effects varying according to the time that has lapsed since daylight.

Under short-day conditions the light treatment will generally inhibit flowering in short-day plants and promote flowering in long-day plants. Keeping long-day plants in the dark for several hours during the day has virtually no effect on the time of flowering, while illuminating the plant at night, even for short periods, can have a dramatic effect on flowering. When a short-day plant, such as *Sorghum bicolor* for example, is subjected to a brief interruption of darkness during the night, normal flowering is prevented. Similarly, it is possible to provoke flowering in long-day plants such as *Tanacetum cinerariifolium* (pyrethrum) during a short-day period. Such manipulations have a number of practical applications, enabling growers to control the time when certain high-value crops, such as flowers, will be ready for the market, as well as enabling the plant breeder to control anthesis (Tootill, 1984; Arnon, 1992). See also Chapter 3 for further discussion.

6.   **VERNALISATION**

Some plants require a cold treatment, i.e. vernalisation, before they will flower. The most effective temperature for vernalisation is *ca*. 6°C and, depending on the species, for a period lasting from 4 days to 3 months. The process can also be reversed by high temperatures immediately following the cold treatment. The vernalisation process may be essential for flowering or merely hasten its onset, and is of particular interest to both the agriculturist and horticulturist. For some plants it has been demonstrated that the sensitive region for vernalisation is the stem apex, which has to reach the required stage of maturity before the cold treatment is effective, e.g. *Beta vulgaris* (beet). In other photoperiodic plants it is the leaves that must reach the appropriate stage of maturity; photoperiodic induction being effective only after the formation of a particular number of nodes. Many of those species that form bulbs, e.g. winter-flowering cultivars of *Hyacinthus orientalis* (hyacinth) or produce rosettes at the end of the growing season, as well as many perennial grasses, are unable to flower until they have undergone a period of vernalisation. Such an adaptation ensures reproductive development and seed production does not commence until the start of the next growing season. Thus, in order to ensure that flower initiation may occur in plants such as *Lilium longiflorum* (Easter or Bermuda lily), it is necessary for a minimum temperature requirement to be satisfied, i.e. vernalisation must occur. For *Hyacinthus* (hyacinths), *Narcissus* (daffodils and narcissus) and *Tulipa* (tulips), the flowers are already fully initiated and differentiated by planting time, notwithstanding cold treatment (vernalisation) still has an inductive effect on flower initiation (Rees, 1972; Fitter and Hay, 1989).

The winter varieties of wheat, barley, oats and rye are all long-day plants (see Section 5), and will normally only flower in early summer if they are sown before the onset of winter. Such winter cereals, which have both a cold and long-day requirement, can achieve vernalisation during the winter months either through the embryo of the immature seed or in the moistened mature seed. This will induce them to flower as rapidly as the spring varieties, although their day length requirement can be met only when the young seedlings have attained the appropriate stature.

By way of contrast, the biennial strain of *Hyoscyamus niger* (henbane), also a long-day species, has an absolute cold requirement and if overwintered at too high a temperature will remain in the vegetative state indefinitely. A cold requirement has also been found in a short-day variety of *Chrysanthemum*, as well as in a number of day-neutral species of *Erysimum*, *Geum*, *Lychnis*, as well as *Saxifraga rotundifolia* (round-leaved saxifrage) and *Tanacetum cinerariifolium* (pyrethrum) (Langer and Hill, 1982; Street and Öpik, 1984).

7.   SYMBIOTIC RELATIONSHIPS

A symbiotic relationship is the intimate relationship between two or more living organisms. This can be a relationship, known as *mutualism*, in which all participants benefit. A lichen is an example of *obligatory mutualism* between an alga and a fungus, where neither partner can survive without the other. *Facultative mutualism* differs in that both partners can survive independently, as occurs among many members of the Leguminosae and the bacterium *Rhizobium* (Tootill, 1984).

7.1   Bacterial Association

Free-living, N-fixing Eubacteria are often found in association with living plants. A *nitrogenase* enzyme system isolated from the bacteria has been found responsible for the nitrogen fixation, and an acetylene reduction assay may be used to confirm the presence of any active nitrogenase activity. Included among the N-fixing Eubacteria are members of the Cyanobacteria, formerly known as Cyanophyta or blue-green algae, where genera such as *Nostoc*, etc. are known to be responsible for symbiotic N-fixation associations with some members of the Diatoms, formerly Bacillariophyta, lichens, Anthocerophyta, Hepatophyta, Bryophyta, Filicinophyta, formerly known as

Pteridophyta (*Azolla* spp.), and Cycadophyta (*Cycas*, etc.), with *Gunnera* the sole genus represented by the angiosperms. Such symbiotic N-fixation provides plants with a competitive advantage over those plants that are unable able to fix atmospheric nitrogen, especially in the arid and semi-arid regions where, as is often the case, soil nitrogen levels are low. Even if there were any nitrates present from mineralised litter, the annual rainfall would generally be insufficient to wash them down to the nodulating zone, where they could inhibit nitrogen fixation (Tootill, 1984; Sprent, 1985; Sprent and Sprent, 1990).

N-fixation can be inhibited by such factors such as desiccation and drought, heat stress, oxygen diffusion resulting from low photosynthate availability, the effect of high soil nitrogen, low soil phosphate and non-availability of an appropriate strain of inoculate. Such failure may occur in arid and semi-arid ecosystems, even in such species as members of the mimosoid legumes *Acacia* and *Prosopis* that are elsewhere either known or expected to fix nitrogen (see Sprent and Sprent (1990) and Waisel *et al.* (1991) for further information).

The lack of nodulation in arid ecosystems is often compensated for by the development of a deep root system in order to obtain nutrients and water, which is considered to be a better utilisation of plant resources than for those required to fix nitrogen. Such **phreatophytes** might even find conditions near the water-table suitable for their roots to nodulate and fix nitrogen to support growth. The presence in arid ecosystems of grass roots bearing rhizosheaths associated with a possible nitrogen-fixing activity has already been discussed in Section 2.3. In a Mediterranean-type climate where the growing season coincides with winter rainfall, the nitrogen cycle is likely to broadly resemble that of the wet seasons in temperate regions, although low temperatures rather than water deficit may limit the biological processes. In summer rainfall regions, such as the tropical savanna woodlands of Australia's Northern Territory, dry season fires may lead to a loss of 90% or more of the available nitrogen from litter, thereby biasing the nitrogen cycle in favour of N-fixing species (Sprent, 1985; Sprent and Sprent, 1990).

The microsymbiont of the Leguminosae are known as **rhizobia**, the important genera being *Azorhizobium, Bradyrhizobium* and *Rhizobium*. Non-leguminous, N-fixing plants are known as **actinorhizal plants**, with the filamentous *Frankia* as the microsymbiont. Although only *ca.* 57% of the legume genera and *ca.* 20% of the species have so far been examined for nodulation, results show N-fixation is least common among members of the subfamily Caesalpinioideae. Some of the reports of nodulation by earlier authors, especially among the Caesalpinioideae, may be suspect, either due to a wrong identification of the species or by the inaccurate identification of root nodules with stubby root outgrowths, including ectomycorrhizas, with which they may be easily confused.

In the Caesalpinioideae nodulation is largely restricted to the tribe Caesalpinieae, which includes such rain forest genera as *Campsiandra* and *Melanoxylon*, and in the genus *Chamaecrista* of the Cassieae. Within the Mimosoideae nodulation is fairly general except for four groups within the tribe Mimoseae, and a few species of *Acacia*. Nodulation is widespread among the Papilionoideae examined, the exceptions being the tribe Dipteryxeae and a number of genera in the tribe Swartzieae, while the tribe Euchresteae has yet to be examined. The successful introduction of many potentially nodulating woody species to various parts of the world strongly suggests that effective rhizobia for woody species are widely distributed between continents (Allen and Allen, 1981; Faria *et al.*, 1989; Sprent and Sprent, 1990). See Skerman *et al.* (1988) for management practices of tropical forage legumes.

The value of N-fixing species is well illustrated in eastern Australia after pure stands of *Acacia harpophylla* (brigalow) were cleared for pasture. There were sufficient nitrogen reserves remaining in the soil to maintain the pasture for about a decade, after which nitrogenous fertilisers had to be applied in order to overcome the rapidly falling productivity.

In most woody and herbaceous perennial legumes there is a more or less constant cycle of root-nodule turnover which may be expected to benefit the plants for most of their life span, whereas annuals may derive no such benefit. In Colombia, for example, *Phaseolus vulgaris* (common bean) appeared to be able to meet only 50% of its inorganic nitrogen requirements through N-fixation, even though it was growing in its natural environment under almost optimal conditions. It is possible that during selection for domestication the ability for efficient N-fixation was unconsciously neglected when the life span during which N-fixation could be effective was reduced (Sprent, 1985; Sprent and Sprent, 1990; Smartt, 1990).

Our knowledge of actinorhizal plants is far less extensive, partly because bacterial research is expensive, so that research has tended to have been concentrated on the more economically important legumes, and partly because of the scattered occurrence of actinorhizal nodules and the technical difficulties associated with their research.

The plant breeders' dream of N-fixing cereal crops is now approaching reality. The cereals maize, rice, pearl millet and grain sorghum inoculated with such N-fixing bacteria as *Azotobacter chroococcum*, *Azospirillum brasilense* and *Azospirolla* have produced statistically significant increases in grain yield. In Egypt it has recently been discovered that rhizobia from legumes grown in rotation with rice can colonise rice plants; *Rhizobium leguminosarum* from clover has already been identified in rice. Interestingly, it is the older, more traditional varieties of rice that appear to respond better to N-fixation than the new varieties bred to respond to fertilisers. From Germany, *Azoarcus,* a new genus of N-fixing bacteria, has been discovered in the roots of *Leptochloa fusca* (kalla grass). Kalla grass can also colonise rice plants, with laboratory results showing a 10-20% increase in growth. It is possible that seed inoculation might have a practical application in those marginal areas where fertiliser applications are considered uneconomic. However, there is a school of thought that believes the introduction of symbiotic N-fixation to non-leguminous plants and the associated higher energy requirements for fixing nitrogen, could result in lower productivity and cancel out any advantage for the developing countries. In the developed countries N-fixation could reduce the need for heavy applications of nitrogenous fertilisers, thereby reducing water pollution. Unfortunately there are indications that the increase in use of glyphosate herbicides in association with glyphosate-resistant GM crops can have an adverse affect on soil mycorrhiza, which could reduce the advantages of such crops (Anonymous, 1997a; Armstrong, 1985; Gruben *et al.*, 1996; Anonymous, 1998)

The ability of certain C_4 tropical grasses, such as *Brachiaria decumbens* and *Digitaria decumbens,* to fix nitrogen with the aid of the bacterium *Azospirillum lipoferum, Leptochloa fusca* with *Azoarcus,* etc. is clearly of considerable economic importance Since it has been demonstrated that the application of nitrogen fertilisers imposes transitory limitations on rhizophere N-fixation, only light fertiliser should be applied. Whether the nitrogen is immediately available to the host grass after fixation or whether it only becomes available after the death and breakdown of the bacterial cells is not yet understood. Neither have there been consistent improvements obtained from the pasture after appropriate inoculation.

The root-microbe association also appears to exhibit a degree of specificity. For example, an association has been found between *Azotobacter paspali* and the tetraploid cultivar 'batatas' of *Paspalum notatum* (Bahia grass) and the fixation of appreciable quantities of nitrogen (Döbereiner *et al.*, 1972, cited by Sprent and Sprent, 1990), but no N-fixing activity with the diploid cultivars. In Queensland Weier (1976) cited by Skerman and Riveros (1990) reported that over a 12 week growing season *P. notatum* fixed nitrogen at the rate of 4 kg N ha^{-1} day^{-1} or 93 kg N ha^{-1} year^{-1}. The N-fixing *Azospirillum brasiliense* was also found to be present in the rhizosphere of 95% of the northern Australian grasses examined. This is a promising source of nitrogen for tropical pastures and requires further investigation in order to assess the extent of N-fixation by grasses, the variation between species and the possibility of improvement through selection of bacteria and grasses (Sprent, 1985; Skerman and Riveros, 1990).

7.2   Fungal Associations

A surprising number of plants occur in the wild which have a micro-fungal association with the roots. Three forms of such symbiotic associations are recognised: ectotrophic, endotrophic and vesicular-arbuscular mycorrhiza.

7.2.1   Ectotrophic Mycorrhiza

Ectotrophic mycorrhiza, also known as *ectophytic mycorrhiza* refers to an association where a well-developed mycelium, frequently by members of the Agaricales (mushrooms and toadstools), forms a sheathing mantle outside the root. A specific fungal association may be involved or, as is the case with *Pinus sylvestris,* more than a 100 different fungi may be involved. The same fungi are also known to form associations with more than 700 other tree species. A characteristic *coralloid* growth is exhibited by the infected roots where, instead of elongating, the lateral roots repeatedly branch to form an intricate mass. Initially the root cortex is penetrated by some of the fungal hyphae and form an intercellular mesh known as the *Harteg net*, although no intracellular connection is made. Subsequently a sheathing mantle develops from this net and replaces the piliferous layer of the root. These mycorrhizal roots have been shown to be more effective than uninfected roots in taking up water and nutrients, especially nitrates and phosphates, while the fungus benefits by obtaining carbohydrates and possibly B-group vitamins from the tree. Some of these agarics are even unable to produce fructifications unless connected to the host tree.

Ectotrophic mycorrhiza are associated with most trees and the presence of the appropriate fungus is often essential for healthy growth. Consequently it is sometimes necessary to inoculate the soil when planting young trees in areas where the species has not been previously grown r where a different strain of the required fungus would be more beneficial (Harder *et al.*, 1965; Tootill, 1984; Fitter and Hay, 1989).

7.2.2   Endotrophic Mycorrhiza

Endotrophic mycorrhyza, also known as *endophytic mycorrhiza*, represents an association where the fungus lives between and within the outer cortical cells, with limited growth outside the root. The hyphae form tightly coiled masses known as *pelotons* within the outer cortical cells. Endotrophic mycorrhiza are found in many herbaceous species, especially among members of the Orchidaceae where species of *Rhizoctonia* are usually involved, and the Ericaceae involving species of *Phoma*. The fungal association is often essential for normal growth. With orchids such as *Corallorrhiza*, *Epipogium* and *Neottia nidus-avis* (bird's-nest orchid), for example, which either lack chlorophyll or are otherwise photosynthetically inefficient, the association with the fungus is an essential source of plant nutrients, although the benefits to the fungus are uncertain (Harder *et al.*, 1965; Tootill, 1984; Fitter and Hay, 1989).

While a number of terrestrial orchid species can be successfully raised from seed under asymbiotic conditions, others have proved more difficult, while others growing well *in vitro* have proved difficult to transfer to a soil-based potting medium. To overcome this problem a successful technique for the symbiotic germination of orchid seed in a mycorrhizal association has now been developed (see Clements et al., 1986 and references therein).

7.2.3   Vesicular-Arbuscular Mycorrhiza

Vesicular-arbuscular (VA) mycorrhizas or *VAM* are where the fungus, often a species of *Endogone*, *Gigaspora* or *Glomus*, or *Pythium* of the Oomycota, live between the cortical cells and forms temporary hyphal projections that penetrate the cortical cells. Such projections may take the form of swollen vesicles or finely branched masses known as *arbuscules* (little trees). VA mycorrhizas may be found in most annual and perennial plants, irrespective of whether they are nodulated legumes or non-legumes. It is noteworthy that when plants are infected with both *Rhizobium* and VA mycorrhizal fungi it has been demonstrated experimentally that the two may interact to produce an effect on a plant greater than the sum of their individual effects.

Like plants associated with N-fixing bacteria, plants infected by VA mycorrhiza can obtain a competitive advantage over uninfected plants. Thus, plants growing in impoverished soils, for example, can exhibit improved phosphorus uptake, especially from an insoluble source, compared with non-infected plants, although the reverse may be true in soils rich in available phosphorous. Indeed, inoculation with effective rhizobia have been found ineffective in a number of tropical and temperate pasture legumes growing in impoverished soils without the presence of VA mycorrhiza. Water uptake too may be increased due to the VA mycorrhiza providing a low resistance pathway for radial water flow across the cortex. This in addition to possible changes in the permeability of the membrane to water due to the improved phosphorus metabolism (Tootill, 1984; Fitter and Hay, 1989; Sprent and Sprent, 1990; Danin, 1996; Jacobson, 1997).

8.   PLANT METABOLITES

The products of the physical and chemical metabolic reactions within the plant cells are known as *plant metabolites*. They provide the food energy for all stages of development during the plant's life cycle, and thereby ensure the plant's survival from one generation to the next. Some, like proline, respond to environmental stress (see Sections 1.1 and 1.3). Many metabolites are of economic importance, and are considered in Chapter 15, while those of ecological importance, such as antifeedants, and other plant defensive metabolites are discussed in Chapter 17. For further information on plant biochemical adaptations to the environment see Harborne (1988).

There are also *secondary compounds*, such as auxins and allelochemicals, which were formerly believed to be the waste products of the primary metabolism. But a pioneer study on the coevolution of plants and butterflies by Ehrlich and Raven (1964) concluded that these secondary plant substances played a leading role in determining patterns of evolution, a role that now appears to be true for all phytophagous groups.

9.   PLANT AUXINS

Plant growth substances are known as *auxins*, the most widely occurring being *indole acetic acid* (*IAA*), while others are based on the indole ring. When present in low concentrations they have the ability to promote apical and root growth as well as having a number of other effects on the growth of plants, including yeasts and other fungi. At higher concentrations auxins may inhibit growth following the auxin-promoted synthesis of ethene (ethylene), which inhibits cell elongation. This inhibitory action has been exploited for the manufacture of synthetic auxins, such as 2,4-D, MCPA and 2,4,5-T, for use as weedicides. Synthetic auxins, such as *indole-3-acetic acid, indole butyric acid* (*IBA*) and *naphthalene acetic acid* (*NAA*), have also been developed for use as rooting compounds, while *steviol*, a derivative of the diterpene glycoside *stevioside* obtained from *Stevia rebaudiana* (the sweet herb of Paraguay) has a similar action to the plant growth metabolite *gibberellin*. The gibberellins are a complex of hormone-like substances from the fungus *Gibberella fujikuroi*, an anamorph of *Fusarium moniliforme*, and were first discovered during investigations into abnormal growth of rice (Bakanae disease) caused by the fungus. The terpenoid *gibberellic acid*, which has similar properties to gibberellin, is synthesised commercially for use in horticulture (Tootill, 1984; Brücher, 1989; Holliday, 1990; Sharp, 1990; Hawksworth *et al.*, 1995).

9.1   Allelochemicals

Probably the least understood of the biochemical interactions between plants is allelopathy, chiefly because of the difficulties in isolating the action of the plant's allelochemicals, consisting mainly secondary metabolites, from competition for light, moisture and nutrients. The term *allelopathy*, from the Greek *allelon*, of each other, and *pathos*, to suffer, was coined by Molisch (1937).

Allelopathy is defined by Rice (1984) as *"any direct or indirect harmful or beneficial effect by one plant (including micro-organisms) on another through production of chemical compounds that escape into the environment"*. Early reports of now known allelopathic effects include that by the Greek philosopher Democritus (460-370 BC) who reported trees could be killed by treating the roots with a mixture of lupin flowers soaked in hemlock juice, while Pliny the Elder (23-79 AD) in his *Historia naturalis* cites numerous examples of apparent allelopathic reactions with chickpeas, barley, fenugreek and vetch (Rizvi *et al.*, 1992).

The allelochemicals effect plants in a number of ways, including the cytology and ultrastructure, phytohormones and their balance, membranes and their permeability, the germination of pollen and spores, mineral uptake and plant productivity, stomatal movement, pigment synthesis and photosynthetic respiration, the synthesis of protein and *leghaemoglobin* (a protein found in root nodules of legumes), conducting tissues, plant water relations, specific enzyme activity, and genetic material. There is also evidence that allelopathy inhibits nitrification. While it appears likely that allelopathy could affect denitrification directly and indirectly through its affect on nitrification in determining the amount of substrate available for denitrifying organisms, no information is yet available of any allelopathic effect on denitrifying bacteria (Rice, 1992).

In the past credit for vegetation patterns was largely attributed to competition, it is now known that allelopathy can play an important role. In agriculture the inhibition effect of many smother crops used to suppress weed growth are now known to be due to allelopathy and not competition for light and nutrients. Current research is now concerned with the effect of weeds on crops, crops on weeds and crops on crops, where the main problem is in distinguishing between the effects of competition and allelopathy. See Chapter 17 fir further discussion.

A number of other topics are also under investigation and include: (1) The importance of allelochemicals in promoting pathogen infestation/resistance development to pests and diseases; (2) The role of allelopathy in providing seed protection prior to germination, e.g. the effect of fire in eliminating allelochemicals in plant litter and resulting in mass seed germination; (3) The implication of allelopathy in nitrification/denitrification; and (4) The use of allelopathy as growth regulators.

In forestry allelopathy plays a crucial role in both natural and man-made forests, and in the management of agroforestry systems for ensuring compatibility between crops and trees (see Chapter 17). There appears to have been relatively little horticultural research into the effects of allelopathy on ornamental plants, although allelochemical-based soil sickness causing replanting problems with apple, peach and citrus plantations, as well as with roses and many other ornamental plants, is now recognised. An accumulation of the cyanophoric glycoside *amygdalin* is responsible for the peach replant problem and *homovanillic acid* for citrus decline. It is probable that the reason why young bulbs of *Fritillaria pudica* and corms of *Gladiolus segetum* do not develop while attached or adjacent to the parent plant is also an allelopathic effect (Rice, 1975, 1984; Gutterman, 1993; Brian Mathew, pers. comm., 1999).

Most allelochemicals are secondary metabolites and include phenyl propanes, acetogenins, terpenoids, steroids and alcohols. Those produced by the higher plants and micro-organisms were classified by Rice (1984) into the following major categories: (1) Simple water-soluble organic acids, straight-chain alcohols, aliphatic aldehydes and ketones; (2) Simple unsaturated lactones; (3) Long-chain fatty acids and polyacetalyenes; (4) Naphthaquinones, anthraquinones and complex quinones; (5) Simple phenols, benzoic acid and its derivatives; (6) *Cinnamic acid* and its derivatives; (7) Flavonoids; (8) Tannins; (9) Terpenoids and steroids; (10) Amino-acids and polypeptides; (11) Alkaloids and cyanohydrins; (12) Sulphides and glucosides; and (12) Purines and nucleolides.

The volatile terpines are more usual among the aromatic plants of the arid and semi-arid habitats, such as *Artemisia californica* and *Salvia leucophylla* of the

Californian chaparral. The volatile compounds released by *S. leucophylla* produce a halo of bare soil adjacent to the shrub, beyond which there is an 'inhibitation' zone where certain species are absent and the stunting of others decreases with distance from the plant. Its essential oil yields appreciable quantities of five terpenoids, *camphor, 1,8-cineole, isothujone, α-thujone* and *artemene ketone*. Of these the camphor is the most toxic, followed by 1,8-cineole, with artemene ketone as the least toxic. The composition and quantity of the essential oil changes with the season, increasing in late summer and autumn and decreasing in the winter, while in late summer and autumn the isothujone largely replaces α-thujone and camphor becomes more abundant.

The phenolic compounds tend to be found among plants of more humid habitats. For example, in the subhumid deciduous forests of South Carolina the bare area within the crown drip zone of *Quercus falcata* (American red oak) has been identified as being due to salicylic acid from the leaf leachate. Water soluble phenolic compounds have also been found among the dominant shrubs of the semi-arid Californian chaparral in the leaves of *Adenostoma fasciculatum* and *Arctostaphylos glandulosa*. They also exert allelopathic effects on the surrounding herbs similar to those of *Salvia leucophylla* (Muller and Chou, 1972; Halligan, 1975; Tootill, 1984; Harborne, 1988).

Chapter 6

Plant Breeding and Propagation

Crop plants are directly dependent upon human management and have evolved in part under the influences of the farming practices of particular cultures. For more than 10,000 years there have been both conscious selection by the cultivator for larger seeds, etc. and an unconscious selection for non-shattering seed heads and other useful characters. To some extent chance has enabled a particular species to be brought into cultivation and to be preferred while an apparently equally deserving species has been ignored. Similarly, the political, military and economic superiority of a particular nation has enabled it to impose its own preferences for particular cultivated species against those of other nations. Thus, the suppression by the Spanish Conquistadors during the 16th century of the cultivation by the Incas of *Chenopodium quinoa* (quinoa) brought about a dramatic decline in the cultivation of the crop. So much so that it is only recently that it has undergone a revival (Risi and Galwey, 1989).

It is not too far-fetched to assume that if many of the so-called 'minor crops' had received even a proportion of the selection and breeding given to 'major crops' that they too could have become 'major crops'. Unfortunately there is a degree of intellectual snobbery attached to 'high tech' plant breeding. This is coupled with the desire to replace local cultivars such as, for example, members of the minor cereals, with more socially acceptable and apparently more productive crops such as wheat. As a result the minor cereals remain under-researched and therefore of largely unknown potential, often requiring comparatively simple 'low technology' selection and breeding for their immediate improvement. On the other hand, the replacement major cereal has, for commercial reasons, been well-researched elsewhere and is consequently more amenable to 'high tech' manipulation and, for plant breeders in the developing countries, helps to demonstrate their scientific achievement. The simple fact that the original crops are already 'adapted to' the environment while the introduction has to be 'adapted for' the environment is largely ignored.

105

For the developing countries especially, plant breeding should also reflect the needs of the rural population; prestigious, government-sponsored, large-scale, development projects rarely raise the living standards of the rural poor. In my opinion the developed countries are doing a disservice to the developing countries by encouraging such projects. For the small farmers in the developing countries local farming practices do not necessarily demand only high yields. For example, at San Pedro de Atacama in the Andean foothills of Chile, I was told by the local farmers that early maturing, dwarf maize had been introduced, but after one season the crop was rejected in favour of their former slower maturing and taller varieties. Although the dwarf variety yielded more grain it failed to produce sufficient straw for overwintering the livestock. Similarly, in many marginal areas, such as the Sahel, local cultivars with a wide variability in maturing and rainfall requirements, while an anathema to the plant breeder and mechanised farmer, suit the needs of the local people because they ensure that at least a proportion of the crop will be harvested annually despite the vagaries of the climate.

Today the breeding techniques include the artificial control of pollination, biotechnology, the generation of variability by artificial hybridisation and mutation, as well as by selection procedures such as pedigree breeding and backcrossing (Dalal-Clayton, 1981; Tootill, 1984).

Biotechnology is defined as 'the application of biological organisms, systems or processes to manufacture and service industries' (Anonymous, 1980, cited by Armstrong, 1985), i.e. any technique that uses living organisms or parts of organisms, to modify products, to improve plants [or animals], or to develop micro-organisms for specific uses (Rexen and Munck, 1984). It overcomes the difficulties encountered in conventional plant breeding, especially the reliance on using natural fertilisation processes to introduce genetic modifications and the limitation of the gene pool available to the range of sexually compatible plants. It embraces a number of technologies based upon an increasing understanding of biology at the cellular and molecular levels, including recombinant DNA manipulation, monoclonal antibody preparation, tissue culture, protoplasm fusion, protein engineering, immobilised enzyme and cell catalysis, sensing with the aid of biological molecules.

Biotechnology is applied to crops for: (1) Improving yields and the ratio of primary and secondary products; (2) Improving the nutritive value of products. The nutritive value of low quality food, for example, can be improved by the introduction of enzymes capable of degrading or modifying the ligno-hemicellulose complexes of the mature plant cell walls; (3) Improving the physical properties. For example, there is the possibility of plant product improvement of cultivars whose seeds are modified to meet the needs of man, such as the gluten and protein content in the seed endosperm of wheat and its effect on bread making; also the alteration of the amino-acid complement of cereal protein present, especially lysine and to a lesser extent, threonine; (4) Improving resistance to insects and pests. Additional improvements in disease resistance are likely to be associated with increased and longer lasting plant resistance to crop pests, and increased resistance by plants to cost-effective herbicides and pesticides. Tissue culture, for example, has already produced maize that is resistant to a family of herbicides; and (5) Improving the capacity to resist stress, such as introducing symbiotic mycorrhizal fungi to facilitate water and nutrient uptake by the host plant, particularly phosphorus and some trace elements, thereby increasing stress resistance (Armstrong, 1985).

1. CONVENTIONAL PLANT BREEDING

Plant breeding is defined as the development of new and improved varieties and strains of agricultural, horticultural and forestry crops by selection, hybridisation or genetic manipulation of individuals or 'varieties' possessing desirable characteristics or qualities. The 'variety' as used here includes both botanical varieties and strains or races. Conventional breeding is carried out within the gene pool available from the genus involved or from closely related genera.

1.1 Plant Breeding Techniques

There are four basic breeding techniques recognised today: clones (*CLO*), open-pollinated populations (*OPP*), inbred pure lines (*IBL*), and hybrid varieties (*HYB*). Of these, CLO and OPP are historically ancient techniques that relied on visual selection and propagation, IBL was developed during the 19th century from primitive heterogeneous crops, while HYB is relatively recent.

A *clone* represents a population of homozygous or genetically identical individuals within a highly heterogeneous outbreeding plant population. CLO breeding involves selection from among the vegetative descendants of the variable F_1 families produced by crossing heterozygous parents. Vegetative reproduction ensures a homozygous crop. Clonal crops grown for seed production must, of necessity, be able to reproduce more or less normally, whereas those for vegetative or fruit products are not so dependent. Notwithstanding all will exhibit some degree of reproductive derangement.

In *OPP* breeding it is necessary to distinguish between population improvement in which the emergent varieties are indefinitely propagated as closed populations and synthetics, which are regularly reconstructed from selected source materials from seed propagated lines or, in the case of perennial crops, clones. The emergent varieties will, in both cases, be somewhat heterozygous and heterogeneous.

The *IBL* method is a plant breeding technique, limited to self-pollinating species, that endeavours to combine the best qualities of selected existing varieties to create entirely new varieties of plants. The selected parents are artificially crossed. A few seeds of the resulting F_1 hybrid are then sown widely spaced in order to encourage a prolific seed set. Assuming that the parents were homozygous, the F_1 plants should all be identical. The seed from each plant is harvested separately and grown, again in widely spaced rows, in separate 'family' plots. The genotype variability becomes apparent in the F_2 generation and the characteristics of each plant noted. Seeds from a small number, perhaps as little as 1%, of promising F_2 plants are then harvested separately to grow on to the F_3 generation. For *pedigree breeding* the seeds from each plant are again grown in separate plots, thus allowing the 'pedigree' of each plant to be recorded. By noting the variability within each of the F_3 plots it is possible to determine which plants of the previous generation depended excessively on heterozygosity for vigour. At this stage whole 'families' of plants may be discarded. Since heterozygosity is approximately halved with each generation, increasing emphasis is placed in succeeding generations on selection between 'families rather than within 'families'. If breeding a crop plant, then at about the F_6 generation, seeds from selected plants are grown under normal field conditions for that particular crop in order to determine yield. From this point onwards selection is entirely between 'families'. By the F_8 generation the remaining selected lines may be assumed to be over 99% homozygous and can be bulked up for variety trials. Instead of the pedigree method, the modern tendency is to use the *bulk method*, delaying selection until lines have attained some individuality. This has the advantage that seed stocks are built up faster that they would otherwise; however, there are some problems with heterogeneity.

The *HYB* method, first developed with maize and now increasingly used for other crops, involves the removal of anthers or emasculation of normally self-pollinating species and the artificial transfer of pollen from another plant. Highly inbred parental lines are used, consequently their F_1 hybrid progeny are virtually homogenous, uniform and heterozygous. Their progeny cannot, therefore, breed true.

Backcrossing is the technique used in traditional plant breeding to introduce a desirable gene into a cultivated variety. Unlike the pedigree method of plant breeding, the objective of the backcrossing programme is not to create entirely new varieties, but to modify existing varieties. The cultivated variety (the recurrent parent) is crossed with the donor parent, which apart from possessing one or more desirable gene or genes, may be entirely useless to agriculture (it is the presence of such desirable genes in the donor parent that is one of the reasons for the conservation of all plants - and other organisms). The progeny of this cross, which contain 50% of the donor genetic material, are screened for the desired character, such as disease resistance, and those possessing the desired character are crossed back to the recurrent parent. The progeny of this cross, the B_1 generation, now contain 25% of the donor's genetic material. The plants are again screened and the disease resistant plants are again backcrossed to the recurrent parent. This process is continued to the B_7 or B_8 generation, by which time less than 0.25% of the donor genetic material remains. At this stage the B_7 or B_8 generation plants are *selfed* (crossed with each other) to produce plants homozygous for the desired character; which may be identified by a test cross. i.e. a cross between an individual of uncertain genetic makeup and a homozygous recessive. The above process assumes that the allele for the desired character is dominant. But should it be recessive, alternate backcrossing with selfing of the backcross generation will be necessary. While backcrossing is more efficient with self-pollinating species, the method is not necessarily limited to self pollinators (Simmonds, 1979, 1997; Tootill, 1984; Driscoll, 1990).

Perennial crops such as sugarcane, bananas, etc. and tree crops are of particular importance in the tropics. They are all outbreeders, heterozygous and show inbreeding depression, the exception being *Coffea arabica* (arabica coffee), an inbreeding allotetraploid and is bred using IBL, although CLO and HYB would be better. Breeding plans for perennial crops can range from OPP from good parents, composites or synthesis from elite clonal parents raised in 'seed beds', manual crosses between superior seedlings from good families, bi-clonal pair crosses, and vegetative clones (Simmonds, 1997).

1.2   Vegetation Propagation

Selection and simple vegetative propagation techniques can also produce significant increases in yield and/or quality without recourse to the genetic manipulation used for the CLO method. It is equivalent to mass selection in a breeding programme and is suitable for some perennial species, such as fruit trees and shrubs, but obviously unsuitable annual crops.

Okafor (1977, 1980a, b) has successfully demonstrated how quite simple grafting techniques using propagating material from selected elite trees can be used to improve the fruit yield, quality, etc. of indigenous trees and shrubs harvested from the wild in the Nigerian forests and savannas and within the traditional farming systems. The idea is not new; for *Ziziphus jubata* (Chinese jujube) and *Z. mauritiana* (Indian jujube have been similarly propagated, the former for at least 4000 years. Numerous cultivars of both species are now available and the latter species widely introduced in the tropics. It is noteworthy that Johnston (1972) clearly states that *Z. mauritiana* has been introduced in East Africa and that the smaller-fruited specimens found in the wild away from cultivation may represent an atavistic behaviour, where recessive or complementary genes cause the reversion from the cultivated to the wild form. Atavism is also known in the olive, where the cultivated olive is known as *Olea europaea* subsp. *europaea* var. *europaea* and the reversion to the wild form is taxonomically recognised as var. *sylvestris*.

1.3   *In Vitro* Propagation

In vitro, i.e. 'in glass', or *micropropagation*, is the use of small pieces of tissue, such as meristem grown in culture, to obtain large numbers of individuals. It requires strictly controlled sterile conditions and appropriate growth media and nutrients. The resulting plantlets are then raised under propagators and, when established, removed for growing on in suitably maintained glass house (Tootill, 1984; Woods, 1985).

The technique is a valuable alternative to the propagation of desirable cultivars or endangered plants from seed, especially for those species that are either difficult to propagate by conventional methods, e.g. *Argania spinosa* (argan), *Melia volkensii* and orchids. The Leguminosae, however, have been found particularly difficult to culture using *in vitro* techniques, although according to Thomas and Wernicke (1978) the perennial forage legumes perform better than seed legumes. The technique may also be used to raise virus-free plants from infected stocks. This is because the viruses do not penetrate meristematic tissues, consequently their culture ensures virus-free explants. It is also invaluable for the propagation of purely male or female lines of monoecious species, such as *Simmondsia chinensis* (jojoba), where mainly female individuals are required for seed oil production (Corley, 1988).

Tissue culture is also a particularly useful technique for the rapid bulking up of limited material from selected genotypes. For conservation purposes plants are preferably raised from seed in order to maintain genetic variability and then multiplied by bud formation through a callus induced on either a seedling or hypocotyl and root meristems. Where no seed is available the *in vitro* culture of vegetative material, e.g. nodal cuttings, may be used. However, tissue culture does have the disadvantage that in many plants, including sugar cane and bananas, some *soma-clonal variation* can occur due to mutation in the callus tissue, which will mean that not all regenerated plants will be identical (Corley, 1988).

Embryo culture has also proved beneficial where, because of differences in the number of chromosomes or their genetic constitution, it has proved difficult to cross normally incompatible plants, such as wild rice with a cultivated variety. Success has been achieved by growing the rescued embryo in culture before abortion can take place. Tissue culture now enables wide-cross embryos containing extra chromosomes to be rescued, and for them to be maintained through several back-crossing cycles until their chromosome number has been reduced to that of the cultivar. Excised anthers may also be cultured on a sterile nutrient medium so that embryos and subsequently haploid platelets may be induced to form from the pollen grains. Colchicine may then be used to raise the ploidy level in order to obtain completely homozygous diploid plants (Tootill, 1984; Anonymous, 1993).

2.   GENETIC ENGINEERING

It is now possible, often with great precision, by procedures known as *recombinant DNA technology* or more popularly as *genetic engineering*, to combine DNA sequences from widely different organisms *in vivo*, i.e. 'in life'. It involves the isolation of useful genes from a donor organism or tissue and their incorporation into an organism that does not normally possess them. For example, genes from a donor may be incorporated into the DNA of a bacterium or yeast which, by its own replication, will produce many copies of the foreign gene and hence may produce substantial quantities of the gene's products. These products, which may be enzymes, hormones, antibodies, etc. may have medical or commercial value. Such genetically engineered micro-organisms are already proving commercially successful in both agriculture and medicine, e.g. frost-resistant tobacco, increased shelf-life of tomatoes, in the production of human insulin, interferon, growth hormones, and a banana which produces a vaccine against cholera.

A common method of introducing foreign genes into higher plant cells is the *Bacterium tumefaciens* (*Bt*) method. This involves using the *tumour-inducing plasmid* (*Ti plasmid*) associated with the bacterium *Agrobacterium tumefaciens*, the pathogen responsible for crown gall and hairy root diseases in plants. Once introduced into a plant cell by the bacterium, part of this plasmid becomes incorporated into the plant genome and disrupts the control of cell division. Cell differentiation as well is also often disrupted, resulting in the development of disorganised tumourous tissue. Fortunately certain benign strains of *A. tumefaciens* plasmids exist that do not cause disorganised cell growth. If new DNA sequences are incorporated into such Ti plasmids and the plasmid is then introduced into a culture containing isolated differentiated plant cells capable of regenerating whole plants (*totipotent plant cells*), a genetically transformed plant may be produced. While all dicotyledons are susceptible to *A. tumefaciens,* most monocotyledons are not naturally affected, and as a consequence the bacterial method cannot always be applied.

In 1982 the first GE organism about to be deliberately introduced to crops was Bt carrying the ice-minus gene. The object was to displace forms of Bt carrying genes specifying an ice-nucleation protein, the cause of frost damage in susceptible plants such as strawberries. But because of the resultant public outcry and legal action by environmental groups in the USA the trial was effectively delayed for several years (Tootill, 1984; Crawley, 1990).

Other natural vectors, such as the caulimoviruses, which include the cauliflower mosaic virus, have also been tried. Unfortunately, methods not employing natural vectors can create problems of identity. If the introduced genes do not cause an identifiable disease in the recipient, there is then the added problem of introducing mutant marker genes in order that the transformed tissue can be distinguished. Such techniques include the removal of the cell walls by enzymatic digestion or *electroporation* (a brief voltage pulse transiently increasing membrane permeability to permit DNA or chromosome uptake), osmotic shock, etc. to produce protoplasts, i.e. the living part of a cell minus its cell wall and vacuole, and thereby facilitate genetic manipulation. The technique has already been successfully used, for example, in the transfer of DNA from bacteria to rice protoplasts. The mature plants raised from these protoplasts have then transmitted the implanted DNA to their offspring. It is also a potentially useful technique for the production somatic hybrids or cybrids from unrelated or sexually incompatible species. The technique, known as *protoplast fusion*, involves the induction of fusion between naked cells under culture conditions.

Because the bacterial method cannot be used for most monocots, including cereals, an alternative technique, popularly known as the *golden bullet technique* is used. This involves bathing microscopic pellets of gold or tungsten in DNA and blasting the pellets into the recipient cells using a particle gun. As the pellets pass through the cells some of the DNA remains behind and mixes with the original DNA (Tootill, 1984; National Research Council, 1996; Land and Farquhar, 1998).

Examples of genetic engineering of wheat include work with the halophyte *Agropyron junceum*, which is believed to have seven pairs of chromosomes carrying genes for salt tolerance. Whole chromosomes have been transferred into Chinese spring wheat in order to identify genes for salt tolerance. Chromosome substitution between normal and dwarf wheat has been used to make genetic maps and markers to identify beneficial genes (Law, 1988). Molecular biologists in England and Spain are also involved in developing transgenic varieties of salt-tolerant barley, melons and tomatoes using a salt-regulating gene from *Saccharomyces cerevisiae* (brewer's yeast). Even so, salt-tolerance is a complex mechanism and far too often success in the laboratory may not survive the greenhouse, let alone reach the field (Knight, 1997).

In general physiological modifications involve a number of genes, whereas herbicide and insecticide resistance usually involves a single gene and are consequently simpler to manipulate. Genetically engineered, herbicide-resistant crops was one of first issues targeted, of which there are two forms of resistance that can be genetically engineered into crops. The first is *metabolic resistance*, where the plant degrades the herbicide into non-toxic products. Such crops, e.g. glyphosate-tolerant oilseed rape, permit the use of pre-emergence herbicide applications for a more effective and flexible weed control and, it is claimed, using less herbicide. However, this procedure would leave root parasites, such as *Orobanche* spp. (broomrapes) and *Striga* spp. (witchweeds), unaffected.

The second is *target-site resistance* where the target enzyme affected by the herbicide is modified to preclude herbicide binding without changing the normal function of the enzyme, thus permitting passage of the enzyme from the crop host to kill the root parasite. This is especially desirable where damage from parasitised crop roots has already taken place before the aerial parts of the parasite are visible. Crops having target-site selectivity, however, permit the translocation of the active herbicide to the parasite. Even so, such herbicide control also carries the risk of the possible development and build up of herbicide-resistant weeds. A strict regime of roguing is therefore recommended in order to prevent any build-up of herbicide-resistant strains of *Striga* spp. as it has been calculated that it would otherwise result in complete infestation of the fields in 3-5 years. Obviously, target-site resistance should only be used where the crop does not interbreed with the local related weeds (Joel *et al.*, 1995; Abayo *et al.* 1998).

One of the dangers of using genetic engineering is the possibility of producing something that cannot be controlled and escapes into the wild. A problem that could arise with such crops as rice, oats, sunflower, chicory, potatoes and tomatoes that have wild interbreeding relatives growing in the same area. Already experimental work in France has demonstrated that herbicide resistance in genetically engineered *Brassica napus* (oilseed rape) can be transferred to wild *Raphanus raphanistrum* (wild radish). Insecticides lack specificity, which is not surprising when considering their numbers. Thus, the larvae of Neuroptera (lacewings), the natural predators of *Ostrinia nubilalis* (European corn borer), died when fed corn borers raised on *Bt* maize. Similarly, in Scotland, the targeting of *Myzus persicae* (peach-potato aphid) using GM potatoes expressing the lectin *Galanthus nivalis* agglutinin (GNA), also affected the life span of *Adalia bipunctata* (two-spot ladybird), the aphid's natural predator (Gressel *et al.*, 1994; Chèvre *et al.*, 1997; Gledhill and McGrath, 1997; Land and Farquhar, 1998).

But what of the effect of insecticides on humans? The investigation by Ewen and Pusztal (1999) on the effect of GM potatoes expressing GNA on the small intestine of rats has been severely criticised as being flawed in design and the results inconclusive. The findings by Fenton *et al.* (1999) of GNA binding to human blood cells are also considered incomplete and inconclusive (Horton, 1999; Kuiper et al., 1999). But until new techniques have been developed to detect and characterise the possible unintentional effects of GM foods there will inevitably remain some doubt by the public in the safety of such foods.

It is important to remember that for any nuclear or plastomic transgene to spread successful hybridisation between sexually compatible cultivar and recipient species must take place. This requires them to flower at the same time, share the same insect pollinator if insect pollinated (bees can forage over a radius of 5 km), or otherwise be sufficiently close for pollen transfer. Future success will depend on the sexual fertility of the hybrid progeny, their vigour and sexual fertility in subsequent generations, and the selection pressure on the host of the resident transgene (Chamberlain and Stewart, 1999).

There is an element of chance regarding wind pollination since pollen, depending on meteorological conditions and grain size and density, presence or absence of air sacs, can be dispersed over long distances. Most pollen grains land within 3 m, less than 1% travels 1 km, although given the right meteorological conditions pollen can travel more than 200 km (Stanley and Linskens, 1974). As an extreme example from West Africa Maley (1972) has reported pollen of *Erica arborea* being transported 800 km from Tibesti to Lake Chad during the dry season by the harmattan winds (Bruneau de Miré and Quézel, 1959). The latter example clearly demonstrates that there is always an infinite possibility for cross pollination. Fortunately it is a risk that can probably be ignored.

As an example of environmental damage GM maize is held responsible in the USA for the poor survival of the larvae of *Danaus plexippus* (monarch butterfly). The

Bt toxin is expressed in the pollen, which can be wind dispersed for at least 60 m and deposited on other nearby plants and presumably then ingested by non-targeted organisms. Laboratory trials have already clearly demonstrated how the larvae of the monarch butterfly given access to leaves of *Asclepias curassavica* sprinkled with the *Bt* pollen had a significantly lower survival rate and poorer growth rate than leaves with or without non-*Bt* pollen. Since the ineptly named *A. syriaca* (common milkweed) is frequently found near maize fields as well as being the primary host plant of monarch butterflies, it is concluded that the monarch butterflies of the mid-western US corn belt are likely to be at risk (Losey *et al*, 1999). It is apparent that the US Environmental Protection Agency and Monsanto both failed to investigate the environmental risk before the maize was released for cultivation. The discovery also raises the question as to whether *Bt* maize pollen may also have an allergic effect.

So far the successful transfer of transgenic chloroplasts has only been achieved with great difficulty. Nevertheless, the technique could be used for other crops and the danger of chloroplast transfer from a cultivar to wild recipients has, therefore, to be assessed. The conclusions from an assessment of such a transfer from *Brassica napus* (oilseed rape) to *B. rapa* was that chloroplast exchange would be extremely rare and scattered (Scott and Wilkinson (1999).

In the past pest and weed control was achieved by spraying the crops with pesticides and herbicides. They were applied later and often at heavier dosages than sprays required for GM crops. The use of earlier and lighter spray application are considered by the manufacturers to be more environmentally friendly, even though there has been very little monitoring of their effect on the environment. Since pesticides and weedicides are by their very nature environmentally unfriendly, and whereas it is legislatively possible to enforce the withdrawal of their use, it would be extremely difficult, if not impossible, to control any undesirable genes that had escaped into the environment.

Genetic engineering has also been used to help solve problems in mineral toxicity. For example, excessive aluminium is a major problem throughout one third of the world's arable lands, especially on skeletal soils in high rainfall areas. Such soils are characteristically acidic and low in plant nutrients. Plant breeders have been able to overcome this problem using traditional inbreeding methods and a number of aluminium-tolerant wheat, maize, etc. genotypes are now available. Recently researchers in Mexico have used genetic engineering to develop a bacterial gene for citrate synthase which causes the plant roots to excrete organo-citrates, the citric acid present tying up the aluminium in the soil. The method has also been used to create aluminium tolerant tobacco and *Carica papaya* (pawpaw). However, further trials will be needed to see whether the photosynthetic cost of organo-citrate production affects crop yields (BBC cassette, Farming Today, 25 Sept. 97).

2.1   The Genetic Engineering Debate

It is impossible to discuss genetic engineering without discussing the moral issues involved. Those against claim that it is unnatural, interfering with Creation, and of the danger of creating uncontrollable super species that will destroy the environment. Those in favour believe that it is the logical advancement on traditional breeding, with the breeder using his God-given skills for beneficial purposes, preserving the environment by using fewer chemical inputs, as well as helping to feed the world's population.

Whereas traditional plant breeding is through selection and the transfer of genes between closely related species and genera, genetic engineering can involve the transfer of DNA material between completely unrelated organisms, such as a gene from a fish in order to provide a longer shelf life for the tomato. Naturally there is considerable consumer suspicion and resistance against the introduction of such genetically engineered foods, especially when they are not labelled.

Genetically engineered crops for herbicide and insecticide resistance is equally controversial. It is claimed that less herbicide and insecticides are required to produce control equivalent to several spray applications. Both methods can affect the desirable insects as well as the unwanted, but whereas sprays can usually be confined to the crop, there is an inherent danger that resistance can be passed from the crop to other plants, creating super weeds. Neither should it should be forgotten that insects too have the ability to develop resistance following poor pesticide applications, resulting in the creation of super bugs.

Whether we like it or not, genetic engineering is here to stay, forced on many of us prematurely by the greed of multinationals. The technique is new and requires far better monitoring than it is now receiving if uncontrollable disasters are to be avoided. Thus, while GM foods may be considered safe albeit not proven (in the Scottish legal sense), there is no room for complacency, especially with regard to genes that have been transferred to a GE plant from a completely unrelated organism with which normal breeding would have been impossible. The proteins for which they would have been coded will have been formed 'out of context' and are, therefore, a completely unknown quantity.

It is also essential to know whether the GM crop is stable and has no adverse environmental effect in the long term. But it should not be forgotten that there is still a great deal that can be achieved by conventional breeding techniques and clonal propagation of selected genotypes. See Straughan and Reiss (1996) and Ingram (1999) for further discussion.

3.   INTELLECTUAL PROPERTY RIGHTS

This is a complex subject and those requiring more detailed information than given here should consult The Crucible Group (1994), who claim to represent the widest international cross-section of socio-political perspectives and agricultural experience ever assembled on the subject. Under the General Agreement on Tariffs and Trades (GATT), now the World Trade Organisation (WTO), signatory states had to adopt either a *patent for intellectual property* (*IP*) or a *sui generis* (i.e. of its own kind, unique) IP system for plant varieties, or follow one of the 1978 or 1991 Conventions on Plant Breeders' Rights (see Section 4.1). It should be noted that after 1995 new member states were no longer accepted under the 1978 Convention, although members of good standing continue to be recognised. The United Nations Educational, Scientific and Cultural Organisation-World Intellectual Property Organisation (Unesco-WIPO) Model Provisions on Folklore or Inventors' Certificates, will monitor IP developments until the next round of WTO, formerly GATT, provisions are reviewed (The Crucible Group, 1994).

Plant Breeders' Rights are a form of intellectual property designed specifically to protect new varieties of plants, entitling the holder to prevent anyone from reproducing, offering for sale, or selling a variety without the breeders' consent, without royalty payment. Prior to 1985 plant breeders world-wide received relatively little protection under Plant Breeders' Rights (Section 3.1) and Trademark Names (Section 3.2) in obtaining royalties for their cultivars. Within the past two decades the development of genetic engineering and the demand for patents on genes has caused considerable disagreement and controversy regarding the ethics, morality and dangers of crop biotechnology. Since 1985 the granting of utility patents for novel plant cultivars in the USA has provided substantially greater protection to plant breeders than was available under previous laws. Breeders who identify useful genes can now obtain a guaranteed 17 years of exclusive protection for the specified gene in any plant or derivation thereof and, under licence obtain royalties from marketed products developed from the patented product. The USA is now exerting considerable pressure for cultivar patent protection to be internationally recognised. It can be argued that product development by the food, chemical and pharmaceutical industries, like plant breeding, is slow and expensive and, since the products can be expected to be marketed for long period, they should be protected.

It is debatable whether a patent system actually benefits society or limits opportunity and growth. It can also be argued that without some form of intellectual property rights there would be no commercial plant improvement. There is a need to achieve a balance between economic and social concerns arising from exclusive intellectual property rights against agricultural research and development (Williams and Weber, 1989).

3.1   Plant Breeders' Rights and Patents

The *International Convention for the Protection of New Varieties of Plants* provided by the International Union for the Protection of New Varieties of Plants (UPOV) was first signed in Paris in 1961 by eight European countries; 24 states are now signatory to the 1978 or 1991 Conventions (Anonymous, 1978; The Crucible Group, 1994).

Plant cultivar names are regulated on the assumption that a named 'variety' is an indispensable feature in trade since it gives purchasers the possibility of effectively choosing the 'variety' of their choice (Loscher, 1986). Thus, "*A new variety shall be designated by a denomination. The denomination of the new variety shall be regarded as the generic name for the variety. Such denominations must enable the new variety to be identified; in particular it may not consist solely of figures. It must not be liable to mislead or to cause confusion concerning the characteristics, value or identity of the breeder. In particular it must be different from every denomination which designates in any member state of the Union an existing variety of the same botanical species or a closely related species. Any person who offers for sale or markets propagating material of a variety protected shall be obliged to use the denomination of that variety even after termination of protection*" (UPOV (1984).

In this rather ponderous extract it is obvious that we are dealing with another use of the terms 'genus' and 'variety'. A '*genus*' is defined here as any group of objects having one or more properties in common, and thus distinguishes it from other groups of objects, and '*variety*' is used in the sense of what the Cultivation Code would term a cultivar (Loscher, 1986). The UPOV 'variety' must not be confused with 'trademark' designations, which will be dealt with in the following Section 4.2.

In the UK the Plant Varieties Act 1997 implemented fundamental revisions to the UPOV Convention adopted in 1991, and brings the UK plant breeders' rights regime into line with the separate EU regime, which came into effect in 1995 and is based on the 1991 Convention (MAFF, 1997a).

After developing a new cultivar, a plant breeder in the UK who wishes to market it as a new variety submits the plant for trial by the Plant Variety Rights Office of the Ministry of Agriculture, Food and Fisheries (MAFF). Provided the variety is novel, truly distinct from other cultivars, uniform and stable in cultivation, it will be granted Plant Breeders' Rights, enabling the breeder to claim royalties from other nurseries or business organisations who wish to propagate and/or market the new variety under licence. A name must be provided for each new variety and the name is then advertised in the *Plant Varieties and Seeds Gazette*. Provided there are no objections the name can be approved. Provided the holder pays an annual renewal fee, Plant Breeders' Rights are granted for a period of 25 years for all species except trees, vines and potatoes, which are for 30 years.

Similar conditions of distinctness, uniformity and stability plus improved agronomic or use value can be applied to mainly agricultural and vegetable species, enabling the new variety to be entered in the National List. The National List system was adopted in 1973 following the UK entry into the EU and ensures that no seed of a prescribed species may be marketed in the UK unless the variety is on a UK list or the EU Common Catalogue. Once entered the plant can only be traded under its registered name (MAFF, 1998a, b).

Plant Breeders' Rights apply to the authorised sale of seed used once in a cycle of production. For example, a farmer is entitled to propagate the seed and produce a crop for the food or industrial markets. The production of a further generation from the seed will then be subjected to Plant Breeders' Rights, but people who propagate the new variety for their own private use, i.e. in their own gardens, will not be breaking the law provided they do not barter or sell the plants.

There is pressure from the USA for breeders to be allowed to patent their new varieties and thereby prevent anyone propagating such plants for their own use. Under the present WTO accord governments still have the option of either adopting patent laws or devising some other form of *sui generis* legislation (The Crucible Group, 1994; MAFF, 1997a). Since seed merchants nowadays consist largely of international cartels, the monopoly imposed by the copyrighting of cultivars under the UPOV Convention is now becoming a matter of considerable contention.

As Plucknett *et al.* (1987) point out, the champions of plant breeders' rights argue that only when patents or royalties can be applied to the finished products of research is there sufficient stimulus for further research and development in agriculture, or any other scientific enterprise. Their many opponents claim genes are a world heritage and in the public domain and cannot be patented. They also claim that patents will encourage greater secrecy in crop research and may even hinder the flow of germplasm; and that crop genetic diversity could be further reduced by the private plant breeders' interest in crop uniformity.

Patents are granted for inventions, i.e. a new process or product; they are granted for discoveries, i.e. new knowledge. Thus, natural genes, as distinct from engineered genes, cannot be patented since they already exist, they are not new, they are discovered, not invented. A patent on a gene refers to a gene isolated from its natural surroundings and products containing the isolated gene. Such gene inventions are patented under **Trade-Related Intellectual Property (TRIPS)** in both Europe and the USA. Similarly, new plant cells, i.e. containing transformed DNA, or the product of cell fusion, or in the form of a cell culture, can also be patented under TRIPS.

With international representation from public and private research institutions, nongovernmental organisations, governments and academia, The Crucible Group (1994, p. xv) comment *"It is certain that no country has cornered the market on biodiversity. No country is even remotely self-sufficient in its needs for genetic resources. Genetic diversity is full of surprises. Some of the most biologically diverse regions in the world may depend upon much less diverse regions for some of their most important foods and medicines. The world needs a strong, multilateral framework within which nation-states can manage their resources and negotiate their access*." They continue (p. 4) with the following significant statement: "*The agricultural research community cannot guarantee the long-term survival of any crop, in any country, if the breeding options for that crop are curtailed through the nonavailability of cultivated or so-called wild germ plasm*".

The morality of private seed companies exploiting the developing countries by utilising germ plasm initially obtained from those countries and then selling it back to them is now being questioned, although in some cases the seed has been so 'improved' that it is no longer suitable for growing in the donor country. Even the research undertaken by international scientists working for the developing countries within the CGIAR (Consultative Group for International Agricultural Research) establishments funded by the World Bank have had their freely available research results patented by private companies! It is an unsatisfactory situation which WTO, and the US in particular, must resolve.

3.2   Trademark Names

In some countries a trademark, i.e. a name or mark identifying a plant which is officially registered and legally restricted to the owner or breeder, is also attached to the cultivar and as such is subject to the country's trademark laws. The same trademark may even be attached to different cultivars, and sometimes even to other objects. In the UK a trademark or trade name can be used in connection with the registered variety name provided that the mark or name and registered name are juxtaposed and the registered name is easily recognisable. This applies both during and after the expiring of plant breeders' rights.

There are two approaches to trademarking names of cultivated plants. The more usual and preferred method is to trademark a word (or words) which is then used in association with a cultivar name, e.g. *Lilium* Enzett (R) 'Orangefeuer', and would be cited in the *International Register of Trademark Names* in the manner shown, Enzett (R) indicating that it is the firm Enzett that is registering the trademark for the cultivar 'Orangefeuer'. In some International Registers a different usage has unfortunately developed, incorporating the trademark with the cultivar name, e.g. *Pelargonium* 'Enzett-Domino (R)', which is not acceptable by the International Commission for the Nomenclature of Cultivated Plants, although national trademark laws may mean that they are legal for that particular country.

A second method of trademarking that has been introduced in recent years is the trademarking of cultivar names, e.g. *Dianthus* 'Alexis' (R), thereby enabling the trademark 'Alexis' to be used for any other plant or product, even soap powder if the owner wishes. A further complication occurs when breeders not only trademark the 'fancy' cultivar name but also add varietal denomination in order to obtain plant variety rights, e.g. *Dianthus* 'Alexis' (R) (Alexis) (C.D. Brickell, pers. comm. 1989).

A trademark provides far greater legal possibilities than a varietal denomination when used for advertising. Trademake names are consequently gaining increasing significance in the marketing of propagating material of certain species. Its use as a cover for various products of the same firm, however, is not suitable means of designation in the context of plant breeders' rights (Loscher, 1986). The owners of trademark names enjoy full ownership rights, and any long-term publicity investment based on the trademark will clearly benefit the owner. The scope of protection afforded a trademark within a given class (International Trademark Classification) of products is very far reaching. Breeders diversifying in several species can use the same trademark for several cultivars belonging to several species. This is both an economical and commercially more profitable use of the trademark. But of course each cultivar must be identified by a different denomination for nomenclatural purposes. For a financial consideration breeders may also licence their trademark for use by other breeders for cultivars belonging to other species. Since the commercial life span of many modern cultivars is becoming shorter and shorter, the trademark may outlive the cultivars for which it was used to promote. Nevertheless it can continue to be used for new, improved and cultivars (Royon, 1986).

Chapter 7

Marketing of Crops and Crop Products

Quantity, quality, availability and economics are the key factors that affect the marketing of a crop or crop product, irrespective whether it is for the domestic, national or international market. Obviously the purchaser will require the product to be available at the time required and for it to be of the desired quantity and quality. While the domestic market may permit some flexibility, where further processing or national and international markets are involved there can be no such flexibility. A processing plant will require a steady supply in order to keep the plant working efficiently throughout the year or season. Importers likewise require a steady supply, if not they will seek supplies elsewhere or find substitutes.

The market in gum arabic from *Acacia senegal*, harvested mainly from wild populations, is a good example of how the interplay of quantity, quality, availability and economics affect the international market. Gum arabic is a major export crop of the Sudan with a the statutory monopoly of the gum arabic trade within the country granted to the Gum Arabic Company whereby it effectively controls *ca*. 85% of the World Market, with Senegal, Mauritania, Mali, Chad, Niger and Nigeria supplying much of the remainder. Deforestation, drought and the poor prices paid to the farmers for their gum has reduced the quantity available for export. Despite the advantages of low cost and superior performance over possible substitutes, gum arabic has in the past suffered from variations in quality arising from the varying composition of each batch as well as the uncertainty of maintaining a regular supply. Supplies dwindled, especially after the loss of many of the gum trees following the Sahelian droughts of 1973-74 and 1982-83. The losses of tree from drought were further compounded by the high prices obtained in selling the trees of firewood. As a result, over the past 15-20 years there has been a steadily decrease in demand by the processors. Gum arabic sales peaked at around 1970 at approximately 70,000 t, of which about 70% went into confectionery products. The high prices and world shortage as a result of the 1973-74 drought led to some major users seeking alternative modified starches. Annual sales fell to *ca.* 40,000 t during 1978-82 and the 1982-83 drought created a further world shortage and loss of markets, with annual sales reduced to *ca.* 20,000-24,000 t. Although world consumption of hydrocolloids by the food industry is fairly stable, the uncertainties created by the fluctuations in the supply of gum arabic, and the increasing availability of modified celluloses and fermentation products such as *xanthin* and *gellan* has certainly led to the major users paying serious attention to cheaper alternatives to gum arabic. Furthermore, the gum importers and users consider that the Sudan, with its almost monopolistic control of the market, is bad for trade confidence. The spreading of sources of supply to give the Sudan control of only 50% of the market would lessen the risk of a major crop failure and improve confidence (Anderson, 1990, 1993).

The ability for a product to be kept or stored without deterioration is also an important requirement. Edible products with a long shelf life will have an advantage over more desirable products with a short shelf life, unless the latter can find a ready market. A good storage life is essential for products required out-of-season, such as Brazil nuts for the Christmas trade. The mature fruits are harvested from November through to August in the following year, the wide harvesting period being governed by the seasons within the distribution range of *Bertholletia excelsa*. The seeds (Brazil nuts) are extracted and dried as quickly as possible to ensure preservation before being packed for sale to the marketing organisations. Consequently the Brazil nuts can be a year old before reaching the consumer (Prance and Mori, 1979; LaFleur, 1992; Wickens, 1995a).

1.   MARKETING CRITERIA AND INFLUENCES

The marketing stages involved from planting to consumer can be most readily illustrated with a new crop or crop product that is 'new' to a country, state or region, or has been developed *de novo* from a previously unexploited plant, i.e. agricultural, horticultural and forestry produce. The introduction of a new crop or crop product is a long and tenuous process and it is essential that all stages from the first identification to consumer consumption or use are rigorously examined and a considered judgement arrived at regarding marketing feasibility and future prospects. While it is not intended to enter into too much detail here, a number of questions have to be asked regarding marketing prospects, crop and product research requirements, and the needs of the producer.

Before marketing the following questions need to be asked: (1) Is the plant or plant product needed? (2) What is the likely consumer demand? (3) With what would it compete or replace, and what would be the social, economic, etc. effects of such competition/replacement for the producer and the world market? (4) Who will grow, buy, store, process, transport and use it? (5) Are the transport services adequate to convey the crop to the market, processing plant or port? (6) What is the minimum yield required for the potential market to be profitable? (7) What are the long term prospects in terms of potential sales, likely price structures and market locations? (8) What are the packaging requirements from farm to consumer? (9) Can phytosanitary precautions be rigorously maintained from farm to consumer/end user? (10) Does the product name appeal to the purchaser? For example, 'cocoa', produced from *Theobroma cacao*, is a pleasanter sounding name than the local vernacular 'cacao'. Similarly, the sales of the fruit from *Actinidia deliciosa* increased significantly when the name was changed from 'Chinese gooseberry' to 'Kiwi fruit'; and (11) Is the product acceptable to the consumer?

The relevant past and present research requirements include: (1) Has all the known information been assessed and examined? (2) Has the new crop been adequately researched from planting through harvesting to storage and processing in the environment where it is to be produced? (3) What are the storage and processing requirements? (4) What existing technology can be used or adapted? (5) Has adequate attention been paid to provenance and genetic diversity? (6) Are there any major problems anticipated from pests and diseases? Note that field trials at the research stage may not always reveal problems that may occur during large-scale commercial production; (7) What are the research costs and who pays? For pharmaceutical products this could run into millions of pounds; (8) Are there any health and safety hazards, especially during processing? Are health and safety regulations adequate? and (9) What are the prospects for future research?

The needs of the producer include: (1) What are the production costs for seed, labour, fertilisers, etc.? (2) What financial support is there for the new venture? (3) Can existing equipment and storage facilities be used or adapted? (4) How will the new crop fit into the existing cropping system in terms of labour, rotations, etc.? (5) Would extension support and advice be available should any problems arise? (6) Will it be profitable? Farmers do not seek maximal yields or gross profits but prefer high net profits, which are generally less than maximal; (7) Can quality and quantity of the product be maintained or improved? (8) Is bank or co-operative financial support available? and (9) What are the field risks from pests and diseases? (Meadley, 1989; Muchaili, 1989; Wallis *et al.*, 1989; Holness, 1995; Simmonds, 1996).

Price and availability can sometimes determine the market. Jojoba oil from *Simmondsia chinensis* currently finds a ready and profitable market in the cosmetics industry. It also offers an alternative substitute to sperm whale oil and could, therefore, save *Physetes catodon* (sperm whale) from the threat of extinction. Be that as it may, there would have to be a dramatic increase in production accompanied by an equally dramatic fall in price. The problem is how to bridge the change in global production from a relatively small but highly profitable acreage to a very large and highly productive acreage producing an oil that can compete on the world market with sperm whale oil.

There is a somewhat analogous problem with palm oil from *Elaeis guineensis* (African oil palm) concerning the relative economic values of its products. The oil can be used for food, pharmaceuticals, biochemicals, and biofuel. If all the palm oil could be sold for food, then the time and money spent in converting it into the less valuable methyl esters as a substitute for diesel fuel would appear inappropriate. However, during the conversion process to produce the methyl esters, a number of pharmaceutical derivatives such as tocopherols (0.05-0.1%) are also produced; the tocopherols being natural antioxidants which also function as vitamin E. The methyl esters can also serve as valuable chemical intermediates for fatty alcohols, surface active agents and other products. Approximately 1% carotene, a precursor of vitamin A, is also present and imparts a distinctive orange-red colour to the oil, but is unfortunately destroyed during processing. If the carotene could be economically extracted it could then have an economic value close to that of the palm oil (Shay, 1993).

Economics, politics and chance play a large part as to whether a plant or plant product has a role to play in the domestic or commercial economy of a community or nation. Commodities can be rendered no longer economically viable due to increased labour or other production costs, or they are superseded by technology. An example of the latter is the production of the chemical analogues of many pharmaceuticals to provide a better control of quality and quantity than would be possible from natural resources.

Politico-economic considerations can also greatly influence the rise and fall of plant products on the local, national and international markets. The following three examples illustrate such influences: (1) The Spanish Conquistadors, in an effort to subdue the Incas prohibited the cultivation of their staple pseudocereal, *Chenopodium quinoa* (quinoa). The crop is now achieving a minor comeback internationally as a health food (National Research Council, 1989; Risi and Galwey, 1989; Lewington, 1990); (2) Brokensha and Riley (1986) and Riley and Brokensha (1988) studied the social and environmental changes in Mbeere, Kenya between 1970 and 1987, documenting the changes in the uses and incidence of wild and cultivated plants. Mbeere is a microcosm of the many changes characteristic of Kenya, tropical areas in particular and typical of the developing countries generally as people try to adapt to the pressures of a dramatically rising population on diminishing natural resources, the change from communal to individual land tenure, and the new values brought about by modernisation. Many of the indigenous economic plants have disappeared and being replaced by alternative plant substitutes or commercial substitutes; and (3) World War II meant that the USA was deprived of its source of *Hevea brasiliensis* (Pará rubber) from the plantations of Southeast Asia, a commodity of strategic importance, e.g. for aircraft tyres and the arresting wires on aircraft carriers, etc. Production of the previously economically nonviable guayule rubber from *Parthenium argentatum* received an immediate, albeit temporary, boost. Production again declined when Pará rubber was once more available on the international market. Nevertheless, guayule is still regarded as a potential crop for the future (Whitworth and Whitehead, 1991).

Public opinion must not be neglected. Markets are becoming increasingly influenced by the consumer with regard to human rights, not only with reference to oppressive governments but also for fair wages, healthy working conditions and, in the case of food, crops containing the minimum of chemical residues. There is concern also about genetically engineered products as to whether they are really safe for both consumer and the living world. Can, for example, genetically engineered food plants solve world hunger or will some hitherto unforeseen circumstance trigger genetic chaos? People have the right to choose and know what they are buying, i.e. full and accurate labelling of all products. This has recently come to the fore with the labelling of food additives such as GM soya flour. The US claim that any plan to label such soya products would involve segregating a bulk commodity, which would not only be impractical but also 'clearly trade-inhibiting' (Anonymous, 1997b). It rather looks as if consumer choice is going to win on that particular issue.

2.   ISO STANDARDS

Industrial, technical and business experts from some 120 national standards institutes world-wide have, through the International Organization for Standardization (ISO), developed voluntary technical standards for products that are internationally marketed. These are designated *[ISO] International Standards*. (Because the International Organization for Standardization would have different abbreviations in different languages, e.g. IOS in English, OIN in French, it was decided that ISO, derived from the Greek *isos*, meaning equal, would be the accepted abbreviation). Although the standards provide quality control, the ISO has no regulating function. It is the individual governments, or multi-governmental organisations such as the European Union,, that are the regulating bodies who decide whether to accept or ignore ISO standards (ISO, 1997).

3.   IMPORT AND EXPORT LEGISLATION

The import and export of plants and their products are subject to regulations according to the countries concerned. For example, endangered species harvested from the wild are banned under the *Convention on International Trade in Endangered Species* (CITES). Other plants, both living and dead, and plant materials may also be prohibited for plant health reasons or require a phytosanitary certificate or license for importation. The regulations governing the importation of plants into Great Britain are outlined below.

Under the Single Market arrangements within the European Union (EU) from 1 June 1993 plants and plant products can move freely throughout the Union, with only a limited range of plants and particular plant products requiring a plant passport to facilitate such movement. Phytosanitary certificates are now only required for plants and particular plant products entering the EU from third countries, i.e. countries other than members of the EU. The exception is for certain areas within the EU which are still free from certain pests and diseases already established elsewhere within the Union. These have been designated ***Protected Zones***. The zones in Great Britain are for *Leptinotarsa decemlineata* (Colorado beetle), *Bemisia tabaci* (tobacco whitefly), beet necrotic yellow vein virus (beet rhizomania) and the forestry insect pests *Dendroctonus micans* (great spruce bark beetle), *Ips amitinus* (smaller eight-toothed spruce bark beetle), *I. typoraphus* (larger eight-toothed spruce bark beetle) and *I. duplicatus* (northern spruce bark beetle). Imported plants and plant products regarded as potential hosts of these protected zone pests must be accompanied by a phytosanitary certificate when entering Great Britain.

The relevant plant health restrictions and requirements regarding plants for planting, fruit, seeds, parts of plants, potatoes, and other plant products are laid down in *The Plant Health (Great Britain) Order 1993* (Statutory Instrument 1993 No. 1320 and summarised in MAFF (1994b).

The controls for wood, wood chips, etc. and bark are laid down by the Forestry Commission in *The Plant Health (Forestry) (Great Britain) Order 1993* (Statutory Instrument 1993 No. 1283), and summarised in Forestry Authority (1994). It should be noted that phytosanitary certificates issued by a state, province, or regional or local government are not acceptable. In the USA, for example, only the Federal Department of Agriculture may issue a certificate, a state may not do so.

Compliance with the above Plant Health Orders for Great Britain does not relieve the importer from obligations imposed by other relevant regulations. These include CITES, EU quality controls on the import of horticultural produce, tariff classification, import duties, CAP levies and charges for goods imported from third countries, forest reproductive materials import licences, hay and straw licences, and general import licensing (other than plant health licensing).

Chapter 8

Human and Animal Nutrition

Nutrition is the process by which a living organism, both human and animal, physiologically absorbs and uses food to ensure growth, energy, reproduction and repair of tissues. Nutritional studies involve the study of both diets and deficiency diseases. A *balanced diet* is one that will provide at least the minimum nutritional requirements for energy, protein, minerals and vitamins required by an organism. An *unbalanced diet*, such as one low in fibre or with excessive vitamins, will have an adverse effect on the organism. Clearly the actual dietary requirements of an organism will vary according to the age, sex, level of physical activity and, in the case of females, the reproductive stage, i.e. pregnancy and lactation. Carbohydrates, proteins and fats are required in relatively large quantities to provide energy and build body tissue while only small quantities of minerals and vitamins are needed to enable the body to function properly. Fibre aids digestion by stimulating muscular activity in the digestive tract. It also facilitates the digestion of concentrates by opening them up to the action of digestive juices. Excess fibre in the diet can reduce food intake

In general pure carbohydrates and proteins provide 4 kcal g^{-1}, fat 9 kcal g^{-1}, dietary fibre little, and vitamins and minerals no energy. In addition. the water content of the actual food consumed will affect the energy value, so that most fruit and vegetables will, because of their high moisture content, have an energy value less than 1 kcal g^{-1}. On the other hand, grains and pulses with only *ca.* 10% moisture, will provide *ca.* 3.5 kcal g^{-1}. *Antinutrients* are those food constituents that interfere with the beneficial effects of food intake and are discussed in Section 8 below (Passmore *et al.*, 1974; Pietroni, 1994; Macpherson, 1995; Vaughan and Geissler, 1997). See also Chapters 9 and 10 for additional information and human and animal foods.

1. FOOD AND FEED ANALYSIS

The term *proximate analysis* refers to a type of analysis that was originally developed by the Weende Experiment Station in Germany in 1895. It is now universally accepted as a means of providing a summary of the major components present in human and animal foods. The laboratory procedures used are given in *Official Methods of Analysis of the Association of Official Agricultural Chemists*, published by the Association of Official Agricultural Chemists, Washington, DC.

All analyses are based on *dry matter* (*DM*), and represents the sum of the various mineral and organic components present in the food. To convert a dry matter value of a component to a fresh-matter basis for the formulation of feeds, the percentage is multiplied by that of the dry-matter content. Six fractions are recognised: (1) *Moisture content* represents the difference between wet and dry weights and is expressed as a percentage of the wet weight; (2) *Crude protein* (*CP*) is approximately the value of Kjeldahl nitrogen x 6.25 and is based on the assumption that protein contains all the nitrogen in the food and that all proteins contain 16% nitrogen. Depending on the nature of the feed, the *true protein* content is derived by subtracting the non-protein nitrogen from the total nitrogen content. The non-protein consists of inorganic nitrogen salts, amino nitrogen, amides, etc. In the majority of green foods only a part of the nitrogen occurs as protein nitrogen; (3) *Ether extract* (*EE*) or *crude fat content* is measured as diethyl ether or petroleum ether extracted material, and includes not only oils and fats but also fatty acids, resins, chlorophyll, fat soluble vitamins, etc.; (4) *Crude fibre* (*CF*) refers to organic matter insoluble in a hot, diluted H_2SO_4 and diluted NaOH solution and consists mainly of cellulose, lignin and related compounds; (5) *Ash* represents the mineral residue remaining after igniting samples until they are free of carbon. A high level of ash in fodder is often an indication of soil contamination; and (6) *Nitrogen-free extract* or *available carbohydrate* (*NFE*) consists of soluble carbohydrates such as sugars, starches, etc. and constitutes the difference between the original dry weight and the sum of the other constituents,

The proximate analysis has a number of advantages: (1) Most laboratories are suitably equipped; expensive and sophisticated equipment is not required; (2) The method provides a good, general evaluation of the food; and (3) Much of the available data is already reported as proximate analyses. The disadvantages are: (1) The method does not define the individual nutrients of the food material, i.e. the fractions represent mixtures of the various nutrients present; (2) The CP, CF and FE are rough estimates; (3) The analysis is time consuming, with little scope for automation; (4) The analysis does not show how much indigestible material is present because the acid-alcohol treatment dissolves much of the CF, making it impossible to indicate how much is indigestible. The nutrient values of some foods are over-estimated while others are under-estimated. It also fails to show how the constituents of the indigestible residues are related; and (5) No information is provided regarding palatability, texture, toxicity, digestive disturbances and nutrient availability.

More specific analyses are becoming available that better reflect the structural constituents of plant materials: (1) The *Van Sorest analysis* separates and classifies the digestible and indigestible parts of plant cells; (2) *Bomb calorimetry* provides complete combustion in the presence of oxygen to give the gross heat or heat of combustion. The method may be used to determine the *gross energy value* in kgCal of a food. The *metabolisable energy* is the gross energy value less body waste products, it represents the proportion of energy in the food used in metabolism; (3) *Chromatography* provides an extremely sensitive and easily adaptable technique for measuring extremely minute quantities of many compounds. It has the advantage of being both rapid and inexpensive, but has the disadvantage of being a complex operation involving sophisticated equipment and requiring careful sample preparation; (4) *Colorimetry* and *spectophotometry* are used for measuring concentrations of certain substances, such as vitamin A, or protein and amino-acids, and thereby replacing the Kjeldahl process. Nevertheless, the various procedures that measure total nitrogen content only provide a rough indication of total protein because many foods contain non-protein nitrogenous compounds. Tests for true protein include the Bluret, Lowry, and Warburg-Christian assays, turbidity measurements, and the peptide bond method; (5) *Biological assays* are required to provide information on the biological availability of nutrients, i.e. whether chemically desirable compounds are actually available to the consumer. Such assays are time-consuming and subject to differences between the test animals in their digestive abilities; (6) *Digestibility trials* are used to determine the nutrient and energy differences between food input and output. The chemical analyses of the percentage of each digested nutrient in a food provides a *coefficient of digestibility*. The *digestible crude protein* provides a more useful measurement than either the CP or TP content when, for example, formulating rumen diets, since both CP and TP can vary considerably in digestibility and composition. The *digestible energy value* provides a measurement of utilisable energy in a food after deducting the energy value of undigested faecal residues; (7) *Microbiological assays* test the presence or absence of a nutrient according to the growth or not of micro-organisms; and (8) *Microscopic analysis* determines the presence of unacceptable quantities of filth, i.e. insect fragments, rodent faeces, hairs, etc., adulterants and contaminants, including dirt, sand, shell fragments, husks, etc. as well as substitute materials (Dalal-Clayton, 1981; Ensminger *et al.*, 1994).

2.   WATER

Water is essential for both plant and animal life. In plants water serves to maintain turgor, in the transportation of nutrients, temperature regulation, etc. and the water content varying with the organ and functions. In dry storage organs, such as nuts and oilseeds, the water content may be as low as *ca.* 2-5% and in cereal grains and pulses 10-12%, while in root crops and leafy vegetables it may be as high as 80-90% and 90-95% respectively. In arid regions the high water content of some plant organs, such as the fleshy fruits of *Acanthosicyos naudiniana* (herero cucumber) and *Citrullus lanatus* (water-melon) may even be the sole source of moisture available for human consumption for several months of the year. Similarly, the succulent ephemeral *gizu* or *asheb* grazing of the northern Sahel was, before the recent protracted drought, capable of supporting camels, and to a lesser extent other stock, for several months without access to other sources of potable water (Wilson, 1978; Göhl, 1981; Arnold *et al.*, 1985; Vaughan and Geissler, 1997).

3.   CARBOHYDRATES

Carbohydrates are naturally occurring organic compounds essential to plant and animal life. In plants they are synthesised from atmospheric CO_2 and water, and function as energy storage molecules and as structural elements in the plant body. In general the carbohydrates have an energy-producing value of *ca.* 4 kcal g^{-1}.

The carbohydrates have the approximate formula of $(CH_2O)_x$, where the various values of 'x' represent the sugars, starches and cellulose. The simple carbohydrates are the *monosaccharides*, *disaccharides*, *polysaccharides*, and *cellulose* (*dietary fibre*), with repeating units usually containing 5 or 6 carbon atoms joined through oxygen linkages. The basic sugar skeleton of carbohydrates, involving hydroxyl groups, gives them such properties as water solubility and sweet taste. When digested the polysaccharides are converted into simple sugars and assimilated by the body tissues (Passmore *et al.*, 1974; Sharp, 1990; Macpherson, 1995; Vaughan and Geissler, 1997).

3.1   Sugars

The sugars represent members of the lower molecular weight carbohydrates, i.e. monosaccharides and the smaller oligosaccharides. The latter consist of short chains of between two and ten polymerised monosaccharide units, and are represented by the disaccharides, trisaccharides, etc.

The dietary *monosaccharides* are sugars with the formula $(C_nH_{2n}O_n)$, where n = 5 or 6, of which the hexose sugars *glucose* and *fructose* are among the few monosaccharides that are found free in plant tissues. Free glucose is not particularly abundant, occurring in small amounts in fruits and vegetables, particularly in grapes and onions, while the more common fructose is widely found in fruits and vegetables.

All monosaccharides have a potentially active aldehyde or ketone group capable of reducing an oxidising agent, they are consequently also known as *reducing sugars*

(For a disaccharide to be a reducing sugar one of the two component monosaccharides must be left intact). The most common hexose sugar is D-*glucose* (dextrose or grape sugar) and is present in many plants. It is the sugar present in the blood, produced by the body tissues, with large quantities excreted by the kidneys in diabetes mellitus. An isomer of glucose, D-*galactose* is also fairly widely distributed among plants; it is a constituent of *raffinose* and *stachyose*, of hemicelluloses, pectin, gums and mucilages, and of some glycosides. Galactose forms half of the lactose molecule (milk sugar) as well as being the sugar found in the brain. D-*fructose* (laevulose or fruit sugar) is the commonest ketose sugar, i.e. a sugar containing a potential keto-(CO) group; it is present in fruit juices in the unstable *furanose* form. Polymers of fructose are known as *fructosans*, the most important of which is the storage polysaccharide *inulin*. The aldohexose sugar *galactose* commonly found in plants does not normally exist in the free state but as polymers such as *galactan*; enzymes in the liver convert the galactose to glucose. In some plants, especially among some members of the Leguminosae, the aldohexose *mannose* largely replaces glucose as the building block for reserve polysaccharides, the chains of mannose units being known as *mannans*. Galactose and glucose may also be present, forming galactomannins, glucomannins and galactoglucomannins. The film-forming characteristics of the glucomannin from *Amorphophalus konjac* (devil's tongue) are used in China and Japan in preparing stabilisers and emulsifiers for food and drinks, as well as for industrial purposes (see Chapter 15). *Mannitol* is a soluble sugar alcohol formed by reduction of the carbonyl group of mannose or fructose to an alcohol. Mannitol is a major photosynthetic product in the brown algae, lichens and some higher plant species, e.g. *Fraxinus ornus* (flowering or manna ash) and was formerly an important sweetening agent (see Chapters 9 and 19). The pentose sugars *ribose* and D-*2-deoxyribose* (*deoxyribose*) are important genetic constituents of ribonucleic acid (RNA) and deoxyribonucleic acid (DNA) respectively.

Disaccharides are sugars derived from two monosaccharides by the elimination of a water molecule, e.g. sucrose and maltose. *Sucrose* (cane or beet sugar) is the major transport sugar in the higher plants, with 50-85% sucrose present in *Saccharum officinarum* (sugar cane) and 18% in *Beta vulgaris* subsp. *vulgaris* (sugar beet). Sucrose is a *nonreducing sugar* since the component glucose and fructose units are linked through a high energy bond between carbon one and carbon two of their respective aldehyde and ketone groups. When ingested sucrose is converted by the digestive juices into glucose before being absorbed by the digestive tract by a process known as inversion. It is a valuable source of heat and energy and, to a certain extent, a tissue-builder in respect to fat. Consequently it should be avoided by people tending to put on fat, also by diabetics. *Cellobiose* is another common reducing disaccharide in plants, being the basic repeating unit of cellulose. *Maltose* (malt sugar), which consists of two glucose units, is produced by the amylase present in the ferments of the saliva and pancreatic juice, and is still further changed by the intestinal ferments into glucose before being absorbed by the digestive tract (see Section 3.2 below for further discussion). The non-reducing *trehalose* (mushroom sugar) is composed of two glucose molecules and generally replaces sucrose in plants lacking chlorophyll and starch, such as the fungi, where it constitutes up to 15% of the dry matter.

Another common oligosaccharide is the long-chain fructosan *inulin* (not to be confused with insulin) forming the storage polysaccharide in the Compositae, with up to 58% in *Helianthus tuberosus* (Jerusalem artichoke). Inulin is flatulent-forming. since it is not broken down by the digestive enzymes but undergoes bacterial fermentation in the larger intestine (Tootill, 1984; Sharp, 1990; Macpherson, 1995; Jansen *et al.*, 1996; Vaughan and Geissler, 1997).

3.2   Starch

Starch is the most abundant and important reserve polysaccharide in plants. It is an early end product of photosynthesis, continuously being formed and then broken down at night into sucrose for transporting to other plant organs, where it is mainly stored as starch granules, especially within the seeds and roots. Starch $(C_6H_{10}O_5)_x$ is composed of varying proportions of the water-soluble *amylose*, consisting of long straight chains of 200-1000 glucose units linked by α-1,4-glycoside bonds, and the mucilaginous *amylopectin*, with relative short chains of *ca.* 20 glucose units cross-linked by α-1,6-glycoside bonds. Starch is insoluble in cold water, but in hot water the granules are gelatinised to form an opalescent dispersion. The characteristics of starch granules in terms of size and shape, and amylose and amylopectin content vary according to the species (see Table 15). To some extent granule size is also a function of age, increasing in size with time. Starch grains, because of their abundance, ubiquity and varied morphological characteristics, can be diagnostically important. Depending on the species, archaeologists have made use of starch grains to identify samples of dehydrated, carbonised or pulverised vegetable materials, even from traces of starch granules found on tools used for grinding and preparing food (Cortella and Pochettino, 1994; Moore, 1998).

Variations in the amylose and amylopectin content can have important applications. The naturally occurring variant **high amylose maize** with 60-70% amylose (instead of the usual 30%) has starch granules which, when heated in water, form strong gels, which are used in the confectionery industry for making jellies and gums. There are also naturally occurring variants of several plant species, such as rice and maize, whose starch consists solely of amylopectin. Such variants are the so-called '*waxy starches*'. Genetic engineering has also be used to develop waxy starches in GM maize as well as several other species. It should be noted that the term 'waxy' applies to the type of starch and must not be confused with those varieties of potato where the texture is also described as 'waxy'. The waxy starches, when heated in water form viscous pastes with a 'stringy' texture, making them highly suitable as all-purpose thickeners for the food industry (Tootill, 1984; Sharp, 1990; Macpherson, 1995).

Using an X-ray diffractometer three types of X-ray diffraction patterns are recognised for plant starch granules. Most cereals are *type A* (cereal type), while potato, canna and tulip are *type B* (tuber type). The classification is not perfect since *Colocasia esculenta* (taro), some *Ipomoea batatas* (sweet potato), *Manihot esculenta* (tapioca) and *Iris* are type A, while high amylose barley, maize and rice starches are type B. Types A and B and believed to be independent, while *type C* represents possible different proportions of A and B and, depending on the proportions of A and B, form Ca, Cb and Cc subgroups (Hizukuri, 1996).

Quantitatively starch is the most important source of available carbohydrates in the diet, as well as being an important source of indigestible carbohydrates. While the smaller starch granules are reputed to be more easily digested than large granules, this is not entirely true. The granule structure and how the starch is subsequently processed can also affect digestibility. Thus, the tubers of *Maranta arundinacea* (West Indian arrowroot) contain starch granules that are only 15-70 μm long, with *ca.* 20% amylose and 80% amylopectin. By way of contrast, the granules of *Canna edulis* (Australian arrowroot) - possibly a synonym of *C. indica*, are very large, 125-145 × 60 μm. Both, because they are readily digested, may be used as an invalid food (Flach and Rumawas, 1996).

The digestion of starch is through the action of two or more enzymes, i.e. stabilising or liquefying enzymes, usually α-amylase, to convert starch to water-soluble dextrins or oligosaccharides, and a saccharifying enzyme to convert intermediate products to glucose or maltose. In humans all the products resulting from starch digestion, i.e. glucose, dextrins and starch granules, that arrive at the end of the ileum (lower part of the small intestine) are subjected, irrespective of their size, to bacterial degradation and fermentation and have potentially the same metabolic behaviour. The sum of starch and products of starch degradation not absorbed in the small intestine of healthy individuals is known as *resistant starch* (*RS*) and provides an important contribution to the beneficial effect previously assigned solely to dietary fibre. RS is defined as being composed *in vivo* of oligosaccharides (including glucose), high molecular weight α-glucans (mainly starch granules), and a crystalline fraction whose size depends on the origin and treatment of the starch. Three categories of RS are identified according to their resistance to digestion: (1) *RS type I*, physically inaccessible starch, enclosed in cell or tissue structure and occurring mainly in partly milled grains and seeds; (2) *RS type II*, resistant native starch granules, present in raw B-type starch granules, e.g. potato and banana; and (3) *RS type III*, retrograded amylose, i.e. chemically or thermally modified starches and found, for example, in cooled, cooked potato, bread and corn flakes.

While starchy foods will contain one or more RS types, their presence is physical rather than chemical, depending on botanical origin, processing, i.e. cooking, cooling, freezing, storage, the nature of the enzymes and resistance of the starch to digestive enzymes, the ratio of starch/enzyme, hydrolysis characteristics, etc. Indeed, it is probable that no starch fraction would fail to be digested by any enzyme provided the concentration is not limiting and the duration of hydrolysis sufficient (Champ, 1994; Björk, 1996).

Edible starch is mainly obtained from milling cereal grain. In the USA the second largest use of maize after animal feed is wet milling for the isolation and recovery of modified and unmodified food grade starch, glucose (dextrose) and fructose syrups, *ethanol* and other chemicals via fermentation processes. *Wet milling* involves steeping, germ, fibre and protein recovery operations, and starch washing (for further details see Eckhoff and Paulsen, 1996). In addition to pure starch, maize gluten feed and maize-germ oil meal are produced as by-products for feeding to livestock (Göhl, 1981). Although a high capital and energy intensive process the products remain economically viable.

As well as extracting starch by wet milling there are three other process known as *dry milling*: (1) Dry milling the whole kernel into a whole grain flour using a stone grinder or roller mill. The ground germ present in the flour decreases the storage life but improves the flavour; (2) A dry grind ethanol process where the kernel is first ground in a hammer mill and then cooked, saccharified and finally fermented to convert the starch to *ethanol*; and (3) Degerminating dry milling, where the grain is initially tempered to facilitate the separation of the germ, pericarp, and endosperm, followed by passing through a mechanical degermination processor before drying and separation.

The products of dry milling are: (1) *Bran*, the flakes from the coarse outer covering of the wheat, rice, maize or rice grains and are removed by sieving. Bran may be used as a useful protein, vitamin and fibre feed for livestock, eaten as a breakfast cereal, or returned to the flour for whole grain flour. The high water-absorbing properties of the fibre has a laxative action; (2) *Flour*, a soft, powdery product derived mainly from the inner portion of the kernels of wheat, barley, maize, oats, rice and rye. It consists mainly of starch and protein and is low in fibre. It is used for baking bread, etc.; (3) *Germ*, the protein-rich embryo of wheat, rice and maize grains and may or may not be returned to the flour for bread making; (4) *Gluten*, the tough, elastic protein remaining after the starch is removed from wheat flour. It imparts elasticity to dough, enabling leavening with yeasts; (5) *Grits* consist of the coarsely ground deskinned grain from which the bran and germ have been removed. It is usually cooked for a breakfast cereal or side dish. In the UK grits, especially oats, are used for livestock feed; (6) *Hominy* refers to the bran coating and the maize germ. It too may be eaten as a breakfast cereal or fed to livestock; (7) *Groats*, grains of barley and oats after the hulls have been removed. It is usually cooked for a breakfast cereal or added to soups; (8) *Meal*, coarser particles of ground wheat, maize, oats or rice grains, It is usually cooked and eaten as a breakfast cereal; and (9) *Polishings*, the soft, fine residue obtained by polishing brown rice. It is added to baby cereals to increase the mineral and vitamin content (Göhl, 1981; Ensminger *et al.*, 1994; Eckhoff and Paulsen, 1996).

There has been some mistaken concern that the term *'modified starch'* may refer to GM starch. In fact modified starch does not include any GM material and is defined in the *Miscellaneous Food Additives Regulations 1995* as 'any substance obtained by one or more chemical treatments of edible starch, which may have undergone a physical or enzymatic treatment, and may be acid or alkali thinned or bleached' material (MAFF, 1999).

3.3   Dietary Fibre

Dietary fibres or *roughage,* as distinct from textile fibres (see Chapter 14), are collectively known as *non-starch polysaccharides* (*NSP*). They chiefly consist of cellulose, plus mainly water-soluble, non-cellulose polysaccharides such as pectin, β-glucans and gums, mucilages such as alginates, etc., and the non-carbohydrate phenyl propene polymer lignin. *Cellulose* is the main component of cell walls of most plant structures. It is colourless, transparent, insoluble in water and composed solely of glucose units linked by β-1,4-glycoside bonds. While adding bulk to vegetable foods, it lacks nutritive value since it is practically unaffected by digestion. The structural polysaccharide present in cell walls, *pectin* consists of a mixture of substances, of which the most important constituent is *methyl pectate* with *araban* and *galactan* also present. *Lignin* is a highly polymeric substance that occurs with cellulose in lignified plant tissues and, like cellulose, is of no nutritional value. The *soluble NSPs* such as pectin, gums, etc. serve to reduce cholesterol levels in the blood and other tissues by preventing the reabsorption of bile acid by the intestine. Barley, oats and rye are important sources of soluble NSP, while *insoluble NSPs*, which do not reduce cholesterol levels, are found in wheat, maize and rice.

It should be noted that the published values for dietary fibres are dependent upon the analytical methods used. The determination of *crude fibre* remaining after extraction with petroleum, dilute NaOH, and dilute HCl is no longer used for human nutrition as a measurement of cellulose and lignin content because it greatly underestimates the dietary fibre in food. The higher values represented by *total fibre* provides a better and broader range of non-digestible substances. Crude fibre is still used for animal nutritional studies (Vaughan and Geissler, 1997).

Fibre, by stimulating muscular activity in the digestive tract, facilitates digestion, probably through the ability of the fibre to hold water in a gel-like form. Although fibre has long believed to be inert in the human diet, it is now known that it is digested and metabolised by micro-organisms in the colon. It is even probable that the fibre fraction has the greatest influence on digestibility. Increasing dietary fibre in the diet of humans and non-ruminants increases the faecal bulk and frequency of elimination.

Fibre may also be associated with inhibiting effects on the absorption and retention of the minerals Ca, Mg and Zn. The action of the *phytic acid* (*meso-inositol hexaphosphate*) component of the outer endosperm of cereals and consequently present in the bran fraction, by forming insoluble salts with these minerals reduces their absorption by the digestive tract (Dalal-Clayton, 1981; Sharp, 1990; Duffus and Duffus, 1991; Macpherson, 1995; Vaughan and Geissler, 1997). See also Section 8.4.

4.   PROTEIN

Proteins are composed of large numbers of amino-acids which differ in both arrangement and in qualitative relationships. Two or more of the constituent amino-acids are linked by the -CO·NH- peptide bond into chains known as *peptides* and may be broken down into smaller chains by partial hydrolysis. The proteins are the major nitrogenous constituent of living organisms. They form an essential part of the diet because: (1) They are essential for growth. Carbohydrates and fats cannot be substituted for protein because they do not contain nitrogen; (2) Proteins provide the essential amino-acids, the building stones for tissue synthesis, making good daily wear and tear; (3) They provide the raw material for the formation of digestive juices, hormones, plasma proteins, haemoglobin, vitamins and enzymes; (4) They are a source of energy, *ca.* 4 kcal g^{-1} protein, albeit a wasteful source; and (5) Proteins function as buffers, helping to maintain the reactions of various media, e.g. plasma, cerebrospinal fluid, intestinal secretions, etc. (Passmore *et al.*, 1974).

Functional, structural and storage proteins are recognised within a plant, each with different nutritional properties. (1) *Functional proteins* are present in the contents of live cells. They are the most nutritious and approach animal proteins in quality; (2) *Structural proteins* are present in the cell walls and although of good quality they are difficult to digest; and (3) *Storage proteins* are rare in plants yielding non-seed carbohydrates, and are usually of inferior quality. Both functional and structural proteins are usually already present in young storage organs and, as the storage organs age and the carbohydrate content increases, the relative value of the protein decreases (Flach and Rumawas, 1996).

The nutritive value of proteins is essentially similar for both humans and animals. During absorption, or digestion, the proteins are broken down into their constituent amino-acids. Those proteins that supply all the essential amino-acids in adequate amounts to meet dietary requirements are known as *biologically complete*, while a *biologically incomplete* protein is deficient in one or more essential amino-acid. Most vegetable proteins are biologically incomplete, although mixtures of vegetable proteins usually meet dietary requirements. Proteins from different sources mutually supplement each other and consequently may have a higher biological value than individual sources. For example, the cereals are low in *lysine* while the legumes are rich in lysine, so that a cereal-legume diet has a significantly higher nutritive value than diets based solely on cereals or legumes. A food should not, therefore, be condemned because its proteins have a low biological value.

The legumes provide 20% of all plant protein in human diets; their contribution is even more important for livestock. Undoubtedly the most significant nutritional contribution of legume protein to food and feed is in balancing cereal proteins, which can be improved by increased availability and by improved protein digestibility and sulphur amino-acid content. The contribution of beans is especially beneficial in those countries where carbohydrate foods such as maize, rice, cassava, etc. form the staple cereal diet. However, legume proteins are relatively indigestible and have also been found to have hypocholesterolemic effects. They also have a poor biological value, which can be significantly improved by the addition of the amino-acid *methionine* (Bressani and Elias, 1980).

Excess dietary protein does not accumulate in the body since it is readily converted into its constituent amino-acids, which are subsequently used for energy. Thus, the traditional almost entirely meat diets of the guachos of South America, the Masai of East Africa and of Eskimos provide a protein diet containing more than twice the normal intake without any apparent ill effect. However, the protein of some unconventional new foods, such as yeasts and *Chlorella*, is associated with large amounts of nucleic acids, which break down to form uric acid, and may cause kidney stones if large quantities of such protein sources are consumed.

4.1 Amino-Acids

The *amino-acids* are a large class of organic compounds containing both the carboxyl, COOH, and the amino, NH_2, group. They are colourless, crystalline substances which melt with decomposition and are mostly soluble in water and insoluble in alcohol. They represent the ultimate products of digestion of protein foods and feed, from which the body's protein materials are rebuilt.

There are ten amino-acids, or keto acids, of which animals are incapable of manufacturing the carbon skeletons. These are known as *indispensable amino-acids* and are essential for growth, maintenance and reproduction. Those amino-acids which an animal is able to synthesise within its own tissues are termed *dispensable amino-acids* (Table 5). Non-ruminants are totally dependent upon a dietary supply of the indispensable amino-acids while ruminants are able to synthesise a substantial proportion within the rumen. However, young animals do not achieve their genetic growth potential if the dietary nitrogen is supplied entirely as indispensable amino-acids; an optimal combination of the dispensable amino-acids must also be provided to ensure maximum efficiency of utilisation of the amino-acids. (D'Mello, 1991).

The range of essential amino-acids required depends partly on the species. For example, the eight considered essential for the adult human body are: isoleucine, leucine, lysine, methionine, phenylalanine, threonine, tryptophan and valine; infants also require histidine (Sharp, 1990; Macpherson, 1995). The potentially toxic amino-acids that may also be present are discussed in Section 8.7.

TABLE 5. Nutritional classification of amino-acids (D'Mello, 1991, reproduced by kind permission of the Royal Chemistry Society)

Indispensable	Dispensable	Toxic (non-protein)
arginine[1]	alanine	β-aminoproprionitrile
histidine	aspartic acid	canavanine
isoleucine?	cystine	β-cyanoalanine
leucine	glutamic acid	3,4-dihydroxyphenylalanine
lysine	glycine	α,γ-diminobutyric acid
methionine	proline	djenkolic acid
phenylalanine	serine	homoarginine
threonine	tyrosine	indospicine
tryptophan		*S*-methylcysteine sulphoxide
valine		*Se*-methylselenocysteine
		mimosine
		β-*N*-oxalyl-α,β-diaminopropionic acid
		selenocystathionine
		selenomethionine

[1]arginine can be synthesised by mammals but at a rate insufficient to meet the demands for growth

5.   FATS

Fats are esters of fatty acids with glycerol. In lay terms fats are solid and oils liquid at 18-24°C. They are essential constituents of the diet, although the conversion of carbohydrates to fats in the animal body also occurs. On a weight to weight basis fats have more energy-producing power than any other food, with each gram of fat having an energy-producing equivalent of 9.3 kcal. The fat intake has to be balanced with carbohydrates or proteins in order that it may be completely oxidised within the digestive system, otherwise harmful *ketones*, e.g. *propanone* (acetone or dimethyl ketone) are likely to be formed in the blood.

From a medical viewpoint the edible fats can be divided into: (1) *Saturated fats*, having no free valency bonds, i.e. their molecules hold the maximum amount of hydrogen ions. They include animal fats and dairy produce, and tend to be solid at room temperature. Cocoa butter, coconut oil and the palm oils are the exception among the vegetable oils in being high in saturated fatty acids; and (2) *Unsaturated fats*, i.e. containing atoms that share more than one valency bond and tend to be liquid at room temperature. They include most vegetable oils such as soya bean, maize, sunflower, etc. (and fish oils). The vegetable oils are also free of *cholesterol*, which is suspected of contributing to atherosclerosis when excessive amounts of animal fats are consumed over long periods by susceptible people. Hence the medical recommendation for replacing saturated animal fats by vegetable oils in order to reduce the tendency for coronary thrombosis and high blood pressure.

The unsaturated fatty acids can be further divided in *monounsaturated* and *polyunsaturated fatty acids*. Oils, such as olive and rapeseed oils and some nuts, are high in monounsaturated fatty acids; they also contain some polyunsaturated linoleic and linolenic fatty acids. While the body is able to manufacture its own saturated and monounsaturated fatty acids, it is unable to manufacture the polyunsaturated.

The *essential fatty acids* (*EFAs*) represent a group of unsaturated acids that play a vital role in fat metabolism and transfer, and as precursors of prostaglandins in mammalian tissue and some simple animals where they have a hormone-like activity. The principal EFAs are *arachidonic acid* (*AA*), *linoleic acid* (*LA*), *α-linolenic acid* (*ALA*) and *γ-linolenic acid* (*GLA*). They are abundant in most vegetable oils and to a lesser extent in the fats of some animals of marine origin.

Two families of polyunsaturated EFAs are recognised: (1) *Omega-6* or *n-6* family based on LA and its derivatives GLA and AA. They are found, amongst others, in safflower and sunflower oils, also in walnuts. AA is present mainly in peanut oil; it is the precursor of prostaglandins E_2 and F_{2a} and also induces clotting. An omega-6 deficiency in both humans and animals leads to profound abnormalities in many body systems; and (2) *Omega-3* or *n-3* family derived from LA and includes soya bean and rapeseed oils among its sources. In the absence of sophisticated experimentation it is extremely difficult to demonstrate omega-3 deficiencies. While both types of EFAs are important, the omega-6 are the more so (Horrobin, 1990a; Sharp, 1990; Macpherson, 1995; Ensminger *et al.*, 1994; McWhirter and Clasen, 1996).

Among the commonly conducted tests for assessing the chemical and physical properties of edible oils are: (1) *Iodine value* is a test that measures the number of grams of iodine absorbed by 100g of a fat or oil and indicates the amount of unsaturated acids present; the higher the iodine value, the greater the unsaturation. The susceptibility to oxidative rancidity also increases, while the melting point decreases with the iodine value; (2) *Saponification value* measures the average length of the carbon chains in the fatty acids present in the fat. It is a measure of the reaction of the fats to alkalis to form soaps, with the potential for soap-making increasing with the saponification value. A high saponification value also indicates a low melting point and a short chain length, and vice versa; and (3) *Melting point* is important in the food industry because consumers expect fats to be solid at ordinary room temperature (18-24°C) and oils to be liquid. This is particularly important for pastry making where most shortenings need to be at least semisolid at room temperature because the flakiness of the pastry depends on the production of layers of solid fat (Ensminger *et al.*, 1994).

Depending on the species, the extraction of the fats and oils usually involves *decortication*, i.e. the removal of the seed coat, followed by either extraction using solvents or the application of screw or hydraulic pressure. The latter method is used for extracting the edible oils; the residual seed cake is then used for stock feed.

6.   MINERALS

Minerals are elements that are required in relatively small quantities for the body to function properly. The six *macrominerals*, so called because they are required in relatively large quantities, are Ca, Cl, K, Mg, Na and P. Calcium, for example, is required to strengthen the bones and teeth, for the proper function of the nerves and muscles and normal clotting of the blood, others are required to maintain a healthy immune system and help vitamins to function properly. There are also a further eleven *microminerals* or *trace elements*: Al, Cr, Cu, F, Fe, I,, Mn, Mo, S, Se and Zn that are required in relatively small quantities (McWhirter and Clasen, 1986; Macpherson, 1995).

Many of these minerals are also essential for plant growth and development, and are extracted by the plants from the soil. The essential plant *macronutrients* from the soil are Ca, Fe, K, Mg, N, P and S, while the plant *micronutrients* include many of the heavy metals such as Co, Cu, Mn, Mo and Zn as well as trace amounts of B and, in some cases, Cl, Na, Si and V (Tootill, 1984). Plants are thus a valuable source of minerals for human and animal nutrition.

7.   VITAMINS

Vitamins occur in minute quantities in natural foods and are essential for normal nutrition, especially for growth and development; in their absence growth will be defective and health impaired. Although the term vitamin was first used in 1912 by the Polish chemist Casmir Funk in the belief that they were vital amines, only a few are actually amines, the majority being chemically unrelated (Passmore *et al.*, 1974). Some are fat-soluble and can be stored in the body, the remainder are water-soluble and quickly excreted. The data given for the vitamins listed below are taken from Macpherson (1995) and refer specifically to humans, but could equally apply to most mammals.

The *fat-soluble vitamins* present in plants are: (1) *Vitamin A (retinol, retinal, retinoic acid)* and is essential for normal growth, reproduction and development, and for the maintenance of the immune system. It is especially abundant in carrots, dark-green vegetables and fruits. It also occurs in palm oil from *Elaeis guineensis* as the precursor *carotene*; (2) *Vitamin D* helps in the absorption of Ca and P, and functions by regulating the concentration of blood Ca and the mineralisation of bones and teeth. It is formed in plants by the action of sunlight on the sterol precursor *ergosterol* to produce *vitamin D_2 (calciferol)* which differs chemically from vitamin D found in fish liver oils as well as probably being less active; (3) *Vitamin E (α-tocopherol)* functions by preventing the oxidation of vitamin A in the digestive tract, preventing red blood cells from haemolysis (breaking up of the blood corpuscles) and maintaining cell membranes. It is a terpenoid-like substance with high concentrations present in certain seeds, notably the cereals. Dietary sources include wheat germ, whole grains, nuts, vegetable oils, legumes and green leafy vegetables; and (4) *Vitamin K (phytomenadione, phylloquinone)* is necessary for the formation of *prothrombin* and other factors necessary for blood clotting. It is present in dark green leafy vegetables and soybean oil.

The *water-soluble vitamins*, i.e. the B group and C, unlike the fat soluble vitamins, are not stored in the body, consequently their deficiencies are more likely to occur. The B vitamins form a complex group of. chemically unrelated *coenzymes*, i.e. organic non-protein compounds which, in the presence of the appropriate enzyme, perform an essential function in the chemical reaction catalysed by the enzyme. Those present in plants are: (1) *Vitamin B_1 (thiamine, aneurine)* is synthesised by plants but not by some micro-organisms or by most vertebrates. It plays a role in carbohydrate metabolism, helps the nervous system, heart and muscles to function properly as well as promoting appetite and the functioning of the digestive tract. Dietary sources include whole grains, wheat germ, legumes and brewer's yeast. It is readily destroyed by heat; (2) *Vitamin B_2 (riboflavin*, formerly known as vitamin G) is essential for certain important enzyme systems in food metabolism. It is not synthesised by animals. Dietary sources include green, leafy vegetables, groundnuts and whole grains; (4) *Vitamin B_3 (niacin, nicotinic acid*, a derivative of *pyridine alkaloids*) is heat resistant and functions as part of two important enzymes regulating energy metabolism, promoting good physical and mental health and in helping to maintain the health of the skin, tongue and digestive system. It occurs in whole grain flour, legumes and brewer's yeast; (5) *Vitamin B_6 (pyridoxine)* functions in the metabolism of proteins, amino-acids, carbohydrates and fat and is essential for growth and health. It is present in whole grains, maize, potatoes and green vegetables; (6) *Vitamin B_{12} (cobalamin, cyanocobalamin)* is important in haemoglobin synthesis and normal cell function. It is not present in significant quantities in plants; commercial production is from bacterial cultures of *Streptomyces olivaceus*; and (7) The *vitamin B complex* including: (a) *folic acid*, important in the formation of red blood cells and the normal function of the gastrointestinal tract. It is found in dark-green leafy vegetables, legumes, whole grains and yeast; (b) *biotin*, a heat-stable enzyme that takes part in the metabolism of amino- and fatty-acids, and occurs in cauliflower, legumes, nuts and yeasts; and (c) *pantothenic acid*, an essential component of coenzyme A, a key factor in many of the body's metabolic functions. It is synthesised by plants and bacteria but not by vertebrates. *Vitamin C (ascorbic acid, dehydroascorbic acid)* provides protection against infection and promotes healing of wounds as well as being important for tooth dentine, bones, cartilage, connective tissue and blood vessels. Important sources include citrus fruit, tomatoes, strawberries, currants, green, leafy vegetables and potatoes (Passmore *et al.*, 1974; Liener, 1980; Nowacki, 1980; Tootill, 1984; Sharp, 1990; Macpherson, 1995).

8.   ANTINUTRIENTS

Antinutrients are those compounds present in a food that adversely affect the digestive processes. The mode of action of a particular toxic compound may vary according the species ingesting. The implication is that a single detoxification mechanism for human food and livestock feed may not have universal application, thus necessitating different strategies (Duffus and Duffus, 1991).

8.1   Protease Inhibitors

Protease inhibitors, especially those produced in soya and pulse legumes, are proteins that interfere with the serine proteases, mainly the pancreatic secretions chymotrypsin and trypsin (formerly known as peptidase), inhibiting their ability to hydrolyse the peptide bonds of proteins. Protease inhibitors, especially *trypsin inhibitors*, are often believed to be the prime case of mortality in experimental animals fed raw soya beans. The trypsin inhibitors are present to some extent in all legumes but may be rendered safe for human consumption by heat treatments, i.e. leaching followed by steaming, pressure cooking, extrusion cooking, by fermentation or sprouting (Liener, 1980; House and Welch, 1984; Duffus and Duffs, 1991; Ensminger *et al.*, 1994).

8.2   Lectins

Lectins, also known as *haemagglutinins* or *phytohaemagglutinins*, are those proteins or glycoproteins of non-immune origin that have the ability to bind specific sugar groups on the surface of the cell membranes or in glycoproteins, each lectin having its own sugar specificity. There are some exceptions that lack a carbohydrate group, such as *concanavalin A* found in *Canavalia ensiformis* (jack bean), and those from *Arachis hypogea* (groundnut) and wheat germ. They may be toxic because of their tendency to substitute for amino-acids in the protein structure. Lectins are wide-spread in the plant kingdom and are noticeably concentrated in legume seeds, including most if not all species of beans and lupins.

Lectins stimulate the agglutination of the erythrocytes (red blood cells), and can cause diarrhoea, gastro-enteritis and nausea in both humans and cattle. Fortunately the lectins are heat labile, with their molecular structure and activity destroyed by cooking and are only a potential hazard when edible beans are eaten raw.

Other lectins, known as *mitogens,* stimulate mitosis in the leukocytes (white blood cells). Mitogens, such as those from *Phytolacca americana* (pokeweed) are now regarded as powerful clinical and research tools for investigating the body's immune response system under pathological states, e.g. Hodgkin's disease (Liener, 1980;

House and Welch, 1984; Blackwell, 1990; Duffus and Duffus, 1991; Ensminger *et al.*, 1994).

8.3   Saponins

Saponins are widely distributed through the plant kingdom and can occur in all plant organs, with the concentrations varying according to stages of growth. They are glycosides with a non-sugar aglycone portion or *sapogenin*. They are classified into two groups according to the nature of their sapogenins: (1) *Steroidal saponins* with 27 carbon ions and are the precursors of steroids for human synthesis; and (2) *Triterpene saponins* with 30 carbon ions.

They are generally harmless to mammals unless consumed in large quantities, although highly toxic to fishes and snails, hence their use as piscicides and molluscicides (see Chapter 17). Their potential antinutritional effects include haemolysis of erythrocytes, cholesterol precipitation in the cell membrane and a decrease in the palatability of the diet (Nowacki, 1980; House and Welch, 1984; Blackwell, 1990; Duffus and Duffus, 1991).

8.4   Phytic Acid

Phytic acid (meso-inositol hexaphosphoric acid) is a strongly acidic syrup which is sparingly soluble in organic solvents. It is present in seeds as its insoluble calcium magnesium salt, *phytin*. Phytin is the principal form of phosphorus storage in plant seeds and is also present in small amounts in potato tubers; it is also present in the blood plasma phospholipids and the erythrocytes of chicken blood. It is reputed to depress protein digestion and to have anti-vitamin D activity, as well as having an anti-nutritional effect on mineral availability, especially zinc. See also Section 3.3 for further details (House and Welch, 1984; Sharp, 1990; Ensminger *et al.*, 1994).

8.5   Dietary Tannins

The *tannins* are widely distributed among the dicotyledons and are consequently ingested by many herbivores (see Chapter 15 for further discussion). They form a diverse group of astringent water-soluble phenolics which are able to bind protein to form soluble or insoluble complexes. Their astringency adversely affects palatability, with above 5% tannin dry weight levels leading to food rejection.

Two classes of tannins are recognised according to their chemical structures: (1) *Condensed tannins (proanthocyanidines)* are flavonoid polymers which, under mild or anaerobic conditions, are stable but can be oxidatively degraded in acid to yield *anthocyanidines*. The binding of the condensed tannins to protein in the gut decreases both protein and dry matter digestibility and can lead to griping diarrhoea or constipation; and (2) *Hydrolysable tannins* represent the gallic acid or hexahydroxydiphenic acid esters of glucose or other polyols. The *gallotannins* are simple esters of gallic acid and are easily hydrolysed to yield *gallic acid* or *hexahydroxydiphenic acid* and the parent polyol. They do not affect digestibility and are degraded in the gut to small phenolics that do not react with the protein.

Dietary tannin may or may not, depending on the mammalian species, diminish protein and dry matter digestibility. Their anti-nutritional effects in mammals are linked to: (1) Inhibition of digestive enzymes; (2) Formation of relatively less digestible complexes with dietary protein; (3) Depressed growth rate; and (4) Inhibition of the microbial flora in the digestive tract.

This diversity of activity is partly due to the different physiological abilities of the ingesting animal to handle tannins and partly due to differences in chemical reactivity of the various tannin types. The binding of the tannins to the digestive tract may sometimes act as a toxin rather than a digestion inhibitor in some animals but not in others. Some mammals produce salivary tannin-binding proteins which apparently protect other more valuable proteins from the effects of tannin, while others do not have this facility. This is possibly correlated with browsing and non-browsing feeding habits respectively. In some insects they may also have little effect on digestibility because of adaptations by the digestive system to tannins (Liener, 1980; Harborne, 1988; Hagerman *et al.*, 1992).

8.6   Amylase Inhibitors

During digestion starch is initially hydrolysed by the action of the salivary and pancreatic *α-amylases* to the disaccharide *maltose*, with cooked starch being more readily hydrolysed than raw starch. The maltose is further degraded by the action of maltases in the brush-border of the intestine epithelium. Amylase inhibitors, which are present in various cereals and dry legume seeds but not groundnuts, interfere with this process, resulting in flatulence (Liener, 1980; House and Welch, 1984; Ensminger *et al.*, 1994).

8.7   Toxic Non-Protein Amino-Acids

Non-protein amino-acids (Table 5) are not normal components of proteins, although they freely occur in a wide range of plants. Among the Leguminosae their presence undermine efforts to fully exploit such otherwise promising fodder sources as species of *Canavalia*, *Indigofera*, *Lathyrus* and *Leucaena*; they are also present in the Cruciferae among the *Brassica*. The toxic amino-acids are usually concentrated in the seeds but can occur throughout the plant. They affect the nervous system, although a range of other symptoms have also been reported. Human diets consisting of 25% or more of lathrogenic legumes, such as *Lathyrus sativus* (blue vetchling) or *Vicia sativa* (common vetch) and consumed for several months can result in *neurolathyrism*, which is associated with spastic paralysis of the legs and pain in the loins. It is a disease that is particularly prevalent in India. In livestock the symptoms are weakness and paralysis of the hind legs and difficulty in breathing, or as *osteolathyrism*, characterised by skeletal abnormalities and, in some animals, haemorrhage. The *lathyrogens* present in the blue vetchling contain the water soluble non-protein amino-acid *β-N-oxalyl-α,β diaminoproprionic acid* (*ODAP*) and *β-(γ-L-glutamyl) aminoproprionitrile* (*BAPN*). While it is possible to breed for reduced ODAP, breeders have unfortunately so far been unable to overcome the associated reduction in yield.

Since amino-acid toxicity is not confined to the nutritional boundaries shown in Table 5, all amino-acids should be regarded as potentially injurious to animals and humans. In addition, the toxicity of many non-protein amino-acids is, to some extent, determined by complex interactions with the nutritionally important amino-acids. For example, the adverse effects of *canavanine* in metabolic reactions are due to competition with *arginine*, while *lysine* exacerbates this toxicity in avian species due to its own antagonism with arginine (Nowacki, 1980; Smartt, 1990; D'Mello, 1991; Duffus and Duffus, 1991; Ensminger *et al.*, 1994).

8.8 Cyanogens

Traces of *cyanogenic glycosides* are widespread in the plant kingdom, with relatively high concentrations restricted to certain plants and generally concentrated in the leaves. Most cases of cyanide poisoning follow the consumption of members of the Gramineae, Leguminosae and Rosaceae. The lethal dose for cyanide poisoning is 2-4 mg kg^{-1} body weight, the effect is very rapid, with death occurring within a few minutes to an hour following ingestion.

Toxicity from cyanogenic glycosides is due to the release of *hydrocyanic acid* (*HCN*) following the breakdown of the glycosides after ingestion or as a result of cell damage before ingestion. In humans the enzyme present in *Escherichia coli*, which is normally present in the intestine, may be responsible for the release of HCN. This may be counteracted by encouraging a reduction in the intestinal *E. coli* population by encouraging the growth of *Lactobacillus* populations with fermented milk products.

In humans and monogastric animals the acidic contents of the stomach inhibit the action of enzymes on the cyanogenic material, although they may become active during subsequent digestion in the duodenum. In ruminants the neutral pH of the rumen encourages enzyme action on the ingested material and is furthered by the mechanical action of rumination and the activity of the rumen bacteria. The bacterial enzymes may also have a role in glycoside hydrolysis, which may explain why ruminants are more susceptible to cyanogenic plants than non-ruminants (Liener, 1980; Nowacki, 1980; Duffus and Duffus, 1991; Ensminger *et al.*, 1994).

8.9   Glucosinolates

The *glucosinolates* (mustard oil glycosides), formerly known as *thioglucosides*, are hydrolysed when wet raw plant material is masticated or crushed, releasing glucose, an acid sulphate ion and the goitrogens *thiocyanate, isothiocyanate* and *goitrin*. Many different glucosinolates have been isolated from cultivated members of the Cruciferae grown for fodder, but only one or two are present in relatively large quantities per species and these are usually concentrated in the seed. The toxic symptoms include diarrhoea, severe gastro-enteritis, salivation and irritation of the mouth from the vesicant action of the isothiocyanates.

The thiocyanate ions may also be present in other families as well as the Cruciferae following the breakdown of cyanogenic glycosides. Goitrogenic substances are also produced by the breakdown of the glucosinolates and can produce hyperthyroidism in mammals due to interference with iodine utilisation by the thyroid gland. This will not respond to treatment with iodine because the goitrogenicity is not caused by an iodine deficiency. As a consequently the eating of large quantities of groundnuts and soya beans should be avoided (Liener, 1980; Harborne, 1988; Duffus and Duffus, 1991; Ensminger *et al.*, 1994).

8.10   Alkaloids

Alkaloids occur in a wide range of plants and plant organs (see Chapter 15) and it is believed that their toxic effect are due to mimicking or blocking the action of nerve transmitters. Symptoms of toxicity include excess or absence of salivation, dilation or constriction of the pupil, abdominal pains, diarrhoea, vomiting. incoordination, convulsions and coma. The effect of poisoning by pyrrolizidine alkaloids, however, differs in that it is chronic and the liver is the principle organ affected. Some alkaloids can produce abnormalities, *coniine* in *Conium maculatum* (hemlock), for example, is teratogenic, causing foetal defects when the plant is ingested by the mother. Others, such as the bitter tasting seeds of the Andean annual *Lupinus mutabilis* (tarwi or blue lupin) contain several toxic quinolizidine alkaloids, mostly *sparteine* and *lupanine*; so that the raw seeds are toxic to livestock except sheep. Tarwi is a potential high protein crop, with seeds containing up to 50% protein, so that there is a need to either select low alkaloid biotypes from the wild or for plant breeders to create them (Nowacki, 1980; Smartt, 1990; Duffus and Duffus, 1991).

Potato tubers also require selection for low levels of the toxic steroidal alkaloid *solanidine*, which is mainly concentrated in the skin and increases in quantity when the tuber is exposed to the light. Although many other factors are also involved, the degree by which alkaloids accumulate and are synthesised in potato tubers is inherited and is thus a varietal character. Most potato cultivars normally contain low concentrations of solanidine. Unfortunately, the variety 'Lenope', which has *Solanum chaoense* in its ancestry and was released for commercial production in 1967, had to withdrawn in 1970 because of its consistently high concentrations of alkaloid. As a result all new potato varieties are now tested for the alkaloid content of their tubers before they can be registered in the national cultivar lists (Burton, 1989; Wagih and Wiersema, 1996; P.Day, pers. comm. 1998).

9.   FOOD SAFETY REGULATIONS

The World Trade Organisation's *Agreement on Sanitary and Phytosanitary Measures* requires that there should be the minimum of restrictions in the movement of goods between countries. But where there are grounds to believe that imported foods may represent a concern to public health, measures can be taken to restrict their importation.

Food standards are subject to government legislation. In the UK the Ministry of Agriculture, Fisheries and Food (MAFF), acting on advice from the Ministry of Health, was the controlling authority. It is now the responsible authority Food Standards Agency established on 3rd April 2000, whose role is *"to protect public health from risks which may arise in connection with the consumption of food, and otherwise protect the interests of consumers in relation to food."* It is also responsible for the safety of animal feed. The Agency has the great advantage of being able act independently of political and commercial interests (Food Standards Agency, 2000). The EU is expected to provide a similar organisation in the not too distant future.

In the USA there are four major sets of standards. The US Department of Agriculture Grade Standards cover canned, frozen and dried fruits and vegetables, and related products such as preserves, also rice, dry beans and peas. Standards for fresh products are voluntary. The Federal Food and Drug Administration of the Department of Health and Human Services provides the three remaining standards. The Food and Drug Administration Standards of Identity are mandatory standards for defining a food product. They also provide for the use of optional ingredients in addition to the mandatory ingredients for canned vegetables, tomato paste, ketchup (catsup), vegetable juices and frozen vegetables. The Federal Drug Administration Minimum Standards of Quality are, unlike the USDA Grade Standards, mandatory and are used to supplement standards of identity by specifying the minimum acceptable characteristics for such factors as tenderness, colour and freedom from defects. Finally, the Federal Drug Administration Standards of Fill of Containers, as its name suggests, provides mandatory standards for the filling of containers, i.e. banning the sale of air or water in place of food (Ensminger *et al.*, 1994).

While it is impossible to control the chemicals present in food so that there is never any risk to health, the Food Standards Agency ensure that any risk of consuming injurious chemicals is minimal. They determine what is termed *NOAEL* or *No-Observed Adverse Effect-Level*, i.e. the maximum quantity that various animals can eat without showing any adverse effects. A large safety margin is then applied to obtain the much lower *ADI* or *Acceptable Daily Intake* for chemicals added deliberately as food additives, or *TDI* or *Tolerable Daily Intake* for chemicals present accidentally, i.e. the small amounts of natural toxins present in the plant or small quantities of polluting chemicals from contact with fertilisers, pesticides, industrial chemicals, radioactivity (taking into account the. proximity to nuclear sites, fallout from the Chernobyl disaster), etc. (MAFF, 1994c, 1998c).

In view of the concern regarding genetically modified (GM) foods, the safety procedures within the EU, and for the UK in particular, are summarised from MAFF (1997b, 1999). Since May 1997 the *EC Novel Food and Food Foods Ingredients Regulation* (258/97) provides a pre-market approval system for all novel foods, including GM foods, throughout the EU. Novel foods are defined as those foods which have not previously been used for human consumption within the EU. The safety of GM foods has to be scientifically assessed by comparisons with similar non-GM foods so that any differences can be recognised and any changes that were not supposed to have occurred, identified. The GM food must be labelled if it is judged not to be equivalent to an existing food, if there is any ethical concern and if it contains a GM organism, even though it is equivalent to an existing food. Food offered for sale in the UK has to meet various requirements, including those in *The Food Safety Act 1990*, which applies equally to both GM and non-GM foods.

Perversely, while the protein sequences in GM foods are matched against known allergies before being released onto the market, GM food additives have so far escaped labelling since the quantities consumed during a meal are regarded as insignificant. There is, however, always the possibility of eating a succession of foods containing a GM additive and the resulting accumulative effect resulting in an allergic reaction. Even so, a distinction must be drawn between the identification of the cause of known allergies and the unknown since, given the variability of the genetic constitution of the human race, it is inevitable that somebody somewhere will always have an allergic reaction to a particular substance.

The regulatory safeguards for GM food in the UK required a breeder, when using premises for the first time, to notify the *Health and Safety Executive* (*HSE*) of the intention in order that an assessment may be made in conjunction with the *Advisory Committee on Genetic Modifications* (*ACGM*) of any risks to the environment or human health. The consent of the Secretary of State for the Environment must then be obtained by the breeder before any GM crop can be grown outside the premises in field trials. The breeder has to supply an up-to-date environmental risk assessment to the *Department of the Environment, Transport and the Regions* (*DETR*), who would then be advised by the *Advisory Committee on Releases to the Environment* (*ACRE*). Following successful field trials, the food safety would then be considered under the *EC Novel Foods Regulation* by the *Advisory Committee on Novel Foods and Processes* (*ACNFP*) who, before acceptance, may consult the *Food Advisory Committee* (*FAC*), the *Committee on Toxicity of Chemicals in Food Consumer Products and the Environment* (*COT*), and the *Committee on Medical Aspects of*

Food Policy (*COMA*). All committees previously under the auspices of MAFF now fall under the Food Standards Agency. In the event of these Committees agreeing that the product is safe, their assessment would then be considered by other EU Member States. If there are no objections, the product could then be marketed provided it was labelled in accordance with the EU Regulation (MAFF, 1999).

However, despite such apparently stringent regulations and control there has been no biological testing on mammals of GM food considered as being safe, i.e. food considered substantially equivalent to non-GM food. But with GM foods following so closely on the UK outbreak of bovine spongiforme encephalopathy (BSE) among cattle and new variant Creutzfeldt-Jacob Disease (CJD) in humans, the public now have understandable doubts about the safety of GM foods and the effect on the environment of GE crops.

Within the EU the responsibility for assessing the safety of additives is carried out by the European Commission's experts on the Scientific Committee for Food and in the UK by experts on the Department of Health's Committee on Food Toxicity. An 'E' number is given when the additive is considered safe (MAFF, 1997c, 1998d). The Food and Drug Administration has similar responsibilities in the USA. But what is approved for one country does not necessarily meet the approved of another.

Chapter 9

Human Food and Food Additives

Food plants may be defined as those plants or plant organs that are used for human consumption. Although the angiosperms are the major source, the gymnosperms, ferns, lichens, fungi and algae are also represented. Human food differs from animal food in the sense that it usually undergoes some form of preparation prior to eating. This may include washing to remove dirt, leaching to remove toxic substances, grinding to reduce excessive mastication and to improve digestibility, and even fermentation to produce a more palatable food or intoxicating beverage. Some foods, such as nuts, fruit and salad vegetables may be eaten raw, more often food is cooked. Food may also be preserved by drying, freezing, canning, pickling, fermentation, etc. and stored for future use. Also, depending on the culture, it may be presented in a manner that is aesthetically pleasing.

The use of plant food sources ranges from total dependence on wild plants by some aboriginal peoples to almost total reliance on cultivated plants by people of developed countries. Ogle and Grivetti (1985), for example, have shown that more than 220 species of wild plants are commonly consumed by the people of Swaziland and that for 39% of the people the wild foods provided a greater share of the annual diet than the cultivated crops. In the Garfagnana region of Tuscany the local inhabitants still use 133 species of wild plants (including 19 species of fungi) for food, with 20 to 45 species included in the traditional soup known as minestrella (Pieroni, 1999).

In many regions, especially in the drier areas, wild food sources provide food security during times of dearth arising from drought, famines and wars, as well as regularly providing an important contribution during pre-harvest food shortages. The nutritional value of these wild food sources has been sadly neglected. This is particularly distressing since, from the little work that has been done by, for example Brand and Cherikoff (1985) and Arnold et al., (1985), many have been found to be nutritionally superior to cultivated crops.

151

Whereas the plant organs that are used for food have defined botanical identities, the lay terminology of the western world is completely illogical. Thus, an apple is a fruit yet a tomato is a vegetable! In an attempt to rationalise the problem the US Supreme Court in 1893 decided '*that a plant or plant part generally eaten as part of the main courses of the meal is a vegetable, while a plant part which is generally eaten as an appetiser, as a dessert, or out of hand is a fruit*' (Ensminger *et al.*, 1994). Jowitt (1989) has also attempted to overcome this problem by referring to the tomato as a 'fruit vegetable'. The status of the petioles of *Rheum* × *hybridum* (garden rhubarb) is also questionable, being consumed as a dessert fruit, in jams or fermented for wine.

With the influx of ethnic foods in recent years food nomenclature is now even more imprecise. *Relish*, for example, in the UK refers to a sharp sauce to accompany savoury food, in south-east Asia it is used to describe side dishes accompanying a rice dish in order to add variety, spice and nutritional value to the diet, while in West Africa it may similarly refer to a side dish or, in one-pot cooking, to pot herbs added to the cooking pot, especially the more highly flavoured ones. In French cooking the term ***pot herb*** traditionally refers to the six vegetables *Atriplex hortensis* (orache), *Beta vulgaris* subsp. *maritima* (seakale-beet, sea spinach), *Lactuca sativa* (lettuce), *Portulaca oleracea* (purslane), *Rumex acetosa* (sorrel) and *Spinacia oleracea* (spinach). They are used in the preparation of soups and broths, but because many other vegetables could also be included, the selection is considered arbitrary (Montagné, 1977). Again in West Africa, the leaves or young shoots of such pot herbs as *Amaranthus caudatus*, *Basella rubra*, syn. *B. alba*, and *Corchorus olitorius*, etc. are commonly referred to as ***spinach*** (Irvine, 1953).

In the lay terminology the range of foods eaten by man include cereals, pseudocereals, pulses, nuts, dessert fruits, green, root and fruit vegetables, leaf protein concentrates, starches, oils and fats, gums and mucilages, sugar, etc. In addition there are food additives used in the preparation of food, including clarifiers, colouring agents, emulsifiers, fermenting agents, rennet substitutes and milk curdlers, fermentation retarders, gelling agents, preservatives, purifiers, raising agents, stabilisers, tenderisers and thickening agents, as well as additives for flavouring, such as non-sugar sweeteners, vegetable salt, herbs, spices, vinegar and essences (Cook, 1995). Their terminology is also often confusing and is dealt with where appropriate.

It is perhaps sobering to consider that some 12,500 edible plants have been listed by Kunkel (1984) and that his list is certainly not exhaustive. Jardin (1967) records over 1400 wild plants gathered in Africa for food, while Yanovsky (1936) recognises some 1000 food plants utilised by the hunter-gatherers of North America. The Australian aborigines recognise some 290 food plants (Lazarides and Hince, 1993), which is almost certainly an underestimate. Yet it is only recently that the non-aboriginal Australians have begun to consider the possibility of developing these plants for food.

TABLE 6. Global food production for 1993; staple foods in bold type (FAO, 1994)

Crop	1000 tonnes	Crop	1000 tonnes
CEREALS		GREEN VEGETABLES	
wheat	564, 457	**cabbage**	404, 414
paddy rice	527, 413	**tomatoes**	70, 623
maize	470, 570	onions, dry	29, 961
barley	170, 364	cucurbits	26, 345
sorghum	57, 667	carrots	13, 997
oats	35, 443	green peppers	10, 603
pearl millet	26, 442	aubergines	8, 682
rye	26, 200	garlic	7, 624
DESSERT FRUIT		cauliflowers	6, 754
dessert grapes	57, 165	green peas	4, 602
oranges	56, 818	green beans	3, 087
other citrus	26, 243	artichokes	1, 137
bananas	50, 596	ROOTS AND TUBERS	
apples	42, 388	**potatoes**	288, 183
watermelons, etc.	40, 039	**cassava**	153, 628
plantains	27, 902	**sweet potatoes**	123, 750
pineapples	11, 740	yams	28, 126
pears	10, 333	taro	5, 639
peaches, nectarines	9, 785	BEVERAGES	
plums	6, 197	wine	26, 349
pawpaws	5, 663	coffee, green	5, 808
dates	3, 823	tea	2, 639
apricots	2, 248	cocoa beans	2, 417
avocados	2, 104	SWEETENERS	
mangoes	1, 774	**sugar cane**	1, 040, 600
raisins	1, 031	**sugar beet**	281, 682
LEGUMES			
soya beans	111, 011		
groundnuts in shell	25, 005	TREE NUTS	tonnes
dry beans	16, 163	almonds	1, 194, 498
dry peas	16, 032	walnuts	1, 006, 547
chick peas	6, 641	hazelnuts	565, 157
dry broad beans	4, 208	cashew nuts	479, 804
lentils	2, 698	chestnuts	437, 403

With such vast food resources it is a disturbing fact that only 100-150 species of the world's flowering plants are now considered sufficiently important to enter world trade. While more than 150 species have been commercially cultivated, the world's non-aboriginal population relies on a mere 20 staple food plants (Table 6) for 90% of their food, i.e. cereals (wheat, rice, maize, millets, sorghum), roots (potato, cassava, sweet potato), pulses (soya beans, ground nuts, beans, peas), vegetables (cabbages, tomatoes, cucurbits), fruit (grapes, oranges, bananas, apples), sweeteners (cane sugar, beet sugar), vegetable oils, etc.

It is also strikingly noticeable how the developed countries rely on Asia, Central and South America and Africa for the genetic origins of their staple foods (Table 1), while Australia has contributed nothing. Among the non-staples North America is responsible for *Helianthus annuus* (sunflower) and *H. tuberosus* (Jerusalem artichoke), while *Asparagus officinalis* (asparagus), *Beta vulgaris* (beet) and *Brassica oleracea* (cabbage) are of European origin (National Research Council, 1975). Schultes, 1980; Prescott-Allen and Prescott-Allen, 1990; Hawkes, 1998).

While perhaps not yet making a significant contribution to international trade, as the developed countries become increasingly multiracial and people travel more and more widely, there is an increasing demand for exotic foods and this has considerably increased the range of foods that are now available in the developed countries.

The main considerations given to the choice of food are: (1) *Palatability*, whether pleasant to taste either raw or processed, alone or as a condiment to other foods; (2) *Quantity satisfaction*, the ability to overcome hunger and supply the energy requirement of the consumer without using up internal reserves or leading to obesity; (3) *Nutritional value*, a balanced diet in terms of protein, fats, carbohydrates, fibre, minerals, amino-acids, vitamins and calories is essential for human well-being. No single plant product will supply all these requirements, but when taken in combination with other plant and/or animal products, a balanced diet can be provided. The energy requirement will, of course, vary according to the age, sex, size, and work being done by the consumer, a heavy manual worker requiring three or more times the amount needed by a lightweight sedentary worker. Although people in a hot climate will eat less food than those in a cold climate, there is no quantifiable basis for correcting the diet for climatic differences. Food production in the tropics is also highly seasonal and malnutrition is likely to be most severe in the pre-harvest wet season before the staple cereal crops are harvested, a period that also coincides with high prices due to a food scarcity value, accompanied by a low food intake and high energy expenditure by the cultivators; (4) *Digestibility* refers to the percentage of food that has been rendered soluble and assimilated by the consumer, the undigested portion being excreted; (5) *Toxic properties*, these may prevent eating the raw plant or its organs, although this may not be too great a disadvantage if the toxins can be readily removed by, for example, cooking. On the other hand, the necessity to thoroughly leach the food in order to remove any toxicity would be a major disadvantage in regions where water is in short supply; (6) *Seasonal availability*, for commercial processing a short harvest season where, for example, all the fruit ripen together, is usually an advantage, whereas domestic requirements may be for fruit that ripen over a long period and thus do not create a sudden and unmanageable surplus. It follows that those plants that provide fresh food out of season when other foods are not readily available will clearly have benefits for the producer; (7) *Storage life*, some fresh plant products, especially leaf vegetables such as lettuce, cabbages, etc. or soft fruit, including bananas, have a very short shelf life and have to be eaten within a short time of harvesting or following removal from cool storage. Other foods may be stored by drying, freezing, preserving, etc. and are thus available when fresh foods are not readily available; (8) *Preparation requirements*, these includes the removal of inedible or unpalatable parts such as the shells of nuts, orange peel, tough cabbage stalks, leaching toxic cyanogens from such foods as cassava, as well as various methods of cooking or preserving: and (9) *Presentation*, the serving of food in accordance to defined ethnic and cultural standards is believed to render the food more aesthetically pleasing and thereby more palatable (Passmore *et al.*, 1974; Poulter 1988; Ensminger *et al.*, 1994).

1.   CEREALS

Cereals, which are named after Ceres, the Roman goddess of agriculture, are the cultivated members of the Gramineae whose seeds or grain are mainly grown for food and drink or animal feed. Usually the thin pericarp is firmly adherent to the grain and the grass fruit is known botanically as a *caryopsis*. If the pericarp is free and soft, as in *Eleusine coracana* (finger millet), it termed a *utricle* and, more rarely, an *achene* if the pericarp is free and hard (Clayton and Renvoize, 1986).

Three groups of cereals are recognised: major cereals, minor cereals and wild cereals, the latter a name applied to the edible seeds of wild grasses. The grain from non-Gramineae species, i.e. members of the Amaranthaceae, Chenopodiaceae and Polygonaceae, are known as *pseudocereals* and are dealt with in Section 2.

As a food cereals have the great advantage of being energy-rich, easily grown, stored and prepared. The ease with which they can be stored also means that they can be available throughout the year. When used for human food cereal grains, apart from rice, are usually ground to a flour for making bread or other cooked products. They may also be processed for breakfast cereals, or fermented or distilled for alcoholic beverages. Interestingly, the only grain to be commonly used as a vegetable in the US is sweet corn. It should also be appreciated that there can be regional differences in cereal use. For example, in Africa and Asia the millets and sorghums are a staple food, while in Europe and North America they are grown chiefly for feeding to livestock and poultry. Their use for animal feed is discussed in Chapter 10 (Ensminger *et al.*, 1994; FAO, 1994).

Archaeological evidence has shown that 30,000 years ago Palaeolithic people were already collecting grass seeds for food, a practice that continues to this day among traditional hunter-gatherers such as the Aborigines of Australia, etc. Traces of barley and wheat have been discovered from the excavations of settlements in the delta of the Nile and in the Fertile Crescent dating from *ca.* 8000-7000 BC. During the Neolithic, 7000-3000 BC, wild forms of wheat, barley, millets, sorghum, maize and rice were being collected for food. Their domestication and cultivation gradually began so that by 4000 BC cereal cultivation was already widespread in the Mediterranean basin, Western Asia and Western Europe (Grubben *et al.*, 1996).

The major cereals, their order of importance in terms of production are shown in Table 6. They are discussed in some detail in order to demonstrate some of the factors that have determined their use and global acceptance as staple foods in preference to other potential candidates.

1.1   Wheat

The genus *Triticum* contains 10-20 species distributed through the Eastern Mediterranean to Iran. The wheats have a complex ancestry and were first domesticated in the Near East some time before 7000 BC. Principally a temperate crop, it is the most important of the temperate cereals. It can also be grown in the tropics during the cool, dry season, or in the highlands, preferably with irrigation. Far too often it is grown in the tropics because it is considered prestigious to eat wheaten bread, while the more agronomically suitable and often more nutritious indigenous sorghums and millets are neglected.

Wheat is noteworthy for providing a classical example of crop evolution through polyploidy, with at least three levels of ploidy. Among the most important ploidy representatives are the diploid *T. monococcum* (einkorn), the tetraploids *T. dicoccon* (emmer wheat), *T. turgidum* (rivet or cone wheat) and *T. durum* (durum, flint, hard or macaroni wheat), and the hexaploids *T. aestivum* (common bread wheat) and *T. spelta* (spelt). Within each level selection has proceeded from wild species with brittle spikes shattering at maturity into separate spikelets to the hulled non-brittle cultivated species with grains tightly invested by lemma and palea (chaff), through to free-threshing naked species. Depending on the cultivar, the lemma may be awnless or awned. It is possible that the preference for awnless cultivars was associated with ease of harvesting by hand while the awns were reputed to deter birds and insects. There is now a trend towards awned cultivars, partly because some selections with desirable yield characteristics are awned and partly because the awns confer an advantage by being photosynthetically active (Langer and Hill, 1982; Clayton and Renvoize, 1986; van Ginkel and Villareal, 1996).

Wheat represents over one third of the world's total cereal production. It is the main cereal for approximately 37% of the world's population and accounts for some 20% of the total food calories consumed by man. By far the most important world-wide are the bread and durum wheats, although the prehistoric cultivated wheats are currently undergoing a revival. Among the first wheats to be domesticated in the Near East, emmer still accounts for 7% of Ethiopia's total wheat production, while einkorn, which is still grown in the southern Alps, is beginning to find a potential in the health food markets. Neither the flours from emmer nor einkorn wheats are suitable for making leavened bread due to their high α-amylase content. The high protein, 14-15(-17)% and mineral content of spelt are, nutritionally speaking, very exciting and are undergoing a come back, especially in Germany where, despite its lower yield, produces more protein per hectare than bread wheat with *ca.* 11% protein. The primitive and wild wheats are also an important gene source for the plant breeder (Langer and Hill, 1982; National Research Council, 1996).

Wheat grains, like other cereals, are usually subjected to milling before being utilised for food. The milling process involves the breaking open of the grain, separating out the embryo and bran (pericarp and testa), and crushing the endosperm to produce granules of *meal* or *semolina*, which further grinding reduces to a *flour*. The hard durum wheat can be milled to produce a coarsely ground gluten-rich *semolina* or *durum flour*. The best macaroni, noodles, spaghetti, vermicelli, semolina pudding, etc., collectively known as pasta, are made from semolina flour, as are also biscuits.

There are three basic flour categories: (1) *Wholemeal* or *whole-wheat flour*, known as *Graham meal* or *Graham flour* in America, with 100% extraction, containing the whole wheat grain with nothing added or removed. But, because of the presence of minute quantities of wheat germ oil, it has poorer keeping qualities than white flour; (2) *Brown flour*, usually with 85% of the grain, although there is a recent demand for 81% extraction to bridge the gap between brown and white flours; and (3) *white flour*, with 75-78% extraction, with most of the bran and wheat germ removed. Commercial millers aim at an extraction rate of 78%, depending on the amount of endosperm in the grain and the ease by which the embryo and bran can be removed. The *bran*, which constitutes 13% of the grain, is an important source of dietary fibre, fat, minerals and vitamins. It may be made into various patent foodstuffs, or reblended with white flour for 'wheatmeal' breads or, more frequently, used for animal feed. The embryo or *wheat germ* represents 2% of the grain and is rich in protein, vitamins and oil. When extracted separately it may be made into various patent foodstuffs, otherwise it is sold for animal feed (David, 1978; Langer and Hill, 1982; Davis, 1993; Grubben *et al.*, 1996).

When water and salts are mixed with wheat flour to form a dough, the protein forms a *gluten complex*, enabling the production of characteristic wheat loaves. The absence of gluten in barley or rye flours makes them unsuitable for bread making, although recent research in India has found that fermenting a mixture of a cereal grain and pulse, such as rice and *Vigna mungo* (black gram) with *Leuconostoc mesenteroides*, produces a thick enough gum to act like gluten and contain the CO_2 for leavening the bread. Loaf volume and crumb texture are closely related to the gluten complex content (consisting of the proteins *glutenin* and *gliadin*) of the grain and its biochemical composition which, in turn, is dependent on genotype and the environment. The glutenin confers elasticity or strength to the dough and the gliadin provides extensibility. The ratio of elasticity to extensibility needs to be high for leavened bread, intermediate for flat breads such as chapatis, and very low for wafers and semisweet biscuits. Warm, dry, continental climates produce the so-called hard or strong wheat with 12-15% gluten, which is so highly suitable for bread making. The higher yielding but softer wheats of the more temperate regions with 8-10% gluten are generally considered better suited for pastries, breakfast cereals, noodles and starch or for livestock feed. Nevertheless, it is possible to improve soft wheat flours for bread making by the addition of a concentrated gluten extract (Ensminger *et al.*, 1994; National Research Council, 1996).

The milling quality of wheat for making bread and cakes is affected by the activity of the naturally occurring enzyme *α-amylase* in the endosperm, which reduces the starch to sugar. Its activity is determined by either the Hagberg or Farrand Tests. The *Hagberg Test* involves a special tube standing in a bath of boiling water; the tube containing a suspension of 7 g of ground wheat in 25 cc of water. The suspension thickens as it is stirred and subsequently becomes thinner if the enzyme is present. After 60 secs the stirrer is allowed to fall under its own weight to give what is termed the *falling time*. A good milling wheat has a falling time of over 200 secs; falling times of less than 100 secs being unacceptable for milling. *Falling numbers*, which consists of the total time in the water bath (60 secs plus falling time), are sometimes also quoted. In the *Farrand Test* an extract of ground wheat is reacted with a solution of 'limit' dextrin (a derivative of wheat starch). After a measured interval it is mixed with iodine solution and the colour produced compared with that of a suitably diluted iodine-'limit' dextrin solution. The loss of colour due to the reaction between the α-amylase in the wheat extract with the 'limit' dextrin is used as an indicator of enzyme activity, which is expressed as units. Sound wheat has an activity of 10 units or less; over 40 units is normally not acceptable for milling. Forty Farrand Units is approximately equivalent to a Hagberg Falling Time of 100 or a Falling Number of 160 (Dalal-Clayton, 1981).

The assessment of flour quality for cooking suitability is based on various laboratory end-product and component tests. The former use small-scale procedures that mimic large-scale industrial processes, while component tests examine protein, starch, water relations and colour. Grain quality is assessed as follows: (1) *Grain hardness* is indicated phenotypically by a particle size index, the grinding of soft wheats producing finer meals and flours and is manifested genotypically by the presence of 15kD proteins. A near infra-red spectrophotometric method and a single kernel crushing device may also be used. Hardiness is attributed to the short arm of chromosome 5D; the absence of the D genome in durum wheats produces a very hard wheat, while hexaploid wheats may be soft or hard depending on the allelic state; (2) *Grain size*, the kernels should be large, uniform and well-filled; (3) *Protein quality* is highly variable and dependent on environmental and cultural factors, including the use nitrogenous fertilisers. It may range from 7-17% within the same cultivar; (4) *Moisture content* needs to be 12-13% or lower in order to prevent moulds during storage; (5) *Sprouted grains* are of inferior quality due to the presence of carbohydrases, proteases and other hydrolytic enzymes normally associated with germination (Morris and Rose, 1996).

For fermented or leaven bread, fermentation is most commonly based on the production of fermented sugars from starch. Maltose is produced by the combined action of α- and β-amylase on starch. Starch grains damaged during milling are more susceptible to enzymatic hydrolysis, hence the preference for hard wheats because of their higher level of mechanically damaged starch. The strain of *Saccharomyces cerevisiae* known as baker's yeast is normally used as the leavening agent for bread making. But for sourdough breads *Lactobacillus* spp. are used, producing CO_2 for the leavening and lactic acid to provide the sour taste.

The flat breads, such as chapati, pitta and roti from the Indian subcontinent, North Africa and the Middle East and are typically produced from a high extraction (75-90%) or whole-wheat (atta) flour. Such loaves have a characteristic high crust to crumb ratio. They are generally baked at high temperatures and are, therefore, relatively insensitive to variations in flour water absorption and α-amylase. Pizza crust, English muffins and crumpets, bagels and pretzels are also considered as flat breads, as are tortillas. The latter were traditionally made from maize flour, although wheat is now increasingly being used. Soda crackers, cream crackers and water biscuits are also produced using a fermentation process; gluten is developed and the product referred to as **hard dough**. Chemically leavened soft wheats are used for hard-dough sweet biscuits or semisweet, such as Marie, also for cookies, cakes, etc. (David, 1978).

1.2   Rice

The genus *Oryza* contains *ca.* 20 species distributed through the tropical and subtropical regions of both hemispheres, growing in humid forests and open swamps. The cultivated rice *O. sativa* (Asian or paddy rice) was probably derived from *O. rufipogon* in the Nepal - Yunnan area some 2500 years ago, possibly with the independent domestication of a tropical race in India and a cool-tolerant race in Japan. The two major life forms recognised are **swamp rice** and **upland rice**. The former requires flooding for 2-3 months during growth, the latter requiring less irrigation. A second cultivated species, *O. glaberrima* (African rice), was domesticated in West Africa from *O. barthii* (Clayton and Renvoize, 1986).

Grubben *et al.* (1996) suggest that the cultivation of irrigated rice was probably developed in China during the Neolithic, *ca.* 5000 BC, spreading from South China to South and Southeast Asia and, more recently, westwards to central Asia, Mediterranean basin, Africa, Mediterranean basin and the Americas.

Rice is the second most important cereal in terms of global production. It is also a crop that is often associated with low incomes and poverty. Rice is the staple food for some 1.6 billion people and, in the humid and sub-humid tropics, it is the primary source of human energy. The farmers in southern and eastern Asia grow 90% of the world's harvested rice, providing the main source of calories for 2.7 million people. Rice differs from the other cereals in that it usually cooked and eaten boiled as whole grain and not milled, although it can be ground into a flour for rice bread, honey cakes, etc. or the grain fermented to produce rice beer known as sake, or distilled for the Japanese alcoholic drink shôchû. In European cooking it may be used as a breakfast cereal, for various puddings as well as accompanying curries, kedgerees and other oriental-type dishes. It should be noted that English cornflour, unlike American cornflour, is often made from rice, not maize (David, 1978; National Research Council, 1996).

The raw rice, complete with glumes, lemma and palea, is known as *paddy rice*. After harvesting the threshed grain, known as *rough* or *cargo*, consists of the caryopsis (*brown* or *husked rice*) tightly enclosed by the lemma and palea (the husk) and usually including remnants of the glumes and perhaps a portion of the pedicel. The rough rice then passes through a series of operations, including cleaning, parboiling to facilitates the remove of the hull and improve the keeping quality and thiamine content of the grain, hulling to remove the husk and expose the usually brown in colour and strongly flavoured pericarp of the caryopsis, pearling, polishing and grading, to free the grain from the hull, germ and bran and produce the familiar *polished* or *white rice* of commerce. The *brown rice* may also be marketed, especially in health food shops. Unlike other cereals, except oats, the endosperm starch consists of compound granules.

TABLE 7. Nutritional composition of *Oryza sativa* (paddy rice) and *O. glaberrima* (African rice) (Göhl, 1981; National Research Council, 1996)

| | paddy rice | | | African rice |
	rough rice	brown rice	polished rice	whole grain
dry matter	86.9-89.5		87.5-90.1	95.5
crude protein	7.8-11.9	7.6-10.8	7.9-9.1	7.6
crude fibre	9.3-11.9	0.9-1.5	0.3-1.8	0.5
ash	4.5-9.3	1.0-1.5	0.6-1.4	3.8
ether extract	1.2-1.7	1.3-1.6	0.1-1.8	1.9
N-free extract	69.4-75.4	85.4-88.4	87.1-89.9	81.0
Ca	0.11		0.05	0.03
P	0.29		0.32	0.26

The fine glossy, translucent appearance of polished rice desired by the European and American markets is obtained by glazing. The processing results in a reduction in the nutritive value (Table 7). Nevertheless, despite popular opinion, polished rice is less nutritionally disadvantaged than brown rice since the latter has a poorer digestibility and poorer protein retention. The protein content of rice ranges globally between 5 and 17%, with a mean value of 10.5%, with a significant correlation between total protein and lysine levels, although the differences in protein retention between high and low protein polished rice is negligible. The colour of milled rice, the rate at which the endosperm yellows with age, are also influenced by the protein content, as well as tenderness and cohesiveness within the same variety - the higher the protein the lower the cohesiveness. Polishing removes the thiamine (vitamin B_1) from the outer layers of the grain, and where polished rice provides the staple food; *thiamine deficiency* may cause the disease beriberi. Brown rice is less popular than polished rice because it requires more time for cooking and consequently more fuel. It also has a reduced storage life due to the tendency of the oil in the bran to turn rancid; anti-nutritional factors are also associated with brown rice. The advantage of brown rice is in the higher vitamin content (Göhl, 1981; Langer and Hill, 1982; Purseglove, 1985; Blakeney, 1996; Greenland, 1996; Grubben *et al.*, 1996).

Grain quality is based on appearance, milling and cooking qualities: (1) *Grain appearance* is assessed in terms of moisture content, grain size, shape, colour, gloss, translucency, uniformity. cleanliness and freedom from empty hulls, straw and other seeds. Traditionally quality is evaluated by eye or by microscopic measurements. Objective methods for assessing colour and translucency using a reflectance spectrophotometer and size and shape using image analyses are now available; (2) *Milling quality* of the grain varies according to length, width, thickness and weight, and is primarily based on the yield of whole grain after the milling process has removed the hull, bran and germ with minimum breaking of the endosperm. There are two major sources of grain breakage, which are referred to as *checking* or *sun cracking*, and *chalking*. Both are attributable to environmental factors, especially immaturity, and genetic factors. Checking is caused by delays in harvesting and threshing or a too rapid drying of the crop. The fluctuations in temperature and moisture cause the outer portion of the grain to expand more rapidly than the centre, thereby creating cracks along the line of the endosperm. Chalking and immature grain is recognised by layers of cells with loosely packed starch granules. Parboiling reduces rice breakages during milling, the improvements being attributed to a greater grain hardness and the sealing of any internal cracks and chalkiness. Inadequate penetration and distribution of water in the grain during parboiling results in opaque spots due to ungelatinised starch granules in the grain. Other factors affecting milling quality are moisture content, infestation and grain shape; and (3) *Cooking quality* in terms of preferred texture and flavour varies according to ethnic and regional preferences. Cooking characteristics are generally associated with grain size and shape. The short and medium grain varieties, which usually become somewhat sticky when cooked, are often preferred in some north Asian countries, while Australia and the western nations prefer the drier and flaky quality of the long grain. In the USA the short grain rice is used for puffed rice, precooked canned rice and for other dry quick-cooking rice products. Both types are used for dry breakfast cereals. The two most important physiochemical cooking properties are gelatinisation temperature and amylose content. Cooking mainly involves the gelatinisation of the starch, the gelatinisation temperature being related to the cooking time, granular size, molecular size of the starch subfraction polymers and their relative amounts. A correlation has been noted between amylose content and eating quality preferences, cooked rice with a high amylose content is dry and flaky while rice low in amylose is sticky and moist when cooked. Cooking time relates to the rate at which water penetrates the grain and since soaking causes cracking to develop in the grain, cooking time is reduced (Blakeney, 1996).

1.3   Maize

The genus *Zea* consists of four species from Central America, including the cultivated *Z. mays* subsp. *mays* (Indian corn, maize, mealies), which is not known in the wild and was apparently domesticated in Central America some 7000 years ago from a species resembling *Z. mays* subsp. *parviglumis* (Balsas teosinte). But because of the major morphological differences between *Z. mays* and the other three species, they were formerly placed in a separate genus, *Euchlaena*. Nevertheless, despite the visually extreme differences they are actually under relatively simple genetic control, as would be expected between species subject to disruptive selection. The selection pressure on the wild species being towards more efficient dispersal, whereas for the cultivated maize it has been for cob retention. Indeed, maize is believed to have the widest range of genetic diversity of any of the major cereals (Clayton and Renvoize, 1986).

The grain is not only a valuable source of starch but also has a higher oil content than most other cereals. Of the various grain types several commercially important cultivars are recognised, of which flint and dent corn are by far the most important: (1) *Flint corn* has a hard, vitreous endosperm surrounding a small area of soft or floury endosperm. The grain exhibits little shrinkage at maturity, resulting in a rounded, undented grain; (2) *Dent corn,* a cross between flint and floury maize, has a soft inner endosperm and shrinks at maturity to give the grain a characteristic indented shape. Dent maize hybrids with a small percentage of floury endosperm can exhibit little indentation, while hybrids with a high percentage of endosperm can have a shrivelled and immature appearance. It is the principal maize of the US Corn Belt; (3) *Soft* or *floury corn* is one of the oldest types to be cultivated by the early Native Americans. The kernels are usually flat and large, without any indentation and with a soft and floury endosperm. It is cultivated in the drier parts of the Americas and South Africa; (4) *Waxy corn* has, as its name suggests, a waxy endosperm which when milled produces a flour lacking amylose and resembling that of tapioca, which is obtained from the root of *Manihot esculenta* (cassava); (5) *Rice corn* has small, pointed grain, the endosperm of which expands rapidly on heating, bursting the grain and exerting the endosperm; (6) *Pearly popcorn* has small, rounded grains and is an early variation of flint. It was selected for its ability to pop during rapid cooking. The popping being a response to the build up of internal pressure from the vaporising moisture which, unable to escape because of the low diffusibility of the thick and dense protein matrix encapsulating the starch granules, bursts the grain and exerts the endosperm. An important snack food, popcorn can be coated to provide a wide range of flavours; and (7) *Sweet corn* is a genetic variant which inhibits the conversion of sugars to starch, resulting in a glossy, sweetish endosperm which is translucent when immature and produces a wrinkled appearance when dry. Three classes of sweet corn are recognised: (a) *Standard sweet maize* arising from a selection of the sugary mutant gene *su*, resulting in an accumulation of sucrose and phytoglycogen in the endosperm. It has the lowest sugar content, with *ca.* 16% sucrose; (b) *Augmented sugary sweet maize*, which has been genetically modified by involving the sugary extender gene *se* to enhance the sugar content; and (c) Other mutants such as shrunken-2 (sh_2), brittle-2 (bt_2) and brittle-1 (bt_1) bred to increase the sugar content to 35%. Sweetcorn is used in South and Central America for producing a potent beer known as *chicha* and elsewhere eaten as corn on the cob; it is also used for canning and freezing (Langer and Hill, 1982; Purseglove, 1985; Eckhoff and Paulsen, 1996).

Maize is now grown on every continent except Antarctica. It is the staple cereal food crop of the tropics as well as being extensively grown as a forage crop; it is also an important source of starch, oil, syrup and alcohol. Maize and maize products are widely eaten in parts of the former Soviet Union, Asia, Africa and Central America as bread (arepa, tortilla, corn bread, hoe cake or blintzes), for fermented or unleavened porridges (atole, ogi, kenkey, ugali, ugi, edo pap, maizena, pasho or asifah), steamed (tamalos, cous-cous, Chinese breads, dumplings and chengu), alcoholic beverages (koda, chicha, Kaffir or maize beer), non-alcoholic beverages (mahewa, magou or chica dulce), or as snacks (empanada, chips, tostades or fritters). Masa, the dough produced from fine grinding of alkali (lime) cooked maize can be made into sheets, cut and baked or fried for tortillas, tacoshells, tortilla or maize chips. It is the major source of nutrition in Mexico and other countries in Central and South America. In the United States corn meal or maize meal is much used for muffins and other breakfast foods. In Italy the grilled or toasted paste or maize meal is used for making 'polenta'. Buttered corn cobs are also widely eaten. The very fine starch, cornflour or cornstarch (not to be confused with English cornflour made from rice flour) is widely used as a sauce thickener and in blancmanges, custards, etc. (David, 1978). The gall produced by an infection of the cob with the fungus *Ustilago maydis* is eaten as a vegetable (Clayton and Renvoize, 1986).

As a staple food maize is deficient in *lycine* and, provided the lycine can be supplemented from other food sources, the maize kernel is a valuable source of protein and energy for human food and animal feed. Breeders are now actively engaged in improving the lycine content, which is genetically controlled. Although the high lycine cultivar 'Opaque 2' can produce 30-100% more lycine than normal dent hybrids, the extremely soft endosperm makes it susceptible to mechanical damage during handling and therefore more susceptible to insect and fungal infestation during storage. ***Quality protein maize (QPM)***, a high lycine maize with a hard endosperm, has now been developed which has better handling qualities.

Despite having a ***nicotinic acid*** content as high as that of wheat, much of it is in a bound form and unavailable. It is associated with the amino-acid ***tryptophan*** and in a normal diet part of the tryptophan is converted into nicotinic acid. A deficiency of tryptophan and thus of nicotinic acid, causes the dangerous skin disease known as pellagra, a disease prevalent in those regions where maize forms the major part of the diet (Macpherson, 1995; Eckhoff and Paulsen, 1996; Grubben *et al.*, 1996).

1.4   Barley

The genus *Hordeum* contains *ca.* 40 species distributed throughout the temperate regions of the world. The major cultivated barley is *H. vulgare*, an important cereal of the temperate regions which, because of its short growing season and tolerance of drought, alkaline and saline conditions, is grown further north than any other major cereal, even to beyond the Arctic Circle. It is also grown at high elevations in Ethiopia and Tibet. Barley may also be grown in the tropics during the cool season.

The domestication of *H. spontaneum* in the Near East prior to 7000 BC gave rise to two major types of barley. The original 2-row cultivated *H. distichon*, where the two lateral florets of the sessile racemes are sterile, from which mutation gave rise to the 6-row *H. vulgare*, where all three florets are fertile, and a 4-row variant in which the central spikelet is sterile. The progenitor, *H. spontaneum*, with its brittle rachis, differs only by one of two linked genes from the cultivated *H. vulgare* with its non-brittle rachis. The barley grain mostly consists of the caryopsis tightly enveloped by lemma and palea. Naked or hull-less forms of both 2- and 6-row barleys also exist, the loose hulls being removed during harvesting, making the grain more digestible and more desirable as a food also occur.

The morphology of the two cultivated barleys affects their handling properties. The lateral kernels of the 6-row barleys tend to be crushed and twisted during grain filling, giving a greater variability in grain size compared to the 2-row barleys, so that they have to be handled differently during processing. The 6-row barleys yield better than 2-row and are consequently often grown for feed, while the uniform grain of the 2-row barleys are preferred by the brewing industry (Langer and Hill, 1982; Clayton and Renvoize, 1986; Edney, 1996).

Natural mutants include: (1) ***High-lysine barleys*** from Ethiopia with 30-40% more lysine in its protein. To date no commercial cultivars of high lysine barley have been released; (2) ***High-amylase barleys*** with an altered starch containing *ca.* 40% amylase and 60% amylopectin (standard barley contains 20-25% amylase and 80-75% amylopectin); and (3) ***Waxy barley*** with starch containing *ca.* 95% amylopectin and offering exciting possibilities to nutritionists and industrialists.

Apart from its major use as animal feed, which accounts for as much as 75% of the world's barley crop, the best quality barley grain is important for the brewing and distilling industries (see Sections 6.1 and 6.2). Malting barley is high in starch (source of maltose) and low in protein, the latter being determined by measuring the nitrogen content. While the use of barley malt for alcoholic beverages is increasing, the use of barley for food is declining in response to the increased availability and access to better cereal food sources, including wheat.

The mechanical removal of successive outer layers of the barley kernels is known as *pearling*. By the removal of excess fibre, which is unsuitable for human consumption, as well as removing inconsistencies in colour and shape of the kernels, a more useful starting material for future processing is produced. Pearling also reduces any adhering hulls and removes any barley bran. The bran can give an undesirable colour to the finished products and also cause problems during milling because it is brittle and shatters, unlike wheat bran, which flakes off.

Barley flour is generally used when wheat is not available. The meals and flour are low in *gluten* and cannot be leavened, although making a good flavoured, rather flat, greyish loaf which quickly becomes dry and unpalatable. Loaves of more volume and a better colour and keeping quality can be made by mixing barley flour with white or brown wheat flours.

Milled barley is used in North Africa, where it is pearled, coarsely ground into semolina and used in cous-cous; coarser grindings are used for soups, while finely ground flour is used for breads. In SE Asia ground barley is incorporated into noodles, or the pearled barley consumed as rice extenders. In Europe and North America pearled barley is mainly used in soups, or as a flour for food thickeners. There is also an increased demand for barley health foods as it is reputed to reduce heart disease by lowering the blood cholesterol, which is probably due to the soluble *β-glucan*, higher levels of which occur in barley than in oats, the previous major source. It also reduces colon cancer and postprandial glucose concentration, the latter being important for diabetics (David, 1978; Dalal-Clayton, 1981; Edney, 1996).

1.5   Sorghum

The genus *Sorghum* contains *ca.* 20 species, mostly in the Old World tropics and subtropics, with one species, *S. trichocladum*, in Mexico. The cultivated cereal *S. bicolor* (sorghum) was probably domesticated in the Sudanian Region of tropical Africa *ca.* 3000 years ago from *S. arundinaceum* (Clayton and Renvoize, 1986). It is now widely grown throughout the tropics and subtropics and is second only to maize among the tropical cereals as a staple food, especially in Africa and Asia, although in the west it is mainly grown for livestock fodder and grazing. Indeed, only rice, wheat, maize and potatoes surpass it in feeding the world's population. The caryopsis (grain) is readily separated from the enveloping glumes on thrashing and the stems (stover) can be fed to livestock or used for building materials.

Sorghum is an extremely versatile crop, tolerating a wide range of environmental conditions, including waterlogging and some salinity. It is second only to pearl millet as a major grain of hot and dry conditions. Sorghum is among the very few major food crops, along with maize and sugarcane, that follow the C_4 synthetic pathway. It boasts of having one of the highest dry matter accumulation rates as well as being one of the quickest maturing food crops. It also has the highest production of food energy per unit of human or mechanical energy used, surpassing even that of maize silage, sugarcane and maize grain. Unfortunately sorghum does suffer from the mistaken prejudice of being a "coarse grain", an "animal feed" and "food of the peasant classes".

Sorghum is unique among the major cereals in that the brown and black sorghum cultivars with a pigmented testa contain bitter-tasting polymeric polyphenols, i.e. *condensed tannins*, although tannic acids do not occur. The tannins have the agronomic advantage of conferring protection to the grain against fungal attack and predation by birds and insects. Weathering, preharvest germination or early sprouting are also significantly lower in most high tannin cultivars. The tannins do have the disadvantage of reducing the nutritional value of the grain by binding with both the grain proteins and with enzymes in the digestive tract. They also pose problems in brewing traditional opaque and lager-type beers, binding with the malt amylase enzymes and inactivating them. Such high-tannin sorghums require careful processing if the tannins are to be eliminated and although decortication is the most effective method available, it does have the disadvantage of excessive nutritional losses (Rooney, 1996; Taylor, 1998).

Popping sorghums, which readily pop (like popcorn) when heated, have an enhanced flavour. They are nutritionally favoured since the popping only slightly denatures or hydrolyses the proteins and vitamins, also little fuel is required for the popping process. Some sorghums have a sugary endosperms so that while the grain is still soft and sugary the entire panicle may be cooked and eaten like sweet corn. Some of the yellow-grained sorghums are relatively rich in vitamin A precursors, albeit at considerably lower concentration than found in yellow maize, nevertheless a useful attribute where the main diet is sorghum. The yellow endosperm types are widely used in breeding modern hybrids, while waxy endosperms cultivars are used for starch manufacture. Despite its huge potential sorghum remains relatively undeveloped, with an array of untapped variability in plant and grain type, adaptability and productivity. Indeed, it is regarded as having a greater undeveloped and under-utilised genetic potential than any other major food crop (Langer and Hill, 1982; Purseglove, 1985; National Research Council, 1996; Rooney, 1996).

There are numerous cultivars world-wide, 3,682 of which are currently maintained at the International Crop Research Institute for the Semi-Arid Tropics (ICRISAT). The grains of the various cultivars may be eaten boiled, ground to a flour for flatbreads (the grain lacks gluten), cracked like oats for porridge, fermented for beer, processed for breakfast cereals, popped for snack foods, etc., while the sugary grains may be boiled at the green stage like sweet corn. Traditionally the hard endosperm cultivars are preferred for thick porridge and cous-cous, the intermediate endosperm cultivars for unleavened breads, boiled rice-like products, malting and brewing, and the soft endosperm cultivars for leavened breads. The absence of gluten has been overcome by what is known as the custard process, where a starch gel or 'custard' is prepared by boiling cassava or similar starch source and then used as a gluten substitute to produce a rather firm and crumbly loaf. The stems of sweet sorghums are harvested at the dough stage, stripped of leaves and crushed to yield a juice which is evaporated to yield a high quality sorghum syrup or sorghum molasses. Clarification of the syrups using starch can cause a problem since the starch may be removed by the action of the enzymes present.

For most traditional African and Indian foods the grains are processed using a pestle and mortar or abrasive dehuller. This removes the outer layer and yields a decorticated or pearled kernel with reduced fibre levels and lysine content. Sorghum for food and feed industries are usually dry milled to produce grits and meals for brewing, baking, snack foods, etc. and malted to produce opaque and lager beers, breakfast foods, etc. Abrasive drums are used for decorticating since the pericarp pigments can give the cooked starch an undesirable light pink tinge, which can be removed by bleaching with $NaClO_2$. When malting for opaque beers, weaning foods and other traditional dishes, pearling results in an 8 to 30% loss of dry matter, decreased levels of prolamins, fats, tannins and starch and increased levels of free amino-acids, albumins, lysine, reducing sugars and most vitamins, including synthesis of vitamins B_{12} and C (Rooney, 1996; Taylor, 1998). For information on brewing sorghum beers see Section 6.2.

1.6   Oats

A genus with *ca.* 25 species, *Avena* is mainly distributed through the Mediterranean and the Middle East, extending into northern Europe and with the semi-domesticated *A. abyssinica* in Ethiopia. There are six to seven cultivated species, including *A. sativa* (oats) which, from archaeological records, appears to have been domesticated in north-western Europe during the Early Bronze Age, *ca.* 2000 BC. The cultivated oats belong mainly to the three, fully cross-fertile, hexaploid species, i.e. *A. sativa* subsp. *sativa* (white or yellow oats) and provides more than 75% of the oat crop, followed by *A. byzantina* (red oats) and *A. nuda* (naked oats), with a small area under the diploid *A. strigosa* (black oats). *A. sativa* subsp. *fatua*, syn. *A. fatua* (wild oats) is now ranked as one of the world's worst weeds.

The grain consists of an elongated caryopsis tightly enveloped by a tough lemma and palea. Attempts at breeding for a naked grain have so far been only partially successful due to poor yields, irregular development and grain shattering. As a crop oats give a poor monetary return per hectare compared to other competitive crops, and is primarily grown as a feed grain. (Langer and Hill, 1982; Clayton and Renvoize, 1986; Webster, 1996).

The commercial oat products are: (1) ***Rolled oats*** produced by flaking whole groats, i.e. the hulled and usually crushed grain; (2) ***Steel-cut oat groats*** produced by sectioning the groats into several pieces; (3) ***Quick oats*** produced by steaming and flaking steel-cut groats; (4) ***Baby-oat flakes***, which have thinner and finer flakes than steel-cut groats; (5) ***Instant oat flakes*** commercially produced by subjecting the groats to special proprietary processes before cutting; (6) ***Oat flour*** obtained by grinding the flakes or groats; and (7) ***Oat bran,*** the fraction remaining after sieving coarsely ground oat flour.

The first references on the use of oats are by Hippocrates *ca.* 460-360 BC, Dieusches *ca.* 400 BC, Dioscorides in 1st century AD and Galen 130-200 AD. All refer to their medicinal uses rather than for food. Dioscorides, for example, refers to the use of oats as a healing agent, as a desiccant when applied to the skin, for relieving coughs and as a natural food for horses and for humans in times of scarcity. A somewhat similar opinion was also expressed by Samuel Johnson (1755) "*Oats. A grain, which in England is generally given to horses, but in Scotland supports the people*". It was not until the 19th century that oats became accepted in England as a staple breakfast food. This followed improved milling and stabilisation technology, and by the development of improved packaging of branded products and associated sales promotion.

As a food oats are mainly used as a whole grain flour or flake, unlike other cereal grains where significant quantities of bran are removed prior to their use for food. The oats are also heat-processed in order to develop the characteristic toasted oat-like flavour and inactivate the complex enzyme systems present. Unless so treated the oat flour rapidly develops soapy and bitter flavours due to the enzyme activities of ***lipase, lipoxygenase*** and ***peroxidase***; the treatment also dramatically reduces the protein solubility.

Of good nutritive value, rolled oats and oatmeal are commonly employed in the food industry and home cooking. Oats are used mainly as a hot or cold breakfast cereal food, as well as being an important component of many infant foods. Oatmeal is low in gluten, but high in fat, and can be used for flat unleavened oatcakes. Small quantities of oatmeal may also be used to give flavour and increase the fat content of other bread flours, especially wheatmeal and wholemeal breads. It is also used as a thickening agent for soups, sauces, gravies, as a meat extender, etc. The antioxidant properties are effectively used to stabilise milk powder, butter, bacon, ice-cream, cereals and frozen fish.

Oats are also reputed to have a very positive medicinal effect on problems of ageing, cholesterol oxidation and cancer incidence. There is also an increased interest in oats as a health food because of its hypocholesterolaemic properties, with the soluble fibre slowing the rise in blood sugar following ingestion, an important factor in the insulin responses of both normal and diabetic individuals, plus the presence of a powerful antioxidant complex (David. 1978; Webster, 1996).

1.7   **Pearl Millet**

Millet is a blanket term embracing the major cereal *Pennisetum glaucum* (pearl or bulrush millet) and a number of minor cereals, which are dealt with in Section 1.10. The name is from the Middle English *milet*, from Old French, from *mil*, millet, from Latin *milium* (Long, 1994). The genus *Pennisetum* contains *ca*. 80 species distributed throughout the tropics. Of these, pearl millet is the most widely grown, ranking sixth among the world's most important cereals. It was probably domesticated in the Sahel of West Africa *ca*. 2000-3000 BC from the wild, ruderal species *P. violaceum*.

Pearl millet is the most drought-resistant of the tropical cereals. It is a staple food crop throughout the drier parts of tropical Africa and Asia, whereas in Australia, South Africa and USA it is largely grown for livestock feed. It also has one of the highest rates of dry matter gain among the major C_4 cereals, yet only a small proportion of the available genetic variation has ever been utilised. There is, therefore, a good potential for improvement, both as a staple cereal for the low rainfall regions of the world and as a high yielding nutritious livestock feed for the developed countries of the tropics and subtropics (Clayton and Renvoize, 1986; National Research Council, 1996; Oyen and Andrews, 1996).

Pearl millet has a higher protein and ash content than either sorghum or maize, as well as having significantly higher energy levels. The protein content and composition does, however, vary with agronomic conditions. The use of nitrogen fertilisers, for example, significantly increases the accumulation of *prolamines*, which form, as in sorghum, the major protein fraction, followed by *glutelins* and are located within the protein bodies and protein matrix respectively. The protein is deficient in the essential amino-acids *lysine*, *threonine* and *tryptophan*, although the lysine content is higher than either those of sorghum and maize proteins, and is probably related to the increased proportion of germ, i.e. embryo, in the caryopsis (Rooney, 1996).

1.8   **Rye**

Secale is a genus of four species distributed mainly from eastern Europe to central Asia, but also present in Spain and South Africa (*S. africanum*). The cultivated *S. cereale* (rye) is noteworthy in being the only outbreeder among the small-grained temperate cereals. It probably began as an inbreeder domesticated from the inbreeding *S. vavilovii* and later became an outbreeder as a result of introgression with the self-incompatible *S. montanum* infesting the rye fields of Turkey. Eventually its cultivation spread northwards into Russia and hence westwards into Poland and Germany; it was introduced elsewhere in Europe via the Balkans. Because of its winter hardness, resistance to drought, insects and pests, and for its ability to grow better in poor soils and under greater adverse conditions than other cereals, rye is an important cereal in both northern Europe and the former Soviet Union. It is also cultivated in the Himalayas, Canada, southern Chile, Africa and Australia. (Langer and Hill, 1982; Clayton and Renvoize, 1986; Weipert, 1996; Pickersgill, 1998).

International trade in rye is mainly as a feed cereal. Global production of rye as a staple grain has declined in recent years, its use for bread in some countries being limited to certain ethnic communities and health shops. Rye is also distilled for alcoholic drinks, e.g. Canadian rye whiskey and Bourbon. The presence of *pentosans* are a disadvantage when brewing but are not necessarily insuperable. Rye is also used for controlling fermentation in fruit juices and in the production of *ethanol (ethyl alcohol), propanone-butyl-alcohol (acetone-butyl-alcohol)*.

Rye differs from the other cereals, particularly wheat, in its susceptibility to pre-harvest sprouting due to the virtual absence of dormancy. Sprouting is accompanied by a reduction in both starch and protein content. There is also a tendency for lodging, hence the need for early harvesting and drying prior to storage. The crop it is liable to bacterial and fungal infections, especially by *Claviceps purpurea* (ergot).

The products from the dry milling of rye are flour, meals and whole-kernel meals and flours. Due to the viscous properties of the kernels, rye flour is generally finer and of a higher pentosan content than wheat flours. The pentosans are carbohydrates but behave significantly difficult from starch in that they do not undergo disintegration and gelatinisation, nor do they release water like proteins when heated during bread-making, thereby providing the crumbs with a softness and longer shelf life. The EU standards for making rye bread are measured by an amylograph and whole kernel meal to show peak viscosity not lower than 200 BU (*Brabender Units*) and peak temperature not lower than 63°C.

The structure of the rye proteins and their inability to form gluten, together with the high water-binding capacity of the pentosans, affect the gummy nature of the kernels during milling. The consistency, yield, stickiness and behaviour of the dough during proofing, as well as the resilience and chewability of the bread crumb, are also affected. Also, because rye starch gelatinises at a lower temperature than wheat starch, it has a longer exposure to α-amylase activity, so that the starch is unable to form an elastic and resilient crumb. Consequently, apart from pumpernickel and one or two Dutch breads, and dry crispbreads, nearly all rye breads are made with a mixture of wheat flour (David, 1978; Langer and Hill, 1982; Weipert, 1996).

1.9   Triticale

The intergeneric hybrids created by plant breeders in the late 19th century by crossing wheat and rye are referred to ×*Triticosecale* (triticale). Tetraploid, hexaploid and octoploid forms now exist, of which the hexaploid is the most successful with yields of recent forms approaching those of wheat but with consistently higher protein and essential amino-acids. The crop is well suited to poor soils, including acid soils and those with high iron and aluminium content. Processing is more difficult than for rye due to its susceptibility to sprouting and intrinsic enzyme activity.

Although the milling qualities of triticale are closer to wheat than rye, this does not permit profitable flour extraction because of the shrivelled and disordered structure of the kernels and their high mineral content. Indeed, whole kernel meal flours are preferred. Triticale protein is to a great extent water soluble and forms a soft and weak gluten. The gelatinisation properties of the starch and the generally high α-amylase activity result in a weak starch gel. Such disadvantageous properties and behaviour do not encourage the use of rye for making bread (Darwinkel, 1996; Weipert, 1996).

1.10   Minor Cereals and Wild Grasses

Wild grasses were the source of grain for human food before domestication and are still so used by some 'primitive' communities, while for others they are an important famine food, especially in the drier regions of the world. For example, 79 species are recognised by Renvoize *et el.* (1992) as being used for human food in the arid regions of the world and the list is by no means exhaustive. Among the more important genera are *Aristida*, *Brachiaria*, *Cenchrus*, *Coix*, *Dactyloctenium*, *Digitaria*, *Distichlis*, *Echinochloa*, *Latipes*, *Oryza*, *Oryzopsis*, *Panicum*, *Paspalum*, *Poa*, *Setaria*, *Sporobolus*, *Stipagrostis*, *Uriochloa* and *Zizania*. Problems with low yields, small grain, uneven ripening, grain shattering, etc. will have to be overcome before they can even be considered for profitably domestication. Not withstanding, they form an important gene source for the plant breeder, even if they unlikely to be brought into cultivation.

Some 50 species of grasses are extensively harvested in the Americas yet only six have been domesticated. They are maize, the only New World cereal of commercial significance and already dealt with above, *Bromus mango*, syn. *B. burkartii* (mango), *Panicum sonorum* (sauwi), *Phalaris caroliniana* (May grass) from south-eastern USA and only known as a source of food from archaeological findings, *Setaria geniculata* (brittle grass) from arid Mexico and used for at least a millennium before being replaced by maize 4000 years ago and *Zizania aquatica*, syn. *Z. palustris* (American wild rice). The latter is the only grass species that has been successfully domesticated as a cereal in recent times. It is also the only New World minor cereal of economic importance, mainly as a health food. There are also possibilities for creating perennial cereals, such as a hybrid between *Zea mays* and the perennial *Z. diploperennis* to produce a perennial maize, or the domestication of the perennial *Distichlis palmeri* (Palmer saltgrass), which is capable of tolerating salinity levels up to 56.0 dS m^{-1}.

Mango is a native of Chiloé Island and the adjacent Chilean mainland south to *ca.* 37°30'S. It is unique among the cereals in being a biennial. It is grazed by livestock during the first year and the grain harvested the following year. It was domesticated and cultivated as a cereal in central Chile by the indigenous population from before the Conquistadors until the mid 18th century, when it was replaced by wheat introduced by European colonists. The grain was used for making unleavened bread and an alcoholic beverage. Domesticated varieties were believed extinct until their rediscovery in the Andeso-Patagonia region of Argentina in 1987 (Muñoz-Pizarro, 1944, 1948; de Wet, 1981,1990, 1992; Matthei, 1986; Brücher, 1989; National Research Council, 1989). However, trials in Tasmania have shown that whilst the seed yields are quite high, the size of the grain (4.5 mm long fide Matthei, 1986) does not warrant further development (R. Reid, pers. comm. 1993).

In Africa some 60 grasses were harvested as wild cereals until recent historical times and some continue to be harvested in times of famine. For example, *Brachiaria deflexa* (Guinea millet), *Oryza barthii* (progenitor of the cultivated *O. glaberrima*) and *Paspalum scrobiculatum* (kodo millet) were extensively harvested in West Africa, while *Cenchrus biflorus* (cram cram) and *Stipagrostis pungens* were harvested by the Saharan nomadic peoples and are still encouraged as useful weeds of cultivation for harvesting in times of dearth. *Eragrostis tef* (tef), a native of Ethiopia, is cultivated in eastern Africa. It is also being investigated at Wye College (University of London), Kent for improved varieties.

There are a number of minor millets that are locally important and grown in Asia, with some introduced elsewhere. They include *Brachiaria ramosa* (browntop millet), *Coix lacryma-jobi* (Job's tears), *Digitaria cruciata* (raishan), *D. sanguinalis* (crab-grass), *Echinochloa colona* (Indian barnyard millet), *E. frumentaceae* (Japanese barnyard millet), *E. oryzoides* (paddy-rice mimic weed), *Panicum miliaceum* (common millet), *P. sumatrense* (little millet), *Setaria italica* (foxtail millet) and *S. pumila* (yellow cat-tail millet), of which foxtail millet is the most widely grown (Harlan, 1989; de Wet, 1990; Renvoize *et al.*, 1992).

2.   PSEUDOCEREALS

The *pseudocereals* are herbaceous dicotyledons with small edible starchy and grain-like seeds and belonging to the genera *Amaranthus* (grain amaranths), *Chenopodium* (grain chenopods) and *Fagopyrum* (buckwheats). Their seed storage proteins are salt-soluble globulins and water-soluble albumins, unlike the cereals where alcohol-soluble prolamines that form the main storage protein fraction. The biological value of their protein is higher than that of the true cereals (Table 8), so that they are sometimes mixed with wheat flour to improve the nutritive value of the bread. The absence of gluten however makes the use of pseudocereals alone unsuitable for leaven breads. The pseudocereals also differ from the cereals in that their seed starch reserves are stored in the perisperm and not in the endosperm (Berghofer and Schoenlechner, 1998; Tatham *et al.*, 1998).

The grain amaranths, like maize, sorghum and the millets, follow a C_4 photosynthetic pathway, while the grain chenopods and buckwheat, like wheat, barley and rice, follow the C_3 pathway. In general those cereals and pseudocereals following the C_4 photosynthetic pathway are better adapted to hot climates and high light intensities than C_3 plants (Grubben *et al.*, 1996).

TABLE 8. Comparative nutritional values of the pseudocereals and cereals (Cole, 1979; Göhl, 1981; National Research Council, 1989; Risi and Galwey, 1989; Jain and Sutarno, 1996)

	grain amaranths	quinoa	buckwheat	maize	wheat
moisture	10.1	6.8-20.7	11.0	9.2-13	0.8-2.4
crude protein	13.6-18.0	7.5-22.1	11.7	9.1-12.5	9.4-17.0
crude fibre	2.9-10.8	1.1-16.3	9.9	1.4-4.6	3.1-3.3
ash	2.5	2.2-9.8	2.0	1.3-2.0	1.9-3.3
ether extract	7.0-8.3	1.8-9.3	2.4	4.2-4.6	1.7-2.1
N-free extract	52.3-60.2	38.7-71.3	63.0	76.8-83.7	77.8-79.4
Ca (mg)	490		114	10-20	30-70
P (mg)	455		282	300-370	480
amylase (%)	5-7	20		1-85	23-27
vitamin A (IU)	0			70.0	0
thiamine (μg)	0.14			0.43	0.57
riboflavin (μg)	0.32			0.10	0.12
niacin (μg)	2,0			1.9	4.3
ascorbic acid (μg)	3.0			trace	0
(amino-acids as % of protein)					
lysine	3.2-8.2	6.6		2.4-4.4	2.6-2.9
methionine	1.8-2.5	2.4		0.6-1.9	1.3-1.7
threonine	3.2-4.3	4.8		3.1-4.0	2.7-3.3
tryptophan	0.2-0.9	1.1		0.6-1.0	1.1
isoleucine	3.3-6.9	6.4		3.1-4.6	3.8-4.1
leucine	5.1-8.0	7.1		10.7-13.1	6.3-6.8
valine	3.7-4.7	4.0		4.2-6.7	4.3-4.9
phenylalanine	7.5-14.7	3.5		4.9-6.1	4.3-4.6
arginine	7.5-14.7			3.5-5.1	4.4-4.5
histidine	2..3-2.9			2.1-3.5	1.9
cystine	1.6-2.9	2.4		1.3-2.0	1.7-2.3
tyrosine	3.2-4.1	2.8		3.7-6.1	2.9-3.5

Three of the grain amaranths have been domesticated in the New World and became important crops of both the Inca and Aztec civilisations. According to the first hypothesis *Amaranthus caudatus* was domesticated from *A. quitensis* in Andean Peru and Ecuador, *A. cruentus* from *A. hybridus* in Mexico and Central America, and *A. hypochondriacus* from *A. powellii* in Mexico. The alternative hypothesis is that *A. cruentus* arose from *A. hybridus* and that subsequent hybridisation led to the selection of *A. caudatus* and *A. hypochondrus*. Unfortunately, neither of these hypotheses are conclusively supported by hybridisation studies. The grain amaranths were introduced to India, Nepal and Sri Lanka during the colonial period, and later to areas of Southeast Asia, and Africa, and are now grown in many countries of the world.

The small, *ca.* 1 mm in diameter, highly nutritious seeds, unlike quinoa and the buckwheats, contain no adverse compounds requiring decortication before use. Their seeds are used in various milled and popped/puffed products, including unleavened bread; because amaranth flour lacks functional amylose it must be blended with wheat meal or flour for leavened bread. Their wholemeal flours can be separated into starch-rich and protein-rich fractions. With up to 30% protein the latter fractions are proposed for the protein enrichment of other foods. The seeds can also be fermented for a soy sauce, beer and brandy. The small starch grains, 1-3 μm in diameter, have potential applications as a fat substitute. The amaranth oil produced as a by-product of starch production contains a high proportion of the acrylic triterpene *squalene*, which is a precursor of cholesterol and other sterols and triterpenoids of plants and animals. Squalene is widely used by the pharmaceutical and cosmetic industries for promoting absorption of drugs applied to the skin and as a non-toxic vehicle for cosmetics (Jain and Sutarno, 1996; Berghofer and Schoenlechner, 1998; Sharp, 1990).

The seeds of *Chenopodium album* (fat hen, white goosefoot, pigweed) were used in prehistoric times for food in both the Old and New World; it is now widely regarded as a noxious weed. Domesticated in the Himalayan region, it is now cultivated in Nepal, northern India, northern Thailand and in the mountains of Java as a subsistence food crop. Elsewhere the leaves are widely eaten as a vegetable (Mastebroek *et al.*, 1996).

The pseudocereal *Chenopodium quinoa* (quinoa) is believed to have been domesticated in the Andes 3000-5000 years ago. High Andean and lowland cultivars occur in Chile to south of the Rio Bio and Rio Maule *ca.* 37°S. The seeds are used mainly for local consumption, especially in soups and porridge, although there is an increased international interest for their sale in health food shops.

The pericarps of the landraces and early cultivars contain up to 4% saponins, which give an unpalatable bitter taste to the seeds, and can be toxic unless removed by vigorous washing and polishing or by decortication, resulting in some nutritional loss. The seeds have a slightly nutty flavour, as well as having an exceptionally high nutritional value, the protein is particularly high in the essential amino-acids lysine, methionine and cystine. The small starch grains, 2-4 μm in diameter, give the starch a fatty taste, and is used to replace fats in diet products.

The seeds (achenes) from the local landraces are small, *ca.* 2 mm in diameter. Higher yielding cultivars with larger seeds, such as cv. 'Blanca de Junín', cv. 'Royal quinoa', have since been developed for the export market. Two sweet saponin-free cultivars, cv. 'Nariño' and cv. 'Sajama', have also been developed and are considered to be most suitable for human consumption (Brücher, 1989; National Research Council, 1989; Mastebroek *et al.*, 1996; Ruales, 1998).

Two other grain chenopods are cultivated on a small scale. *Chenopodium berlandieri* subsp. *nuttalliae*, syn. *C. nutalliae* (huauzontle) in Mexico and is now more important as a vegetable than a grain crop. The highly variable *C. pallidicaule* (cañihua, kaniqua, quananhua) is cultivated up to 4500 m in the marginal regions of the Andean plateaux of Peru and Bolivia. Possibly few other crop are as resistant to the combination of frost, drought, salty soils and pests, or are so easy to grow and require so little care in cultivation as cañihua. The plants are variable in their ripening and free shedding of their seeds (achenes), thus making harvesting difficult. The black or dark brown, *ca.* 1 mm in diameter seeds are rich in the essential amino-acids lycine, isoleucine and tryptophan and contain no saponins. The seeds are usually toasted and ground to a flour and used with wheat flour in baking, or made into a hot beverage (Gade, 1970; Brücher, 1989; National Research Council, 1989; Mastebroek, *et al.*, 1996).

There are two species of cultivated buckwheat originating in Central Asia. *Fagopyrum esculentum* subsp. *esculentum* (common buckwheat) was probably domesticated in China from the wild and readily seed shattering *F. esculentum* subsp. *ancestralis* and *F. tataricum* (Tatary buckwheat) from Sichuan and Yunnan provinces respectively; both probably have *F. dibotrys* in their ancestry. The former requires a longer growing season, but has the higher yield.

The cultivation of buckwheats were first described from China in the 5th century. They are important subsistence and cash crops in the Himalayan region, from northern India to Mongolia, China, Korea and Japan, as well as parts of the former USSR. Common buckwheat was introduced into Europe and North America where, during the 17th century it was an important food crop. The increased use of artificial fertilisers in the Western World during the 20th century led to its decline in favour of higher yielding cereal crops. The crop also suffered from being considered a food identified with poverty and peasant culture. Not-withstanding, buckwheat remains an important crop for the health food market. It is eaten in food preparations similar to those for wheat. Buckwheat groats, i.e. flaked grain, for example, are eaten as a porridge in both Europe and North America.

The high biological value of the protein is attributable to the high essential amino-acid content, especially lysine. The grains require decortication before use, the hulls containing the phytophotodermatitis-active *fagopyrine* and *filloerythrine*, which can cause cutaneous eruptions and, in severe cases, toxicosis (Grubben and Siemonsma, 1996; Grubben *et al.*, 1996; Aubrecht *et al.*, 1998; Ohnishi, 1998).

3.   FRUITS AND NUTS

Botanically a *fruit* may be defined as the structure that develops from the ovary wall (pericarp) as the enclosed seed or seeds mature. A fruit may be succulent (drupes, berries, etc.) or dry and either dehiscent, indehiscent or schizocarpic, their origin monocarpic or polycarpic. A berry may be simple or compound, true or false, the latter includes the apple, of which the swollen receptacle is eaten. Botanically a *nut* is a dry indehiscent fruit that is usually shed as a 1-seeded unit formed from more than one carpel, but with only one seed developing, the remainder aborting. The pericarp is usually lignified and is often partially or completely surrounded by a '*cupule*' formed from fused extensions of the pedicel (Tootill, 1984).

In lay terminology fruit is applied to those botanical fruits that may be eaten as an appetiser, as a dessert or out of the hand. Other botanical fruits, e.g. aubergines, cucumbers, gourds, marrows, pumpkins, olives and tomatoes, are eaten as 'vegetables'; pickled cucumbers and pickled olives, for example, may also be eaten as a 'condiment'. The seeds and sometimes pods of the dry, dehiscent, monocarpic fruits (legume) of the Leguminosae are also treated as a vegetable, e.g. beans, and even some as 'nuts', e.g. groundnuts. The term 'nut' include a range of fruits, seeds and tubers (Montagné, 1977; Ensminger *et al.*, 1994).

While the botanical distinctions are clear and the lay terminology relatively easy to understand, food legislation terminology can sometimes be somewhat bizarre. For example, in its wisdom the EU has ordained that the carrot, which the Portuguese use to make carrot jam, should be designated a fruit in order to suit the regulations regarding jams and preserves!

3.1   Edible Fruits

In general edible fresh fruits are succulent structures with pleasant aromas and flavours, often perishable, with poor transporting characteristics and a short shelf life. While the production of fresh fruit for local consumption may be satisfactorily spread throughout the growing season, commercial production usually requires a relatively short harvest season. Depending on their ripening characteristics, the disadvantages outlined above may restrict their availability as fresh fruit for overseas markets without careful temperature control, disinfestation, post-harvest ripening, etc.

Two classes of fruit are recognised, depending on the stage at which they attain optimum ripeness, acceptable colour, flavour, aroma, texture, etc.: (1) *Climacteric fruits* are picked when mature and ripen after picking. Examples include *Carica papaya* (pawpaw), *Malus* × *domestica* (cultivated apple), *Mangifera indica* (mango) and *Musa* spp. (banana); and (2) *Non-climacteric fruits* do not ripen significantly after being detached from the parent plant and have to be harvested when ripe. They include *Citrus maxima* (pummelo, shaddock), *C. reticulata* (mandarin) and *Fragaria* × *ananassa* (cultivated strawberry).

Some fruits, such as most citrus fruits can be stored for a considerable time in the fresh state. They are high in vitamin C although the level declines gradually during storage, with the rate of decline depending on the temperature. An unusual exception is *Adansonia digitata* (baobab), where the already high vitamin C content (six times that of citrus) of the pith surrounding the seeds actually increases during storage.

The commercial processing of fruit may also be undertaken, either to obtain added value as juices and jams or to utilise otherwise unsaleable fruits that are perishable and surplus to requirements or small and ill-shaped. Processing methods include: (1) *Fermentation* to produce wine and vinegar; (2) *Pickling* of green mature fruit in brine, e.g. olives, or vinegar, e.g. gherkins, with or without bacterial fermentation; (3) *Drying and dehydration*, e.g. dried sliced banana, prunes and raisins; (4) *Juice extraction*, e.g. citrus and apple juices; (5) *Glacéing* by 'pickling' whole fruit in a sugar syrup; (6) *Preserving* as jams, jellies, marmalades and conserves.

The terminology for some of these products can be confusing. *Jams* are preparations of whole fruit boiled to a pulp with sugar. The legal definition in France is: 'Products constituted solely of refined or crystallised sugar and fresh fruits or juice of fresh fruits, or preserved in some way other than by drying' (Degree of 25 September 1925). They must contain a maximum of 40% moisture and of the resulting 60% plus dry extract 55% must be sugar (fruit already contain 5-7% sugar). *Jellies* are preserves made from fruit juice and sugar and set with pectin. They are used as a jam. The term is also applied to a fruit-flavoured dessert set with gelatine. In the UK and North America the word *marmalade* applies to a clear, gelled product in which citrus fruits and quinces, together with their pulp and peel or skin, are suspended. The word *marmelade* is French and is believed to be from the Portuguese *marmelada* for quince jam, from *marmelo* for *Cydonia oblonga* (quince). In France marmalade applies to any fruit stewed for a very long time until it has been reduced to a thick purée. They must conform to the same definition as jams, although the dry extract can fall as low as 55% and the use of brown sugar, or sugar that has already been used for crystallising fruit, is permitted. *Conserves* are jams made with a mixture of fruit, with or without nuts and raisons. It is also applied to fruits preserved with sugar; and (7) *Canning* whole or sliced fruit in a light syrup, e.g. pineapple. At the domestic level bottling offers a similar process (Montagné, 1977; FAO, 1988a; Verheij and Coronel, 1991; Ensminger *et al.*, 1994; Mabberley, 1997).

Popular wild fruit sources in the temperate regions include the aggregated fruitlets of *Rubus* spp. (blackberries and raspberries) and berries of *Vaccinium* spp, (blaeberries, etc.), some of which may also be cultivated. In the tropics wild fruit supplement largely starchy diets based on subsistence crops or provide food when crops fail or during pre-harvest food shortages. For examples see Mabey (1972); FAO (1982, 1983, 1984, 1986, 1988a), Burkill (1985-1997 and in press) and Verheij and Coronel (1991). Okafor (1973, 1980a, b) describes how many fruits harvested from wild trees can be improved in yield and quality by using grafts from elite trees.

3.2   Edible Nuts

True nuts include the fruits of species of *Corylus* (hazelnuts), *Fagus* (beechnuts) and *Quercus* (acorns). The term is also loosely applied to any woody fruit or seed, including the drupes of *Juglans* spp. (walnut), also the seeds of *Bertholletia excelsa* (Brazil nut) and *Arachis hypogea* (peanut). It is even incorrectly applied in the vernacular names for non-fruiting bodies, such as the root tubers of *Conopodium majus* (pignut or earthnut) and *Cyperus esculentus* (tiger nut, earth almond) where, in the latter example, the root tubers are eaten as a form of 'dessert nut'. A further complication is that the seeds of *Helianthus annuus* (sunflower) and several members of the Cucurbitaceae are referred to as 'dessert nuts' and are so included in the literature on edible nuts, yet they are always referred to both botanically and in the vernacular as seeds.

Previous authors have also had problems in defining an edible 'nut' and have provided their own apparently arbitrary limits. Menninger (1977) provides the widest definition, defining a nut *'as any hard-shelled fruit or seed of which the kernel is eaten by mankind.'* This definition is so broad that even grasses and a number of herbaceous species have been included, although the former are generally regarded as cereals and the latter more appropriately listed under edible seeds. More succinctly, Woodroof (1979) states *'Tree nuts are the edible kernels of seeds of trees'*. Moreover, all the authors consulted apart from Verheij and Coronel (1991) have accepted the popular or everyday use of the term 'nut', including its conservative use by Howes (1948) to include *'any seed or fruit consisting of an edible, usually oleaginous kernel, surrounded by a hard or brittle shell'*. Interestingly, the major authors consulted, Howes (1948), Menninger (1977) and Rosengarten (1984) have, like the present author in Wickens (1995a), all made their own interpretation as what to include or exclude as an edible nut. Verheij and Coronel (1991), however, have wisely declared *'it is not possible to define the edible fruits and nuts in such a way that clearly sets them apart from species in other commodity groups'*.

Most edible 'nuts' contain concentrated reserves for future generations of plants and thereby provide valuable sources of energy, protein, oils, minerals and vitamins suitable for human consumption. These include: *Canarium* spp. (pili), *Corylus* spp. (cob), *Juglans* spp. (walnuts) and *Prunus dulcis* (almond). Instead of protein other nuts may have their food reserves in the form of starch, including acorns, chestnuts, *Cordeauxia edulis* (ye'eb), *Nelumbo* spp. (lotus seeds), *Trapa* spp. (water chestnuts), and *Vigna subterranea* (Bambara groundnuts). Among the gymnosperms species of *Pinus* (pine nuts) are rich in protein while other members, such as *Araucaria* spp. (Chile nuts) and *Ginkgo biloba* (ginkgo tree) have starchy food reserves (Melville, 1947; Wickens, 1995a).

Nuts have a reputation for being indigestible, especially if eaten in large quantities or are poorly masticated. In general nuts are a highly concentrated food and low in water and fibre content, requiring thorough mastication if they are to be properly digested. From a dietary point of view they are preferably eaten with other foods. Nuts have the additional value in that their fats are, in the main, highly unsaturated and consequently do not raise the blood cholesterol. Peanuts, *Caryocar nuciferum* (souari nuts), *Glycine max* (soynut) and sunflower seeds are specifically mentioned in this context (Howes, 1948; Rosengarten, 1984).

Most nuts now sold on the international market are grown in plantations, the major exception being the Brazil nut, which are still largely harvested from wild trees. Pure stands of Brazil nuts outside the forest may fail to yield because their natural pollinators are inhabitants of the forest. Furthermore, it is an obligate outcrosser with significant differences between donor and receptor compatibilities (Mori and Prance, 1990).

The post-harvest treatment of nuts is mainly aimed at reducing the moisture content and liability to rancidity. Nuts such as Brazil nuts may be preserved in-shell, others, such as macadamia and pili nuts, are decorticated. Others, such as chestnuts may be preserved under refrigeration or, like walnuts, either decorticated or pickled. The processing of cashew nuts, from *Anacardium occidentale*, involves roasting to remove the caustic **cardol** and **anacardic acid** (**cashew-nutshell liquid** or **CNSL**) in the pericarp before removing the kernel. The CNSL by-product is used industrially in brake-linings, plastic resins, etc. (Wickens, 1996; Mabberley, 1997).

4.   VEGETABLES

A vegetable is here defined as the edible part of a wild or cultivated plant which is traditionally not classified as a grain, fruit or nut and is eaten either cooked or raw with meat or other articles of food. Within the angiosperms vegetables are represented by a wide range of morphological forms. Examples include: bulbs - *Allium cepa* (onion); corms - *Colocasia esculenta* (taro), *Eleocharis dulcis* (Chinese water chestnut); swollen taproots - *Daucus carota* (carrot); rhizomes - *Canna indica* (Queensland arrowroot), *Curcuma longa* (tumeric); root tubers - *Ipomoea batatas* (sweet potato), *Manihot esculenta* (cassava); stem tubers - *Solanum tuberosum* (potato), *Cyperus esculentus* (tiger nut); swollen hypocotyle, i.e. portion below the cotyledon - *Raphanus sativus* (radish); hypocotyle plus cotyledon - *Lepidium sativum* (garden cress), *Sinapis alba* (white mustard); stem - *Asparagus officinalis* (asparagus); petiole - *Apium graveolens* var. *dulce* (celery); leaves - *Spinacea oleracea* (spinach); peduncles - *Arenga pinnata* (sugar palm), *Borassus flabellifer* (toddy palm); floral bracts - *Cynara scolymus* (globe artichoke); calyx - *Hibiscus sabdariffa* (roselle); inflorescence - *Brassica oleracea* Botrytis Group (cauliflower); fruit - *Phaseolus coccineus* (runner beans), *Pisum sativum* (mangetout peas), *Cucurbita pepo* (marrows), *Lycopersicon esculentum* (tomato); seeds - *Phaseolus lunatus* (Lima bean) (Brouk, 1976; Montagné, 1977; Flach and Rumawas, 1996; Vaughan and Geissler, 1997). The algae and fungi used as vegetables are discussed in Chapters 18 and 19 respectively.

With the development of cheap international transport and improved technology in handling and preservation, there is an increasing demand for both exotic and out-of-season fruits and vegetables, particularly among the developed countries. As a result crops normally grown in the temperate summer can now be supplied in winter by growers in tropical climes. While the USA is able to meet much of its demands internally, other developed countries with high labour costs can often obtain labour-intensive crops more cheaply from the developing countries, e.g. Europe importing Brussels sprouts from Kenya. Transportation costs too are increasingly competitive. For example, it costs more per kg to send a lorry load of apples from Taunton to London than to transport by air from Mombassa to London. See Robbins (1995) for further details.

4.1   Legumes

Legumes are second only to the cereals in importance for food. The *legume*, which is also the botanical term for the fruit, as distinct from pulse or seed, is from the French *légume*, derived from the Latin *legumen*, pulse or bean. The term *pulse*, or *edible grain legume*, is from the Middle English *pols, puls*, from Old French *po(o)ls*, porridge, from the Latin *puls*, pottage made of meal and pulse, possibly from the Greek *poltos*, porridge (Long, 1994).

Despite their value as food, the pulses are unfortunately largely regarded as non-prestigious foods in both the New World and Africa south of the Sahara, their consumption being related to the social status of the consumer, with consumption being highest mainly among the poorer classes. As a crop they have the disadvantage of often being subject to insect attack and, in the case of groundnuts in particular, being susceptible to the mould *Aspergillus flavus*, producing the highly carcinogenic toxin *aflatoxin*.

Pulses may be utilised in a number of ways: (1) As mature dried seeds, which will store well provided they are protected against insects, although during prolonged storage there will be some decline in nutritional value and tenderness when cooked. They may used whole, like *Phaseolus lunatus* (butter or Lima bean) and *P. vulgaris* (haricot or kidney, bean), or split as dhal, like *Cajanus cajan* (pigeon pea) and *Pisum sativum* (garden pea); (2) They may be consumed as green, fully mature or immature seed. The garden pea may be eaten either mature or immature, the immature seeds are also popular when frozen; the Lima bean too may be eaten at the green mature stage. The fresh green beans of pigeon pea are a popular vegetable in the Caribbean region, while *Vigna subterranea* (Bambara groundnut or ground bean) may be consumed either as green mature seeds or boiled in their pods; (3) Immature pods that remain fleshy for 2-3 weeks after setting may be picked green and used as a vegetable. These include haricot bean, *Phaseolus coccineus* (runner bean), *Pisum sativum* var. *macrocarpum* (mange-tout and sugar peas), also *Canavalia ensiformis* (jack bean) and *C. gladiatas* (sword bean). Such immature fruits contain lower protein but are relatively richer in vitamins and soluble carbohydrates; (4) Germinated seeds and young seedlings are used as a fresh vegetable, especially in the

Orient, with both the field crop and stored seeds ensuring a constant supply throughout the year. Highly nutritious, 48 h after germination the vitamin C content rises from negligible levels in the seed, to 12 mg 100 g^{-1} seed; the riboflavin and niacin contents also increase significantly and the insoluble carbohydrates are mobilised. This mobilisation of the cotyledonary reserves improves the general digestibility, especially for soya bean, whose sprouts may be used either as a salad or a vegetable. Historically germinated seeds were used on sailing ships and more recently in prisoner of war camps as a source of vitamin C against scurvy. *Vigna radiata* (green gram) is the most widely used pulse for bean sprouts; *Vicia faba* (broad bean), *Cicer arietinum* (chickpea), etc. may also be used; (5) Fermented pulse products are particularly popular in Southeast Asia, where a range of micro-organisms, including *Aspergillus oryzae, Rhizopus oryzae* and *Bacillus subtilis* are used. Fermentation certainly improves both the digestibility and palatability of soya bean; and (6) The separation of readily digestible seed protein from the less readily assimilated protein is a common practice in the Far East, especially for soya bean (FAO, 1988a; Smartt, 1990).

Unfortunately various deleterious substances are present in many of the edible legumes, e.g. antivitamin A, B$_{12}$, D and E factors in soya beans. Other antinutritional factors among the legumes include cyanogenic glucosides, haemagglutinins, trypsin inhibitors, lathrogenic neurotoxins, goitrogens, metal binders, and favism (Chapter 8). *Favism* is a hereditary haemolytic anaemia present in people of African, Mediterranean and Asian ancestry susceptible to eating broad beans; in extreme cases even inhaling the pollen can be damaging. This is due to the absence in the body of the essential enzyme glucose-6-phosphate dehydrogenase, which is believed to be activated by the plant glucosides *vicine, convicine* and the amino-acid *3,4-dihydroxyphenylalanine (Dopa)*. With careful selection it may be possible to reduce the presence of these compounds in *Vicia faba*. For the present those susceptible to favism must avoid eating broad beans. Fortunately, most of the antinutrients mentioned above can usually be overcome by leaching and cooking. Fermentation or sprouting can also be used for trypsin inhibitors and for improving protein digestibility.

Soya bean is unique among the grain legumes in being primarily an industrial crop, requiring processing before the seed can be exploited. One method of separating readily digestible seed protein from the less readily assimilated protein involves fermentation with *Rhizopus oryzae* or bacteria to produce soya bean 'milk', 'cheese', 'curd' or 'tofu', or soya sauce. The fermentation process also neutralises any anti-nutritional factors, which would make the otherwise untreated utilisation of soya bean difficult.

The soya bean milk can be used as a milk substitute by those who are allergic to cow's milk. A useful protein extract may also be obtained from soya bean meal after oil extraction, which can be either used alone or further processed to produce *textured vegetable protein (TVP)*. The TVP may then be used to enrich cereal products, as meat extenders and in the preparation of low-fat cream substitutes. There is certainly a potential for introducing this technology to other legumes where the utilisation of potentially valuable protein is inhibited by the presence of toxic materials such as for example, alkaloids and essential amino-acid analogues. The alternative possibility of prolonged leaching or cooking in order to render seeds safe for local consumption is an unacceptable requirement in countries such as Africa where potable water and fuelwood are in short supply (Smartt, 1990; Ensminger *et al.*, 1994).

4.2   Green Vegetables

Green vegetables refer to leafy plants that are usually eaten either cooked, e.g. *Brassica oleracea* Capitata Group (cabbage), or raw in salads, e.g. *Lactuca sativa* (lettuce). Because of their high moisture content green vegetables are perishable and have a short shelf life unless preserved. Wilting by exposure to the sun can result in high losses in both ascorbic acid and carotenoids.

In practice there is no clear distinction between green cooked and salad vegetables, both can usually be eaten either cooked or raw. In French cooking terminology *salads* refer to dishes seasoned with oil, vinegar, pepper, etc. and made up of herbs, plants, vegetables (plain salads, cooked or raw), or with eggs, meat or fish (mixed salads) (Montagné, 1977; FAO, 1988a).

Little use is made of wild vegetable sources by the developed countries of the temperate regions. In the UK members of the Chenopodiaceae such as *Atriplex* spp. (orache), *Beta vulgaris* subsp. *maritima* (sea spinach, seakale-beet) and *Salicornia europaea* (common glasswort, marsh or sea samphire), not to be confused with the less popular *Crithmum maritimum* (rock samphire), may be eaten. See Mabey (1972, 1996) and Vickery (1995) for further examples. Rare examples of edible leaves from temperate trees and shrubs are *Vitis vinifera* (grape vine) in Europe and *Rubus spectabilis* (salmonberry) by the native Americans of north-west America, while in north-eastern China the leaves of *Morus alba*, *Salix* spp. and *Ulmus pumila* eaten.

A greater use is made of wild sources in the tropics, especially by the developing countries. Weeds such as *Cleome gynandra*, syn. *Gynandropis gynandra*, and *Portulaca oleraceae* (purslane) may even be allowed to grow among the cultivated crops and harvested when required. The leaves of many shrub and tree species may also be eaten, including the apical buds of a number of palms as palm cabbage and baobab leaves (Baranov, 1962; Hedrick, 1972).

4.3   Root Vegetables

Root vegetables are those plants whose roots, tubers or stolons are used for food. The general trend in the developing countries, and for Africa in particular, is towards a diminishing consumption of root crops and of *Musa × paradisiaca* (plantain). This decline is attributed to the rapidly increasing urban populations and the associated demand for convenience foods, the latter often being provided from non-indigenous sources and even encouraged by food pricing policies and cereal subsidies. The plantain, although a fruit, is included here because its culinary treatment is similar to that of root vegetables.

Root crops are capable of producing double the dry matter yield of cereals. Whereas the moisture content of cereals is *ca.* 12% so that they can be readily stored, root crops are bulky, containing 60-70% moisture. They can only be reasonably stored if, like certain *Dioscorea* spp. (yams) and the Irish potato, they have a dormant period although some crops, such as *Manihot esculenta* (cassava), can be left in the ground until required. Once roots have been harvested they become highly perishable and high losses can be incurred during transport and marketing as a result of poor handling and inadequate transport infra-structures. Some root crops, such as potatoes and yams, contain considerable amounts of vitamin C, which is reduced during storage and lost completely when dried.

The processing of root crops can transform a perishable product into a more stable or preserved form, thereby improving storage security and food distribution, as well as making the product more acceptable to the consumer, easier to use and, in the case of high cyanide cultivars of cassava, safer for consumption. The traditional processing of yams, *Ipomoea batatas* (sweet potato), *Colocasia esculenta* (cocoyam, taro) and low cyanide cassava, for example, is often no more complicated than peeling followed by boiling or roasting. The cleaned and grated cassava roots may also be fermented in heavily weighted jute sacks to expel excessive moisture. The fermentation increases the protein content and the resulting product, known as **gari**, may then be stored for some time. Boiled roots as well as plantains, are often traditionally pounded in mortars to produce a glutinous mass known in West Africa as *fufu*. Such roots may also be preserved by peeling, chipping or slicing, par-boiling and sun-drying. Unfortunately, because roots are often irregularly shaped, peeling is a difficult process to mechanise if wastage is to be avoided (FAO, 1988a; Poulter, 1988).

4.4 Other Vegetables

Edible bamboo shoots are an important vegetable in tropical Asia. They represent the new growth from the rhizome apices and consist of young culms with their internodes protected by sheaths. After removal of the sheaths the sectioned or shredded shoots are cooked in boiling water. Although a number of species produce edible shoots, the preferred species of south-east Asia are *Dendrocalamus asper*, *Gigantochloa albociliata*, *G. levis* and *Thyrostachys siamensis*, and in China *Bambusa oldhamii*, *Dendrocalamus latiflorus* and *Phyllostachys pubescens*; the latter species is regarded by some as a synonym of *P. edulis*. China and Japan are the main export source of canned shoots, which are obtained from *Phyllostachys edulis* (Brücher, 1989; Dransfield and Widjaja, 1995; Mabberley, 1997).

The pith from the felled trunk of the palm *Metroxylon sagu* (sago palm), native to Indonesia, is the source of **sago**. The grated pith is kneaded over a strainer, through which the starchy sago passes and settles in trays, which are then heated in rotating drums where the sago is rolled into small balls known as **pearl sago** or slightly larger **bullet sago**. Containing *ca.* 90% starch, sago is widely eaten in the eastern Pacific countries. The EU health regulations only permit sago that has been recleaned in Europe to be accepted by the retail industry, where it is used to make puddings, as a sauce thickener, and as a textile stiffener (Robbins, 1996).

The Cycadophyta are also an important source of edible starch. The pithy stems of *Cycas circinalis* (sago palm) from tropical Asia, *C. beddomei* from India, *C. revoluta* from Japan and *C. media* from Australia, all yield edible starch. The boiled seeds of the latter species are a staple diet of the Australian Aborigines of Arnhemland. In southern Africa the pithy stems of *Encephalartos* are also a source of an edible starch, as are the seeds from the enormous, up to 92 cm long strobules of *E. caffer* (hottentot bread-fruit, kaffir bread). Similarly, the starchy seeds of *Dioon edule* from Mexico yield a flour, which is used to make tortilla pancakes. Also from Central America the rhizomes of *Zamia integrifolia*, syn. *Z. floridana* are the source of a starch known as Florida arrowroot (Brouk, 1975; Lazarides and Hince, 1993; Mabberley, 1997).

5.   FATS AND OILS

Fats and oils are the storage triacylglycerols of plants and are especially abundant in some seeds. **Fats** are solid or semisolid at 20°C, whereas **oils** are liquid at that temperature. The predominant fatty acids of the former are **palmitic** or **stearic acids**, and of the latter unsaturated **linolenic** and **oleic acids**. Linolenic acid is of particular importance nutritionally since it cannot be synthesised in the human body; particularly rich sources are corn (maize), cottonseed, soya bean and sunflower oils.

Vegetable fats and oils are becoming increasingly important in the global economy. This is due to their ease of production, plus reasons of health and the possibilities for increased future production. The tropical palms especially are a major untapped resource, the Amazon basin being particularly important in this respect. On the other hand, animal fats and oils are decreasing in popularity, especially among the more health conscious developed countries (Hill, 1952; Tootill, 1984; Balick, 1988; Ensminger *et al.*, 1994).

5.1   Vegetable Fats

The more important commercial sources of vegetable fats are: (1) **Cocoa butter** from the beans (seeds) of *Theobroma cacao* (cacao). It is an important fat for the chocolate industry where the sweet or eating chocolates have a high cocoa butter content than the plain or cooking chocolate; (2) *Shea butter* is obtained from the crushed kernels of mainly wild trees of *Vitellaria paradoxa*, syn. *Butyrospermum parkii*. It is an important substitute for cocoa butter in the chocolate industry; small quantities are also used for soap and pharmaceutical products, while in the producing countries it is an important oil for cooking; (3) *Brazilian palm oils* which, despite their name are solid vegetable fats, are obtained from both the kernels and mesocarps of *Astrocaryum tucuma* (tucuma) and *A. vulgare*. Other palms with very thin and mealy mesocarps only yield palm kernel oil. These include *Astrocaryum murumuru* (murumuru palm), *Oenocarpus bataua*, syn. *Jessenia bataua* (pataua), *O. distichus* (bacaba palm), *Orbignya oleifera* (cohune palm) whose the kernels contain 63-70% oil, *O. phalerata*, syn. *O. martiana* (babassu palm) and *Syagrus coronata* (nicuri or ouricuri palm); (4) *Dipterocarp butters* are expressed from the solitary seeds of the dry indehiscent fruits of various members of the Dipterocarpaceae. Species of *Madhuca* are the source of the original illipe nuts used for margarine, cocoa butter substitute, etc. as well for local use. They include *phulwara butter* from *M. butracea*, *mowra butter* from *M. latifolia*, *illepe butter* from *M. longifolia* (mahua) and *catian butter* from *M. motleyana*. The kernels of *Shorea seminis* and *S. splendida*, syn. *S. seminis* var. *martiana* yield the *Borneo tallow green butter* of Indonesia and from Borneo *S. palembanica* is the source of *Borneo tallow*. They too are used as substitutes for cocoa butter (Brouk, 1976; Balick, 1988; Robbins, 1995; Mabberley, 1997).

5.2   Vegetable Oils

The edible vegetable oils are mainly used in the food industry for margarine, cooking and salad oils, etc. World trade in edible vegetable oils is shown in Table 9.

TABLE 9. Global production of raw materials and import/ export of edible vegetable oils for 1992 (FAO, 1993a, 1994)

Oil	Production (tonnes)	Imports (tonnes)	Exports (tonnes)
palm oil	12, 836, 386	7, 778, 668	7, 881, 699
soya bean oil	113, 682, 000	4, 114, 085	4, 208, 356
sunflower seed oil	21, 979, 000	2. 500, 915	2, 495, 099
rape + mustard seed oils	26, 239, 000	1, 945, 106	1, 934, 961
coconut oil	43, 385, 000	1, 486, 992	1, 562, 971
palm kernel oil	3, 799, 915	829, 312	786, 217
olive oil	1, 961, 000	685, 448	717, 400
maize oil	527, 715, 000	444, 457	475, 450
groundnut oil	24, 601, 000	334, 132	320, 752
cottonseed oil	33, 296, 000	289, 253	289, 995

Palm, soya, sunflower and corn oils are interchangeable for many of these uses, with their relative global availability determining their cost and consequent utilisation.

The refinement and thickening of the fatty oils used in food production involves a process known as **hydrogenation** where, in the presence of a catalyst such as nickel, hydrogen is added under high pressure. For economic oilseed production it is essential that there should be markets for both the oil and the residual oilseed cakes; fortunately the present long-term market prospects favours both commodities (Sharp, 1990; Smartt, 1990).

The more important vegetable oils are: (1) **Palm oil** and **palm kernel oil** from *Elaeis guineensis* (African oil palm). They represent *ca.* 16% of global vegetable oil production and 40% of the world trade in vegetable oils. They are rich in saturated fatty acids, especially lauric acid, and are widely used in margarine and cooking oils, although considered by some as a potential health risk. The development of genetically engineered *Brassica napus* cv. 'Canola' (from Can-adian oil-seed) with seeds containing 40% **canola oil** rich in lauric acid and low in erucic acid and glucosinolates, now poses a threat to the oil palm and palm-kernel oil market; (2) **Soya bean oil** from the seeds of soya contain 18-22% of an edible, semi-drying oil, which is mainly used in margarine and as a salad and cooking oil; (3) **Sunflower seed oil** is obtained from the light brown, striped sunflower seeds, the oil from the dark brown seeds being used as an industrial oil. The largely polyunsaturated oil is variable in composition, usually with 58-67% linoleic acid, although this is sometimes as low as 20%. It is also rich in vitamin E. The oil is used mainly for cooking and margarine; (4) **Rape** and **mustard seed oils** are obtained *Brassica napus* (oilseed rape) and *B. juncea* (brown or oriental mustard) respectively. Rapeseed has a higher oil content than soya bean, 45% compared to 20% for soya, and is mainly used by the food industry for margarine and salad oils. The linolenic, oleic and erucic acid content were formerly too high and the linoleic acid too low for good quality food oil. The rapeseed cake also had a lower feeding value than soya cake due to the presence of such antinutritional factors as glucosinolates and high fibre content. Plant breeders have aimed at producing seeds with no erucic acid or glucosinolates, low in fibre, and with a 40:3 linoleic:linolenic acid relationship; the first two objections have already been achieved, the third is progressing satisfactorily. Nevertheless, oil millers still prefer soya bean oil because of its better crystallisation properties due to the lower C_{18} fatty acids content, and use only a proportion of rape seed oil for margarine. Breeding of improved rape is hampered by the ease with which cross-fertilisation can take place, creating cross-breeding problems between old and new cultivars.

Brown or oriental mustard is also grown as an oilseed crop in the Indian subcontinent, China and southern Ukraine. In drier climates the high levels of glucosinolates (*ca.* 200 μmol g^{-1}) of oil-extracted meal plus significant levels of erucic acid (25-45%) renders the seed oil unsuitable as an edible oil in some countries, although strains with low levels of glucosinolates and almost erucic acid-free are now known; (5) *Coconut oil* is usually extracted from the dried endosperm, known as *copra*, of the fruits of *Cocos nucifera*. Extraction in the country of origin from fresh endosperm is now increasing. The copra contains 60-68% oil, with an average extraction rate of 64%. Below 24°C it is a white to yellowish solid fat, forming a colourless to pale brownish yellow oil at higher temperatures. The high lauric and myristic acid content gives the oil a high saponification value, while low unsaturated acids provides resistance to oxidative rancidity and enhances the keeping quality for baked foods and fillings; the oil is also widely used in the manufacture of margarine; (6) *Olive oil* is extracted from the fleshy mesocarp of *Olea europaea* subsp. *europaea* (olive). It is almost unique among vegetable oils in that it may be consumed without any refining treatment. The oil is fragrant and with a delicate flavour, high in nutrients and health value. In recent years the dilution of poorer quality olive oil with the cheaper hazelnut oil has been causing problems since the two oils have rather similar chemical contents which, at present, can only be identified with difficulty; (7) *Maize oil* from the embryos of maize kernels contains 4.2-5.6% of a semi-drying oil, the embryo being separated following milling (see Chapter 8). The oil, which is rich in essential fatty acids, is refined and used for cooking and salad oils; (8) *Groundnut oil* is extracted by expression and solvent extraction from groundnut seeds. The golden-yellow, non-drying oil may be used without further processing, or it may be refined and bleached to an odourless, pale-coloured liquid. It is used for salad and cooking oils and in margarine manufacture; (9) *Cottonseed oil* is a by-product of the cotton industry, the edible, semi-drying oil is obtained from the cleaned and usually decorticated seeds of *Gossypium* spp. The seeds contain 18-24% oil, which is extracted using hydraulic or screw presses. The toxic, bitter tasting, phenolic pigment *gossypol* (a possible oral male contraceptive) present in the oil is rendered harmless by heating and solvent extraction. Breeding for low gossypol cultivars has unfortunately increased susceptibility to insect pests; and (10) *Other edible oils* include the high quality semidrying *sesame* or *gingelly oil* from the seeds of *Sesamum orientale*, syn. *S. indicum* (sesame, simsim) and is mainly used in the countries of origin for cooking (in the developed countries the seeds are often used to decorate breads and cakes). Sesame oil contains the phenolic substances *sesamol*, *sesamolin* and *sesamin*, which is sometimes used as an adulterant of olive oil. *Argan oil* is extracted from kernels of the fruit of the evergreen tree *Argania spinosa*, syn. *A. sideroxylon* (argan), a monotypic Moroccan endemic. The unsaturated fatty acid fraction resembles that of olive oil but with a higher linoleic and lower oleic acid content. A gave disadvantage for commercial production is that the kernels are extremely hard and difficult to crack, also the oil has a strong flavour, although it is actually preferred by Moroccans to olive oil. The oil was exported to Europe during the 18th century but trade ceased in favour of the milder olive oil. *Asparagus pea oil* from *Psophocarpus tetragonolobus* (asparagus pea, winged bean) is basically similar that of soya bean oil and could be developed as a substitute crop, except for the fact that it would require sound economic establishment and marketing which, at present, are not forthcoming. *Tarwi oil* is from the seeds of the Andean *Lupinus mutabilis* (tarwi) and contains 14-24% of a light coloured oil which is roughly equivalent to groundnut oil, being relatively rich in essential fatty acids, including linoleic acid. **Avocado oil** is usually obtained from surplus fruits of *Persea americana* (avocado). Over-production and the relatively short shelf life of the fruits, led to it being utilised to produce an edible oil. The oil is similar but slightly inferior to olive oil. It can also be used in cosmetics and as a sun-screen oil. At present there is only a limited demand for this oil on the international market (Hill, 1952; Dogged, 1970; Pothook and Saxon, 1979; Göhl, 1981; Languor and Hill, 1982; Rexene and Muncie, 1984; Morton and Voss, 1987; Purseglove, 1987; Kiritsakis and Markakis, 1991; Hemmingway, 1995; Robbins, 1995; Wickens, 1995a, 1998; Nerd *et al.*, 1998).

6.   BEVERAGES

Beverages are any liquid used or prepared for drinking. They may be classified as follows: (1) *Aromatic or stimulating infusions*, of which the major commercial beverages are: (a) *China tea* from *Camellia sinensis* var. *sinensis* and *Assam tea* from var. *assamica*. For black tea, which forms the bulk of the world market, the withered young leaves undergo an enzymatic fermentation of the polyphenols and then dried. The polyphenols, derivatives of gallic acid and catechin, but not tannins in the accepted sense, are oxidised to produce *o*-quinones, which are polymerized to produce the astringent condensation products that are partially extracted in brewing. For green tea the enzyme is destroyed by steaming or by rapid drying before fermentation can take place. Oolong tea from Taiwan is semi-fermented; (b) *Coffee*, the three major sources are *Arabian coffee* from *Coffea arabica*, **robusta** or *Congo coffee* from *C. canephora*, which is the preferred bean for instant coffee, and the rather bitter *Abeokuta* or *Liberian coffee* from *C. liberica*. The harvested berries (beans) are first pulped to remove the skin, fermented by natural enzymes to remove the mucilaginous mesocarp adhering to the coffee parchment enveloping the bean, and then marketed; (c) *Cocoa* is prepared from the fermented seeds of *Theobroma cacao*. It is the fermentation by yeasts and other microfungi plus lactic and ethanoic acid bacteria that develop the essential oil responsible for its peculiar aroma, conversion of the bitter-tasting *cacaoal* compounds and the liberation of the alkaloid *theobromine* to give cocoa its tonic and stimulating qualities; (d) Of more local importance are the herbal teas, such as *maté*, also known as *Brazilian* or *Paraguay tea*, a hot water infusion of the dried leaves of *Ilex paraguariensis* (yerba), and *rooibos tea* from *Aspalanthus linearis*. There are also a number of roasted grain drinks; (2) *Fruit juices* include lemonade, fruit-flavoured drinks and kerkedeh from the ripe accrescent calyx of *Hibiscus sabdariffa* (roselle, Jamaica sorrel) and is a drink rich in vitamin C; (3) *Fermented beverages* include beer, wine and spirits (see Section 6.1). Such drinks are generally regarded as being aesthetically pleasing and stimulating social drinks when taken in moderation but can become an addictive drug when taken in excess. The beers can, however, provide an increased protein and vitamin content to the diet; (4) *Distilled liquors,* such as whisky, gin, brandy and liqueurs (see Section 6.2) also need to be taken in moderation; (5) *Soft drinks or carbonated beverages* are non-alcoholic and fizzy drinks and usually contain soda water plus up to 0.02% by weight of caffeine. *Caffeine* occurs naturally in such plant products as coffee, tea, cocoa, kola nuts from *Cola* spp., maté, and guaraná paste from the seeds of *Paullinia cupana*. It is obtained commercially as a by-product from the manufacture of decaffeinated coffee, by extraction from waste coffee beans and tea leaves, or by the methylation of theobromine from cacao waste; and (6) *Water* is essential for life. Plants, especially succulents, may store substantial quantities of water that can be tapped to provide a welcome source of clean, potable water, especially in the arid and semi-arid regions. The emergency water sources of the Seri of Baja California include the sap of *Agave cerulata, Ferocactus wislizenii* and the fruit juice of *Stenocereus thurberi*. In the deserts of Israel and Sinai drinkable water can be obtained from the roots of *Emex spinosa* and *Erodium crassifolium,* syn. *E. hirtum*. The cut stems of the scandent shrub *Tetracera potatoria* of tropical Africa provide a clear and potable water, as do the stems of *Sterculia setigera*. In south-western Africa an important source of water for the San people is the flesh from the fruits of the cucurbits *Acanthosicyos naudinianus* (herero cucumber) and *Citrullus lanatus* (tsamma, wild watermelon), containing *ca.* 90 and 94 % water respectively. In Australia potable water can be squeezed from the roots of *Brachychiton* spp. (kurrajong), while the root pulp can be cooked and eaten; *Heteropogom triticeus* (giant speargrass) and *Nymphaea macrosperma* are also used as thirst quenchers. In tropical forests the stem sections of the rattan palms and the internodes of the bamboos are well-known sources of drinking water, as are the pools of water that collect at the leaf bases of many bromeliads (Irvine, 1952; Corner, 1966; Ko, 1982; Danin, 1983; Arnold *et al.,* 1985; Felger and Moser, 1985; Purseglove, 1987; Kobayashi and Kawakamis, 191; Ensminger *et al.,* 1994; Robbins, 1995; Yunupinu *et al.,* 1995).

6.1   Fermented Beverages

The fermentation of the sugar present either naturally in plants or produced by the transformation of starch to yield alcohol has been practised from the earliest times. The designation 'beer' is generally applied to alcoholic drinks obtained from cereals or roots and 'wine' to fermented beverages from fruit, flowers and less commonly other plant organs. The cereal grains for beer are the major source of alcohol today, including sake or Japanese rice wine (a strong beer rather than a wine) from steamed rice grains by the action of the mould *Aspergillus oryzae* following the breakdown of the starch to fermentable sugars.

Again the terminology is not clear cut as the following examples show. For example, the sap of *Betula lenta* (black birch) for birch beer, the tuberous roots of cassava for cassava beer and the roots of *Pastinaca sativa* for parsnip wine. The fermented juice of apples produce cider and pears perry. The characteristic bouquet of cider from Normandy is attributed to the yeast *Kloeckera apiculata,* an anamorph of *Hanseniaspora uvarium,* although Dr J.A. Barnett (1998, pers. comm.) suggests that the evidence for this is inadequate. The flowers and berries of *Sambucus nigra* produce elderflower and elderberry wines respectively,. the petioles of *Rheum* × *hybridum* rhubarb wine, while palm inflorescences and the sugary palm sap yield palm wine and toddy respectively (Batra and Millner, 1974; Brouk, 1975; Postgate, 1992).

6.1.1 Beers

The term *beer* is from the Middle English *ber(e),* from the Old English *bëor,* from a West Germanic word from Late Latin *bibere,* a drink, from Latin *bibere,* to drink (Long, 1994). The brewing of beer has a long history. Archaeological evidence has shown thar the ancient Babylonians of *ca.* 6000 BC brewed beer. Beer residues from emmer wheat and barley have been reported from Ancient Egypt as early as *ca.* 3500-3400 BC. and is believed to have been similar to the present-day Egyptian beer known as *bouza,* which is prepared from a flour and water dough to which yeast is added. It is then lightly baked and the broken bread placed in a jar of water to ferment. *Bouza* is an ancient Coptic word, from which I suggest the English slang *booze* could have originated, although according to Long (1994) *booze* is from the 12-15th century Middle English *bousen,* to carouse, from Middle Dutch *büsen.* Sprouted spelt and bread wheats, barley grains and rye reported from Roman Britain, Sweden and Jutland are also considered as evidence of early brewing (Täckholm *et al.,* 1941; Renfrew, 1973; Manniche, 1989; Postgate, 1992; Maksoud *et al.,* 1994).

The word *beer* includes a variety of fermented cereal beverages. The original malt liquors, which were fermented without hops, are known as *ales,* from the Middle English *ale,* Old English *alu, ealu* (Long, 1994). The introduction of hops from France during the 14th and 15th centuries from the dried female inflorescences of *Humulus lupulus,* imparted a characteristic bitter flavour to the ale, which then became known as *beer.*

Today the two terms are now regarded as more or less synonymous, with ale being regarded as a beer made with very little hops, although Ensminger *et al.,* 1994) make the distinction of ale being top fermented in the US and beer bottom fermented. The dark and often almost black drink made from a wort containing dark roasted barley or malt is referred to as a *stout.* In the UK ales and stouts are also made using 'top-fermenting' or 'high yeast' strains of *Saccharomyces cerevisiae,* syn. *S. ellipsoideus* (brewer's yeast). Similarly, the *lagers* and *light beers* made in continental Europe use a low temperature 'bottom-', 'low-' or 'sedimentary-yeast' fermentation in a closed fermenter with *S. carlsbergensis*, which is sometimes considered to be a strain of *S. cerevisiae* (Barnes and Barnes, 1978; David, 1978).

The 2-row barley cultivars are preferred because they malt better and more slowly and produce lower enzyme levels, the malts also yield greater amounts of extract. The 6-row malts are favoured for adjunct brewing and grain whisky distilling because they have a higher enzyme content, which is necessary for breaking down the adjunct starch. Although most of the world relies on 2-row malting barley, the 6-row is used in North America for historical reasons. The 2-row barley did not perform well when originally introduced to eastern North America but the 6-row flourished in California, and achieved eventual dominance. However, successful breeding programmes in the USA have now increased the use of 2-row malting barleys (Edney, 1996).

Barley, wheat and rye are the most commonly used grains for malting. The process is described in some detail since the principles can be applied to other grains. Classical brewing involves the following stages: (1) *Malting*, the steeping of the cleaned grain in cold water until it has absorbed 42-45% moisture, i.e. sufficient to induce germination. The grain is then drained and spread out as a long, 0.6-1 m high heap known as a *couch*. The couch is left until the rising temperature induces germination, when it is evenly spread over the malting house floor to a depth of *ca.* 14 cm and left for about 10 days. The sprouting grain, known as a *piece*, is regularly sprinkled with water, ploughed and turned to control the temperature and enable the *acrospires* (embryo green shoots) to start growing under the testa. The malster aims for the acrospire to grow to approximately two-thirds of the length of the grain with minimum rootlet growth. It is during this germination process that the enzyme *amylase* (*diastase*) breaks down the grain's stored starch to *maltose*. The piece is again made up into a heap and left for 24 hours. The green *malt* is then transferred to a drying kiln and there dried with progressively warmer air for 3-4 days to check germination and create the characteristic sweet, malt flavour. The final flavour is determined by the degree of curing and takes approximately 8 hours; (2) *Preparation of the grist* by grinding the grain or malt after removing the *malt culms* (roots and shoots); (3) *Mashing* by steeping the grist in water, where it undergoes further enzyme action and the maltose extracted into solution, or *wort*. (4) *Lautering*, the process where the spent grains are separated from the liquid wort. The residual grain, known as *brewers' grain*, is used as an animal feed; (5) *Brewing*, boiling the wort with hops; and (6) *Fermentation* of the cooled liquid by *Saccharomyces cerevisiae* to produce a bitter beer (David, 1978; Dalal-Clayton, 1981; Hawksworth *et al.*, 1995).

The popularity of barley for malt is due to the husk protecting the growing kernel during malting, which is essential for successful lautering. It is also popular for its ability to produce large quantities of enzymes during germination and for the firmness of the steeped barley when handling during the malting process, plus the contribution of both the barley and barley malt flavour to the alcohol.

Malted is used by the brewing and distilling industries, predominantly for the brewing of beer but is also an important ingredient for the distilling industry, and to a lesser extent the food industry. Unmalted barley, known as **barley adjunct**, is similarly used in brewing and distilling but to a much lesser extent. It is an inexpensive source of carbohydrates and requires the gelatinising of malt amylases in order to inactivate the enzymes breaking down the starch into maltose and other sugars. Because the gelatinising temperature of barley starch is below the inactivating temperature of many of the malt enzymes, barley can be mushed with the malt and its starch degraded by the malt enzymes. Other cereals, such as rice and maize, have higher gelatinising temperatures than the inactivation temperature, hence these adjuncts need to be heated to gelatinisation in a separate cereal cooker, cooled and then added to the malt mush.

The physical and chemical characteristics of a malting barley are subject to standards set by the American Society of Brewing Chemists, European Brewing Convention, Institute of Brewing, American Association of Cereal Chemists and the Association of Official Analytical Chemists. Malting quality based on appearance, requiring a uniform, plump grain and good colour; a homogeneous grain sample is also essential. Plump grains have a lower protein content, and since there is a significant negative correlation between barley protein and the amount of extract produced, high protein barley will consequently results in a poor quality malt. It is possible, however, for the high protein malt to be diluted in the brewery with an adjunct. A clean, bright grain colour is indicative of good ripening and harvesting conditions and suggests a low moisture content. On the other hand, stained grain often indicate microbial infection, which can cause problems during malting and in the finished beer. In-depth protein and germination tests are also used to indicate malting potential. In addition, farming practices and environmental factors can affect the quality of malting barleys, so that there is a need to match the appropriate cultivar to the geographical area (Edney, 1996).

Sorghum is used for traditional *opaque*, *sweet* and *sour beers*, and clear beer (lager) from sorghum or maize-sorghum blends. It is also used in European beers as an adjunct with barley malt. Because of their low β-amylase activity, sorghum malts are inadequate for the production of lager or clear beers. Nevertheless, in South Africa and Nigeria, industrial processing for sorghum malt supplemented with commercial amylase is currently totally replacing barley malt for lager beers. The brown, bird-resistant sorghums with their bitter tannins are considered undesirable for brewing. But if they must be used, they need to be treated with methanal.

Opaque beer is a soured, viscous, effervescent and opaque beverage, pale buff to pink in colour and containing 2-4% alcohol. During brewing the various micro-organisms present in the germinating sorghum grain multiply and produce hydrolytic enzymes accompanied by a doubling of the vitamin B content. The malt is then dried, ground and mixed with cold water. Either more pounded grain or a grain other than sorghum steeped in boiling water is then added. The enzymes present in the malt begin to saccharify the starch, bacteria act on part of the sugar to form lactic acid and vitamin B. The resulting sweet wort, with nil or very little alcohol content, may be drunk after less than 1 day's fermentation. More often the mixture is heated and the various fungal spores present germinate, resulting in the production of even further hydrolytic enzymes. The fermentation process continues with the production of alcohol, more vitamin B and more protein. Opaque beer is thus a good source of vitamins, minerals, protein and carbohydrates solubilised during the malting and brewing processes (Doggett, 1970; Rooney, 1996).

An unusual beer is *kvass* or *quass* (Russian rye bread beer). It is prepared by pressure-cooking brown rye bread, treating the cooked mash with rye malt, and then fermenting with a combination of yeast and *Streptococcus lactis,* syn. *Bacillus lacta,* and finally, filtering to remove any solids (Ensminger *et al.,* 1994)

6.1.2   Wines

Wine, the fruit of the vine, *Vitis* spp., has a long history. According to the Bible "And *Noah began to be an husbandman, and he planted a vineyard: And he drank of the wine and was drunken;.....*" (Genesis 9, verses 20, 21). Archaeological evidence of earthenware pots with traces of wine discovered from a site in western Iran confirm that it has been known for at least 5000 years (Atkin, 1996). Renfrew (1973) suggests that wine making possibly arose as a result of attempts in preserving the perishable grape, a wild food source that was certainly used for food in the Early Neolithic and Bronze Ages. By 1000 BC vineyards had spread from Greece and Italy to France and in 280 AD were introduced by the Romans into Britain. Viticulture declined in Britain following the departure of the Romans but has undergone a revival in recent years, producing wines that can compete favourably with continental wines (Beech and Pollard, 1970; Bianchini and Corbetta, 1975).

The traditional grape vines of commerce are believed to be derived from cultivars of *Vitis vinifera* subsp. *sylvestris*. Following the 1867 outbreak of *Phylloxera vitifoliae* (vine rootlouse) in Europe, resistant stocks of American species were introduced. These, especially *V. labrusca,* gave rise to the American hybrid cultivars, some of which were crossed with *V. vinifera* to produce the French hybrids (Mabberley, 1997).

For *white wines* the ripe grapes are harvested and are usually crushed and destemmed before pressing, although for *sparkling wines* the bunches are usually pressed whole without crushing. After crushing and removal of the tannin-rich stems, the resultant mixture of juice, pips and skins is known as *must*. After relatively gentle pressing in order to avoid extracting the bitter tannins from the pips, the resultant *juice* is left for 12 hours or more to allow the residue of pips and skins, known as *solids,* to settle in a process known as *débourbage*. The juice is then transferred, or *racked,* to a covered vessel for fermenting, using either the natural yeasts present on the grape skins or cultured yeasts. Following fermentation the wine may be left on the yeast sediment or *lees* for up to 2 years to acquire more richness. The stirring of the lees, or *batonnage,* is an optional addition. Secondary fermentation, or *malolactic fermentation*, where the tart *malic acid* is converted to the softer *lactic acid*, may occur after the primary fermentation, or even simultaneously. It adds a buttery richness albeit at the cost of fruitiness. In cool areas the malic acids are naturally high, whereas in warmer areas a little malic acid may even be added to give a bit of bite to the ripe wines. The wine may then be either allowed to mature in oaken casks or immediately bottled. Before bottling the wine is clarified, or *fined*, usually with bentonite (an assemblage of clay minerals from weathered volcanic rocks) in order to flocculate any suspended particles, and then filtered.

While the skins are removed at the débourbage stage for white wines, for *red wines* the grapes are fermented with their skins and pips and then either pressed or not pressed. Some stalks may also be included in the must. During fermentation a cap of pips, skins and stalks rise to the surface covered by a blanket of CO_2, thus preventing excessive oxidation and permitting the use of open vats rather than enclosed vessels. The cap is kept wet by pumping the wine from the bottom to the top of the vat, or by *remontage*, where the cap is forced down into the wine. After fermentation the *free-run wine* is run off and the *pomace*, consisting of a residue of skins and pips, pressed to produce *press-wine*. The press-wine is thicker and contains more tannin than the free-run wine, some of which may be blended with the free-run wine to provide colour and tannin. The wine is then racked and matured before being fined, often with bentonite but more traditionally with egg whites. The wine may or may not be filtered before bottling (Atkin, 1996). The removal of the skins of red grapes during fermentation produces a pink, light wine known as *rosé*. See also Chapter 12 regarding wine barrels, their construction and their effect on wine.

Sparkling wine is the result of secondary fermentation by *Saccharomyces bayanus* and *S. cerevisiae*, syn. *S. oviformis* within the sealed bottle and the dissolving of the resultant CO_2 by the wine. *Champagne* is a sparkling wine where the debris of the controlled fermentation is removed by laborious *rémuage* or *riddling* of the individual bottles. This involves the gradual twisting and turning of the bottle until it is inverted. The neck of the bottle is then frozen, the bottle turned upright, the cap removed and the ice and debris abstracted. The space, or *spave*, is topped with fresh wine, a process is known as *dégorgement*.

Sweet wine is where the sugar content of well ripened grapes is too great for the yeasts to convert all the sugar to alcohol. This results in a stable non-sparkling wine with residual grape sugar, the result of the yeasts ceasing to function when the alcohol level reaches 13-14%. Late harvesting enhances the grape sugar content; sun-drying the harvested grapes, known as the *Passito method*, will also enhance the sugar content.

Under warm, damp autumnal conditions accompanied by plenty of sunlight the fungus *Botrytis cinerea* (grey mould) may infect the grapes, producing what appear to be rotten grapes. But, depending on the weather, the effect can be beneficial. The fungus penetrates the skins of ripe grapes and, if followed by warm, dry weather, can cause the grapes to loose some water, thereby concentrating the sugar content. The

Botrytis also metabolises some of the sugar into glycerol instead of alcohol, so that only a proportion of the sugar will undergo alcoholic fermentation; the levels of malic and tartaric acids also decrease. Such **botrytised wines** are sweet, rich and honey flavoured, such as the Tokay from Hungary. But, should the weather following infestation be cold and humid, the crop will be spoilt. Harvesting grapes after they have frozen solid will also concentrate the juice; the method is used to produce the German *eisweins*. The characteristic resinous flavour of the Greek wines known as *retsina* is obtained by using pine casks made from *Pinus halepensis* (Aleppo pine).

Fortified wines are made by adding extra alcohol during the fermentation stage in order to stop fermentation and maintain the sweetness. For **port** the fermentation of the red wine must is stopped when 4-5% alcohol is reached. This is achieved by adding grape alcohol to increase the alcohol content to *ca.* 20%, after which it is allowed to mature for a few years. **Madeira** is made in a similar fashion to port but heated to 35-45°C for a minimum of 90 days up to a year, after which it is allowed to mature for 10 years. **Sherry** is also fortified. At 15-16% alcohol it is known as *fino sherry* with a naturally occurring yellowish yeast, known as *flor*, covering the surface of the wine. Fortified to 18% alcohol the flor is killed to produce *oloroso sherry* which, after maturing for 8 years, is then known as *amontillado sherry* (Ainsworth, 1994).

Flavoured wines such as **vermouths** and **martinis** (an Italian trade name for a vermouth) are mainly white wines flavoured with herbs and spices, such as *Artemisia absinthium* (wormwood), *Cinnamomum verum* (cinnamon), *Pimpinella anisum* (anise), etc. (Montagné, 1977).

Wine vinegar is produced by the ethanoic (acetic) fermentation of wine by *Acetobacter aceti*, syn. *Mycoderma aceti*, which first appears as a light veil and increasingly penetrates the liquid, forming a thick folded, sticky skin known as *mère de vinaigre*. A good wine vinegar should have an acidic taste, be clear and transparent, colourless if made from white wine, pinkish if from red wine, and with an aroma reminiscent of the parent wine. Other culinary vinegars may be made from beer, cider, perry, etc. (Montagné, 1977; Rexen and Munck, 1984).

6.2   Distilled Beverages

Spirits are produced from the distillation of an alcoholic liquid, usually from the fermentation of cereal grains and other fruits, especially the former. The result is a potable alcoholic liquid with a higher alcohol content than that of the original fermented beverage. There are two categories of grain spirits recognised, **whisky** and **neutral spirits**. For whiskys care is taken to ensure the retention of colours and flavours introduced during production, while the introduction of colours and flavours are avoided in the production of **neutral spirits**, i.e. ethanol distilled at or above 190° proof and used in blended spirits. The cheapest available cereal can be used for neutral spirits; it is also more economical to use enzymes derived from micro-organisms rather than from malt (Grubben *et al.*, 1996). *Proof spirit* is defined as an alcohol-spirit mixture or alcoholic drink containing 49.28% ethanol by weight, or 57.1% by volume and having a relative density of 0.92 at 15.56°C. In the United States proof spirit contains 50% ethanol by volume at 15.56°C.

Whisky, as spelt by the Scots, Canadians and Japanese, and as *whiskey* by all others nations, is an English corruption of *uisge* from the Gaelic name *uisge beatha* or 'water of life'. Whether whisky originated in Scotland or Ireland during the early Middle Ages is uncertain, but what is certain is that it is now being produced and drunk in very in many countries around the world. Scotland is by far the largest producer and boasts of 125 brands of malt, 9 of grain, 58 vatted and 370 of blended whisky (Gabány, 1997).

Scotch whisky and Irish whiskey are produced from barley, Canadian whisky from rye and USA whiskeys from maize (corn) or rye. *Malt whisky* is distilled from fermented barley malt with no adjuncts, consequently the malting barley should contain high levels of starch and low levels of protein (see Section 6.1). Particular emphasis is placed on taste and character of the local spring water used; other factors determining taste are the malt sugar and the peat smoke employed during preparation. For tax purposes HM Customs and Excise in Scotland accept an annual 2% loss from the vats through evaporation, which is known as the 'Angel's share'. *Grain whisky* from a distillate of the grain, produces a purer but more flavourless whisky than malt whisky. Limited amounts of barley malt are used to provide an enzyme source, with cereal grains as the main source of starch. The barley for grain whisky should have higher levels of protein than that required for malt whisky in order to produce increased levels of enzymes. *Vatted malt* contains at least two single malts from different distilleries, which are mixed in a process known as *vatting*. A vatted malt whisky is almost never referred to as such, the label usually refers to 'malt', 'pure malt' or 'all malt' whisky. Pure malt can also mean a single malt, although any malt whisky not definitely labelled as a single malt is considered vatted. *Blended Scotch whisky* consists of a mixture of malt and grain and is the world's most popular whisky. *Blended American whiskey* must contain at least 20% straight whiskey, the remainder consisting of neutral grain spirits. *Bourbon* is distilled from a mash containing between 51 and 80% maize and must have matured for a minimum of 2 years. *Tennessee whiskey* differs from bourbon in that the fresh spirit is filtered through a 3 m thick layer of maple charcoal. *Pot still whiskey* involves distilling the grain mash in onion-shaped copper stills, the first distillation producing low wines with *ca.* 28% alcohol, followed by a second distilling to yield baby whiskey with *ca.* 70% alcohol (see Gabány, 1997 for further information).

Gin, sometimes known as *Hollands*, is a strong alcoholic drink obtained by distilling rye, barley, or other grains and flavouring with juniper berries. The name is a shortened form of the Dutch *jenever*. Other distilled drinks are *brandy* from wine, *Calvados* from cider, *slivovitz* from fermented plum juice, *rum* from sugar cane juice or molasses, *mescal* from *pulque*, the fermented juice of *Agave* spp., etc.

Liqueurs are strong, alcoholic fruit liquors sweetened and flavoured with aromatic herbs and spices and usually drunk in small quantities. Formerly liqueurs were often referred to as *cordials*, a term now usually used for liqueurs produced by infusion, where distillation would destroy or modify the fugitive flavours (Fowles, 1977).

7.   FOOD ADDITIVES

Food additives are those small quantities of plant, animal (e.g. cochineal), mineral (e.g. salt) and chemicals that are added to food in order to: (1) Make the food more attractive both visually and in flavour; (2) For food preservation; (3) To help food preparation and/or processing; and (4) Maintain or improve the nutritional value, including the replacement of vitamins destroyed during cooking. A number of these additives may even perform several functions. This is not a new technology, sugar, vinegar and salt, for example, may have been used in prehistory. The safety of food additives for use is, of course, a paramount consideration. Those additives approved for use for food and pharmaceuticals within the EU are designated by an 'E' number; those used in the UK are listed in MAFF (1998d).

Not all food additives are necessary or even beneficial. Among the undesirable and inappropriate uses of food additives are: (1) To disguise faulty or inferior processes; (2) To conceal damaged, inferior or spoiled foods, although it should be remembered that before refrigeration and other sophisticated preservation processes many of our herbs and spices were used for precisely such a purpose; (3) To gain some functional property at the expense of nutritional quality; (4) Used in excess of the effective minimum; and (5) Used to replace economical, well-recognised manufacturing processes and practices (Ensminger *et al.*, 1994; MAFF, 1997c).

7.1   Food Attractives

Additives for *food attraction* are used to make the food more appealing in appearance and taste. These include colouring agents, flavouring agents, and sweeteners.

7.1.1   Colouring Agents

Colouring agents, such as bixin and crocein, are used either to restore colours lost during food processing or to make food visually more attractive. The plant kingdom possesses an abundant supply of suitable plant pigments, of which four main groups suitable for colouring food, can be recognised: (1) *Anthocyanidines* are intensely coloured water-soluble orange, red, violet and blue *flavonoid pigments* commonly found in flowers, fruits and vegetables. For stability they are usually kept in a rather acid medium; (2) **Betalains** are a small group of red and yellow pigments sensitive to pH, heat, and light, the most common of which is **betanin** from *Beta vulgaris* (beetroot); (3) **Carotenoid pigments** are red, orange and red in colour and sensitive to oxidation, hence their usefulness in the food industry necessitates limiting their exposure to air; and (4) **Chlorophyll pigments** are green and sensitive to acidity and light. See Table 10 for examples, also Chapter 18 regarding food dyes from lichens.

TABLE 10. Examples of plant pigments used for food (Brouk, 1976; Lemmens *et al.*, 1991; Green, 1995)

Class	Colour	Species	Phytochemicals
anthocyanin	burgundy red	*Vitis* spp. (grape)	apigenin
betanin	red	*Beta vulgaris* (beetroot)	betanin
carotenoid	red to orange	*Bixa orellana* (annatto)	bixin
carotenoid	orange-red	*Capsicum annuum* (paprika)	capsanthin, capsorubin
carotenoid	yellow to orange-red	*Curcuma longa* (tumeric)	curcumin, etc.
carotenoid	yellow	*Crocus sativas* (saffron)	crocein
chlorophyll	green	*Urtica* spp. (nettle)	chlorophyll

In the developed countries food colourings are subjected to a rigorous screening before they are permitted to be used, although such scrutiny does not necessarily bestow widespread acceptance. Chlorophyll, for example, is permitted for use in food and drinks within the EU but is not approved for use in the USA. Similarly, although widely used in Japan, extracts from the fruits of *Gardenia jasmnoides* (Cape jasmine) are not permitted for use in the USA.

The fungal discoloration of food can sometimes be acceptable, e.g. rice fermented by *Monascus purpurens* produces red-coloured grains known as **angkak**, which is considered to give an attractive colour to oriental foods (Ko, 1982).

Although generally more expensive to produce, the future of vegetable food colourings does rather depend on the fact that they tend to be safer than synthetic colourings. A few of the latter have even been suspected of causing hyperactivity and learning difficulties in children. The vegetable dyes are also more environmentally friendly. Hence, with the increasing suspicion that certain synthetic food dyes may be harmful, natural food colourings are increasing in popularity. The burgundy red dye obtained from grape skins, for example, is increasingly preferred to synthetic alternatives. The major component, **apigenin**, is currently obtained as a by-product of the wine industry. Furthermore, apigenin has recently been found to be an important component of the anthocyanins present in the leaf sheaths of *Sorghum bicolor* race *caudatum* and not only is it readily extractable, it is also present in four times the quantity found in the grape. (Lemmens *et al.*, 1991; Burkill, 1994; National Research Council, 1996).

Production costs of natural colorants are a problem, especially with such a highly labour-intensive crop as saffron. It requires 500 individual stigmas to produce 1 g, or 10^6 flowers to produce 10 kg of the dried spice. Indeed, labour costs are so high that the more-or-less Spanish monopoly of the saffron trade is now in the process of being taken over by countries where labour costs are very much lower (Brouk, 1976; Robbins, 1995; Mabberley, 1997).

7.1.2 Flavouring Agents

Flavouring agents or *food aromatics* embrace all forms of food seasoning that emit flavours and odours of varying degrees of sweetness and pungency and are used to stimulate the sense of taste and/or smell. The terminology, however, tends to be imprecise. The term *seasoning* is applied to those substances added during cooking, e.g. herbs and spices. In general, *herbs* refer to aromatic vegetative organs, usually leaves, e.g. *Laurus nobilis* (bay leaf), *Mentha* spp. (mint), *Thymus vulgaris* (thyme), etc. *Spices* refer to aromatic seeds and fruits used either whole or ground e.g. *Cinnamomum verum* (cinnamon), *Coriandrum sativum* (coriander), *Myristica fragrans* (nutmeg and mace), *Piper nigrum* (pepper), *Syzygium aromaticum* (cloves), etc. A spicy or savoury condiment, such as pickles or chutney is known as a *relish*. *Condiments* refer to substances added after the food has been prepared for eating, e.g. *Armoracia rusticana* (horseradish); *Brassica juncea* (brown, Indian, or oriental mustard), *Sinapsis alba* (white or yellow mustard). The term vegetable salt is sometimes applied to a condiment such as celery salt, where the mineral salt is flavoured with dry and powdered celery. Nowadays the terms seasoning and condiment tend to be regarded as interchangeable (Montagné, 1977; Ensminger *et al.*, 1994).

Salt is such a common seasoning agent and condiment that people forget that there are parts of the world where neither mineral nor sea salt are readily available, and alternative sources have to be found. For example, in North America the ash from the leaves and stems of *Petasites palmata* (sweet coltsfoot) was formerly the only source of salt available for some native Americans. In the Israel and Sinai the salt secreted by the stem or leaf glands is still extracted from *Avicennia marina*, *Cressa cretica*, *Reaumuria* spp. and *Tamarix* spp. by soaking the leafy stems in water and then evaporating the water. The leaves of *Atriplex halimus* can even be used directly during the summer months to season food (Saunders, 1976; Danin, 1983).

Cheap artificial flavourings are available but are not necessarily preferred. Vanilla, for example, from the pods of *Vanilla planifolia* is used to flavour ice-cream, biscuits and custard, of which the odoriferous principle is *vanillin* (4-hydroxy-3-methoxybenzaldehyde). A cheap synthetic vanillin involving the 4-allyl-2-methoxyphenol *eugenol*, which also occurs naturally in clove, bay and cinnamon oils, is manufactured commercially from the *lignosulphonic acid* obtained from the waste sulphite liquor during paper manufacture or from tar extracts. Despite this, or perhaps because of, there is an increasing preference by the ice-cream industry for the natural product (Brücher, 1989; Sharp, 1990; Robbins, 1995; Mabberley, 1997).

Agronomic practices can also affect the economics of production. For example, both brown and white mustard are widely cultivated in North America, Europe, Australia and Argentina for their seeds, which are used for the condiment *mustard*. The former species, whose ripe pods are non-shattering has now replaced B*rassica. nigra* (black mustard), which was unsuitable for combine harvesting and had to be hand-harvested green to prevent loss of seed through shattering. Brown mustard is also grown as an oilseed crop.

The name mustard is from the Middle English *mustarde*, a condiment and later applied to the plant, from the Old French *mo[u]starde*, from Common Romance *mosto*, from the Latin *mustum*, i.e. *must* or juice of new wine, because the mustard paste was originally made by mixing grape juice with mustard powder (Long, 1994). The use of mustard seeds as a spice are recorded in Sumerian and Sanskrit texts from *ca.* 3000 BC, from Egyptian texts *ca.* 2000 BC (*Sinapsis alba* fide Manniche, 1989) and in the Chinese literature from before 1000 BC. The *B. juncea* is responsible for providing a pungent olfactory flavour and *S. alba* a buccal hotness, flavours which may be used either singly or in combination in the various mustards. Today, the mustards of the developed countries include *English mustard* (a blend of brown and white mustards), *American mustard* (a cream-salad mustard from white mustard with ground turmeric, *Curcuma longa*, to intensify the yellow colour), *German mustard* (predominantly white mustard plus a small amount of brown mustard to add slight pungency) and *French mustard*, e.g. *Dijon mustard* which is restricted by French law to brown mustard (Hemmingway, 1995). Other formulations are to be found in Hazen (1993).

The dry seeds and dry ground products of both mustard species are essentially odourless. They differ in the nature of the pungent isothiocyanates (essential oils) released on wetting and the glucosinolates present come into contact with the enzyme *myriosinase*. The chief constituent of brown mustard is the 2-propenyl (allyl) glucosinolate *sinigrin,* which hydrolyses to produce allyl isothiocyanate or 'volatile oil'. This 'volatile oil' provides a strong olfactory pungency; it is also lachrymase. White mustard contains mainly the 4-hydroxybenzyl glucosinolate *sinalbin* which, on hydrolysis yields the non-volatile 4-hydroxybenyl isothiocyanate known as the '*white principle*'. It is this white principle that produces the 'heat feeling' in the mouth, while its instability results in a sensation of sweetness and warmth. The ISO (1981) mustard seed specification requires no more than 2% damaged or shrivelled seed, or 0.7% extraneous matter, and ground seed free from any odour of mustiness and rancidity. *B. juncea* and *B. nigra* must yield a minimum of 1.0% and 0.7% allyl isothiocyanate, respectively, and *S. alba* a minimum of 2.3% 4-hydroxybenzyl isothiocyanate (Hemmingway, 1995). The mustards have been discussed in some detail because they serve to illustrate many of the principles discussed earlier in Chapter 7.

7.1.3   Sweeteners

Sweeteners are those substances that are added to food and drink in order to make the product sweet to the taste. Sweeteners should: (1) Possess a clean flavour without any aftertaste; (2) Their cost on a sweetness basis should be competitive with the price of sugar; (3) They should be adequately soluble and stable; and (4) Be subject to rigorous health safety testing before marketing. Two categories are recognised, *nonnutritive sweeteners* with less than 2% of the calorific value of sucrose per equivalent unit of sweetening capacity and *nutritive sweeteners* with more than 2% of equivalent unit of sweetening capacity (Montagné, 1977; Ensminger *et al.*, 1994).

The main sugars present in plants are glucose, fructose and sucrose. It is the latter which is normally regarded as 'sugar' and is used as the sweetness standard, i.e. 1, against which other sweeteners are measured. *Saccharum officinarum* (sugar cane) and *Beta vulgaris* subsp. *vulgaris* (sugar beet) are the world's major sources of *sugar*, their refined products being virtually indistinguishable. It should be noted that for sugar cane there is a distinction between what is known as a sugar syrup and molasses. A *syrup* is produced by evaporating the plant juices so that all the sugar is present. *Molasses* are the residue remaining after the juice has been concentrated until much of the sugar has crystallised out and removed.

In the United States enzyme transformation of maize starch is now used for the large-scale production of *glucose* (*dextrose*) and *high fructose syrup*. In the 1980s it was the widespread use by the beverage industry of this 'High Fructose Corn Syrup' as a sugar substitute in aerated drinks and fruit juices that became the major factor in causing the collapse of the sugar cane industry in the Philippines. (Hill, 1952; Brouk, 1976; Robbins, 1995; Grubben *et al.*, 1996).

Among the minor sources of sugar are sweet sorghum for sorghum syrup and the commercial tapping of *Acer nigrum* (black maple) and *A. saccharum* (silver, striped or sugar maple) for maple syrup. Several palms also yield a sweet sap when tapped, including *Arenga pinnata* (gomuti palm), *Borassus flabellifera* (palmyra palm), *Caryota urens* (toddy palm), *Cocos nucifera* (coconut palm), *Phoenix sylvestris* (wild date palm). The resinous exudate obtained from incisions in the trunks of *Fraxinus ornus* (flowering or manna ash) yields a sweet substance containing up to 80% of the sugar alcohol *mannitol* (see Chapter 8), which was formerly used in southern Europe as a sweetener. The tree is now cultivated in Sicily and Calabria and the manna used for sweetening medicines (Hill, 1952; Brouk, 1976; Mabberley, 1997).

In the deserts of the Middle East sweet exudations are obtained during the summer months from *Alhagi maurorum*, syn. *A. pseudalhagi*, *Anabasis setifera*, *Capparis* spp., *Haloxylon salicornicum*, syn. *Hammada salicornia*, *Tamarix mannifera* and *T. nilotica*. The scale insect *Coccus manniparus* punctures the bark, from which a sweet fluid exudes and solidifies. In Iran the psyllid *Cyamophila astragalicola* feeding on *Astragalus adscendens* is responsible for the exudate or 'gaz', known as the 'Gaz of Khunsar' after a town in the producing area. It is produced in sufficient quantities to be marketed as a popular sweetmeat known as the 'Gaz of Isfahan' (Brouk, 1976; Dannin, 1983; Grami, 1998).

A number of non-sugar sweeteners have now been identified but are relatively little used. The leaves of the composite herb *Stevia rebaudiana* (caa-ehe, kaahée, the sweet herb of Paraguay) contain the diterpene glycoside *stevioside*, which is 200-300 times as sweet as sucrose, i.e. a fifth as effective as saccharine. It has long been used by the indigenous peoples of Paraguay for sweetening drinks and is now being used as a sweetening agent in Japan. It is the only *Stevia* species so far found among the 150 New World species with a sufficiently high content of stevioside for commercial extraction.

Three other sources of non-sugar sweeteners have been discovered in West Africa. The berries of *Synsepalum dulcificum*, syn. *Richardella dulcifica* (miraculous berry) contain the glycoprotein *miracularin*, which affects the taste buds and causes sour and salt foods to taste sweet, although the effect wears off within a few hours. The fruits of *Dioscoreophyllum cumminsii* (serendipity berry) contain the *monellins*, proteins which are 800-1500(-3000) times sweeter than sugar and have been used for low calorie foods and drinks. The arils of *Thaumatococcus daniellii* (miraculous fruit) contain the protein *thaumatin*, which is 1600 times as sweet as sucrose. Although a promising sweetener, thaumatin unfortunately breaks down when the food is heated. Despite production attempts using tissue culture and genetic engineering, dipeptides synthesised from aspartate are now preferred to the natural product (Brouk, 1976; Harborne, 1988; Brücher, 1989; Flach and Rumawas, 1996; Mabberley, 1997). See also Chapter 18.

Any sense of sweetness can be destroyed and bitterness partly repressed by chewing the leaves of the asclepiad *Gymnema sylvestre*, a scrambling shrub of the Old World tropics. The pentacyclic triterpenes present, the *gymnemins* are even capable of overcoming the sweetness of the miracularin in the miraculous berry (Irvine, 1952, Harborne, 1988; Mabberley, 1997).

7.2   Food Preservation

Additives for *food preservation* are used to maintain freshness, prevent discoloration, and retard spoilage by micro-organisms. *Antimicrobials*, e.g. *propanoic acid*, are used to retard or prevent spoilage by bacteria, yeasts, moulds and fungi. *Antioxidants*, e.g. *ascorbic* and *citric acids*, are used to prevent discoloration and loss of flavour due to oxidation, delay or prevent fats and oils becoming rancid, as well as preserving bakery products, cereal-based baby foods, soup mixes, sauces and preserved meat and fish products (Ensminger *et al.*, 1994; MAFF, 1997c).

7.3   Food Texturing

Additives used for *food texturing* help in the processing or preparation by giving the food body and texture: (1) *Emulsifiers* are used to ensure the even distribution of particles of one liquid in another, e.g. oil and water. Plant-based emulsifying agents include seaweed sources of alginates and carrageenan (Chapter 19), gums (Chapter 15.) and *lecithin* obtained from soya bean oil and produces a thick yellow emulsion with water; (2) *Clarifiers* are used to remove impurities from food products, especially cloudy liquids, by causing the cloudiness to settle out upon standing. They are particularly important in the brewing industry and for water purification. The cloudiness in beer, for example, may be reduced by *peptain* from *Carica papaya* and *gum arabic* from *Acacia senegal*. The clarification of water is of major importance in many rural areas of the developing world, especially in times of disaster. Depending on the quality of the water 30-200 mg l^{-1} of the powdered seeds of *Moringa oleifera* (horse-radish tree) will clarify cloudy water to tap water quality within a few hours, the process thereby eliminating 98-99% of indicator bacteria (Jahn *et al.*, 1986). See also Chapter 20 and Jahn (1981) for further information; (3) *Leavening agents,* such as yeasts or potassium hydrogen tartrate (cream of Tartar), affect the results of cooking or baking. Cream of Tartar occurs in grapes and is precipitated from wine lees as *argol* during fermentation and is used in baking powders to liberate CO_2 from the $NaHCO_3$. The acidic pith of baobab can also be used in baking as a substitute for cream of Tartar (Wickens, 1982); (4) *Humecants* are hygroscopic substances, such as glucose (dextrose), glycerol and sorbitol, that are used to retain moisture; (5) *Stabilisers* and *thickeners* are used to render a compound, mixture or solution resistant to changes in form or chemistry, to create smoothness and prevent caking or lumping. In the food industry stabilisers are used to impart body, to keep pigments and other compounds in an emulsion form, and to prevent particles in colloidal suspension from precipitating. Plant sources include algin, carrageenan and 1,2-dihydroxypropane [propylene glycol] alginate from the seaweed industry (see Chapter 19), pectin, and various gums; (6) *pH control* to change or maintain acidity or alkalinity, including buffers, acids and neutralising agents, e.g. citric acid, tartaric acid and alkalis; and (7) *Enzyme modification* by proteolytic enzymes derived from plants that are capable of breaking down proteins or peptides into smaller units. Probably the best known plant source is *Carica papaya* (papaya, pawpaw), whose unripe fruit contain the protein *papain*, the only natural plant antibacterial *protease* used commercially as a meat tenderiser; even meat wrapped in the leaves become tender. The potent proteolytic enzyme *bromelin* obtained from the stems of *Ananas comosus* (pineapple) depolymerises fibrin matrix and can also be used as a meat tenderiser as well as a *vegetable rennet* with the property of coagulating the phosphoprotein *casein* in milk. Other rennet sources include *ficin* from *Ficus carica* (common fig), *Galium verum* (ladies or yellow bedstraw), *Pinguicula vulgaris* (butterwort) and *Ranunculus flammula* (lesser spearwort). In the Sudan the seeds from either *Solanum incanum* or *Withania somnifera* are used to curdle milk; *W. coagulans* is similarly used in the Indian subcontinent (Montagné, 1977; Ensminger *et al.*, 1994; Vickery, 1995; Mabey, 1996; Mabberley, 1997).

8.   MISCELLANEOUS FOODS AND FOOD PRODUCTS

8.1   Fermented Foods

Micro-organisms, especially the yeasts and bacteria have been used since prehistory for the preparation of fermented foods and drinks. Every country in eastern Asia has indigenous fermented foods prepared on a scale ranging from individual households to commercial production. The fermented foods result from the action of specific microbial enzymes, selected strains of which can be produced as pure cultures. Usually a combination of two or more micro-organisms are used, working together to produce the desired product (Hawksworth *et al.*, 1995).

The controlled fermentation of food is an inexpensive, simple and proven technique with little waste. It may be used to: (1) Preserve food for future use through ethanoic acid, lactic acid and alcohol fermentation by bacteria, e.g. cabbage, radish, turnips and cucumber by *Lactobacillus brevis*, *L. plantarum*, *Leuconostoc mesenteroides* and *Pediococcus* sp. (*P. cerevisiae* Baloke cited by Ko (1982) is a *nomen confusum*, i.e. of uncertain identity). Alternatively, an initial fermentation by moulds (microfungi with an obvious mycelial or spore mass on a substrate) may be followed by a mixture of yeasts and bacteria; (2) To achieve detoxification during fermentation by destroying undesirable factors in the raw product and rendering it safer to eat, e.g. toxic glycosides in kawal prepared from *Senna obtusifolia*; (3) Biological enrichment of the food with protein, essential amino-acids, fatty acids and vitamins, e.g. porridges from fermented cereals, (see also Section 6.1 regarding sorghum beers); (4) To improve the digestibility, e.g. sauerkraut (pickled cabbage), albeit with a reduction in nutritive value; (5) Improve the flavour, aroma, pH, texture and appearance of the food; (6) To stabilise some components; (7) Produce an acceptable product which, in the case of fermented beverages, may be regarded as convivial; and (8) Reduce the cooking time and fuel requirements (Ko, 1982, Dirar, 1993; Steinkraus, 1995; National Research Council, 1996).

In Indonesia a highly palatable and readily assimilated protein food known as *tempeh* is produced by the fungal fermentation of boiled pulse substrates with *Rhizopus* spp., especially *R. oligosporus*. Other possible inoculums are *Alternaria solani*, syn. *Macrosporium solani*, *Cochliobolus miyabeanus*, syn. *Helminthosporium oryzae*, *Gibberella zeae*, *Penicillium chrysogenum*, *P. notatum* and *Stemphylium solani*. Tempeh, especially that made from soya beans, is an important food for local consumption; it is also marketed as a soya meat substitute in the health food shops of developed countries. In Japan the soya beans are wrapped in rice straw and fermented by *Bacillus subtilis* to produce *natta*.

In the Indian subcontinent *Candida* sp., *Saccharomyces bayanus*, *S. cerevisiae*, *Trichosporon pullulans*, etc. have been identified as the fermenting agents in a number of fermented cereal and gram foods, including the black gram bread *idli*. An inoculant composed of small balls of rice flour containing *Mucor*, *Rhizopus*, yeasts and bacteria and known as *ragi* is used to induce fermentation of starch-rich raw materials, e.g. steamed rice, cassava or sorghum, and as a starter for the alcoholic beverage *arrak* from the sugary sap of palms. In Micronesia pit fermentation (rather like pit silage making) is used to preserve starchy crops such as *Artocarpus altilis* (breadfruit) for use in times of scarcity (Batra and Millner, 1974; Ko, 1982; Atchley and Cox, 1984; Kronenberg, 1984; Hawksworth *et al.*, 1985).

In the Sudan more than 80 different fermented foods are recognised, of which 55 are of plant origin. The best known, apart from the fermented cereals, is *kawal* from *Senna obtusifolia*, syn. *Cassia obtusifolia*. Although formerly a little known food of the poor in Darfur Province, it is now widely used in many parts of the Sudan as a meat substitute and as a flavouring agent. In descending order of importance, the fermenting agents are the bacteria *Bacillus subtilis*, *Lactobacillus plantarum*, *Propionibacterium* spp. and *Staphylococcus sciuri* subsp. *lentis*, and the yeasts *Candida krusei* and *Saccharomyces* spp. (Dirar, 1984, 1993; Dirar *et al.*, 1985).

8.2   Leaf, Seed and Single Cell Proteins

In temperate countries primary production resulting from photosynthesis by fodder crops can yield 20 t dry matter and 3 t leaf protein or more ha^{-1} yr^{-1}. For tropical countries the yields can be up to 80 t dry matter and 6 t leaf protein, with production from C_4 plants being more efficient than from C_3. Protein production from such crops can exceed by three- or fourfold that obtained from cereal crops, and clearly have a potential role as a source of human food and livestock feed.

The wet-fractionation of the green biomass can be used to produce *leaf protein concentrate* (*LPC*) and *leaf nutrient concentrate* (*LNC*). The preparation involves crushing as finely as possible, pressing to extract the green juice, heat or acid protein coagulation or precipitation, followed by washing, decolorisation, drying and concentration. The dried LPC contains *ca.* 40% CP, of which *ca.* 75% is TP. Surprisingly the amino-acid content is remarkably constant regardless of the plant source. The cellulose-rich cake obtained as a by-product can be used as a ruminant feed. The *seed protein concentrates* (*SPC*) are mainly extracted from legumes, especially soya, field beans and peas, as well as from sunflower. Rapeseed has also been tried but so far efforts to remove the toxic antithyroid substances have not been completely successful. *Single cell proteins* (*SCP*) include food yeasts, such as species of *Candida, Saccharomyces, Torula*, using a sugar, starch, molasses, etc. substrate as a growth medium, and the moulds *Fusarium* and *Penicillium notatum* grown on starch. In terms of efficiency, 1000 kg of livestock can produce a maximum of 1 kg of protein in 24 hr; during the same period 1000 kg of yeast can increase five-fold, of which half is edible protein. Other SCP sources include species of the Chlorophyta *Chlorella* and *Scenedesmus,* and the Cyanobacteria *Arthrospira* and *Spirulina.* (Ferrando, 1981; Göhl, 1981; Carlsson, 1989).

8.3   Sodium carboxymethylcellulose

Sodium carboxymethylcellulose (CMC) is a water-soluble ester of cellulose used in the food industry as a substitute for gelatine in ice-cream, meringues, jellies, pie-fillings, etc. It is extracted from wood pulp or cotton linters (short fibres rejected by the textile industry) by treating with sodium hydroxide (Brouk, 1975).

8.4   Pectin

Pectin, consisting of protopectin, pectin and pectic acid, is a mucilaginous substance present in mature fruits, root vegetables, etc. It is present in the middle lamella and in cell walls of soft plant tissues and their tissue juices. Pectin consists of a mixture of three polysaccharides, the most important constituent of which is methyl pectate, which is a methyl ester of pectic acid, a high molecular weight polymer of D-galacturonic acid. Small quantities of araban and galactan are also present. Immature fruits contain *pectose*, which is insoluble in both alcohol and water, and is easily converted into soluble pectin by heating with dilute acid, or by the addition of the enzyme *pectase*. Commercial pectin is mainly obtained from the dilute acid extract of apple pomace produced during cider making, the inner rind of citrus, or sugar beet roots. It is marketed either as an aqueous solution or as a dry powder. It is important for both the home and industrial preservation of fruit, 1% of pectin being capable of converting the fruit and sugar into a gel and is so used in jellies, jams and marmalades (Brouk, 1975; Sharp, 1990).

8.5   Rice Paper

Rice paper is a thin, edible paper which, despite its name, is obtained from the pith of *Tetrapanax papyrifer* (rice paper tree), a rhizomatous, clump-forming shrub of southern China and Taiwan (Mabberley, 1997).

8.6   Pollen Flour

The indigenous North American have traditionally shaken maize pollen tassels to harvest the pollen. The pollen is then kneaded like dough, sun dried, ground to a fine powder and used as a flour. The pollen from *Typha* spp. has similarly been collected and eaten, sometimes mixed with flour. Evidence from ancient coprolites suggest species of *Brassica*, *Salix*, etc. may also have been used (Linskens and Jorde, 1997).

Chapter 10

Feed for Livestock

Livestock feed embraces all the food requirements of not only domesticated livestock, including poultry and fishes, but also wild life. The latter concerns the management of game and other wild life habitats, a complex subject which will be only briefly considered here. The needs of bees and other useful invertebrates such as silkworms, lac insects, dye insects, manna-producing insects, edible insects, etc. are dealt with in Chapter 11.

Three major food classes are recognised: forage, fodder and concentrates. *Forage* refers to all *browse* (the tender shoots and fruits of shrubs and trees) and herbaceous animal feed, including silage and green feed. The term *forage utilisation* is used to define the proportion of the current year's forage production that is consumed or destroyed through soiling and trampling by grazing animals. *Fodder* refers to dried cured material, particularly dry cured roughage high in fibre, e.g. hay, straw and stover (maize stems). The two terms are often confused, with fodder often being loosely applied to fodder. *Concentrates* have a high food value relative to volume; they are low in fibre and, depending on their origin, are usually rich in protein, carbohydrates or fat. They include the cereal grains and their by-products, pulses, oil seeds and their by-products or cake. The formulations of concentrates supplied by feed manufacturers will often contain supplementary minerals, trace elements and vitamins in order to provide a balanced ration (Ibrahim, 1975; Dalal-Clayton, 1981; Glossary Revision Special Committee, 1989).

The ability of an animal to ingest and digest feed is subject to a number of anatomical and physiological adaptations. The major factors affecting actual ingestion are dentition, lip and tongue action; the stance and eye position are also relevant. Grazing methods can vary widely, even among domesticated stock types. For example, grazing cattle are unselective apart from soiled grazing. They are unable to graze closely, using their tongues to tear and draw grass into their mouths; they also supplement their grass and herbage diet with some browse. Sheep are mainly grazers but will eat some low browse. They are extremely selective, grazing closely but unable to tackle stemmy grasses and herbs. Horses also use their teeth to graze closely, concentrating on certain areas and neglecting others and, like sheep, sometimes tackling low browse. Under free range conditions pigs will graze and use their snouts to root in the ground for food. Camels both graze and browse, using their mobile lips to pull leaves onto their teeth. Goats also use their lips and mainly browse, their access to browse being aided by their ability to climb trees and shrubs. Both camelids and goats are capable of eating a wider range of herbaceous and ligneous vegetation than bovines and equines. It is also my impression that goats and camelids are able to tolerate a higher intake of toxic feeds than other stock. Among the game animals, *Loxodonta africana* (elephant) will eat both grasses and ligneous plants, while *Giraffa camelopardalis* (giraffe), *Taurotragus oryx* (eland) and *Gazella granti* (Grant's gazelle) are mainly browsers. *Connochaetus gnou* (wildebeest), *Equus* spp. (zebra), *Syncerus caffer* (buffalo) and smaller antelopes feed mainly on grass. Swamp and aquatic vegetation supply the feed requirements of *Hippotamuses* spp. (hippopotamus). The seeds and fruits from swampland and riverine trees and shrubs may also provide food for a surprising number of fish species.

Such wide variations in eating and digestive habits make any generalisation on the nature of feed for livestock and wild animals virtually impossible. It also serves to emphasise the necessity for plant collectors to define the grazing animal and not just write 'grazed by livestock' on their labels (Miller and West, 1956).

Digestion, absorption and assimilation are the three processes by which feed is incorporated into the living body. Beginning with salivary digestion as soon as the food enters the mouth and continuing in the stomach, digestion is an ongoing process during which food is softened and converted into a form that is soluble in the body fluids. The soluble substances are then taken up and carried by the blood- and lymph-streams of the bowels to where they are to be assimilated for the growth and repair of the body tissues.

Ruminant and monogastric animals will obviously differ in their digestive ability, there are also variations between species. In non-ruminants, such as the horse, pig and dog, for example, there are variations in food retention by the stomach. That of the horse is relatively small, so that when the stomach is two-thirds full the food has to be passed to the small intestine to make room for further food entering via the mouth, thus permitting food two or even three times greater than the stomach's capacity to pass through while feeding. On the other hand the stomach digestion of the pig and dog is closer to human digestion and the food retained in the stomach for a variable time, depending on the state in which it was swallowed (Miller and West, 1956).

Ruminants do not have to rely solely on expensive true protein since the microbial activity of the rumen flora is also able to utilise the cheaper non-protein nitrogen present in the feed. The rumen bacteria thrive on the non-protein nitrogen and incorporate it into their own body proteins, which is then digested in the intestinal tract and absorbed. Rumen flora and efficiency also varies between species. Thus, the temperate breeds of cattle introduced to the tropics have a less efficient rumen flora and are unable to digest the high fibre diet of indigenous tropical breeds, while the buffalo is able to thrive on native grasses that could not even maintain the weight of native cattle.

The relative proportion of fibre to starch in ruminant diets influences the constituent members of the rumen flora. Roughage diets, high in cellulose and other non-starch polysaccharides, will produce high concentrations of ethanoic (acetic) acid in the rumen. As the proportion of starch and protein increases, the production of ethanoic acid falls and that of propanoic acid rises. In lactating ruminants this leads to a depression of the milk fat content (Cloudsley-Thompson, 1969; Dalal-Clayton, 1981; Göhl, 1981; Duffus and Duffus, 1991).

1.   FEED REQUIREMENTS

In many respects the feed requirements for livestock are similar to those for human consumption, and those aspects of plant food analysis and general constituents applicable to both human and animal foods have already been dealt with in Chapter 8. Livestock are raised primarily to provide meat and milk and secondly as a means of transport. It should be noted that cattle only convert 10-15% of the vegetable protein they eat into animal protein. Meat production is clearly an inefficient source of protein for human consumption. Theoretically it should be possible to reduce such feed wastage by feeding *leaf protein concentrate* (*LPC*) or *single cell protein* (*SCP*); see Chapter 9 for further discussion (Ferrando, 1981; Göhl, 1981). Even so, in low rainfall areas raising cattle can be a more efficient utilisation of land resources than crop production provided potable water is available.

The main considerations given to livestock feeds are: (1) *Palatability*, whether pleasant to taste either raw or processed, alone or as a condiment to other foods. However, palatability is very difficult to define in terms of the biological processes involved, i.e. the stimulation of the taste buds, aroma and mouth feel. The term commonly implies acceptability but not necessarily desirability. A palatable food may be essentially neutral with regards preference, neither attractive not repellent to the taste, while in terms of nutritional needs a large proportion of the feed may consist of plants that are non-attractive or bitter to the taste: (2) *Quantity satisfaction*, the ability to overcome hunger (appetite) and supply the energy requirement of the consumer without using up internal reserves. The appetite may be affected by the CP content of the feed, since it has been observed that when the CP content of tropical grasses falls below 6-8% the appetite will also be depressed due to a CP deficiency in the animal. While feed high in fibre may be made more appetising by grinding or pelleting, this will have no effect on CP deficient feeds. Since the CP level of legumes is higher than those found in tropical grasses, the inclusion of legumes in a pasture can supplement the low CP level of the grasses. The energy requirement of the animal will, of course, vary according to the age, size, breeding state and work being done. The nutritional requirements for flushing prior to mating, during pregnancy and lactation are clearly higher than for non-breeding animals; working animals employed for transport, cultivation, etc. will also have a higher food requirement than non-working animals; (3) *Nutritional value*, i.e. a balanced diet in terms of protein, fat, carbohydrates, fibre, minerals, amino-acids, vitamins and calories, is essential for animal well-being. No single plant product will supply all these requirements but when taken in combination with other plant and/or animal products can provide a balanced diet. The amino-acid balance of cereal protein, for example, is poor and is particularly deficient in lysine but can be balanced by the use of a lycine-rich supplement such as soya bean meal; (4) *Digestibility* is the percentage of food that has been rendered soluble and assimilated by the consumer, the undigested portion being excreted and will vary according to the digestive system and breed; (5) *Toxic properties*. Some plants are toxic, others possess toxic organs or may be toxic during some stage of their life cycle (see Section 7 for further discussion). Sometimes the toxins may be rendered safe following wilting and in other cases animals reared in areas containing toxic plants may acquire a local immunity; (6) *Seasonal availability*. Ideally there should be no major changes in the level of nutrition throughout the year, i.e. when one fresh food source has finished another should become available. In practice seasonal changes create periods of dearth and the necessity of relying on conserved feed. In non-domesticated animals the seasonal availability will affect their breeding cycle, migration pattern, etc.; (7) *Preservation*, or the ability to conserve fodder either as hay, silage, foggage, etc. for use when green feed is not available, e.g. during the non-growing season or periods of prolonged drought; and (8) *Crop residues* for feed include stubble, straw, stover, sugar beet tops, etc. remaining after harvest (Minson, 1988, 1990; Molyneux and Ralphs, 1992; Morris and Rose, 1996).

While the nutritional value of the cultivated grasses and fodder crops has been reasonably well investigated, the wild grasses, herbs and browse species have been largely confined to an occasional analysis or entirely neglected, especially those of the developing countries. The need to consider changes in nutritional value throughout the growth cycle, complicated by whether considering pure or mixed stands, the influences annual climatic variations and the environment, etc. makes such neglect understandable but regrettable, since good livestock management is dependent on such information.

The units used to measure feed energy value are measured differently in the various countries and can often cause confusion. For example, the *starch equivalent* (*SE*) measures the quantify of pure starch required to produce as much fat as 100 kg of the feed. In Russia the *oat unit* (*OU*) is similar to the SE but is based on oats, while the Scandinavian *feed unit* (*FE*) is based on barley and refers to milk production rather than fattening. The SE units are gradually being replaced by *metabolisable energy* (*ME*) values, defined as the gross energy content of a feed less the gross energy of the animal faeces and digests, and is expressed in kcal. *True metabolisable energy* (*TME*) is used for poultry feeds where the ME is corrected for the endogenous loss of energy from the digestive tract. Göhl (1981) should be consulted for the necessary conversion calculations.

2.   BROWSE

In many parts of the world, especially in the arid tropics and subtropics, the foliage and fruits, known as *browse*, of many trees and shrubs provide an important source of crude protein, albeit often of low digestibility and deficient in phosphorus. Indeed, *"It is a humbling fact for grass pasture experts that probably more animals feed on shrubs and trees, or on associations in which shrubs and trees play an important part, than on true grass-legume pastures"* (Whyte, 1947). The browse plants can often provide nutritious feed at periods of the year when the ground herbage is either low in nutrients or absent, with the spring flush of the ligneous vegetation often occurring before that of the forbs. In tropical Africa alone it has been estimated that at least 75% of the 7000-10000 tree and shrub species are browsed to a greater or lesser extent. Leguminous browse is particularly nutritious, especially those species with N-fixing capabilities, species of *Acacia* and *Prosopis* being particularly valuable in the tropics of the Old and New Worlds, respectively as a source of leafy twigs and nutritious pods and seeds. Browse from *Faidherbia albida* is of particular value as the tree has the peculiar distinction of producing leaves and flowering and fruiting during the dry season when other sources of green feed are unavailable (Whyte, 1947; Wickens, 1969; Göhl, 1981; Skerman *et al.*, 1988).

Browse can also attract wild animals. In Brazil *Agouti paca* (pacas) and *Dasyprocta punctata* (agoutis) are attracted to the fruit of *Orbignya phalerata*, syn. *O. martiana* (babassu palm) and feed on the starchy mesocarp. The pacas provide an important source of game meat for the local people and because the animals are concentrating around the babassu palms their hunting is made easier (May *et al.*, 1985).

3.   GRASSES

While the cereal grasses are a major source of carbohydrates for humans, their main source of protein and fat is obtained from grazing livestock, whose major food source are the grasses. Of the *ca.* 10 000 species of grasses in the world only *ca.* 40 are commonly cultivated as pasture grasses. Furthermore, in the tropics less than half of the available pasture grasses are cultivated, which is due to the heavy reliance on natural grasslands for grazing. Indeed, *ca.* 50% of the world's cattle population utilise natural grasslands.

Natural grasslands reflect the native vegetation of the region, albeit often modified by the activities of man and his domesticated livestock. Such grazing lands are termed *rangelands*. They include a variety of ecosystems, from pure stands of grasses, grass and herb mixtures, grass and tree savannas, dominance by shrubby chenopods, the *Kobresia* alpine meadows of Mongolia grazed by *Bos grunniens mutus* ((yak) and *Ovis aries* (sheep), and the mosses, lichens and ericaceous shrub grazing of the arctic tundra where *Rangifer* spp. (caribou and reindeer) represent important wild and domesticated livestock.

The composition of such natural grasslands can vary greatly throughout the growing season and present quite different impressions of the vegetation, so much so that an inexperienced observer can obtain a completely wrong evaluation of the grazing potential. For example, in the *Acacia senegal* savanna dominating the sandy soils of the Sudan, the early dominance at the start of the rainy season by the up to 45 cm high annual *Sporobolus* spp. provide a purple hue to the rangeland, masking all other species. The *Sporobolus* is succeeded by the dominant 1 m tall perennial *Aristida sieberana*. Late in the growing season the appearance of the even taller *Eragrostis tremula* provides a quite false aspect of dominance.

The *semi-natural grasslands* are where tree clearances within the forest zone have led to dominance by a perennial herbaceous cover. They differ from rangelands by their existence being based on management, especially stocking rates and fire, rather than climate. They also differ from pastures by being composed of indigenous rather than exotic species.

Artificial grasslands are those grasslands where the native species are either strictly controlled or eliminated. Selected and often exotic grasses are either sown alone or in a legume or herb mixture. Such *pastures* are laid down in any climatic region, either as permanent pasture or forming part of the crop rotation as a short duration *ley*. They are generally intensively managed and highly productive, their productivity often extending to times of the year when the native species are either unproductive or unavailable. The term *meadow* is generally applied to semi-natural grasslands and pastures that are reserved for hay or silage, or to rich waterside fields that are frequently flooded by river water, either naturally or via sluices. Such *water meadows* provide rich grazing. The botanical composition of all these grasslands can dramatically changed by grazing pressure, fertiliser applications, mowing, irrigation, fire, etc. They all require careful management for sustainable productivity.

Not only does the chemical composition of grasses vary between species, there are also variations with stage of growth, climate and soil; the latter especially affects the mineral content. Furthermore, due to selective grazing by the animals, especially sheep and horses, there can be as much as 25% difference between the crude protein and crude fibre content of the pasture and the value of what is actually consumed. In intensively managed grasslands it is possible to reduce losses from defecation and trampling, especially on heavy or wet soils, by a practice known as *soilage* (not to be confused with soiling by excreta) or *zero-grazing* where the daily ration is cut and transported to the livestock (Dalal-Clayton, 1981; Göhl, 1981; Coupland, 1992).

3.1   Grass Conservation

The grasses can be conserved for future use as hay, standing hay or foggage, silage, and dried grass. *Hay* made from grasses cut early in the flowering stage have lower yields and higher nutritive value and palatability than hay made from more mature grasses. The high moisture content of the grasses can make curing difficult, while periods of unseasonable rain during haymaking will lower or even ruin the nutritive content. Unfortunately, in the wet tropics haymaking is often impractical due to the optimum haymaking season coinciding with the peak of the rainy season. In the arid tropics the low yields make haymaking uneconomic and it is usually more practical to leave the grasses uncut as *standing hay*, or *stem-cured grass* for dry season grazing. Naturally, such standing hay will have a very much lower nutritive value than properly made hay, but the method does have the advantage of requiring no labour input beyond maintaining firebreaks and ensuring freedom from pirate grazing. A common temperate variant of standing hay is to cease grazing from midsummer until the onset of winter. Known as *foggage*, it has the advantage of providing low quality pasture grazing, although there is the risk of possible frost damage or destruction by trampling in wet years.

Silage, however, is more or less independent of the weather and silage making is widely practised in both tropical and temperate regions. It is a high moisture plant product and there is very little loss in the nutritive value when it is well made. The principal of silage making is the anaerobic bacterial fermentation of the carbohydrates in the plant material to organic acids and of the proteins to amino-acids, which act as preservatives. It does, however, require a comparative large amount of lactic and ethanoic (acetic) acid fermentation to ensure sufficient acidity, *ca*. pH 4.2, to prevent adverse fermentation and putrefaction.

Unlike silage, *haylage* is a low moisture grass or crop silage, where the mown crop is wilted in the field to 50-60% moisture. It is then chopped and blown into an airtight silo, where it undergoes considerable less fermentation than required for silage. The CO_2 released by cell respiration inhibits any bacterial activity and rotting. Haylage does have the disadvantage of being spontaneously combustible and requires storage in airtight silos to limit access to the atmosphere. The method is regarded as expensive and is rarely used in the tropics. An almost identical technique is used for *drillage*, in which wilting is to 60-70% moisture.

Artificially *dried grass* reduces nutrient loss but it is very expensive and can only be justified for high-quality grasses. The dried grass is either pressed into cobs (cubes) or wafers, or milled and mixed with molasses to form cubes or wafers. Dried grass is usually incorporated in mixed feeds as a source of protein and A and B complex vitamins (Voisin, 1959; Dalal-Clayton, 1981; Göhl, 1981; Skerman and Riveros, 1990).

Appropriate range and pasture management to prevent overgrazing, maximise productivity, prevent wastage, encourage desirable species and discourage the less desirable, can also be considered as a form of grass conservation.

4.   **LEGUMES**

Legumes are a high protein and mineral rich livestock feed. They help balance the amino-acid deficiencies, etc. in cereals and grasses, as well as their presence improving soil fertility and productivity. They may be cultivated as a: (1) *Pulse crop* as part of a pulse feed concentrate, e.g. *Vicia faba* (field or horse bean). However, beans are not very palatable to stock, tend to be indigestible if fed raw, particularly by pigs. Bean meals also have the disadvantage of not storing well, developing a bitter, rancid taste after a few weeks in store; (2) *Fodder crop* for conservation, e.g. *Medicago sativa* (lucerne); (3) *Pasture crop* with grasses in pastures for grazing, e.g. *Trifolium* spp. (clovers); and (4) *Agroforestry crop* to provide browse, e.g. *Leucaena leucocephala* (Dalal-Clayton, 1981; Göhl, 1981).

5.   **CONCENTRATES**

The term *concentrates* refer to a variety of animal feeds with a high food value relative to volume, low in fibre and high in protein, carbohydrates or fat. Trace elements, vitamins and minerals are often added to provide a balanced ration (Dalal-Clayton, 1981).

5.1   Oilseed Cakes and Meals

Oilseed crops, such as linseed, oilseed rape and sunflower in the temperate regions, and cotton, groundnuts, soya beans and oil palms in the tropics, are grown to produce edible oils for human consumption. The residue following oil extraction can usually be fed to livestock as a high protein/low starch feed, with any residual oil content depending on the method used to extract the seed oil. Some seeds, such as cottonseed and sunflower, possess seed coats high in fibre, the empty pod, hull, husk or shell of the groundnut is also high in fibre. The fibre must be removed before processing for oil by a process known as *decortication*. Undecorticated cake, which has a lower the feed value because of the higher fibre content, is also lower in protein than decorticated cake but richer in residual oil because the fibrous seed coats and hulls obstruct the removal of the oil.

The oil is removed either by pressing or with solvents. Of the former there are two methods used, hydraulic and screw pressing. For *hydraulic pressing* the ground seed is heated and wrapped in cloth before placing in the press. The residual oil in the resulting press cake is slightly higher than from screw pressing. *Screw pressing*, also known as the *expelling process*, is a continuous process. The seeds are pressed by a screw through a tubular cage, which has a smaller pitch towards the discharge end. The pressure and temperature consequently increases as the seed mass passes through the cage, causing the oil cells to rupture and the fluidity of the oil to increase and escapes through a small aperture in the cage wall. It is the most drastic of the separation methods since the heat treatment causes some damage to the protein. The resultant *press cake* usually contains 5-10% oil; if ground the cake is then known as *oil meal*.

For *solvent pressing* the ground seed is placed in an extraction pot, through which the solvent is pumped. The process is repeated until further extraction is deemed uneconomical. The solvent is recovered by distillation and reused. The ground flake residue, i.e. *oil meal*, is treated with steam to remove any traces of the solvent and then dried. More thorough than pressing, the solvent process results in an oil meal containing less than 2% oil. The oil meals are less palatable than the oilseed cakes and, because of their low oil content, have a lower energy value. Oilseed cakes, however, unlike oil meals, are liable to turn rancid (Göhl 1981).

5.2   Cereals

Cereal grains are generally more important as livestock feed in the temperate regions than in the tropics, where they are mainly grown for human consumption. However, the introduction of high-energy feeding systems is gradually changing the situation, especially where the introduction of high-yielding cereal varieties has created local grain surpluses that can be fed to livestock. The relative feeding value of the various cereals is dependent on a number of factors, the most important being digestibility, fibre and lysine content. These, together with the type of grain, tannin content (depending on the cultivar) and hydrocolloids (influenced by maturity), affect the rate of livestock growth.

Cereal grains are essentially energy concentrates that can be used to supplement protein-rich feeds. Which particular cereal is favoured for use is usually dependent on the lowest cost per unit of metabolisable energy. The cereal proteins, however, are mostly deficient in the amino-acid *lysine* and while this deficiency is unimportant when cereals are fed to ruminants, careful ration formulations are required when feeding cereals to monogastrics. In addition, all the cereals except maize have low levels of the essential fatty acid *linoleic acid*. Cereals only contain small amounts of vitamin E, chlorine and some water-soluble vitamins. Apart from K and P, all are deficient in most of the minerals required for growth and reproduction. There are also differences in mineral availability. While the K in wheat is almost completely available to pigs and poultry, only about half of the P is available to monogastric animals (Göhl, 1981; Morris and Rose, 1996).

The processing of cereals in order to increase their consumption and avoid wastage by improving their palatability, digestibility and nutritive value, is both expensive and time consuming. The processing used, whether it involves dry or wet milling (see Chapter 8), will depend on a number of factors, including whether the animal is a ruminant, or monogastric and the stage in its life cycle. Rice bran and other milling by-products have traditionally been used for feeding livestock rather than humans because an enzyme mixes with the oil in the bran during its removal and will eventually produces undesirable rancid odours and flavours (Grubben *et al.*, 1996).

The major use of maize grain is for animal feed, either directly or as part of a pre-processed feed which, because of its low lycine content, is often supplemented with soya bean meal. New high lycine strains with the opaque-2 gene are now available, although some supplementation is still required for quality performance. While maize can be fed whole to livestock, feed efficiency is improved with processing, either by hammer or roller milling or by steam flaking, to reduce particle size. The co-products of maize after milling are gluten meal, gluten feed, germ meal, and heavy steepwater; all of which have considerably higher levels of lycine and other essential amino-acids than the grain.

In the US the preferred source for poultry is corn gluten meal, with 60% protein. Poultry farmers greatly value this source of the colour pigment ***cryptoxanthin*** as it is the only cereal grain with enough carotenoids for the consistent production of eggs with yellow yolks. The carotenoid content varies according to the type of maize, being virtually absent in white maize and with high concentrations present in the endosperm of yellow dent maize. The cryptoxanthin can, however, adversely affect the colour of pig fat by producing an undesirable yellow tinge. (Göhl, 1981; Langer and Hill, 1982; Eckhoff and Paulsen, 1996).

Some 75% of the world's barley crop is used as feed for ruminants and monogastrics, the latter mainly for pigs and poultry. The barley protein levels are significantly greater than those of maize and are often lower than those of rye, wheat and oats. The grain is usually rolled or flaked before feeding to cattle, a process that requires a uniform grain since rollers set for thin grain will crush plump grains and form fines, which can cause digestive problems. Alternatively, if set for plump grains any thin grain will escape uncrushed. The 2-row barleys are therefore preferred; they also have a higher feeding value than 6-row. The hammer mill is normally used for monogastric feed, consequently plumpness is not important.

The hull is relatively digestible by ruminants and doesn't affect the digestible energy, although indiscriminate feeding can cause acidosis or bloat due to upsets in the rumen flora caused by the rapid digestion of the barley starch. The barley hulls also cause less acidosis and bloat in ruminants than wheat because the barley hulls assist in diluting the effects of rapid starch digestion. The barley hulls, however, are not digested by monogastrics, resulting in a reduced energy content of the feed, and requiring higher levels of barley feed in compensation. There is currently an increased interest being shown in the hull-less forms of both 2- and 6-row feed barleys because they can produce energy levels for monogastrics approaching that for feed wheat. On a dry weight basis hull-less barleys also have higher levels of essential amino-acids than covered barleys and generally better than those of wheat.

The polysaccharide *β-glucan* present in both hulled and hull-less barleys used for poultry feed reduces the availability of energy and protein and needs to be corrected by the addition of supplements containing enzymes with β-glucanase activity to hydrolyse the β-glucan. The β-glucan appears to have little effect on pig nutrition.

The remains of malted barley, known as *distillers' grains*, can also be fed to livestock, either wet or dry, the latter being preferred because the wet grain retains some raw alcohol and could intoxicate the animal (Dalal-Clayton, 1981; Edney, 1996).

Sorghum has 95% of the feeding value of yellow dent maize. Because of their relatively poorer protein digestibility, especially the brown sorghums, they require proper processing before feeding. Although sheep are capable of thoroughly masticating the grain, the digestibility for ruminants generally can be improved by a processing, including steam flaking, micronising, exploding, popping, reconstitution and grinding processes. For pigs and poultry the grain is passed through the hammer mill and possibly pelleted for poultry and pet food.

The waxy kernels flake more readily than the non-waxy cultivars, they also show improved feed efficiency when fed to ruminants, less for pigs and none for poultry. Sorghum diets are too low in carotene for laying hens and broilers and require supplementing with lucerne, marigold meal from *Tagetes erecta* (Aztec or African marigold) or yellow corn gluten. Sorghum distillers grain, an important by-product of the brewing industry and containing 30% protein, can also be fed to livestock (Göhl, 1981; Rooney, 1996).

Oats is primarily a feed grain, with some 75% or more of the world's production being fed to livestock. Oats are the preferred cereal food for horses, being highly palatable and easily digestible as well as having excellent nutritive qualities. Its popularity as a feed for other livestock is declining with the advent of alternative, low-cost feed ingredients, such as maize, soya meal, cottonseed meal and canola meal (low erucic acid oilseed rape), despite the obvious benefits of oats as a feed (see Webster, 1996 for further information).

Rye is regarded as a cheap, high energy, farm-produced feed, but should not be fed alone. It is often preferred for fattening pigs and poultry despite having a lower palatability and feed conversion ratio than wheat. It should not be fed to young stock as it may cause digestive disorders (Göhl, 1981; Weipert, 1996).

5.3   Pulse Legumes

The legume seeds generally contain various antinutritional substances that exert adverse physiological effects when ingested by livestock. The protease inhibitors in soya bean are heat-labile, as are also the lectins in kidney beans. The coloured seeds of *Phaseolus* spp. also contain more cyanogen than white seeds. Others toxic ingredients, such as the bitter-tasting quinolizidine alkaloids present in some strains of lupins, are heat-stable; reduced toxicity can also be obtained through selection and breeding for 'sweet' cultivars. Such antinutrional disadvantages have to taken into account when considering their valuable protein contribution to the diet of livestock (Liener, 1980; Nowacki, 1980; Göhl, 1981).

6.   ROOTS AND TUBERS

Generally fresh roots and tubers contain 80-90% water and are regarded as succulent feeds. Their feed value lies in their high sugar and starch content, and in their palatability. The dry matter is highly digestible, low in fibre and crude protein, the latter consisting largely of nonprotein nitrogen. The Ca, P and vitamin content are also low, the exception being carrots, which are particularly high in vitamin A.

Fresh roots tend to be laxative and are of particular value when the rest of the feed is dry. However, if ruminant feeds are low in crude protein, starchy feeds such as roots may depress the digestibility of the cellulose since the rumen flora will consume the crude protein during the fermentation of the more easily digested starch rather than undertake cellulose fermentation. The feed value of roots and tubers deficient in crude protein can be enhanced when fed with urea.

Depending on the economics of root *versus* grain crops, roots and tubers could become a major energy source for livestock in many parts of the world. Many root crops are capable of high yields in the tropics and could be important where grains are difficult to grow (Göhl, 1981).

7.   TOXINS

The susceptibility in animals to toxins is related to differences in their feeding behaviour, diet and the anatomy of their digestive tracts. In ruminants toxic substances become diluted in the first stomach and rumen activity may even destroy the poisonous principles or promote the release of toxic constituents. For example, the **mimosine** present in *Leucaena leucocephala* (lead tree) is toxic to monogastric animals and causes hair loss, but is apparently non-toxic to ruminants, although enlarged thyroid glands have been observed in calves at birth. The detoxification mechanisms present in animals are also species specific. For example, the rabbit possesses the enzyme *atropinesterase* and is therefore not adversely affected by the alkaloid *atropine* when eating *Atropa belladonna* (deadly nightshade). The enzyme is also found in goats, making them less susceptible, although contrary to popular belief, they are not immune to all toxins. Other examples of alleged immunity are deer, which are reputed to feed on *Taxus baccata* (yew) and *Rhododendron* spp., while *Sciurus carolinensis* (grey squirrel) are reported to be able to eat *Amanita* spp.

The age and health of the animal too can be important. Species of *Amaranthus*, *Atriplex*, *Chenopodium*, *Oxalis*, *Portulaca*, *Salsola* and some grasses have been responsible for oxalate toxicity, especially in young stock. Because of their high Ca requirement young stock may be adversely affected by the oxalates locking up the body Ca. Oxalates can also be a problem with horses as it too leads to Ca deficiency, although ruminants can acquire a degree of immunity arising from changes in the composition of their rumen flora so that the oxalates are metabolised, and problems with Ca deficiency are consequently extremely rare.

Many forage plants can be toxic to animals at certain times of the year. For example, the tannin content of *Quercus* spp. (oak) can increase quite dramatically from *ca.* 0.5% of leaf dry weight in early May to 5% in September. In the UK, the acorns and oak leaves contain relatively low molecular weight hydrosoluble tannins and these are reputed to cause liver and kidney lesions in cattle and sheep.

Toxicity is sometimes manifested when the animals are under stress. Browse plants, for example, that are not normally toxic under free-range grazing because they not consumed in sufficiently large quantities by resident stock, may be avidly eaten by hungry stock newly introduced or passing through the area. The previous experience by animals of toxicity may also be important since livestock introduced to unfamiliar grazing lands will often eat greater quantities of toxic species than locally bred stock. It would appear that immunity can even be acquired since, in South Africa species of *Moraea* are apparently safely grazed by livestock reared in the area, but are toxic to those that are introduced. Immunity is a complex subject and requires further investigation.

Plants containing *alkaloids* are generally bitter and are normally avoided. Where present the alkaloids are generally widely distributed through the plant organs so that any part ingested can be dangerous to stock; the nervous system usually affected. Climatic and seasonal conditions generally have little affect on the alkaloid levels, but not necessarily so. For example, concentrations of the tryptamine-based alkaloids *gramine* and *hordenine* present in the temperate grasses *Phalaris aquatica*, syn. *P. tuberosa* (canary grass), and *P. arundinacea* (reed grass) have been found to increase with soil nitrogen, shade and following rain, as well as having a diurnal rhythm with a much lower concentration early in the afternoon. The alkaloid concentrations are also under genetic control and preliminary investigations suggest that the character for low concentrations are recessive. Furthermore, experiments with sheep have shown that low levels (0.0.1%) of gramine actually stimulate grazing while higher levels up to 1% lead to rejection (Parodi, 1950; Jackson and Jacobs, 1985; Harborne, 1988; Skerman *et al.*, 1988; Skerman and Riveros, 1990; Cooper and Johnson, 1998).

Cyanogenic glycosides occur in a number of grasses, legumes and fodder plants. While toxicity from grasses is generally not considered to be of major importance, HCN can be a serious problem in *Sorghum* spp. and *Stipa subaristata*. Toxicity tends to be higher on soils high in nitrogen and in young plants and regrowth.

The glycosides are regarded as stable and non-toxic until hydrolysed by the appropriate enzyme, although the manner in which the enzyme is expressed in not always clear. Thus, some *Acacia georginae* (Georgina gidyea) trees in Australia are considered toxic and others non-toxic. Hall *et al.* (1972), Askew and Mitchell (1978), Everist (1981) and Baumer (1983) believe that the toxicity is due to differences in the capacity of the acacia to take up fluorides and convert them into fluoroacetate. The toxic principle is probably **monofluoroethanoic acid**, of which the seeds contain more than the pods, and the pods more than the leaves. Up to 4.5 kg of whole pods fed with bran and chaff were found to be lethal to cattle and 0.9 - 6.1 kg lethal to sheep. Toxicity in members of the Australian legume genus *Gastrolobium* has similarly been attributed to fluoroethanate.

But, according to Dayton (1948) and Jackson and Jacobs (1985) toxicity is due to the cattle browsing both *A. georginae* and *Eremophila maculata*, syn. *Stenochilus maculatus* (spotted berrigan). The acacia pods are harmless, but contain an enzyme that liberates prussic acid from a glucoside present in the berrigan browse. Thus, both species may be grazed separately with impunity, and they are only toxic when grazed together. Clearly, a more critical investigation is required to determine whether both or only one process is correct.

Chapter 11

Food for Bees and other Desirable Invertebrates

The flowering plants require either self- or cross-pollination in order to reproduce. For cross-pollination the main pollen dispersal agents are *anemophyly* (wind pollination) and *entomophily* (insect pollination), the latter usually involving the accidental or deliberate carriage of pollen by insects attracted to the flowers for food, i.e. pollen and/or nectar, by scent, colour, honey guides, or to flowers with pseudocopulation camouflage, e.g. many orchids (see Harborne, 1988) for further discussion). Insect pollination is often of vital importance for good crop production and hives of bees may be introduced in order to maximise pollination in orchard and legume crops. In cases of specialised pollination systems, e.g. *Ficus* spp. (figs), their needs have to be carefully considered when cultivating plants outside the environment of their natural insect vectors.

Pollinating insects normally vary in abundance from year to year. Their numbers can also be adversely affected by habitat destruction resulting from mechanised cultivation, while the use of agrochemicals can be lethal. So much so that suitable pollinating insects may be poorly represented or even absent in large-scale agriculture and irrigation schemes. They may also be absent or scarce in arid and semi-arid regions due to a lack of suitable environmental conditions. Special arrangements may also be necessary for greenhouse crops. See Crane and Walker (1984) for further information on crop pollination.

Plant vegetative organs can also provide food and shelter for both economically desirable insects and undesirable insect pests, although the distinction between desirable and pestiferous insects is not necessarily clear-cut. Silkworms, for example, chewing mulberry leaves will be welcomed by their breeder but not by the gardener proud of his/her mulberry hedge!

1.   BEE FOOD

Bees feed on pollen, nectar and water, especially nectar. The pollen provides the protein and nectar the carbohydrate requirements of the bee. However, not all pollen and nectar are desirable, there are a number of sources that are toxic to bees - see Crane *et al.* (1984). The best known examples for the UK are *Tilia* spp. (lime tree), beneath whose canopy the ground may be littered in dry years with dead or drunk bees, probably suffering from excessive mannose after nectar ingestion. ***Nectar*** consists of between 15-75% by weight of glucose, fructose and sucrose solutions, secreted in floral and extrafloral nectaries. After partial digestion by the honeybees the nectar is converted into the familiar sweet yellowish or brownish viscid fluid known as ***honey***, with much of the sucrose broken down into glucose and fructose and stored in the comb for future use. While waiting to store the nectar a portion of the nectar is assimilated by the bee and converted into wax which, space permitting, is used to extend the comb area. The identity of the pollen present in the honey may sometimes be used by prospectors to indicate the presence of plants associated with soil minerals. See Chapter 20 for further discussion.

Honey has been used as a sweetening agent and for food and drink since prehistoric times. When fermented honey produces the alcoholic beverage ***mead*** (from Middle English *mede*, Old English *medu*, *meodu* (Long, 1994). However, not all honeys are edible, some are toxic to humans. The classic 5th century BC example is from 'The Retreat of the Ten Thousand' when the Greek soldiers led by the Athenian Xenophon were encamped near the Black Sea at Trebizonde and ate the *meli maenomenon* or 'mad honey' from *Rhododendron ponticum*. Considering how the heather honeys from *Erica* spp. (heathers) and *Calluna vulgaris* (ling) are so greatly esteemed, it is perhaps surprising to find how many other members of the Ericaceae yield toxic honeys, including *Kalmia latifolia* (calico bush, mountain laurel), *Rhododendron* spp. (including *Ledum*), etc. Other families are similarly implicated, including the Coriariaceae, e.g. *Coriaria arborea* (tutu) in New Zealand, and Loganiaceae, e.g. *Gelsemium sempervirens* (yellow jasmine) of south-eastern USA (Howes, 1949). See Ott (1998) for further examples.

Some bees, including honeybees, will occasionally collect a resinous exudation known as ***propolis*** (Greek *pro* = in front of, *polis* = city or community, i.e. hive entrance). Propolis is collected from tissues that secrete sticky lipophilic substances and flavonoids from the epidermal cells of the leaf buds of *Populus* spp. (poplar), the flavonoid aglycones, terpenes and mucilages secreted by the glandular trichomes on the buds of *Alnus* spp. (alder) and *Aesculus hippocastanum* (horse chestnut), and from wounds to secretary cavities and ducts of other plants, including *Casuarina equisetifolia* (beefwood) and *Mangifera indica* (mango). It is used by the bees either alone or with beeswax in the construction and adaptation of their nests to seal the cracks and crannies in the hive. The use of the resinous and glue-like properties of propolis by humans was recorded as long ago as 300 BC. It was later used in folk medicine, and more recently used pharmaceutically for its antibiotic properties. Propolis has also been used in polishes and varnishes, and possibly by the Ancient Egyptians for embalming (Butler, 1959; Ghuisalberti, 1979; Harborne, 1988; Crane 1990). Honeydew (see Section 6) may also be a source of bee food.

The increasing monoculture of agricultural crops, e.g. *Vicia faba* (broad beans) and apple orchards, may require the seasonal introduction of hives of honeybees to ensure adequate pollination. Honeybees are particularly useful since the foraging bees are present in substantial numbers at flowering time, whereas solitary bees and bumble bees are seldom present in sufficient numbers in spring because the overwintering females are only just beginning their nest development. There are two 'domesticated' species of honeybees that have been induced to establish their colonies in hives. They are *Apis mellifera* (western honeybee) and *A. indica* (eastern honeybee). Notwithstanding, it should be appreciated that all honeybees are wild and that 'domesticated' is used in the sense of managed rather than tamed (MAFF, 1971).

Useful global directories of pollen, nectar, honey and honeydew sources are provided by Crane and Walker (1984) and Crane *et al.* (1984) and for propolis by Crane (1990).

2.   FOOD FOR SILKWORMS

Silk is produced by several insect larvae and spiders, although it is the fine, lustrous fibre from the cocoon spun by the silkworm, the larva of *Bombyx mori* (Chinese moth), that is the principle source of commercial silk. According to Long (1994) 'silk' is from Middle English *silk*, *selk*, Old English *siolic*, *seolec*, from Late Latin *sericum* (noun), Latin *sericus* (adjective), from *seres*, from Greek *Sēres*, 'an oriental people' (probably meaning 'the silk people'), from Chinese *sï*, silk. Despite its ancient lineage, early examples of silk textiles rarely survive; well-preserved examples are reported from cemeteries at T'u-lu-fan (Turfan) in Sinkiang dating from the 5th and 6th centuries AD. The silks have been dyed using the Batik technique of masking portions of the fabric with wax to prevent any dye spreading beyond the designed pattern (Ridley, 1973).

Commercially exploited silkworms are raised in groves and plantations of *Morus alba* cv. 'Microphylla' (white mulberry), a native of central and eastern China but now naturalised in Europe and North America. China is by far the largest exporter of silk with *ca.* 90% of the world's production of raw silk and 40% of the world's silk fabrics (Robbins, 1995).

3.   FOOD FOR LAC INSECTS

Lac is the resinous secretion secreted by female lac insects, especially *Laccifer* spp. The name is from the Dutch *lak* or French *laque*, from the Hindu *lākh*, from the Prakrit *lakkha*, from the Sanskrit *lākshā* (Long, 1994). There are a number of lac insects, of which the most important is *Laccifer lacca*, syn. *Kerria lacca*, whose distribution ranges from northern India to Indo-China. It feeds on more than 160 host trees, including *Acacia catechu*, *Butea frondosa*, *B. monosperma* and *Caesalpinia crista*, while in China the host trees are mainly *Cajanus cajun* and species of *Dalbergia* and *Hibiscus*. The scarlet pigment known as *lac dye* is found in the live pre-emergent lac insects that develop in a resinous cocoon, known as *sticklac*, on the twigs of the host trees. The dye is obtained by aqueous extraction from the resinous cocoon and is used both as a cosmetic and for dyeing wool, silk and leather.

The resinous residue remaining after extracting the lac dye is further processed via *seedlac* to *shellac*. Shellac is a natural thermoplastic used for a variety of purposes, mainly as the definitive ingredient of French polish; it is also used as a resin binder for mica-based insulating materials, as an ingredient of some hair sprays, paints. varnishes, printing inks, sealing wax and adhesives, and as a glaze in confectionery and pharmaceuticals. Former commercial uses of shellac included gramophone records, stiffening felt hats and making buttons, all of which have now been superseded by plastics. India is the largest producer of shellac, with Thailand the only important rival. There is now concern in both producing countries that the demand for wood will place the trees providing lac food in jeopardy (Sharp, 1990; Lemmens *et al.*, 1991; Saint-Pierre and Ou, 1994; Green, 1995; Robbins, 1995).

In south-western North America an orange lac from the excretions of the scale insect *Tachardiella* spp. is collected from the twigs of *Coursetia glandulosa* (resilient plant), *Encelia farinosa* (brittlebush) and *Larrea divaricata* subsp. *tridentata* (creosote bush). The lac is used by the Seri as an adhesive for hafting arrows, spears and harpoons and as a sealant; it is also used in Sonora for treating colds, fevers and tuberculosis (Felger and Moser, 1985; Turner *et al.*, 1995).

4.   FOOD FOR DYE INSECTS

The deep red dye known as *cochineal* is made from the dried and pulverised bodies of pregnant females of the tropical American scale insect *Dactylopius cocceus*, which feed on *Opuntia cochenillifera*, syn. *Nopalea cochenillifer,* and other prickly pears. The deep red colour is due to the presence of *carminic acid*.

Cochineal is produced commercially in Peru (*ca.* 200 tonnes yr^{-1}) and the Canary Islands (*ca.* 30 tonnes yr^{-1}). The cochineal insect was also introduced into the Mediterranean region to feed on *Quercus coccifera* subsp. *coccifera* (Kermes oak) to provide a European source of cochineal. Thus. the three sprigs of the Kermes oak that form the crest of the Dyers' Company, one of the City of London's trade guilds, is evidence of its former importance. Cochineal was formerly used to dye cloth but has now been superseded by synthetic dyes. Nowadays the main use for cochineal is for colouring food, drugs and cosmetics, and as an indicator. Although cheaper synthetic substitutes are available, there is an increasing preference by the public for the natural product (Sharp, 1990; Green, 1995; Robbins, 1995; Mabberley, 1997).

Kermes is the crimson dye obtained from the gravid female scale insect *Kermes ilices* infesting the twigs of oak trees in the Mediterranean basin and Middle East, especially the evergreen *Quercus coccifera* subsp. *coccifera* and subsp. *caliprinos*, syn. *Q. pseudococcifera*, *Q. ilex* (holm oak) and *Q. suber* (cork oak). The etymology of the name kermes has an interesting pedigree, the anglicised name originating from the French *kermès*, which is a shortened form of *alkermès*, from the Spanish *alkermez*, from the Arabic *al-qirmis*, from the Sanskrit *krmi-ja*, i.e. [red dye] produced = *ja*, by the worm = *krmi*. The dye has been used since antiquity and is believed to be the 'scarlet' of the Bible (*Genesis*, 38, 28, 30). The trade in kermes finished in the 19th century following the introduction of cochineal and the subsequent development of aniline dyes (Moldenke and Moldenke, 1952; Meikle, 1985, p. 1486; Long, 1994; Stearn, 1992, p. 232; Green, 1995).

5. FOOD FOR EDIBLE INVERTEBRATES

Edible invertebrates and their host plants are an under-researched non-wood forest product. They form an important seasonal source of food of unknown quantity and quality in many regions of the world, especially among the developing nations. My own experience in a rather seedy restaurant in Mozambique was of a vegetable soup liberally garnished with green caterpillars, which I considered a rather tasteless and over-rated delicacy. However, opinions differ since the roasted 5-7.6 cm long larvae of *Gonimbrasia belina* (mopane moth) which live on *Colophospermum mopane* (mopane), are eagerly eaten by the indigenous peoples of southern Africa. The mopane worms are reputed to be highly nutritious, with a higher protein content than beef! In central Australia the larvae of the witchetty grubs that dwell in the roots of *Acacia kempeana* (witchetty bush) are traditionally eaten by the Aborigines, as are also a number of other unspecified edible grubs associated with *Acacia* seeds.

Similarly, in the New World the Papago and Seri Indians of the Sonoran Desert gather the edible caterpillars of *Hyles lineata* (white-lined sphinx moth) that feed on desert ephemerals, especially *Boerhaavia erecta*. Examples from South America include the edible larvae of the beetles *Rhynchophorus palmarum* and *Caryobruchus* spp., the former harvested from deliberately felled logs of *Oenocarpus* spp., syn. *Jessenia* spp. the latter from felled stems left *in situ* of *Orbignya phalerata* (Babassu palm); the larvae from fallen fruits of *Scheelea* spp. infested with the bruchid *Pachymerus nuclearum*, are also collected and are either cooked and eaten or used for fish bait (Palmer and Pitman, 1972; Beckerman, 1973; Ruddle, 1973; Felger and Moser, 1985; May *et al.*, 1985; Dufour, 1987; Devitt, 1988, 1992).

6.   HONEYDEWS

Honeydews and *manna*, known as *lerps* in Australia, are the sugary and often edible viscous secretions secreted by certain herbivore members of the Hemiptera, especially aphids, and excreted through the anus onto the leaves of plants (see also Chapter 9).

The dried exudate from *Fraxinus ornus* (manna ash) was once believed to be the manna that fed the Children of Israel escaping from Egypt, despite the fact that species is a native of southern Europe and western Asia and does not occur in Sinai. A more probable source of manna is the crystallised honeydew produced by the scale insects *Trabutina mannifera* and *Najacoccus serpentinus* feeding on species of *Tamarix*. In Iran an edible sweet exudate known as 'gaz of Khunsar' is similarly collected from *Astragalus adscendens* (gaz), with the last instar nymph of *Cyamophila astragalicola* being responsible for the exudation (Baum, 1978; Grami, 1998). See Chapter 9 for further details.

From Australia, a lerp is collected from scale insects feeding on *Acacia aneura* (mulga) by specialised workers, known as repletes, of the honey ants *Melophorus bagoti* and *Camponotus* spp. The Aborigines excavate the ant chambers, harvest the repletes and carefully suck the honey stored in their grape-sized swollen abdomens. However, not all honeydews are edible. In New Zealand, that produced by the vine hopper *Scolypopa australis* feeding on *Coriaria arborea* (tutu) is poisonous, as is also the honey (Howes, 1949; Devitt, 1986, 1989).

Chapter 12

Timber and Wood Products

Trees are an extremely valuable and an often over-exploited and endangered natural resource of timber, fuel and numerous non-wood forest products (Wickens, 1991). Following the primary conversion of the felled trees into manageable forms, they provide timber, i.e. wood other than fuelwood, for construction, carpentry and joinery purposes. Timber uses include heavy and light structural and marine timbers, i.e. beams, pilings, planks, poles, props, stakes, struts and rails, for building houses, ships and boats, vehicle bodies, agricultural implements, bridges, sleepers, fencing, handles, ladders, boxes, crates, food containers, matches, vats, battery separators, pattern making, etc. The more decorative woods are used for veneer, parquetry, turnery and carving, including interior decoration, furniture, cabinet work, musical implements, toys and novelties, sporting goods and precision equipment. Other wood uses, either as timber or as a by-product, are for wood pulp, sawdust, wood wool, plywood, wood chips for particle board, blockboard, chipboard, hardboard, laminated wood, etc. Trees also provide cork and cork substitutes, gums and resins, latexes and rubbers, petroleum substitutes, alcohols, etc. (Keating and Bolza 1982; Cook, 1995).

1.   TIMBER

Two major types of commercial timber are recognised, *softwood* obtained from the gymnosperms and *hardwood* from the angiosperms. The terms are misleading since some softwoods, such as *Dacrydium elatum* (sempilor, yaka) and *Taxus baccata* (yew) are 'hard', while certain hardwoods, such as *Gonystylis bancanus* (ramin) and *Triplochiton scleroxylon* (obeche), are 'soft', with the 'hard' woods having a higher density than the 'soft' woods.

The value of trees as timber for commercial utilisation depends not only on their relative availability and wood properties but also on tree habit, with a long, straight, not twisted, clear, wide bole being particularly desirable. Depending on the species and its use, the minimum trunk dimension requirements are generally 4 m high and 7.5 cm in diameter. The forester uses various formulae to calculate the quantity of timber produced by a tree. The more important of these are: (1) The **standing volume** of timber in a tree and is calculated from its length, basal diameter and taper, i.e. decrease in diameter with length. The basal diameter is measured at 1.3 m above ground level, and is referred to as the **diameter at breast height** (**dbh**), while **taper** is assessed by measuring the length of trunk to a point where the diameter decreases to an agreed figure, often 7 cm for softwoods while temperate and tropical hardwoods are crosscut at much higher top diameters. The timbers from trunk diameters below these minimum top diameters are not utilised; and (2) The **felled volume** is either calculated by measuring the mid diameter and multiplying the mid cross-sectional area by the length, or by averaging the top and bottom cross-sectional areas and multiplying by the length.

Some speciality woods are milled from very small stems, e.g. *Dalbergia melanoxylon* (African blackwood), which used chiefly for making wind instruments and carvings, and *Zanthoxylum flavum* (Jamaican or West Indian satinwood) for cabinet work, inlays, turnery, etc. Other timbers come from trees 45-60 m high (including crown) with stem diameters of 2.5 m or more, e.g. the West African endemic *Aucoumea klaineana* (Gabon mahogany, okoume) and *Dryobalanops* spp. (kapur) from Southeast Asia. Some trees may have buttresses up to 4.5-7.5 m above ground level, e.g. *Triplochiton scleroxylon* (obeche) from tropical Africa and *Mora* spp. from tropical America. Such trees are felled above the buttresses, since they represent root adaptations (Chapter 5) and would present anatomical problems in sawing and utilisation (Chudnoff, 1979; Matthews, 1989; Soerianegari and Lemmens, 1993).

The following example illustrates how the total biomass of a coniferous tree of 30 cm dbh is broken down after felling into its various components. Some 55% of the total woody biomass is taken to the sawmill, 23% is left as stump and roots, and 22% as top hamper and branches. The sawmill losses from debarking and sawing result in only 18% of the original tree becoming sawn timber, the remaining sawmill residues being processed into fibre and particle board, pulp and paper, fuelwood and other products.

It is the skill of the sawyer that determines the conversion of a log into the optimum quantity of sawn timber with the minimum of knots and other defects. The outer zone provides cleaner, knot-free wood than that from nearer the centre, with the heartwood producing the heavier dimension timber and large beams. The angle of the cutting in relation to the growth rings determines the nature of the wood surface. **Flat-sawn** surfaces are defined as those cut so that the growth rings meet the face at less than 45°; **quarter-** or **radial-sawn** surfaces are produced when the cut is at 90° to the growth rings, thereby maximising the contrasting effect of the growth rings, as with *Entandrophragma cylindricum* (pencil striped sapele); **rotary-cut** wood is peeled from the log for plywood and veneers. However, not all timbers are sawn. Pole timbers for telegraph poles, fencing posts, etc. are debarked and used in the round (Bramwell, 1982; Matthews, 1989).

Ultimately, as with other products, it is economics, politics, availability and fashion that decide what species will be used for which purpose. These factors are demonstrated by the changes in major woods used for furniture in the UK. Oak was widely used until the depletion of the oak forests by the end of the 16th century. This led to the use of local and imported walnut (from our American colonies) between 1660 and 1720. In 1721 the heavy import duties on timber imported from the British colonies in North America and the West Indies were abolished, and mahogany became the fashion. Economies then followed with the use of mahogany veneers on pine carcasses and, for the poorer classes, cheap pine furniture. In the 1850s a factory industry was started in the High Wycombe area of Buckinghamshire using local beech for the Windsor chair (Joy, 1962).

Today there is increasing pressure by conservationists to conserve the world's forest resources by only using timber from sustainable forestry practices; clear felling and replanting with exotic species is no longer acceptable. Greater use is also being made in finding suitable alternatives to endangered species and, where possible, making greater use of made-made boards for both internal and external work.

1.1   Timber Properties

It is the wood properties that ultimately determine timber suitability and how it is best utilised. The timber properties recognised by the US Forest Products Laboratory and other forestry organisations include the characteristics of the wood, its density, mechanical strength, drying and shrinking properties, workability, durability and preservability. The following summary of wood characteristics is from Chudnoff (1979), Hart (1991), and Soerianegari and Lemmens (1993).

2.   WOOD CHARACTERISTICS

The wood characteristics refer to the general appearance of the wood; (1) *Natural splitting* or *shake* is obviously undesirable. Both *Castanea* (sweet chestnut) and *Quercus* (oak) are prone to shake. In oak this has been linked to soil type, especially stony soils; (2) *Knots* too are undesirable, the size being largely species-linked. To a lesser extent knots reflect forest management, with the knot diameter influenced by initial tree spacing and density, the greater the competition the smaller the knots. Whereas live knots do not affect the strength of the timber, dead knots will have an adverse effect; (3) *Colour* of the sapwood and heartwood and whether the latter changes in colour on exposure to the atmosphere. The dark-coloured woods will often show good resistance to fungal attack due to their heavy impregnation by extraneous toxic substances. Fungal infection weakens the wood and is generally undesirable.

However, brown oak resulting from staining by the mycelia of *Fistulina hepatica* (beefsteak fungus) and the blue-green oak stained by the mycelia of *Chlorociboria aeruginascens*, syn. *Chlorosplenium aeruginascens* (green wood cup) are respectively highly prized for furniture and parquetry, especially Tunbridge ware (see Section 1.2). Some woods may have a desirable *high lustre* or *golden cast* due to the way the light is reflected; (4) *Texture* is largely determined by the size and arrangement of the vessels. Wood with large and irregular anatomical features are recorded as having a coarse and uneven texture, while those with small and even features are fine and even textured; (5) *Grain* or the surface pattern of worked wood will also vary according to the arrangement and alignment of the wood tissues and is defined as straight, spiral or interlocked. The latter can be found in many tropical timbers, while other tropical hardwoods have a pale yellow or white, easily worked timber with a uniform and straight grain, especially among members of the Euphorbiaceae, e.g. *Endospermum macrophyllum* (kauvula) and *Croton* spp., and in the Sterculiaceae *Triplochiton scleroxylon*. An *interlocked grain* is due to an alternating left- and right-hand spiralling of the grain when quarter-sawn and produces a ribbon or roe-like figure, i.e. configuration. Other grain irregularities, which can be enhanced by sawing or slicing techniques, can develop such figures as curly, feather, fiddle-back, stripe, etc.; (6) *Figure* is the decorative appearance of the wood arising from such structural features as grain, rays, growth rings and colour, and will vary according to the orientation of the saw cut. Attractive variegated wood with attractive markings are generally much favoured for cabinet work; (7) *Distinctive scents and taste,* noteworthy examples are the fragrant wood from *Santalum album* (Indian sandalwood), the stench of rotting cabbages from freshly cut *Petersianthus macrocarpus*, syn. *Combretodendron macrocarpum* (essia) and the odourless but bitter tasting *Carapa grandiflora* and *C. guianensis*, syn. *C. procera* (crabwood, bastard mahogany); (8) *Silica*, the percentage of silica bodies can be important, above 0.5% will generally result in blunting of the cutting tools. *Chrysophyllum maytenoides*, for example, has a silica content of over 0.84% and is noted for having a hard and abrasive wood, while that of *Dicorynia guianensis* (angelique, basralocus) ranges from 0.2 to 1.7, with 2.9% is not unknown, and in *Licania* spp. it can be as high as 3-4%! Silica does, however, make wood more resistant to marine borers; (9) *Gums and resins*, where present, may also make the wood difficult to work. For example, the gummy sawdust obtained when sawing green timber of *Baikiaea insignis* clogs the saw teeth, while a high tannin content may cause staining of moist wood in contact with iron. However, the extremely hard and resinous wood of *Guaiacum* spp. (lignum-vitae) with its diagonally opposed fibre-layers provide self-lubricating for bearings and bushes of propeller shafts of ships; (10) *Wood allergies*, although most people are unaffected most woods possess constituents to which somebody somewhere will be allergic or even find toxic, e.g. *Betula papyrifera* (paper birch) and *Nauclea diderrichii* (opepe). The wet sawdust from *Milicia excelsa*, syn. *Chlorophora excelsa* (iroko), and *Maclura regia*, syn. *Chlorophora regia,* can cause dermatitis, and the dust from *Autranella congolensis, Erythrophleum suaveolens,* syn. *E. guineense,* and *E. ivorense* cause irritation of the mucous membranes.

2.1   Density

The density represents a measurement of the dry wood content per unit volume of wood. This naturally varies with the moisture content, which must always be specified. Density calculated from weight and volume when air-dry is usually at 12 or 15% moisture content, sometimes expressed in the US as moisture content of 15 pcf. The *moisture content* (*mc*) is determined by measuring the weight differences before drying and the dry weight, i.e. approaching 0% moisture, as a percentage of the dry weight. The significance of this measurement is that green wood, especially lightweight timbers, can have a mc greater than 100%. The *relative density*, formerly measured in terms of *specific gravity*, may reflect such important wood attributes as mechanical strength, shrinkage, paper-forming properties and the cutting forces required when machining. In general slow growing trees are likely to have denser woods than fast growing trees. Relative densities may range from less than 100 for *Ochroma lagopus,* syn. *O. pyramidale* (balsa), to *ca.* 1100 for *Guaiacum* spp. (lignum-vitae) and as high as 1400 for *Krugiodendron ferreum,* which is one of the densest woods known. Interestingly, following World War II Malaysia, instead of marketing named species has successfully used groups of hardwoods with similar densities and mechanical properties as criteria for identifying timber exports. When correctly applied the method could be a valuable conservation tool by encouraging the felling of non-endangered species with similar properties to those that are endangered. *Basic specific density* is the ratio of wood density to the density of water at 4°C and is calculated from the oven-dry weight and green volume.

2.2   Mechanical Properties

The mechanical properties (strength properties) largely depend on timber density and mc, and are tested when green and dry. Most of these properties may be tested using either the British Standard No. 373 for 2 cm^2 specimens or their equivalent, or the American Society for Testing Materials (ASTM) standard D 143 for a 5 cm^2 or 2.5 cm^2 (2 in^2 or 1 in^2) specimens of various lengths. While these two systems are roughly identical, the test samples may sometimes differ in size and shape, e.g. for testing shear strength and the results are then not completely comparable.

The tests are: (1) M*odulus of elasticity* (*stiffness in bending*), which measures the stiffness of beams or long columns; (2) *modulus of rupture* (*bending strength*) provides the load carrying capacity at 12 % mc on bending until breaking occurs; (3) *Compression strengths* are obtained by testing comparatively small specimens of wood parallel to the grain for *maximum crushing strength,* and perpendicular to the grain to give the *stress at limit of proportionality*; (4) S*hear strength* measures the wood resistance caused by forces producing an opposite but parallel sliding motion to the grain; (5) *Cleavage* or *resistance to splitting* is important when nailing or bolting, and may differ considerably with the plane of cleavage, with radial splitting being more common than tangential; and (6) *Janka side hardness and toughness* measures the resistance of wood to indentation and the ability to withstand abrasion.

2.3   Drying and Shrinkage

Drying and shrinkage refer to the response of the timber on drying from the green condition to 15% and/or 12% mc and/or oven dry. The cell moisture has to be extracted slowly and gently in order to avoid such adverse defects as *splitting*, *checking* (fine reticulated cracking), *warping* or *collapse*. Air drying is effective but requires 2 years to complete and only reduces the water content to the ambient mc, which is *ca.* 20-30%. Kiln drying can reduce the mc to 6%, with each species requiring different drying schedules of slowly increasing temperatures. The process takes *ca.* 2 weeks, depending on the thickness of the timber. Wood used in centrally heated houses for interior fittings and furniture should always be kiln dried since the central heating will bring the mc down to an unscheduled 6%. The usual cause of splitting is due to differences in the radial and tangential shrinkage of the wood cells. *Albizia saman,* syn. *Samanea saman* (rain tree) is noteworthy in that there are no differences between radial and tangential shrinkage; its wood can be turned into salad bowls and not split.

2.4   Working Properties

The working properties of timber are highly subjective and are not quantifiable. It describes the ease of working with hand and machine tools, tendencies to torn or chipped grain, presence and size of knots, smoothness of finished cut, dulling of tools and ease of veneering. Interlocked grain, as in *Julbernardia globiflora*, syn. *Isoberlinia globiflora* for example, may cause tearing when worked. The ability to take nails and screws and hold them firmly without splitting is essential, especially for boxes and crates. Gluing characteristics may be also be important since dense and oily wood, such as that of *Dipteryx odorata* (tonka bean) will not adhere well. Suitability for polishing, painting and steam-bending are also recordable.

2.5   Durability

Durability refers to the resistance of the wood to attack and decay by fungi, insects (especially *Lyctus* spp. (powder post beetles), termites) and marine borers are important, as well as weathering characteristics. The sapwoods are always less durable than heartwoods, while the presence of gums, resins, tannins and other phytochemicals, e.g. the isomer of dihydroquercetin or dihydrorobinetin in *Robinia pseudoacacia* (black locust, false acacia) can inhibit decay. Seasoned wood, because of its lower moisture content, is also less liable to decay than unseasoned. Provided the heartwood or sapwood is not in contact with the ground and is kept dry, they could be free of rot and have an extended service life. The grades of average service life for timbers in contact with the ground are shown in Table 11.

TABLE 11. Approximate service life in years based on decay resistance of heartwood in contact with the ground in temperate and tropical regions (Chudnoff, 1979; Hart, 1991)

Service grade	Temperate regions	High rainfall tropical regions
very durable	>25	>10
durable	15-25	5-10
moderately durable	10-15	2-5
non-durable	5-10	0-2

Insect resistance can also differ between species. The heartwood of *Terminalia superba* (afara), for example, is not durable and is susceptible to members of the Isoptera (termites), *Trypodendron lineatum* (ambrosia beetle) and powder-post beetle; the heartwood is also extremely resistant to preservatives and the sapwood moderately resistant. By way of contrast, *Baillonella toxisperma* (djave) is resistant to both termites and marine borers.

2.6   Wood Preservation

Both sapwood and heartwood may be treated with creosote, zinc chloride or other chemicals against decay and insects using either an open tank or pressure-vacuum process. There is no standard treatability test. Ratings may range from 'permeable' where the absorption of the preservative is 0.19-0.25 kg m^{-3} (15-20 pcf) or more with complete or deep chemical penetration, to extremely resistant with only 0.025-0.038 kg m^{-3} (2-3 pcf) or less absorption and only superficial lateral penetration (Chudnoff, 1979; Hart, 1991; Soerianegari and Lemmens, 1993).

3.   VENEERS AND INLAY

The art of *veneering*, i.e. the bonding of an attractive layer of figured wood onto a plainer or cheaper wood, permits a more economical use of expensive decorative woods as well as granting greater artistic freedom in design, especially in furniture. The use of veneers dates back to pre-Egyptian times. The veneers were formerly cut using multiple-bladed saw frames; modern machined veneers are thinner and usually sliced from the log on the flat, quarter, or half-round. The extremely attractive and highly prized *burr veneers* are cut across the grain of trunk out-growths of hardwood species such as ash and walnut. *Curl veneers*, which have an attractive curl figure, are cut from the junction a branch with the trunk, or from the main root members of some trees. Veneers from *rotary cutting*, where the log is rotated against the knife to produce a continuous sheet, are generally used for plywood as the method produces a rather wild and unnatural grain. Today veneered boards are mass-produced using thermo-setting adhesives under heat and pressure, although traditional techniques are still required for the restoration of antique furniture.

Parquetry is a regular repeating pattern and built up from thin veneers when used in furniture, or of thick wood blocks for parquet flooring. *Marquetry* is a delicate form of veneer inlay using contrasting woods cut to form a pattern. *Inlay* is the art of recessing a number of contrasting veneers into solid wood; *banding* and *string decoration* are a specialised form of composite inlay where blocks of wood are sandwiched between sheets of veneer. They are often used to decorate the edges of tables, desks, etc. *Tunbridge ware* is a popular mosaic decoration developed in Tunbridge Wells, Kent, during the 19th century. Small, square-section rods of different coloured woods are tightly glued together to form a block from which chequerboard patterned veneers may then be cut and used to decorate furniture and various small articles (Bramwell, 1982).

4.   PLYWOOD AND OTHER BOARDS

Plywood, fibreboard and other man-made boards have the great advantage over ordinary timbers in that they can be supplied in almost any size, whereas the maximum size of timber is limited by the size of the trunk. They have the additional advantage in being free of such common defects as shakes (splits), knots, warps, etc. They also expand and contract less with changes in humidity than ordinary timber, with weather- and water-proof grades also generally available. Finally, with a few exceptions, they are, generally cheaper than solid wood. Fibreboard, chipboard and blockboard are especially economical in their use of timber resources in that they can be made from waste wood, offcuts, etc.

Plywoods are usually rotary cut from logs selected for their symmetrical shape. They are constructed from several veneers tightly glued together, with the grain of alternate layers at right angles to each other. Special water-resistant adhesives may be used for outdoor use, the most weatherproof grade being WBP (water- and boil-proof).

Plywood has the great advantage over natural timbers in that it can be manufactured to precise engineering standards. Knots can be removed and the holes patched, small slits filled with synthetic filler; small pieces of veneer may also be sown together and used as core veneer without affecting the strength of the panel;

particleboard and lumber cores may also be used. Any waste material during manufacture from the trimming and sanding operations, etc. can be used for paper-making or in the manufacture of other man-made boards.

Hardboard is made from heated and compressed softwood pulp, the natural resins present usually making any adhesives unnecessary; they usually have one smooth, shiny surface, the other embossed with a mesh pattern. *Chipboard*, also known as *particle board*, is, as its name implies, made with wood chips and, unlike hardboard, contains adhesives and can, therefore, be manufactured in much lower densities. It is probably the most adaptable of the man-made boards. It is possible to replace 10% of the wood used in the core of particle board with coarsely ground straw to act as a filler. It is also technically possible to produce particle boards solely from straw chips that are of as good a standard as wood-based particle board. The straw chips also have the advantage of being able to provide a range of board characteristics and surface textures.

In the past the tubular structure and waxy surface of the straw had created difficulties in obtaining good adhesion between the straw and the bonding material, which had inhibited earlier attempts to produce high quality boards. This problem has been overcome by using abrasive grinders to break the tubular structure and etch the waxy surface, plus the use of a urea-methanal (formaldehyde) glue modified with isocyanate as a bonding agent. Rye and wheat straws are the preferred cereal straw for particle board production, while non-cereal straws, such as rape and flax straws, which lack the waxy surface of cereal straws and are more compatible with wood chips, may be preferred for wood and straw composite boards. Flax may also be used as a filler in wood particle board and alone for flax boards (Rexen and Munck, 1984).

Blockboard, also known as *coreboard*, is a thick, strong board made from wood strips, 19-25 mm thick, glued together and sandwiched between thick veneers. Blockboard is stronger than chipboard and has greater load-bearing properties; it is consequently suitable for shelving. *Laminboard* is the strongest and most expensive of the man-made boards, similar to blockboard but made from thinner strips, 3-6 mm thick, with no unsightly gaps or other internal flaws. It is used for high-quality cabinet work (Hutchinson, 1973; Bramwell, 1882).

5.   RATTANS

Rattan, from the Malay 'rotan', probably from 'raut', trim (Long, 1994) is considered to be the second most important forest product after timber. The bare rattan stems are light, strong, flexible and uniform in diameter. They can be bent into the desired shape by steaming or by the application of a hot iron or blow torch. They are consequently a valuable and versatile resource. Rattans are widely used commercially either in the whole or round form for furniture frames, or in splits, peels and cores for woven chair seats, matting and basketry. They are also extremely important in the domestic economy for cordage, basketry, matting, blinds, thatching, traps, brooms, carpet beaters, walking sticks, furniture, construction purposes, etc.

The true rattans are Old World members of the Palmae subfamily Calamoideae, tribe Calaminae, and include such important genera as *Calamus, Daemonorops* and *Plectocomia.* Their distribution extends from equatorial Africa through the Indian subcontinent to southern China and south to Australia and the western Pacific, with the greatest centre of diversity in western Malesia. In the New World the stems of *Desmoncus* spp., which belong to subfamily Arecoideae, are also used for weaving and furniture but are inferior in quality to the true rattans and only of local commercial significance. The bamboo-like stems of *Chamaedorea* of subfamily Ceroxyloideae may be used for house construction but the stems of the only climbing member of the genus, *C. elatior,* are too soft to be of use as rattans.

Rattans are mainly harvested from the wild since growing them in plantations tends to produce thin and relatively short canes. Production in the wild, however, is being endangered by the increasing destruction of its forest habitat and as a consequence foresters are increasingly seeking for better plantation management systems. Commercially viable plantations of *Calamus* spp. were first established in Indonesia as early as 1850. Plantations of mainly *Calamus,* especially *C. caesius* and *C. trachycoleus* have since been established in a number of other countries in Southeast Asia for domestic, commercial and experimental production.

The large, 25-80 mm in diameter canes without leaf-sheaths, such as the single-stemmed *Calamus manan* and the multi-stemmed *C. merrillii, C. ornatus, C. ovoideus,* etc. are usually cured within 1-2 days of harvesting in troughs containing a hot oil mixture of diesel, kerosene or coconut oil. This is in order to prevent deterioration and increase the durability of the canes by removing gums, resins and water. The cured canes are then air-dried for 1-3 weeks, bundled and stored in covered sheds until marketed. Good quality canes may be fumigated with SO_2, while the poor quality canes are decorticated and stained before sale.

The smaller, 4.5-15 mm in diameter canes without leaf-sheaths, such as those from *Calamus caesius, C. egregius, C. optimus* and *C. trachycoleus,* are also sun-dried before sale. They may also be stored under water prior to being processed, which involves scraping the nodes, fumigating with SO_2, and sun-drying, followed by splitting and coring. Three or four splits are usually obtained from a length of cane. In coring the outer '*peel*' is usually removed by machine; the core is also sold. In a process known as '*runti*' or '*lunti*' the highly silicified epidermis of *Calamus caesius* and *C. trachycoleus* are removed by rubbing with sand or pulling through a series of bamboo bars or wooden pulleys before splitting. When stripped of their leaf-sheaths rattan canes can sometimes be confused with bamboo (Section 3). Bamboo stems are usually hollow and, even when solid, not easily bent, unlike rattans which are always solid and readily bent (Dransfield and Manokaran, 1993).

Indonesia commands 75-80% of the world's production of raw rattan, with prospective earnings of *ca.* US$ 600 million, while Malaysia, Indonesia and

Philippines have imposed bans on rattan exports except as finished products. The bans are intended to encourage the development of local rattan-based industries and to counter the increasing destruction of the wild resource for the export of raw rattan (Robbins, 1995).

6.   BAMBOO

The name bamboo is from the Dutch '*bamboes*', a variant of the Portuguese '*mambu*', which is probably a trade corruption of the Malayan '*samámbu*' (Arber, 1965, Long, 1994). Members of the Gramineae subfamily Bambusoideae, the woody bamboos are represented by *ca*. 1000 members of the tribe Bambuseae. They are essentially pantropical in occurrence, distributed throughout the tropical and subtropical and temperate regions of the world except Europe and western Asia.

The bamboos are perennial trees, shrubs and climbers, usually with woody, hollow culms divided into cylindrical segments by the nodes. Bamboos with solid and hollow stems are sometimes referred to as 'male' and 'female' respectively. The term 'male bamboo' is now restricted to *Dendrocalamus strictus* (Calcutta or male bamboo), whose solid stems were formerly used for spear shafts, they are now used for construction, paper pulp, charcoal, etc. Bamboos are also a source of edible shoots (see Chapter 9), as well as being used for windbreaks, hedges, ornamentals, etc. (Arber, 1965, Clayton and Renvoize, 1986; Dransfield and Widjaja, 1995; Mabberley, 1997).

The mechanical properties of bamboo are generally similar to those for wood. But unlike wood, shrinkage starts immediately after harvesting and does not continue uniformly. The moisture content of the culms also has an important influence on the mechanical properties (see Dransfield and Widjaja, 1995 for details). In some species the silica content of the wood is so high and the wood so hard that it can be used as a whetstone. A silica residue may even be deposited in the hollow internodes, e.g. *Bambusa bambos*, syn. *B. arundinacea* (spiny bamboo), and has long been used in Indian and Arab medicine (Arber, 1965).

With species up to 40 m tall and 29 cm in diameter, they are among the most versatile of timbers. According to Kurz (1876) their numerous attributes make bamboos the tropical economic plants par excellence. The strength, straightness, smoothness and lightness of the culms combined with hardness and hollowness, the range of sizes, lengths and thickness of the internodes, plus the facility and regularity with which they can be split, provide ready-made material suitable for a wide range of purposes. The large diameter and thick walled culms with relatively short internodes of species of *Bambusa, Dendrocalamus, Gigantochloa*, etc. are used for constructing pillars, bridges and scaffolding. They are also used in Asia in the furniture industry. Medium diameter culms with relatively thin walls, such as *Gigantochloa levis*, *Schizostachyum brachycladum* and *S. zollingeri* can be split and used for floors, roofing tiles and walls. The smaller diameter culms with relatively thick walls can also be split and used for basketry, gabions and for a large range of handicraft industries, including the manufacture of table mats, handbags and hats. Unsplit culms or rhizomes can be used for engraving. Domestic applications include the use of culms for making chopsticks, containers, cooking pots, fish traps and fishing rods, heads and shafts for spears and arrows, blowpipes, fencing, pipes, poles, rafts, troughs, irrigation channels, agricultural implements, axles and springs for carts, sticks and poles, fencing, boats, masts, fibre for ropes and cordage and biofuel. Depending on their diameter, bamboos can also be used for a range of percussion, wind and stringed instruments. The leaves can be used for wrapping paper and hats. Bamboos are also a source of activated charcoal (see Chapter 13) for water purification, bamboo mosaic board, plywood and laminated veneer, pulp and paper.

For centuries *Phyllostachys edulis*, syn. *P. pubescens*, has been used in China for making paper, while in Thailand, Indonesia and the Philippines *Bambusa bambos*, *B. blumeana* and *Dendrocalamus strictus* are used by the paper-making industry; the latter species is also the principal species used in India. The sulphate method (see Section 4) is used for making wrapping and bag papers, as well as paperboard from the chipped culms; the bleached pulp is also widely used for writing, printing and wrapping papers (Hidalgo, 1974; Farrelly, 1984; Dransfield and Widjaja, 1995).

The ready availability and its relative low cost, light weight, pliancy, ease of transporting and handling of the culms make them a versatile product for construction purposes. A bamboo house is easily constructed and maintained, comfortable and well-ventilated, resistant to earthquakes and strong winds. The disadvantages are difficulties in making joints, the need to predrill before nailing, its inflammability and susceptibility to insect and fungal attack. The natural durability of bamboo in contact with the soil is *ca.* 2 years. With preservation treatments the life of an untreated house can be extended from 7 years to 15 years or more (Latie *et al.*, 1987; Siopongco and Munandar, 1987).

7.   WOOD PULP

The major source of fibre for the pulp industries is wood. Both softwoods are hardwoods are utilised, the main differences between them being the presence in the latter of a mixture of large vessel elements of greater diameter than the fibres but shorter in length; they are absent in softwoods. In general softwoods consist almost entirely (97% by weight) of fibres, known as tracheids, 2-8 mm in length, while hardwood fibres are from 0.5 to little over 2 mm long (Table 12). The softwood fibres are preferred for paper making since their length and flexibility allows them to be packed closely together into non-porous, tightly bonded sheets, whereas the hardwood fibres pack less tightly and produce an inferior paper.

TABLE 12. Dimensions and types of wood fibres (Eames and MacDaniels, 1947; Esau, 1953; Kirby, 1963; McDougall *et al.*, 1993)

Species	Fibre length (mm)	Fibre type
SOFT WOODS	1-8	tracheids and xylem
Picea abies (spruce)	3-4	
Pinus sylvestris (pine)	2-3	
HARD WOODS	1-3	tracheids, xylem and vessel elements
Eucalyptus spp. (eucalypts)	0.5-1	
Populus spp. (poplars)	0.6-1.5	
Robina pseudoacacia	0.9	
Fraxinus americana	0.9	
Quercus alba (white oak)	1.0	
Betula papyrifera	1.3	
Carya ovata (hickory)	1.3	
Ulmus americana (white elm)	1.5	

The principal pulpwood sources are from the historically preferred species of *Abies, Picea, Pinus* and *Tsuga.* Other major sources are the conifers *Araucaria angustifolia, Larix, Pseudotsuga menziesii* and *Thuja,* and hardwood species of *Acacia, Acer, Alnus, Betula, Broussonetia papyrifera, Eucalyptus, Fagus, Fraxinus, Gmelina arborea, Liquidambar, Nyssa, Populus, Quercus, Salix, Ulmus,* etc. (FAO, 1973).

More than 80% of the world's pulp is produced in North America and Scandinavia, chiefly from softwoods, especially species of pine and spruce. However, as more and more countries begin to develop their own natural resources, the range of species utilised increases. Thus, in the UK more than half of the imported sulphate paper pulp are from *Eucalyptus globulus* (blue gum) and *E. grandis* (flooded gum), imported not from their native Australia but cultivated in Spain, Portugal and Brazil. Other hardwood sources include species of beech, birch, chestnut, popular and willow, plus a number of mixed tropical hardwoods from Southeast Asia (Lewington, 1990).

Depending on the species, the debarked trunk consists of 65-85% cellulose fibres bound together by 15-35% lignin. The presence of usually at least 1% of resin in most softwoods can cause problems arising from plasticity and stickiness or tackiness during pulping and paper making. Although there is generally less than 1% of resin in hardwoods, some woods, such as those of *Betula, Populus* and *Tilia,* contain relatively large amounts of ether-soluble substances which can also cause problems for pulping and paper making (Hillis, 1962; Carruthers, 1994).

The most important use of the wood pulp fibres is for the manufacture of paper (see Chapter 14) and board. Other uses include a viscose process where the cellulose fibres are regenerated to form rayon for textiles, tyre-cords, and cellophane. Cellulose esters and ethers also have a wide range of industrial applications (see Chapter 15). The various mechanical, chemical and semi-chemical pulping processes used to separate out the fibres prior to their recombination in the final product are shown in Table 13.

TABLE 13. Pulping processes and their products (adapted from Bramwell, 1982)

Pulp source	Process	Bleaching	Products
softwood (long fibres)	mechanical	mostly unbleached	newspaper, magazines. paperbacks, carton board and wallpaper
	sulphite	unbleached	hard tissue paper, white paper bags and wrapping paper
		mostly unbleached	soft white tissue paper
	sulphate	unbleached	strong brown wrapping paper
		bleached	hardback books, duplicating paper and typing paper
hardwood (short fibres)	semi-chemical	unbleached	corrugated paper and corrugated cases
	soda	bleached	filler in printing paper

The pulp produced by mechanical pulping is weak such undesirable substances as hemicelluloses, lignin and resins. are also present in addition to the desired cellulose fibres. These pollutants resist bleaching and cause yellowing of the paper. Such pulp is consequently used for short-life products such as newspapers and magazines. Chemical processing is more expensive, but has the advantage of dissolving the lignin and other softer wood components without damaging the cellulose, and produces a far stronger pulp.

The *mechanical pulping* of wood was first developed in Germany in 1840. The process involves passing the debarked, washed bolts (short lengths of timber) over revolving giant wheels or pulp stones, upon which water is played to wash away the ground fibres into underlying pits. Sieves are used to remove any extraneous matter such as knots, lumps, etc. The resulting pulp will also contain all the encrusting substances of the fibres. With the *cold method* large quantities of cold water are used to prevent heating, resulting in a fine, even grade of fibre. Alternatively, the *hot method* uses very little water and produces coarser, longer fibres. The material is then screened to both grade and remove any impurities, after which it is rolled to squeeze out as much water as possible. Finally, a lapping machine turns out sheets of pulp which can then be made directly into paper or further dried prior to shipping.

For *chemical pulping* the wood is first cut into chips in order to bring the lignin into contact with the chemicals as quickly as possible, using acid, alkaline or neutral processes.

Invented in Germany in 1883, the *sulphate (Kraft) process* is the most common chemical pulping process for conifers with a high resin content. It is also the preferred method since it involves fewer unfavourable side reactions. The wood chips are first reduced to pulp by steaming and then subjected at very high temperatures to a solution of $NaOH$, Na_2SO_3 and a little Na_2SO_4. The product is then washed, screened, flash dried and pressed, resulting in fluffy, crumb-like particles suitable for a strong, brown, wrapping paper, i.e. *kraft paper*. It may also be bleached for a white, softer and more pliable paper than that produced from sulphite pulp. It involves boiling the wood chips or fibres under pressure at very high temperatures in either stationary or rotating digests with alkalis such as caustic soda, which is sometimes admixed with sodium sulphate or calcium bisulphate. The sulphate process also yields a valuable by-product known as *tall oil* (see Section 5.4).

The *sulphite process* is used for hardwoods, especially tropical hardwoods. Carefully selected chips are cooked with steam in an acid Na_2SO_3 solution. The digested pulp is then washed, screened, lapped and dried. The pulp is bleached and loading material, e.g. China clay and calcium sulphite, and sizes, e.g. rosin, are added to give the required strength and other paper properties. Such pulp is used for high grade papers and, after further purification, for artificial fibres. The *soda process* is used for hardwoods, especially species of poplars, and pine, digesting the wood under pressure in an $NaOH$ solution at *ca.* 95°C, followed by washing, bleaching and lapping. The pulp produced in a *semi-chemical* process combining both mechanical and chemical methods is used for producing corrugated infill and casing of cardboard boxes.

Mechanical pulping produces wood pulp with short fibres and impurities which, for paper making gives a weak paper of poor colour. It is usually used as a 15-25% addition to chemically processed pulp. Chemical wood pulp is of much better quality than mechanical pulp, and with many more applications. It also fetches a higher price. Because the cellulose fibres are isolated from the resinous substances it produces a '*high alpha pulp*' of cellulose fibre with a neutral pH and a longer fibrous structure. Even so, mechanical wood pulp is still by far the most widely used. Pulp known as '*dissolving*' or '*special alpha*' pulp derived principally but not solely from wood and with a highly purified cellulose, is used in the manufacture of rayon, artificial silks, cellophane and a wide variety of other fibres, transparent photographic films, cellulose plastics and nitro-cellulose for explosives and lacquers.

Because of the higher densities of the hardwoods the requirement for digester space for pulping is lower and is usually associated with lower processing chemical and steam requirements. The disadvantages are a slower draining and weak paper, requiring longer working and sheet-forming equipment. However, the heat content of hardwood spent pulping liquor is lower than for softwoods, due to reduced wood solids. In those hardwoods with longer fibres the extremely high densities in which the fibres occur can cause serious problems in wood preparation and pulping. There are, for example, widely differing pulping, bleaching and chemical requirements for

Pinus spp. and *Liquidambar styraciflua* (gumwood) and the heat content of spent pulping liquors (Hill, 1952; Kirby, 1963; FAO, 1973; Bramwell, 1982; Turner and Skiöld, 1983; Lewington, 1990).

7.1   By-Products from Pulping Processes

The ligno-sulphonates obtained by the digestion of lignins in the sulphite process, are used in tanning, in rubber formulations, as a component of cements and as a source of organic chemicals, including the commercial production of a less desirable *vanillin* than that obtained from the orchid *Vanilla planiflora*. Large quantities of ligno-sulphonates are also used in drilling muds. Neutral sulphite liquors from the semi-chemical method yield *ethanoic (acetic)* and *methanoic (formic) acids*, also sulphate liquors used as dispersing agents and as components of foundry core-binders, ceramics, dyes and printing inks.

Although the acid hydrolysis of wood can be used to yield sugars, mainly glucose, for fermentation to *ethanol*, there are other more economic sources of feed materials, such as starch. The wood hydrolysates are more important commercially for the production of special yeasts for human and animal foodstuffs, and in the fermentation of ethanoic and lactic acids, butanol and propanone (acetone). The more drastic hydrolysis of wood cellulose produces *laevulinitic acid*, which is used in printing cotton, and its derivatives as solvents and plasticisers. The hemicellulose from the wood provides *furfural*, which is used as the starting material for the manufacture of moulding resins, also as a solvent in the extraction of mineral oils and for decolorising rosin (Bramwell, 1982; Sharp, 1990).

8.   NAVAL STORES

The term *naval stores* originally referred to the pitch and rosin obtained from pine trees to caulk ship timbers and preserve the rigging of sailing ships. It is now the generic term applied to turpentine, fatty acids and rosins recovered from tall oil, gum and wood rosin, as well as their derivatives.

Three distinct classes of naval stores are recognised: (1) *Gum naval stores* obtained from the tapping of living pine trees, the distillation of the resin yielding gum rosin and gum turpentine; (2) *Sulphate naval stores* produced as a by-product from the *black liquors* recovered during the sulphate (kraft) pulping of softwood chips, with *sulphate turpentine* condensed from the cooking vapours. The crude *tall oil* produced from the alkaline liquors is then fractionated to yield tall oil, oil rosin, tall oil fatty acids, etc.; and (3) *Wood naval stores* solvent extracted from old, resin-saturated pine stumps and yielding wood turpentine, wood rosin, dipentene and natural pine oil (Hoffmann and McLaughlin, 1986; Prescott-Allen and Prescott-Allen, 1986; Coppen, 1995; Coppen and Hone, 1995).

Rosin and its chemical derivatives, together with the polyterpene resins from turpentine, are widely employed as tackifiers for adhesives used, for example, in pressure-sensitive sticky price labels and printed food labels, paper sizing agents, printing inks, solders and fluxes, a range of surface coatings, insulating materials for the electronics industry, synthetic rubber, chewing gums, and soaps and detergents, as a soldering flux, also in varnish, lacquer, and size manufacture. See Chapter 15 for further discussion.

8.1   Resin

TABLE 14. Resin quality and yield characteristics rated from very good (+++) to poor (-) of some *Pinus* species and the major producing countries. Countries in parentheses are relatively minor producers (after Coppen and Hone, 1995)

Species	Quality	Quantity	Producing country
P. caribaea	+	+++	Venezuela [South Africa, Kenya]
P. elliotii	++	++	Brazil, Argentina, South Africa, [USA, Kenya]
P. kesiya	+	+/-	China
P. massoniana	+	+	China
P. merkusii	+	+	Indonesia [Viet Nam]
P. oocarpa	+/-	+/-	Mexico, Honduras
P. pinaster	++	+	Portugal
P. radiata	+++	+	[Kenya]
P. roxburghii	+	+	India [Pakistan]
P. sylvestris	+/-	+/-	Russia

Crude resin, which is obtained by tapping living pines, is an opaque (due to occluded moisture), milky-grey, thick, sticky and usually a still fluid material. It is invariably contaminated during the tapping with forest debris, insects, etc.

Although there are 93 species of *Pinus,* only a few dozen have had their resin tapped commercially for turpentine and rosin. The principal sources are shown in Table 14. In the past the crude resin was not exported but processed within the country of origin. The recent acute shortages now being experienced by some traditional producers has led to some producers, such as India and Portugal, to import crude resin for processing (Coppen and Hone, 1995; Mabberley 1997).

8.2   Turpentine

Turpentine is a clear liquid with a pungent odour and bitter taste, consisting of light, volatile essential oils obtained either as exudates or by distillation from conifers. Among the better known turpentines are ***Canada balsam*** from *Abies* *balsamea* (balsam fir), **Strasbourg turpentine** from *A. alba* (European silver fir), **Venetian turpentine** from *Larix decidua* (European larch), **Jura turpentine** from *Picea abies* (Norway spruce), **Bordeaux turpentine** from *Pinus pinaster* (maritime pine), and **Oregon balsam** from *Pseudotsuga menziesii* (Douglas fir).

Turpentine is traditionally used as a thinner for paints and varnishes, and as a solvent. The specifications for **'gum spirit of turpentine'** include those of the American Society for Testing and Materials (ASTM D 13-92) and the Bureau of Indian Standards (IS 533:1973), both of which are based on a quality assessment using relative density or specific gravity, refractive index, distillation and evaporation residues. They are designed for solvent use rather than for a chemical feedstock in which the composition is of prime importance. The International Organization for Standardization standard (ISO 412-1976) includes similar physical data.

For the chemical industry the main demand is for turpentines with a high total pinene content for use as a source of isolates for conversion to pine oil, fragrance and flavour compounds, and other derivatives. When fractionated, the **α-** and **β-pinene** constituents especially, are the starting materials in the synthesis of a wide range of fragrance, flavouring and vitamin formulations, as well as polyterpene resins. The α-pinene, for example, is used as the raw material in the manufacture of synthetic camphor chlorinated insecticides. The largest use of a turpentine derivative is for synthetic **pine oil**, which is widely used in disinfectants, cleaning agents and other products requiring a 'pine' odour. Other derivatives, such as **isobornyl acetate**, **linalool**, **citral**, **citronellal**, **citronellol** and **menthol** are used either alone or in the formulation of other fragrance and flavour compounds, while some of the minor constituents, such as **anethole**, are employed for fragrance and flavour without modification.

Suitable sources are *Pinus elliotii* (slash pine) with *ca.* 60 and 30% α- and β-pinene respectively, and *P. radiata* (Monterey pine) with 90% total pinene, over 50% of which is β-pinene. The presence of relatively small amounts of **3-carene** may be undesirable for some applications When the 3-carene present in significant quantities the turpentine is of little value, such as those from *P. roxburghii* (chir pine) and *P. sylvestris* (Scots pine). While turpentines with up to 50% or more of **β-phellandrene** from *P. caribaea* Caribbean pine) may be useful as solvents for paints, they would be unattractive as a source of pinenes for derivatives. A number of North American pines, including *P. ayacahuite* (Mexican white pine), *P. coulteri* (big-cone pine), *P. jeffreyi* (Jeffrey pine), contain the aliphatic hydrocarbon **n-heptane** used for testing motor fuels and consequently commands a higher price despite its presence in ordinary turpentine lowering the quality of the product (Hill, 1952; Mirov, 1954; Bramwell, 1982; Lewington, 1990; Sharp, 1990; Coppen, 1995; Coppen and Hone, 1995).

8.3   Rosin

Rosin is a brittle, transparent, glossy, faintly aromatic, yellowish to dark-brown solid that is insoluble in water but soluble in many organic solvents (see Chapter 15). Rosin is a glass, not a crystalline solid, and on heating softens to what is known as the softening point (rather than melting point), generally in the range 70-80°C, the higher the softening point the better the quality. In North America rosin is normally extracted by the steam distillation of aged pine stumps, by the labour intensive tapping the gummy exudate (*oleoresin*) of live trees, or as a by-product of the sulphate pulping process (Hoffmann and McLaughlin, 1986; Prescott-Allen and Prescott-Allen, 1986; Lewington, 1990; Sharp, 1990; Coppen and Hone, 1995).

It is used to increase the sliding friction of the bows used for various stringed instruments, as well as a variety of products, including varnishes, inks, linoleum and soldering fluxes. The fractionation of rosin yields *rosin oil*, also known as *resinol* or *retinol*. It is used in lubricants, electrical insulation and printing inks. The criteria for rosin quality and acceptability for the various applications based on colour and softening point will usually suffice when the rosin is obtained from proven sources. It is graded according to colour; the two most commonly traded rosins are the palest and most desirable grade WW (water-white), with WG (window-glass) as a slightly lower grade (Coppen and Hone, 1995).

8.4   Tall Oil

Tall oil (including *tall oil rosin*) is a major by-product of the kraft or sulphate pulping process. It consists mainly of fatty acids, together with some unsaponifiable material. It is recovered from the spent cooking liquors and distilled to yield rosin, fatty acids and an intermediate material consisting of a mixture of rosin and fatty acids known as *distilled tall oil*. The tall oil fatty acid production is used in the manufacture of polyamide resins for adhesives, coatings, inks, paints, varnishes and other protective coatings, soaps and detergents, hard floor coverings, esters and plasticisers. During distillation the volatile 'heads' fraction is used in ore floatation, especially by the phosphate industry, while the 'pitch' fraction at the bottom of the still may be used for boiler fuel (Prescott-Allen and Prescott-Allen, 1986; Sharp, 1990).

8.5   Pine Needle Oil

Pine needle oil is obtained by the distillation of freshly cut foliage, principally from *Pinus mugo* (mugo pine) and *P. sylvestris*. It is not normally undertaken at the same time as turpentine production. The oil is used in pharmaceuticals for flavour and perfume, and as an expectorant. The market is rather specialised and is vastly smaller than that for turpentine (Prescott-Allen and Prescott-Allen, 1986; Coppen, 1995).

8.6   Wood Tar and Creosote

Wood tar, or *pix liquida*, is the thick, oily, strong-smelling, dark liquid condensate resulting from the destructive distillation of certain *Pinus* spp. Exports from Sweden and Russia are respectively known as *Stockholm tar* and *Archangel tar*. The tar is applied externally in both human and veterinary medicine for its germicidal action and stimulating properties against chronic skin diseases. It is also one of the most efficient preservatives of animal and vegetable tissue, and is used by veterinarians for treating hoof and horn problems. The antiseptic and preservative properties are due to its creosote, carbolic acid and methanol content.

Creosote is produced by the fractional distillation of tar, especially coal tar, and is used for wood preservation, for fluxing pitch and bitumen, and as fuel. The almost colourless and aromatic *medicinal creosote* is obtained from wood tar and consists of a mixture of phenols, chiefly guaiacol and creosol (4-methyl-2-methoxyphenol). It is a powerful antiseptic and disinfectant, and is used as an ingredient of some disinfectant fluids (Miller and West, 1956; Sharp, 1990; Macpherson, 1995)

9.   OTHER WOOD PRODUCTS

9.1   Cooperage

Cooperage is the art of making wooden barrels, casks, tubs, etc. *Slack* or *loose cooperage* refers to containers used for transporting solid substances, i.e. food and hardware. Such containers are manufactured from cheap, light, easily-worked and elastic woods free from warping and twisting. *Tight cooperage* applies to containers used for liquids, especially alcoholic beverages; they consequently need to be skilfully constructed. Large-sized tanks and vats are known as *heavy cooperage*. Slack cooperage as an industry has now largely disappeared following the increasing use of wooden pallets. The use of tight cooperage has also reduced considerably except where the intrinsic qualities of the wood are indispensable, e.g. oak barrels, which are traditionally used for ageing whisky and wines.

9.1.1   *Whisky Barrels*

Until the 19th century lengthy ageing of whisky was regarded as a luxury and new barrels an unnecessary expense. Hence the custom of charring or *toasting* the inside of barrels to remove all traces of any previous contents. Such toasting is considered to give the whisky a stronger colour as well as conferring vanilla and caramel flavours to the whisky. In the United States the producers of straight whiskeys are even legally obliged to use charred barrels. Bourbon barrels, however, are only used once, they are then exported to Scotland and Ireland. Sherry casks are also exported to Scotland from the Mediterranean region for ageing whisky. Despite not being resistant to decay, the heavy, hard and strong timber of *Quercus coccinea* (scarlet oak), is also imported from America into Scotland for making whisky barrels. Stainless steel barrels with oak lids are now being used in some parts of the world because they help to reduce costs, are easier to clean, and can be repeatedly reused. The 'angel's share' of evaporation losses are also prevented. To produce the desired flavour charred wood can be placed in the steel container.

9.1.2 Wine Barrels

The species and possibly provenance of the oak, its manufacture, size and age, all affect the flavour imparted to wine. The smaller the barrel, the closer the contact of the wine with the wood and the greater the uptake of the oak flavour. Also the younger the barrel the greater the flavour imparted to the wine, the flavour becoming minimal after 5 years. The French or European oak, *Q. petraea*, is generally preferred to the American oak as it imparts more tannin to the wine. In Europe the wood is split into staves, following the natural grain of the wood. Barrels from American oak, especially *Q. alba*, give a more pungent, vanilla-oaky character, partly attributable to the looser grain and partly due to the wood being sawn into staves. The sawing action is believed to rupture the wood cells and release the aggressive flavours. In Greece the characteristic resinous flavour of retsina is due to wine casks made from *Pinus halepensis* (Aleppo pine). Steel casks in which a 'tea-bag' of oak chips or oak staves are suspended are now being used for some of the cheaper American wines. it is even claimed that expert tasters are often unable to distinguish whether a wine has been fermented and aged in or on oak (Hill, 1952; Prescott-Allen and Prescott-Allen, 1986; Davison, 1994; Atkin, 1996; Gabányi, 1997).

9.2 Cork

The commercial production of **cork** is mainly from *Quercus suber* (cork oak), an evergreen tree of the Mediterranean basin. The cork is first harvested when the trees are at least 20 years old, and the operation is repeated every 8-10 years. Slabs of cork up to 7.5 cm thick are removed with an axe, care being taken not to damage the inner bark. Because cork is both compressible and resilient without any lateral spread, it makes a perfect seal as a stopper, liner in crown bottle caps and for gaskets. It is impervious to water, due to air-filled cork cells in a natural resinous binder. Its buoyancy is due to its low relative density (0.20-0.25). It is used in life belts (known in North America as life preservers), fishing floats, etc. The fine air spaces in the cork provide low thermal conductivity, thereby giving excellent thermal insulation as well as sound absorbing properties. Its resiliency absorbs vibration, a property utilised for shock absorbers for vibrating machinery and for the inner soles of shoes. The partial vacuum caused by the microscopic cups formed by the cut surface cells provide a high coefficient of friction, which is utilised in non-slipping floor tiles, etc. A mixture of heated linseed oil, rosin and ground cork pressed onto a canvas or burlap backing is used for the manufacture of linoleum (Cooke, 1961; Davison, 1994; Mabberley, 1997).

Unfortunately cork can be attacked by the phellophagous fungus *Melophia ophiospora*, which is responsible for the unmistakable taint of 'corked' wine, and probably affects up to 5% of wine bottles world-wide. It has become an increasing problem and plastic corks are now being used instead. Up to 3% of the wine bottles in the UK now have plastic corks (Holliday, 1989; Simon, 1994).

Chapter 13

Fuel

The woody biomass from trees, shrubs, herbs and crop residues are an important source of fuelwood, charcoal, petroleum substitutes/alcohol and tinder, especially in the developing countries. The energy released can be used to fuel railway engines, electricity generators, dryers for fish, tobacco, lumber, grain, copra, etc., sugar factories, pottery, brick, charcoal, and limestone kilns, metal smelters, etc., as well as meeting domestic requirements (BOSTID, 1980; Cook, 1995).

1.   FUELWOOD AND CHARCOAL

World-wide *ca.* 1.4 billion tonnes of biomass, mainly as fuelwood and charcoal, were used in 1990 to provide domestic energy. This accounts for just over half of the total wood harvested for all uses, including timber, board and pulp. Yet it represents less than 14% of total world primary energy, a frugal consumption compared with that of fossil fuels, but still of major importance for the future, especially in the developing countries (Lamb, 1995).

In 1980 marked difficulties in fuelwood supplies were detected in 39 countries of Africa, 18 in the Middle East and Asia, and 14 in Latin America. For 1.148 million people their energy needs could only be met by an over-utilisation of the woody vegetation, while for *ca.* 100 million, half of them in Africa, were unable to meet their minimum needs. By the year 2000 it was estimated that 1.49 million people in parts of Southeast Asia will be living in zones where fuelwood supplies will be totally inadequate for meeting minimum energy needs. Similarly, there will be 500 million people in Africa, 160 million in North Africa and the Middle East, and 340 million people in Latin America, who will also be facing serious problems in meeting their minimum energy needs by the year 2000. To meet this energy crisis in the rural areas of the developing countries a greater effort must be made in managing the remaining resources, by increased afforestation and agroforestry practices, and improvements of

251

charcoal conversion techniques, more efficient cooking stoves, etc. (Montalembert and Clément, 1983). It has been suggested by FAO (1993b) that the alarming predictions for increased energy consumption, particularly in the developing countries, did not occur, and that energy consumption growth rate actually fell by approximately 50% between 1983 and 1993. Rising oil prices and a declining economy had contributed to a decrease in non-domestic energy consumption and energy resource improvements. Even so, the domestic requirements in the densely populated and arid areas remain susceptible to a serious fuelwood deficit in the not too distant future. The problems of deforestation and desertification are prevalent in the developing countries and that there still is a major environmental crisis, with the destruction of the natural forests for fuel, timber and farming continuing at a faster rate than reafforestation is able to meet the requirements for fuel.

1.1   Fuelwood

It should be noted that whereas foresters generally measure timber as a solid cubic metre of wood, fuelwood is usually measured as a *stacked cubic metre* or *stere* and represents a 1 m^{-3} stack of branches or split wood, which can very roughly be taken to weigh 200-300 kg. It must not to be confused with a solid cubic metre which, for hardwoods typically used for firewood or charcoal in the developing countries, are usually in the range of 600-900 kg m^{-3}. The contents of the stere will obviously vary with the diameter of the wood and how closely it is packed, with straight, unbranched limbs packing tighter than crooked or forked limbs.

Fuelwood is the major source of energy for a great many rural people in the developing countries. In Burkina Faso and Senegal, for example, the average consumption for cooking purposes alone is estimated at 360 kg person^{-1} year^{-1}, although the use of more efficient stoves instead of open fires could halve the annual consumption. For local industries, e.g. bakeries, breweries, blacksmiths, and brick making, the requirements are 500-750 kg per capita (Montalembert and Clément, 1983; Maydell, 1986). These requirements exceed the rate by which the natural wood resources can be maintained.

There is much anecdotal evidence regarding fuelwood preferences, but because availability rather than suitability is often the major factor affecting choice, a more critical evaluation of fuelwood suitability based on proper standards of comparison is clearly needed. It should be appreciated that in some societies it is the women who cut and collect the fuel and attend to the cooking, consequently their opinions regarding suitability should be sought when evaluating domestic fuelwood. In many nomadic societies, especially those in the arid and semi-arid regions, people may, in the absence of trees, rely on woody herbs and other combustibles as a source of heat for cooking, etc.

Other criteria to be considered are: (1) For domestic purposes fuelwood should be relatively easy to cut, and capable of being split into acceptable sized pieces, without

damage to the axe or saw from excessive calcium oxalate raphids or silica crystals;
(2) Ideally the wood should be straight and easy to handle i.e. reasonably free of
thorns; crooked or branched wood is difficult to carry and stack; (3) Low water
retention on drying is desirable. Soft woods with a high water content, such as
members of the Bombacaceae, require a long time to dry and are generally unsuitable
as fuel; (4) For domestic purposes the wood should burn slowly, with a steady heat
and little smoke, e.g. *Celtis* spp. A fierce heat, however, would be desirable for metal
workers and the smelting of ores, e.g. wood from *Hyphaene thebaica* (dom palm) and
Prosopis spp. Highly inflammable resinous woods are desirable for kindling; (5)
Fuelwood with obnoxious smells, e.g. *Maerua crassifolia*, toxic or excessive smoke,
e.g. *Boscia senegalensis* and *Cassia sieberana*, are undesirable; (6) Resistance to or
protection against insect and fungal damage can avoid unnecessary losses. *Acacia
seyal*, for example, is susceptible to the bostrychid beetle *Sinoxylon senegalense* and
stacks of fuelwood are quickly reduced to a pile of sawdust (J.K.Jackson, pers, comm.
1995b); (7) If possible the source should be renewable and production sustainable, i.e.
from stem coppicing or root suckering, with rapid growth and reasonably dense
wood; and (8) If grown in woodlots the species should be suited to the environment,
cheap and easy to establish, and amenable to management.

1.2   Charcoal

Charcoal is an impure form of graphitic carbon and is defined by Emrich (1985)
as '*the residue of solid non-agglomerating organic matter, of vegetable or animal
origin, that results from carbonisation by heat in the absence of air at temperatures
above 300°C.*" Charcoal is a major fuel source in some regions or, in the developed
countries, used for charcoal grilling and barbecues, for black gunpowder and case-
hardening of metals. It was formerly used a source of propanone (acetone), methanol
(methyl alcohol) and ethanoic (acetic) acid.

Fuelwood is both bulky and heavy. By converting to charcoal it is possible to
produce a lighter and more easily transportable fuel. However, despite the charcoal's
thermal efficiency, which is double that of wood, conversion losses makes it an
inefficient use of fuelwood resources. Ideally charcoal manufacture can only be
economically justified where there is an excess of available wood of suitable
dimensions and high relative densities (850-1000 kg m^{-3}), and markets not too far
distant (Maydell, 1986).

The desirable criteria for charcoal are similar to those for fuelwood with the
additional desirable requirements of large, uniform stands producing wood of suitable
dimensions and density, and capable of being split into acceptable sized pieces. The
characteristics of the finished product are <7% moisture, <3% ash, >75% fixed
carbon and >300 kg m^{-1} apparent density. High wood densities are needed to achieve
charcoal densities of 270-280 kg m^{-3} required by the metal industries (FAO, 1962).

Charcoal production on a weight to weight basis is economically of very low efficiency and quality, with a maximum efficiency of 30-40% possible. In the developing countries, such as those of the Sahel, the slow burning of wood in simple earth or pit kilns, or in locally manufactured metal kilns, can only achieve a 12-20% weight conversion. The by-products, such as tars and ethanoic acid are generally lost. There is clearly an urgent need for improving existing charcoal conversion techniques for the developing countries. Even with more advanced techniques, by 'cooking' the wood within special reactors known as *pyrolysers* and under controlled temperature and atmospheric conditions from which oxygen is excluded, it is only possible achieve 30-40% conversion of wood to charcoal. For every 1 tonne of charcoal produced and using modern carbonisation techniques, *ca.* 0.6 tonnes of dehydrated charcoal products and 7 MJ of energy in the form of poor quality gas can be recovered. Efficiency depends not only on the type of kiln but also on the wood characteristics. The latter including straight branches for ease of packing and avoiding air pockets, moisture content, wood density and diameter. The eventual charcoal characteristics must also be considered. Species such as *Stereospermum kunthianum*, for example, which is widely distributed through the Sudano-Zambesian Region of Africa, may be rejected as unsuitable because the wood does not make good charcoal, disintegrating into ashes during manufacture (Emrich, 1985; Davis and Eberhard, 1991; Lamb, 1995; Rosillo-Calle *et al.*, 1996).

On a weight basis all woods will produce approximately the same amount of charcoal, although a dense wood will clearly produce a greater volume. The temperature attained during pyrolysis will also affect the final yield, while the original vessel and pore structure of the wood by affecting heat and gaseous transfer can affect charcoal quality.

Charcoal quality can be expressed in terms of '*fixed carbon content*', and ranges from *ca.* 50-95%, strongly influencing the final yield. Quality is important, but producing good quality charcoal with a high fixed carbon content can only be achieved by an overall loss in productivity. For iron smelting the fixed carbon content is the most important factor because it is the fixed carbon that is responsible for reducing the iron oxides in the ore to produce metal. Nevertheless, for optimum blast furnace operations there is a need to balance the fireable nature of high fixed carbon charcoal with the greater strength and lower fixed carbon and higher variable matter content. The energy efficiency of the carbonisation process can be much improved if the fixed carbon content of the charcoal is reduced. It is estimated that as much as 51% of energy losses occur if the fixed carbon is kept at 85%. The gains in energy contained in the volatiles can then be transported to the steel mill (Davis and Eberhard, 1991; Rosillo-Calle *et al.*, 1996).

Treating charcoal at high temperatures with steam, air or CO_2 produces highly porous '*active*' or '*activated*' charcoals with surface areas of 300-2000 m^2 g^{-1}. These are widely used to absorb odorous or coloured substances from gases and liquids, e.g. in the purification of water, sugar and rubber, solvent recovery and other volatile materials. They are also used in gas masks for the removal of atmospheric toxins, in dry cleaning and as catalysts in the chemical industries (BOSTID, 1980; Sharp, 1990).

1.3 Fuelwood and Charcoal Combustion

The weight of wood and amount of energy obtained when burnt will both vary according to the moisture content and density of the wood. The *moisture content on a dry basis* expresses the quantity of water present as a proportion of the weight of oven dried wood. The alternative *moisture content on a wet basis* expresses the amount of water as a proportion of the combined weight of the wood and water. The two calculations are often confused or even ignored. When the wood is relatively dry there is little significant difference between the two. For example, *air-dry wood*, i.e. wood that has been allowed to dry out for a few months could have a moisture content of 16.6% on a wet basis or 20% on a dry basis. However, newly cut or green wood with a moisture content of 60% on a wet basis could have a moisture content of 150% on a dry basis.

Wood generally is composed of 49-50% carbon, 6% hydrogen, 43-44% oxygen, and 0.5-1% nitrogen, sulphur and ash. Because of this remarkably small variation in elemental composition there are only slight differences between the calorific values of oven-dry woods. The slight differences found between average values for hardwoods and softwoods are the result of differences in the constituent resins, cellulose, hemicellulose, lignin and mineral water. With calorific values of 35-40 MJ kg^{-1} resins will have a noticeable affect on the gross calorific value, consequently softwoods will have a slightly higher calorific value than hardwoods. This higher calorific value does not necessarily compensate for certain disadvantages when softwoods are used as a fuelwood. Pine, for example, burns rapidly and does not make good coals. The formation of coals may depend on the hemicellulose content, which is generally *ca.* 35% for hardwoods and *ca.* 28% for softwoods. Slight differences can also occur between individual woods within the same species. However, since most oven-dry woods have a gross calorific value within 5% of 20 MJ kg^{-1}, this value is considered a satisfactory approximation for most fuelwoods. The calorific value is not, therefore, a very satisfactory method for evaluating fuel, other parameters, such as reactivity or 'combustibility' might prove to be a more reliable measure (Eberhard, 1990).

The density of wood varies widely between species, with the density of air-dry hardwoods such as *Swietenia* spp. (mahogany) and *Diospyros* spp. (ebony) *ca.* 1000 kg m^3, while those of lower quality hardwoods typically used for fuelwood and charcoal in the developing countries are *ca.* 600-900 kg m^3. The heavier woods are more desirable since they minimise the low heat value to mass/volume relationship on combustion. While the presence of moisture in the fuelwood does not alter its higher heating value, the available heat per unit mass is reduced. This reduction is due to a lower combustible substance per unit weight of fuel plus the use of some of the available energy to vaporise the water. Consequently fuelwoods from species with low green moisture content are generally desirable since they require less drying effort (Table 15).

TABLE 15. Comparison of net calorific values of woods at different moisture contents (FAO, source unknown)

Fuel type (1 kg)	% Moisture (wet basis)	Net calorific value (MJ)
oven-dry wood		20
air-dry wood	17	15
green wood	60	8
charcoal		10
air-dry millet stalks		15
kerosene		54

The *calorific value* or **heat of combustion** represents the quantity of heat released per unit weight of fuel following complete combustion. It is determined according to British Standard BS 526:1961 standard conditions, usually by explosive combustion in a bomb calorimeter under an initial pressure of 23-27 atm at constant volume. The **high** or **gross calorific value** is where all the water formed by combustion is condensed to liquid form and is defined as: *"The gross calorific value at constant volume of a solid or liquid fuel is the number of heat units measured as being liberated per unit quantity of fuel burned in oxygen in a bomb under standard conditions in such a way that the materials after combustion consists of the gases oxygen, carbon dioxide, sulphur dioxide and nitrogen, liquid water in equilibrium with its vapour and saturated with carbon dioxide, hydrochloric acid in solution and solid ash."* The *low* or **net calorific value** is obtained by subtracting the latent heat of evaporation of the water from the high calorific value and corresponds more closely with the conditions under which a fuel is burnt.

Because combustion is an oxidation reaction, the quantity of heat released is related to the reduction state of the fuel, i.e. the chemical make up of the fuel at molecular and atomic levels, hence the extremely high calorific values for aliphatic hydrocarbons (carbon atoms in chains, not rings) with high H:C ratio and absence of oxygen. For the various components of wood, the reduction states are: resins, lignin, cellulose and hemicellulose. Thus, the calorific value of woody material is generally modified according to the lignin content and to a much greater extent by the quantity of resinous extractives. The ash resulting from non-combustible inorganic minerals will adversely effect the calorific value, reducing the heating value per unit weight of the fuel. While wood ash is not usually a problem with domestic fuelwoods, it is undesirable in industrial fuelwood where boiler furnaces achieve high temperatures, producing slag and clinkers from the melting and fusion of the ash.

The size of the fuelwood burnt is of considerable importance. Woodchips used in industrial processes have a critical size for optimum combustion, larger samples having a lower surface area to volume ratio, thus affecting the penetration of oxygen and heat, and the outflow of gaseous products. Similarly, the size of wood used for domestic fires will affect the size of the coals formed, the flammability of the wood and heat retention. The devolatisation and combustion of small pieces will be sequential and degree of overlap for the different phases of combustion restricted. Large pieces will burn in layers and the outer layer may be turned to ash while the inner layers are still undergoing char combustion.

During combustion heat is transferred into the wood through and across the grain. Under the influence of heat the hydrocarbons present, consisting mainly of cellulose (320-370°C), hemicellulose (220-320°C) and lignin (200-500°C) undergo a complex series of decomposition reactions termed *pyrolysis*, also referred to as *wood carbonisation* or *dry wood distillation*, which eliminates much of the volatile components, with a loss of *ca.* 80% of the wood mass and is accompanied by an increase from *ca.* 50% to *ca.* 75% in carbon content of the *charcoal* or *char*, partly as a result of the reduction of the hydrogen and oxygen in the wood.

The following physical-chemical changes occur during pyrolysis: (1) Between 100-170°C, most of the water content is evaporated; (2) Between 170-270°C gases consisting of CO and CO_2 are emitted, plus considerable oils and other extractives, These, after scrubbing and chilling form *pyrolysis oil*; (3) Between 270-280°C the resultant gases from the pyrolysis migrate through the wood structure to the outer layer, where they combust in the presence of oxygen, the surrounding flames providing a major source of heat. The remaining porous solid, or char, consists chiefly of carbon plus small amounts of ash and other compounds; (4) At 300°C the yield is *ca.* 50% char; (5) At carbonisation temperatures of 500-600°C, volatiles are lower and retort yields *ca.* 30% char; and (6) At 1000°C the volatile content is almost zero and charcoal yield falls to *ca.* 25% (Emrich, 1985; Davis and Eberhard, 1991; Lamb, 1995; Rosillo-Calle *et al.*, 1996).

Charcoal combustion involves the diffusion of oxygen across the boundary layer and its penetration into the charcoal, where it reacts with the carbon to produce CO and CO_2. A light charcoal will burn more rapidly because the greater area to volume makes oxygen penetration easier. A weak charcoal that crumbles easily will likewise burn quickly. It is probable that the cohesion of the char is determined by the density of the original wood and the degree of splitting and cracking during combustion. The complex factors that determine cracking are related to stresses during the drying and devolatisation phases, and the ability of the wood to withstand these stresses (Davis and Eberhard, 1991).

Using wood as an industrial energy source does produce atmospheric pollution problems from soot and nitrogen oxides together with traces of carcinogenic chemicals, although these emissions are insignificant when compared with those from fossil fuels. The SO_2, CO and hydrocarbons emissions from wood combustion are negligible or of limited extent, and are generally considered as not causing any serious air pollution problems. On the other hand, the contribution to atmospheric pollution from charcoal processing and combustion are higher. A study in the Philippines has shown that 1 kg of wood emits 450 g CO_2 while a similar weight of charcoal produced more than 2000 g (Wang *et al.*, 1982; Lamb, 1995).

2.   TINDER AND KINDLING

The two terms tinder and kindling are almost synonymous, with kindling being used to describe readily ignited fuelwood and dry herbaceous stems, and tinder for non-fuelwood sources. As used here they refer to readily combustible material used to ignite fires other than paper or commercial 'firelighters'. The term *tinder* is from the Middle English *tinder*, Old English *tynder*, from the Germanic *tund* (unattested), past participle form of *tend-* (unattested) to burn, to kindle (Long 1994).

Tinder has a long history dating back at least to the 8th millennium BC. *Punk* or *touchwood*, rotten wood caused by the bracket fungus *Fomes fomentarius* (tinder fungus), etc. has been found at the Mesolithic site of Star Carr, Yorkshire; the Neolithic 'iceman' from the Hauslabjoch also carried a portion of tinder fungus in his pouch. Much later the tinder fungus, together with flint and steel, formed the contents of a 16th century AD soldier's tinder box. Other prehistoric tinder sources include the puffballs *Bovista nigrescens* and *Calvatia utriformis*, the washed and dried woolly perigonous hairs from the fertile flowers of *Typha* spp. (bulrush, reedmace), thistle-down from members of the Compositae tribe Cynareae, the pith from the stems of *Juniperus* (juniper), *Typha* and *Ferula* (fennel), the hammered bast from *Salix* spp. (willow), and various dried mosses and grasses. The smouldering burning pith within the fennel stems were used by the Ancient Greeks for transporting fire from place to place and as such used for the original Olympic torch. Rolled birch bark and similar materials were used to carry fire in North America (Dimbleby, 1978; Watling and Steward, 1976; Spindler, 1994; Mabberley, 1997; Rudgley, 1998). A number of lichens with a high oleoresin content, such as the oak mosses *Evernia prunastri*, *Pseudoevernia furfacea*, syn. *Evernia furfuracea*, and *Ramelina fraxinea* subsp. *calicariformis*, syn. *R. calicaris* (orchil, cudbear), were also used as tinder.

Even today, particularly in the rural communities of developing countries, tinder is still an important and often overlooked commodity without which, and in the absence of paper, survival could be extremely difficult. for example, in the Negev the Bedouin use the dried inner bark (bast fibres) of *Thymelaea hirsuta* as a tinder by placing a small quantity on a flint stone and igniting the fibres by striking with an iron striker, while in the Sahel charcoal from *Calotropis procera* is used as tinder (Schmidt and Stavisky, 1983; Maydell, 1986).

Kindling is from the Middle English *kind(e)len*, from Old Norse *kynda*, to catch fire, akin to Middle High German *künden*, to set on fire), and refers to easily ignited material (Long, 1994). Good kindling should be light, porous, ignite readily even when damp, and burn with an intense flame. The presence of resins and other extractives improving flammability and are obviously desirable. The ease of ignition has been shown to be directly related with density, the lower the density the easier the ignition (Davis and Eberhard, 1991). Many years ago as a Boy Scout I can recall mastering the art of lighting a fire without using paper, collecting dry pine needles from the under branches (those lying on the ground were usually too damp) for kindling.

A wide range of dry plant material may be used for kindling, often from species that appear to have no other ethnobotanical value. Examples include the dry stems of *Amaranthus* spp., *Opilia campestris* and the wood from *Eucalyptus* spp. and *Phyllanthus* used by the Mbeere of Kenya. In the Sonoran Desert the Seri Indians use fire drills made of *Bursera* spp., *Ficus petiolaris* subsp. *palmeri*, etc. for kindling, while the Yaqui used, among other species, *Baccharis salicifolia*, *Coursetia glandulosa* and *Populus* spp. Interestingly, although both *Baccharis* and *Coursetia* also occur in the Seri region, their use by the Seri for fire drills was not recorded. In Brazil the Waimiri Atroari use the resinous wood of species of *Protium* (also a member of the Burseraceae) for kindling (Felger and Moser, 1985; Riley and Brokensha, 1988; Cotton, 1996).

3.   BIOFUEL

Fossil fuels are an exhaustible resource, consequently there is considerable interest in developing suitable renewable alternatives. The vegetable biofuels offer such an alternative and may be used either as substitutes or extenders of fossil fuels and, although some may have the same energy content as diesel oil, they do have the disadvantage of being much less volatile and more viscous, causing clogging of the injection nozzles and the formation of deposits. However, engine modifications and correct setting can help in overcoming these problems. When used in the pure form, their esterification may even be advantageous since the volatile and viscosity properties of the esters are similar to those of diesel oil.

Biofuels have a number of advantages over fossil fuels. They include: (1) Low sulphur content, giving little or no SO_2 emissions, the precursor of acid rain; (2) Oxygenated biofuels emit little NO_x, which is a smog-forming reactant; (3) Biofuels can reduce the quantity of fossil fuels being used and their associated NO_x and SO_2 emissions; (4) Biofuels are a renewable resource and their production can provide jobs and economic development; (5) Biofuels can replace fossil fuel imports; (6) Biomass growth from biofuels is also a practical way of removing atmospheric CO_2 and ameliorating global warming; and (7) Many petroleum-based organic chemicals can be derived from biomass sources.

Promising sources of biofuel plantation crops include the coconut and African oil palms, *Jatropha curcas* (physic nut), *Orbignya* spp. (Babussu palm), *Sapium sebiferum* (Chinese tallow tree), and oilseed agricultural crops such as groundnut, oilseed rape, olive, soya bean and sunflower. Among the less conventional crops are *Crambe hispanica*, syn. *C. abyssinica* (crambe), *Cuphea* spp. (cuphea), *Lesquerella* spp. (bladderpod), and *Limnanthes alba* (meadowfoam), while trials with sea water irrigation of the annual halophyte *Salicornia bigelowii* (pickleweed) compared favourably or even exceeded oil yields from both soya and sunflower grown under freshwater irrigation (Kochhar and Singh, 1989; Glenn *et al.*, 1991; Shay, 1993; Heller, 1996).

The UK's first biofuel power plant to be fuelled by woodchips was commissioned in Northern Ireland in 1998. The technology is based on the gasification of wood chips obtained from willow and poplar plantations. The gas produced is then burnt in a conventional gas engine to produce heat and/or electricity, the gasification process producing a cleaner and more efficient fuel than direct combustion (Hough, 1998).

A great deal of work has also been carried out on the direct use of vegetable oil triglycerides as diesel fuel. Considerable attention has been given to the conversion of these glycerides into molecular species similar to petroleum-based diesel fuel. Thermal cracking, catalytic cracking, Kolbe electrolysis and transesterification with low molecular weight alcohols, have been used to improve their performance as diesel fuels. A number of seed oils, e.g. sunflower, olive, groundnut, soya, palm, rape and sesame oils are also capable of being mixed with diesel oil and used in diesel engines, with sunflower and groundnut oils being only 14% less efficient than diesel alone.

In 1979 it was discovered that the liquid balsam tapped from *Copaifera langsdorffii* could be used alone to fuel a diesel engine; the related *C. multijuga* has a similar capability. Experimental plantations in Japan have shown the trees to have a potential for yielding 25 l of fuel oil over a 6 month period. Similarly, the seed oil from *Pittosporum resiniferum* (petroleum nut tree) contain *ca.* 30% terpenes, which can also be used directly as a high octane fuel (Wang *et al.*, 1981; Cross 1984; Kochhar and Singh 1989; Lewington, 1990; Shay, 1993; Mabberley, 1997).

Palm oil or palm kernel oil mixed with petrol in proportions of 51:49 and 45-55 respectively, produce a biofuel of the same specific gravity as diesel oil, and have been tested in a Petter Diesel AAI 4-stroke, air-cooled, 220 cc, vertical stroke, single cylinder engine. The rated power with diesel fuel was 3.3 kW at 3500 r.p.m. (*ca.* 4.5 h.p.). Testing against a hydraulic dynamometer and magnetic tachometer at engine speeds ranging from 1500 to 3500 r.p.m. produced similar brake-power for all three fuels up to 3000 r.p.m., above which the palm kernel/petrol blend gave slightly more power. As mentioned above, further improvements may be possible with tuning the injection timing to suit each fuel mixture (Anonymous, 1994).

Microalgae, with their high reproductivity rates (up to 12.5 kg m^{-2} yr^{-1}) have also been examined as a source for methyl ester diesel fuel. Solvent extraction has been used to isolate the storage lipids in *Botryococcus braunii*, *Chaetoceros muelleri* and

Monoraphidium minutum. It has also been found that the yeast *Lipomyces starkeyi* produces an oil similar to that of palm oil. The production of such **single cell oils** has the great advantage of being under controlled, sterile environments, although at present economic production can only be considered for those oils in the US$5000 ton^{-1} price range (Suzuki and Hasegawa, 1974a, b; Kochhar and Singh, 1989; Shay, 1993).

Considerable advances have already been made in the USA in producing **ethanol** from the hydrolysis of maize grain to dextrin, maltose, and finally glucose, which is then fermented to produce ethanol. This, in turn can be converted to **ethene** (**ethylene**) and **butadienes**. Similarly, in Brazil the controlled fermentation of cassava and of sugar cane is used to produce ethanol. France plans to produce methanol commercially from *Helianthus tuberosus* (root artichoke), while in China banana leaves and sugar cane are used to produce biogas.

The fermentation in a simple digester of the tropical aquatic weed *Eichhornia crassipes* (water hyacinth) has been found capable of producing biogas containing 60-90% **methane** and 10-40% CO_2 with a mean calorific value of 5292 Kcals m^{-3}. In a warm climate 1 ha of water hyacinth under nutrient enriched conditions is capable of yielding 58400 m^{-3} of biogas in a 7 month season. The aquatic *Cyperus papyrus* (papyrus) can also be harvested to produce compressed, air-dried briquettes (Kochhar and Singh, 1989).

Chapter 14

Vegetable Fibres

The use of plants for fibres are regarded as second only to food in their usefulness. At a conservative estimate there are well over 2000 species with usable fibre; more than 1000 of which are known from America, 750 from the Philippines and over 350 from East Africa (Hill, 1952). Vegetable fibres have been used by man for cordage, clothing, basketry and matting since time immemorial, although archaeological evidence of their use by early man is often inadequately represented because they do not preserve well. Fortunately ice preserved the cloak of plaited, unspecified grass, etc. and other crude plant fibre accessories worn by the Neolithic 'Iceman' of *ca.* 3300-3200 BC from the Hauslabjoch in the Austrian Alps. The more sophisticated use of flax, whose processing involved retting, was also known to Neolithic man. In Britain impressions of flax seeds in Neolithic pottery were found at Windmill Hill, Wiltshire, but the earliest preserved archaeological evidence of flax fibre are from Jutish and Saxon sites *ca.* 5th century AD (Appleyard and Wildman, 1969; Renfrew, 1973; Spindler, 1994).

Even today, in the age of synthetic fibres, vegetable fibres are still of considerable economic and commercial importance. In order to compete commercially with animal (wool, hair, bristle, etc.) and synthetic fibres an adequate and regular supply must be guaranteed and a suitable market niche available. Plant fibres, including wood pulp, are used for a wide range of products, including cardboard, fibreboard, non-wood board, paper, paper substitutes, cord/string/twine, thread/yard, woven material such as cloth and sacking, packing/stuffing, filling materials, matting, netting, basketry, thatch and tow, also for their cellulose derivatives such as *cellulose ethanoates* (*acetates*), cellophane, plastics, rayon, etc. (Cook, 1995).

It was the pioneer work on lignin chemistry in 1898 by C.H. Stearn and C.S. Cross at the Jodrell Laboratory, Royal Botanic Gardens, Kew that produced the first man-made textile yarn capable of being woven and dyed. Their viscose rayon, later known as rayon, pioneered the development of the artificial fibre industry where cellulose fibres are forced through fine spinnerets and the resultant filaments

263

solidified. Following the development in 1935 of nylon, the first synthetic fibre from coal, a wide range of polyester fibres, e.g. Terylene and Dacron, and acrylic fibres, e.g. Acrilan and Courtelle, are now available. But not all synthetic fibres are cellulose-based, Ardil for example, is manufactured from groundnut protein, while maize and soya proteins are used for other synthetic fibres. Although some synthetic fibres may be inferior in quality to natural fibres, others are either cheaper or, in some instances, more adaptable.

Cotton, continues to dominate the natural fibre market (Table 16), finding an outlet either alone or mixed with synthetic fibres. Natural fibres are biodegradable which, depending on their use, can be an advantage or disadvantage. Polypropylene baling twine, for example, if accidentally ingested by livestock could be fatal, while a natural fibre would be digested (Purseglove, 1987; Robertson, 1994; Robbins, 1995).

TABLE 16. Global fibre production for 1993 (FAO, 1994)

Crop	1000 tonnes
Gossypium spp. (cotton lint)	16,805
Corchorus capsularis, C. olitorius, Hibiscus spp., *Urena lobata* (jute and allied fibres)	3,391
Linum usitatissimum (flax fibre and tow)	610
Agave sisalana (sisal)	291
Cannabis sativa + Crotalaria juncea (hemp fibre and tow)	121
Boehmeria spp., *Ceiba pentandra, Furcraea* spp., *Neoglazovia variegata, Phormium tenax, Samuela carnerosana* (other fibre crops)	408

1. FIBRE CLASSIFICATIONS AND CHARACTERISTICS

Fibres can be classified botanically according to their anatomical and morphological origins, or by their commercial use.

1.1 Botanical Fibres

Botanically fibres are: (1) *Hairs*, cells known as *trichomes*, borne on the seeds or inner walls of the fruit, and consist of elongated, unicellular or multicellular, and nonconducting, epidermal outgrowths. These are referred to as *ultimate fibres*, free of any extraneous plant tissue. Cotton, for example, from the seeds of *Gossypium* spp., consists of long and narrow unicellular hairs, two types of which occur in the cultivated cottons: (a) *Fuzz*, consisting of short hairs which eventually become almost solid from internal cellulose deposits and are firmly attached to the seed; they cannot be spun; and (b) *Lint*, which are relatively long and readily detachable hairs with greatly reduced cellulose deposition, giving a hollow lumen so that the hairs collapse on drying to form a ribbon. The cellulose is deposited spirally, enabling the ribbon to twist and giving the characteristic convolutions that enable the cotton to be spun. Others examples are akund floss from the seeds of *Calotropis procera* and *C. gigantea*, and kapok from the inner capsule wall of *Ceiba pentandra* (silk cotton tree); (2) *Extra-xylary* or *bast fibres* include fibres of the cortex, pericycle and phloem. Among the more commercially important sources of bast fibres are *Boehmeria nivea* (ramie), *Broussonetia papyrifera* (paper mulberry, sa), *Corchorus* spp. (jute), hemp, flax, etc.; (3) *Leaf fibres* are obtained from the lamina and petioles of certain monocotyledons, e.g. *Agave sisalana* (sisal), *A. fourcroydes* (henequen), *Furcraea foetida* (Mauritius hemp), *Musa textilis* (abacá or Manila hemp), *Phormium tenax* (New Zealand hemp). Three types of leaf fibre are recognised: (a) Crescent-shaped median bundles running through the middle of the leaf and are consequently the longest; (b) Fibre bundles from the periphery of the leaf; and (c) Fibres in-between; (4) *Wood* or *xylary fibres* from trees and shrubs and include the fibre tracheids, i.e. xylary tracheids that resemble tracheids by the possession of bordered pits. Wood fibres are widely used in paper-making (see Chapter 12); and (5) *Miscellaneous fibres* obtained from other parts of the plant, such as piassaba or piassava fibre or bass from the leaf bases of *Attalea funifera* (Bahia piassaba or piassava palm), coir fibre from the mesocarp or husk of *Cocos nucifera* (coconut), Italian whisk from millet stems, Mexican whisk from the roots of the grass *Muhlenbergia macroura*, raffia from the lower epidermis of young leaves of *Raphia farinifera*. Also, according to Smartt (1990) a strong fibre from the erect peduncles of *Vigna unguiculata* cv. 'textilis'; these fibres are restricted to the peduncle and are not found elsewhere in the plant.

Plant fibres, as distinct from hairs, consist of relatively long and tapering sclerenchyma cells. They are usually formed directly from meristematic cells and often have inconspicuous simple pits. Such fibres may occur in nearly all parts of the plant, being most abundant in the cortex, pericycle, phloem and xylem. The presence of bordered pits distinguish *xylary fibres* and fibre-tracheids from the simple pits of the *extra-xylary* or *bast fibres*. Short fibres, such as those of Manila hemp and species of *Agave* and *Sansevieria*, have all parts of the cell at the same stage of development. In longer fibres, such as those of *Cannabis sativa* subsp. *sativa* (hemp) and *Linum usitatissimum* (flax), the cells elongate apically, keeping pace with the surrounding cells, with secondary thickening developing in part of the growing cell (Eames and MacDaniels, 1947; Tootill, 1984; Purseglove, 1987; McDougall *et al.*, 1993).

In dicotyledons, as a result of secondary growth, two age classes of fibres are produced, *primary* and *secondary fibres*. The primary fibres are thicker and more compact, with thicker walls and narrower lumina than the secondary fibres, they are also coarse, hard and lustrous. Because the secondary fibres are produced by cambial activity after the herbaceous plant have reached maximum height, plants with thick stems will contain a higher proportion of secondary fibres than those with thin stems. The secondary fibres are finer, softer, weaker and less brittle than the primary fibres. Despite such important differences, it is virtually impracticable to separate the two fibres classes during retting.

TABLE 17. Dimensions of individual vegetable fibres and their uses (Cross and Bevan, 1900; Kirby, 1963; Shaltout, 1992; McDougall *et al.*, 1993; Jarman, 1998)

Species	Length (mm)	Diameter (μm)	Use
Gossypium vitiifolium (cotton)	24-50	19-28	textiles, clothing
Cocos nucifera (coir)	20-150		cordage, brushes, mats, mattress filling
Linum usitatissimum (flax)	2.1-40	12-30	clothing, household linen, canvas, sacks, cordage
Boehmeria nivea (ramie)	39-150	25-75	textiles, canvas, cordage, netting
Cannabis saitiva subsp. *sativa* (hemp)	5-55	22	cordage, tarpaulins, textiles
Musa textilis (abacá)	4.3-6.2	18	cordage, textiles
Crotalaria juncea (sunn hemp)	3-12	13-50	cordage, canvas, paper
Phormium tenax (New Zealand flax)	2.5-5.6	0.8-1.9	cordage, netting, fabrics
Thymelaea hirsuta (mitnan)	1.6-8.9	13-22	paper
Agave sisalana (sisal)	1.5-7.4	19-30	cordage, sacking, floor covering, reinforcement of cement sheets
Hibiscus sabdariffa (roselle)	1.2-6	10-30	cordage
Hibiscus cannabinus (kenaf)	1-31	12-36	cordage, sacks, hessian
Corchorus capsularis (jute)	0.8-8	16-32	cordage, sacks, hessian
Stipa tenacissima (esparto grass)	0.5-1.9	9-15	cordage, sails, mats, paper

Note: cordage is a general term for rope, string, twine, thread, etc.

The fibre bundles, consisting of ultimate cells bound together by the middle lamella or pectinaceous matter, form a lattice structure within the stem which decreases upwards in both number of bundles and ultimate cell content. The ultimate cells vary according to the species in size, shape, wall and lumen thickness, the lumen varying with the origin of the cell and presence of absence of transverse dislocations or nodes. The ultimate cells are generally too short to be used on their own except for paper-making (Table 17) although strands of ultimate fibres bound together by pectins can be used for spinning. Exceptions include the ultimate cells of cotton and ramie. In general the fibre dimensions reflect their potential industrial use. For textiles the requirements are for the length to width ratio to be in excess of 10^3 while for paper making the ratio will generally be less than 10^2 (Kirby, 1963; McDougall *et al.*, 1993).

1.2   Commercial Fibres

Fibres can also be classified according to their use: (1) *Textile fibres*, e.g. cotton, ramie and flax, are capable of being spun into threads and are used for clothing, bags and fabrics; (2) *Cordage fibres* are used for ropes and twines, and include abacá or Manila hemp, flax, hemp and sisal. Most leaf fibres fall into this category; sisal and Mauritius hemp are also used commercially for sugar- and coffee-bags; (3) *Brush and mat fibres*, e.g. piassaba or piassava fibre or bass, palmyra fibre from *Borassus flabellifera*, kitool from *Caryota urens* (jaggery palm), coir, referred to in the brush trade as bristle fibre or coco fibre is from the coconut; the latter can also be spun into a yarn and is much used in the manufacture of matting; (4) *Stuffing and upholstery materials*, including vegetable fibre or crin végétal from *Chamaerops humilis* (dwarf fan palm), and the short fibres obtained as the combings from the husk of the coconut and sisal tow; (5) *Paper-making fibres* from wood, especially soft woods, and *Stipa tenasissima* (esparto grass), the latter being important in the UK for the manufacture of speciality paper; and (6) *Miscellaneous fibres* such as raffia, from *Raphia farinifera*, which widely used by horticulturists for tying plants, and fibres, usually as strips from palm leaves or leaf-sheaths, for basketry and Panama hats (see Section 7).

Commercially the term fibre refers to the fibrous strands or filaments of the numerous ultimate cells remaining after retting. Two categories of fibres are recognised: (1) *Hard fibres* from the leaves of monocotyledons, e.g. Manila hemp (abacá) and sisal; and (2) *Soft fibres* from the dicotyledons. These include the seed fibres of cotton and bast fibres of hemp, jute, flax, etc. (Kirby, 1963; Tootill, 1984).

1.3   Fibre Properties and Classifications

It is the fibre properties that determine their industrial application, and these may be summarised as: (1) *Morphological characteristics*, e.g. length to width ratio for suitability for paper or textiles. Less important are cell wall thickness and lumen diameter; (2) *Physical properties*, e.g. absorption of water and chemicals; swelling and drainage, which are vital for pulp and paper manufacture as well as suitability for brushes and textiles. For some applications colour, refractive index and birefringence (difference between the refractive index of a fibre measured parallel to the fibre axis and perpendicular to it as an expression of the orientation of the molecules within the fibre) are also important; (3) *Mechanical properties*, i.e. stiffness, tensile and shear strength determine the preferred use, e.g. softness is essential for paper tissue, elasticity for brushes and resiliency for stuffing mattresses; and (4) *Chemical constituents*, i.e. crystallinity and molecular size of the cellulose (for details see McDougall *et al.*, 1993) and other chemical constituents. The distribution of reactive chemical groups on or near the fibre surface, for example, will also affect its reaction to dyes or adhesives, while a minimal ash content is required for paper used for electrical purposes.

The fineness of a fibre can be expressed as the diameter of the cross-section of a single strand or filament, and is obviously dependent on the number and diameter of the ultimate cells. In the textile industry the *yarn count* or *yarn number* provides either the mass per unit length or length per unit mass of yarn. A selection of terms for which are given in Jerrard and McNeill (1992), including the formerly widely used and now obsolete measurement of the *denier* based on the gram weight of 900 m of yarn. The ISO unit is the *tex*, expressed in g 1000 m^{-1} of fibre, filament or yarn. It is more or less internationally accepted, apart from North America where the unit is the *drex*, expressed in g 10 km^{-1}, i.e. equivalent to 10 tex.

2.   HARVEST AND POST-HARVEST TREATMENTS

The economics and ease of harvesting, plus the need for and any problems with post-harvest treatments, have to be considered. The seed and fruit fibres are the simplest to harvest. For example, the cotton bolls (fruit) split open while still on the plant and the seed cotton can readily be harvested by hand or machine Raw cotton contains *ca.* 90% cellulose and requires the minimal of post-harvest treatment and suggests that cotton is unlikely to be replaced as the most utilised natural fibre. This contrasts with the leaf fibres, which have to be separated from non-fibrous matter by hand or machine beating to soften the non-fibrous tissue, followed by scraping the fibres clean. Care is also needed in order not to damage the fibres from excessive beating. Where the leaf fibres contain high levels of pectin, this can be extracted industrially with alkali, or in domestic industries, after burial in the soil or by retting.

Retting is a process where the action of water, fungi and bacteria entering the leaf via the stomata, decompose the mainly parenchymatous matter surrounding the fibre bundles and thereby free the fibres from the other tissues. Bast fibres are usually separated by this means and the following description for flax is representative of the process.

For poor quality flax straw *dew retting* may be used where the straw is spread out over the ground. The process requires no capital outlay but has the disadvantage of being laborious and uncontrolled. *Cold water retting* in ditches, rivers, etc. is the most basic form of water retting and may take 2-3 weeks to complete. **Warm water retting** in concrete tanks can be controlled and is more efficient and also produces a finer and more uniform fibre. The preliminary operation is to wash and leach the flax straw in cold water, whereby the air is driven out of the straw and any water-soluble matter removed. The operation requires a plentiful supply of cheap, neutral water with minimal mineral content. Good quality straw may even be double retted. For large-scale operations where water is limited and effluent disposal a problem, **aerated retting** by pumping a controlled air input into the retting liquor reduces the amount of unpleasant and usually volatile substances. With this system the retting liquor can be re-used several times, which is not possible with other systems. This aerated process is not used for kenaf and jute because large-scale, commercial retting is not usually carried out in a central rettery and effluent disposal is consequently not regarded as a serious problem.

Green flax, produced without retting, provides a very fine yarn, but is not popular because the extraneous matter adhering to the fibres can gum up the milling process unless severely scutched, a process that can create an excessive amount of short tow fibre. **Scutching** is where the retted sheaves are air or artificially dried and the fibres separated from the straw using fluted rollers to break the bond between the fibres and the inner pith. The fibres are then buffed and cleaned. Some of the fibres are inevitably broken down into short coarse fibres known as **scutched straw**, which are used for ropes and coarse fibres, etc. (Lock, 1962; Kirby, 1963; McDougall *et al.*, 1993; Jarman, 1998).

3.   SPINNING

Because fibres differ in their dimensions, mean values are required for any one class of fibre, the strength and fineness of which will largely determine the fineness of the spun yarn. For technical reasons a minimum number of individual strands or filaments are required for machine spinning. This is in order to prevent breakage during the actual spinning and in the resulting yarn. For jute 150 fibres in the yarn cross-section are desirable, i.e. a tex value of 25.2, equivalent to the now obsolete spindle value of 8 lb per 14400 yd, whereas kenaf would require a *ca*. 50% thicker fibre with a tex value of 37.8 (12 lb per spindle) for an equivalent strength. Consequently kenaf cannot be spun as finely as jute.

During spinning the initial mass of partly tangled fibres are **carded** to form a mat of parallel fibres and sometimes **combed** to align the fibres parallel to one another; these processes also help to clean the fibres. The fibres are then spun into a continuous untwisted rope, *ca*. 20 mm in diameter and known as a **sliver**. The slivers are then drawn into the required thickness for twisting into a yarn.

The strength and bending properties of the fibre are important during spinning, as they must be strong enough to withstand twisting and bending. Hard fibres in the sliver can loose up to 25% of their strength, consequently the stronger the fibre before spinning the better. The finer fibres can also be given more twist in the yarn. The fine fibres, by presenting a larger surface area for a given Tex value, provide greater insulation or water absorption properties. The fibres have also to be free of any wax coating which would inhibit their spinning; kapok fibres have such a coating (Kirby, 1963; Brücher, 1989; Jarman, 1998).

4.   CORDAGE AND FABRICS

Cordage is the general term for all ropes, packing cords, string, threads, lines and twines (Middle English *twinen*, from *twin*, a rope of two strands). The term *rope* is generally applied to a composition of three or more *strands*, each consisting of two *threads* or *yarns*. Three common types of rope are recognised: (1) *Hawser laid*, consisting of three strands, each containing several yarns, and is the strongest and most commonly used rope; (2) *Shroud laid*, consisting of four strands, each containing a number of yarns. Such ropes are used mainly for running gear and working over pulleys; and (3) *Cable* or *water laid*, made up of three complete hawser-laid ropes to form a single rope. Such ropes are used where elasticity is required (Robbins, 1995).

Fabrics and netting are manufactured from flexible fibres that are twisted together into a thread or yarn and then woven, spun, knitted or otherwise utilised. Fabrics include cloth for wearing apparel, domestic use, awnings, sails, etc. and also coarser materials such as gunny and burlap. Netting fibres are used for lace, hammocks and various forms of nets, include many of the commercial fabric fibres and a host of native fibres.

In addition to their use for cordage, jute and sisal are the only fibres specified for coffee sacks used in international trade. This is because the sacking not only allows the coffee to 'breathe' but also allows samples to be extracted without ruining the bag. Trade also demands that only new bags are used for coffee, cocoa, spices, etc. and while the sacking industry is very much in decline in the developed countries, jute and jute products are still the major export from Bangladesh (Hill, 1952; Kirby, 1963; Robbins, 1995).

5.   PULP

Pulp is the cellulose material formed after mechanical or chemical treatment of wood and other plant material for use as a feedstock for the paper, particle board and chemical industries. The relevant BSI and ISO terminology are contained in BSI (1979). For wood pulp sustainable forest pulp resources depend upon the availability of the required quantity of raw material, the technical suitability of various competing forest industries for timber, poles, plywood, etc. and the presence of acceptable production economics (FAO, 1973). See Chapter 12 for details of pulping processes and utilisation of wood pulp.

5.1   Straw Pulp

Within the EU there are 100 million tonnes of cereal straw available annually as a potential source of cellulose fibre. Their exploitation has been under discussion for the at least 50 years and the recent ban on the burning of cereal straw in the UK has emphasised the need for seeking alternative ways of disposing of any straw surplus to livestock requirements.

Cereal straws consist of leaves, nodes, internodes and ears, and in the case of wheat straw, contains 51-54% cellulose, 26-30% hemicellulose, 16-18% lignin and 7-8% ash. While the internodes have approximately the same α-cellulose and insoluble ash content as wood and are, therefore, well-suited for paper and board production, the leaves and nodes have a lower α-cellulose and higher ash content and are of little value as a fibre source. Furthermore, the presence of the leaves, nodes and ears not only increases the demand for cooking chemicals used in the digester, they also result in a lower fibre yield. In addition, micro-organisms often discolour the leaves, so that the fibres will require additional bleaching. It is now technically possible to separate the internodes from the leaves, sheaths and ears, which may then be used for fodder or industrial extraction. Thus, not only is the efficiency of the cellulose extraction increased, pollution from the extraction process is also reduced.

The smaller diameter of the straw fibres (7-35 μm) compared to those from wood (2-90 μm), present no problems regarding paper quality since good fibre quality is generally characterised by a high ratio between fibre length and width. The lignin fraction may be used in the production of *phenol* or hydro-cracked to form low molecular hydrocarbons for fuelling the process. Phenol is used in the straw cellulose plant as a delignification solvent, resulting in separation into lignocellulose fractions, i.e. *celluloses, lignins* and *pentoses* (see also Chapter 15). The pentose fraction may then be used for the production of *single-cell protein* or *furan* products. (Rexen and Munck, 1984; Mcdougall *et al.*, 1993; O'Brien, 1997).

There is at least one pulp mill in Denmark that processes straw from temperate cereal crops. Straw from wheat and rye are favoured, although oat straw may also be used. Barley straw is regarded as unsuitable for paper making because of the poor drainage and strength of the pulp. The straw is cut into 5 cm lengths, boiled under pressure in 20% caustic soda for 4 hr. The straw is then passed through grinders to crush the more persistent nodes. The yield is *ca.* 30% cellulose. The fibres are *ca.* 1 mm long and 13μm in diameter, consequently it is necessary for the pulp to be mixed with longer fibres from wood or rag pulps for making thin and hard writing papers.

Rice straw, after the silica present has been removed, is widely used in China and Indonesia for writing, printing, wrapping and cigarette papers. Extensive use is also made in China, India, Mexico and Brazil of *bagasse*, the fibrous raw material from the sugar cane after the sugar has been extracted. The best pulp is from cane from which the pith is first removed. The fibres from the bagasse pulp are 1.5-2 mm long and 20 μm in diameter. The disadvantages are that sugar production tends to be seasonal and the bagasse bulky to handle. Despite the problems of depithing and storage, the potential of bagasse as a fibre source is recognised as being increasingly important.

5.2   Bamboo Pulp

Bamboos, especially *Bambusa bambos* and *Dendrocalamus strictus*, are widely used in China, Thailand, Taiwan, Indonesia, Burma, Bangladesh and especially India, for the manufacture of large quantities of wrapping, writing and printing papers; India alone has some 35 factories producing *ca.* 1.75 million tonnes of bamboo pulp annually. Depending on the species, the ultimate fibres are 3-4 mm long and 14-23 μm average diameter and, because they are more slender than those from wood, produce a smooth, flexible paper. Unfortunately processing costs are high due to the many impurities present, including silica. The grass *Eulaliopsis binata* (sabai grass) is also widely used in India for paper-making (Kirby, 1963; FAO, 1973; Lewington, 1990; McDougall *et al.*, 1993).

6.   PAPER

Archaeological evidence from China suggests that paper was invented in the 2nd century BC. These early paper makers prepared their paper from beaten plant fibres, a tradition that still survives in the Orient, although paper as we now recognise it can be attributed to Tsai Lun, the Chinese Minister of Agriculture in 105 AD. In ancient Egypt the sliced and beaten pith of the sedge *Cyperus papyrus* was similarly used to make a paper known as *papyrus* as early as 2400 BC; papyrus is still being made by this method in Egypt for the tourist trade. Indeed, it is from the Greek *papuros* via the Latin *papyrus* that the word 'paper' is derived (Hill, 1952; O'Brien, 1997).

World-wide there are over 400 types of paper utilising a wide diversity of fibres. Indeed the UK Paper Federation alone manufactures some 115 different classes of paper (O'Brien, 1997). Two broad groups of paper are recognised: (1) 'Cultural' paper for newsprint, magazines, books and other printing and writing papers; and (2) 'Industrial' paper for bag, sack and wrapping paper, corrugated and folding box paper, paperboard, fibreboard, structural paper such as building and wallpaper. Also tissue, towel and similar crepe papers, moulded pulp products (e.g., fruit and plant containers), metal layer and plastic coated paper and capacitor paper, non-woven textiles or 'disposables'. As yet there is no universally accepted classification.

The annual paper consumption per capita for India is *ca.* 3 kg and has remained stable over a number of years, while for China's emerging economy it has increased from *ca.* 6 to 20 kg during the past decade, and for the UK 1995 it was 295 kg in 1995 and still increasing. Thus, it is sobering to realise by how much paper is related to education and economic status, by how much consumption by the developing countries could rise and by how much the developed countries rely on the developing countries for their raw materials. Should China alone reach the UK level of consumption it would cause a dramatic fibre shortage. It is worth quoting *"Give them what you take for granted, a Daily Paper, a Take Away Pizza and a Toilet Roll and you would have a world fibre crisis"* (O'Brien, 1997).

Wood pulp, especially from softwoods, is a major source of fibres for paper making, for which the market requirements are for long strong fibres, high in cellulose and free from contamination by resins, gums and tannins. The average length of softwood fibres is 3.5 mm and for hardwood they are (0.7-)1.2(-1.8) mm long. It is because softwood fibres are able to intermesh over a greater area that they make a stronger paper than those from hardwoods. Although the shorter hardwood fibres give a weaker wet sheet and a lower paper strength, small quantities added to long-fibre paper have no deleterious effect on paper strength and will even improve paper smoothness, opacity and printing quality. A typical sheet of writing paper manufactured in the UK could well consist of 55% hardwood pulp (largely from *Eucalyptus* spp.), *ca.* 20% softwood pulp and 15% filler, chiefly chalk and special clays. In general 20 to 33% of long fibres are used in most grades of writing, printing and tissue papers, while linerboard, wrapping and multiwalled bag papers may contain up to 50% of short fibres. Hardwood pulp alone is used for certain kinds of paper, e.g. corrugating medium, household and toilet tissues (Hill, 1952; FAO, 1973; Prescott-Allen and Prescott-Allen, 1986; Lewington, 1990; Robbins, 1995).

The value of plant fibres for making paper depends on the length, texture, strength, softness, natural colour, absorbency, pliability, etc. of the cellulose present in the cell walls, and whether the fibres occur alone or in combination with lignin or pectin. The cellulose content of the fibres must also be sufficiently high to ensure economically feasible extraction. Natural fibres vary from a fraction of a mm to over 3 mm in length, their dimensions being of greater practical importance for paper making than they are for textiles (Table 17), with the ultimate fibres of necessity being sufficiently long for proper felting. The ratio of fibre length to diameter, and the ratio of lumen or cavity diameter to fibre diameter are particularly important since they affect the tear-resistance, flexibility, etc. of the product. The fibres must also be reasonably cheap with consistent and sufficiently abundant supply. In practice surprisingly few plant materials are actually used for commercial paper-making (Kirby, 1963; Turner and Skiöld, 1983; Rexen and Munck, 1984).

6.1   Paper-making

It is the ultimate cells that are used for paper-making and this requires the cellulose to be as pure as possible and, if necessary, reduced in length. The cellulose fibres are suspended in a large volume of water in the ratio of 1% fibre to 99% water, where they swell and form a layer of pulp. The fibres are then filtered from the suspension through a sieve or screen, *mould* or *wire*, to form a uniform layer of drained pulp, i.e. a *wet sheet* of paper. In a process known as *couching* the screen is separated from the paper in such a manner as to leave the wet and therefore fragile sheet unwrinkled and undisturbed. The couched paper is then placed in contact with a woven cloth or *felt* and pressed to remove excess water. The moist paper is removed from the felt and the cohering mass of fibres with their gelatinised and fibrillated surfaces dried. When dried out in sheet form the flexible fibres, as a result of hydrogen bonding, form a semi-rigid sheet of paper containing 8-10 layers of fibres (Kirby, 1963; FAO, 1973).

More modern methods involve the use of mechanised Hollander beaters, an oval tub in which the fibre (or rags) are circulated in water and lacerated by a roll of metal bars revolving over a metal or stone bed plate. The machine is so constructed that the slurry is kept in constant motion by both the backfall and the rotation of the roll. The cutting action can be regulated by the distance between the roll bars and the bed plate, and by the duration of the beating, thereby determining the strength and texture of the paper.

For handmade paper the paper maker dips his *mould* (screen stretched over a frame) and *deckle* (a removable upper frame) into the vat to scoop up sufficient pulp or *stuff* for a sheet. Raising the mould and deckle to a horizontal position above the vat, it is deftly shaken from side to side to distribute the fibres evenly over the mould. The water drains through the wire mesh of the mould, leaving behind a sheet of water-logged fibres. The deckle is then removed and the paper sheet laid with a rocking motion (*couched*) onto a board or damp felt to dry. A pile of couched sheets is built up to form a *post* of *ca.* 25 sheets, which is then forcefully pressed to expel as much water as possible. The sheets are then carefully removed from the felt and dried; partial air-drying is followed by pressing between blotters under pressure to give a flat, smooth appearance.

In traditional oriental paper making very thin sheets of paper are prepared from the pounded bast of *Broussonetia* spp., producing a particularly tough and strong fibre, the sinewy, 5-10 cm long, 0.018 mm wide threads neither shrinking nor expanding. In Japan the *nagashizuki technique* is used, where a viscous agent, *neri*, prepared from *Abelmoschus manihot*, syn. *Hibiscus manihot,* is added. The mucilage slows down the passage of water through the mould mesh, allowing the paper maker more time to form the sheet, as wave after wave of fibre is laminated onto a flexible bamboo screen held in place by a mould and hinged deckle. The sheets are then couched to form a post without any intervening sheets of felt. Overnight the excess water is slowly and gradually pressed out of the post, a procedure that bonds the fibres into a strong paper. Finally, the moist sheets are separated from one another and dried in the sun.

The scale of paper making in the orient varies greatly. In Thailand, for example, paper making is still very much a cottage industry, while in Japan large-scale production uses 1000 tons of bast annually, much of which has to be imported. The wood fibre from wasp's nests, especially *Polistes* spp. (paper wasp) consisting of macerated plant fibres mixed with fluid from the wasp, may also be used for making hand-made paper. Such 'wasp paper pulp' is believed to have given the Chinese the idea of how to make paper (Hunter, 1947; Schmidt and Stavisky, 1983; Turner and Skiöld, 1983; Indrbhakdi, 1989; Lewington, 1990).

Nowadays, in commercial paper production, the pulp is first fed onto a fast-moving, wire-mesh screen, where it drains and the fibres become interlocked. Most of the remaining moisture is then removed and the sheets pressed flat by passing through a series of heavy rollers and heated cylinders, with the resultant paper emerging in a continuous roll. The quality of the paper very much depending on the type of pulp or mixture of pulps used. Chemical pulp processes result in bleached, fluffy, crumb-like particles suitable for all kinds of writing, printing and drawing papers. Mechanical processing produces a coarser pulp containing additional material to the cellulose, i.e. lignin, hemicellulose and resins. Such pulp is unsuitable for high quality paper and, in time, tends to turn yellow when exposed to heat and light (Lewington, 1990).

High quality, speciality papers often contain a proportion of *Stipa tenacissima* (esparto grass, alfa grass) or *Lygeum spartum* (albardine, false alfa) fibres imported from wild grasses in North Africa and from both wild and cultivated sources in Spain. Two commercial varieties of esparto grass are recognised: (1) Long and very fine leaves with regular diameter and great flexibility. They are used for cordage, as are the long fibres of albardine; and (2) Coarser, more variable leaves, which are used for paper-making. For paper-making the leaves are boiled under pressure in caustic soda for 4-5 hr to yield 45-50% cellulose. Although the ultimate fibres are extremely short, 1.5 mm long and have the smallest diameter, 13µm of all the commercial paper-making fibres, they provide bulk and opacity to the paper, plus a unique close texture and smooth surface with excellent printing qualities, which is especially valued for magazine artwork. Such paper also produces very clear water-marks; it also expands less when wetted than other papers.

Although the stem fibres of *Thymelaea hirsuta* (shaggy sparrow wort) have traditionally been used for cordage, their use for paper is a recent development. In 1979 a paper mill was established in Israel for using the bast fibres to produce a high quality, hand-woven paper known as Mitnan paper. The paper is used primarily by artists, but also by paper conservators for repairing documents and works or art. The xylem fibres are comparable to those of hardwoods but with a higher length:width ratio, yet lower than those from rice straw (El-Ghonemy *et al.*, 1974; Schmidt and Stavisky, 1983).

Very strong yet flexible papers are required for bank notes and some legal documents. For such papers a mixture of hemp, cotton and flax fibres are used, all of which are longer than wood fibres and need far less processing. Cotton fibres, after a special beating treatment, are also used for blotting paper. The short fibres or tow of cotton, flax, hemp, ramie, Manila hemp (abacá), etc. are mainly used for textiles, with a secondary use in paper-making. Manila hemp produces a very strong paper, which is used for Manila envelopes, wrapping papers and stencils. Also, because the fibres maintain their strength when wet, *ca.* 40% are used in the paper for tea bags. Amongst other uses, ramie and hemp are used for cigarette papers.

For the finishing of paper a number of additives are added to the pulp in order to give the required colour, smoothness and opacity. They include size, dyes, brighteners and fillers (chiefly china clay or chalk). Although synthetic sizes are available, a number of plant-based starches, gums and resins are still being used as sizing materials, including potato starch, guar gum from *Cyamopsis tetragonolobus*, locust bean gum from *Ceratonia siliqua*, **methyl cellulose** derived from cellulose, rosin (see Chapter 12) and alginates (see Chapter 19) (Lewington, 1990).

7.   FILLING FIBRES

Filling fibres were used in upholstery and for the stuffing of mattresses, pillows and cushions, e.g. kapok and the seed floss of *Asclepias* spp. and *Calotropis* spp., leaf fibres such as crin végétal, henequen, *Copernicia prunifera* and *Yucca elata*, the bast fibres of *Asclepias* spp. and *Tillandsia usneoides* (Spanish moss), and coir. In recent years fire regulations have restricted the use of such fibres for stuffing.

Kapok or *kapoc* fibres are interesting because of their lightness. They are the silky fibres obtained from the fruit of *Ceiba pentandra* (silk cotton or kapok tree). It is a native of tropical America but now grown in commercial plantations, especially in southern Asia, with *ca.* 90% of the world's production coming from Indonesia. The fibres are *ca.* 2.5 cm long but too brittle for easy spinning. Formerly kapok was used to stuff mattresses, pillows and upholstery but because the fibres are highly inflammable, its use for such purposes has declined in favour of less inflammable and mainly synthetic fibres demanded by fire-safety regulations. However, the fibres are very light and water-resistant, due to the walls of the air-filled cells being impervious to both air and water and are thus able to support 30 times their own weight in water. Kapok is consequently widely used in water-safety equipment; it is also used as a thermal insulator. (Kirby, 1963; Robbins, 1995; Jarman, 1998).

In the past, both hemp tow and sunn hemp were widely used as *oakum* for caulking the seams in boats, casks and barrels; coir too was used as an oakum substitute. However, the development of steel and plastic boat hulls have limited the need for such caulking fibres to decking, while steel kegs have largely replaced the traditional wooden casks. Fibres are also used as a stiffening in plaster, and as a general packing material (Hill, 1952).

8.   BRUSH FIBRES

Vegetable fibres have been used for making brushes, brooms and whisks for countless years, probably dating back to early man, although many countries, such as Italy, now rely almost entirely on synthetic fibres. Fibre suitability for brush making is governed by such factors as bend recovery, wear, loss of stiffness when wet. The fibres must be very strong, stiff yet elastic and with a high degree of flexibility, capable of holding moisture if required, and able to withstand any scrubbing action. From a production point of view the available lengths are also important. There are no international or EU Standards for quality, although quality can be tested using a "brushing" machine that simulates the brush in action on various surfaces (P.W.Coward, 1997 pers. comm.).

The fibres used in the brush making industry today fall into three groups: (1) *Leaf fibres*; (2) *Palm fibres* (petioles, ribs and seed fibres); and (3) *Grass fibres* (root fibres and culms). While there was a decline in the use of vegetable fibres in the 1970s and 80s, this trend has reversed during the 1990s with the replacement of the higher quality fibres by cheaper imports. A number of vegetable fibres are used by the Hill Brush Company, the major European broom and brush manufacturer and the following examples illustrate how the fibre qualities are utilised.

The major leaf fibre is from *Agave lophantha* and is known as **Mexican fibre**, *istle* or **Tampico** (named after the port from which it is exported). The Tampico fibres are noteworthy for being highly elastic and resistant to temperature changes, acids and alkalis, and their fineness for polishing and grinding. They are also very water absorbent, retaining 65% more water than its polypropylene synthetic replacement, as well as being non-electrostatic, so that the brushes, which are mainly used for grooming horses, remain dust free.

The major palm leaf fibres are: (1) **Bahia bass** from the leaf bases of *Attalea funifera*, also known as **Bahia piassava, piassalba** or **coquilla**. The fibres have good water retention, do not rot when damp and are very resistant to distortion. Bahia bass is used for the very best yard and street brooms, and in some industrial platform brooms; (2) **African bass** is obtained from the retted petioles of *Raphia hookeri*; that exported from Sierra Leone is known as **Sherbro bass** or **piassave**. **Sulima bass**, also from Sierra Leone, is no longer obtainable. The fibres do not lose their stiffness when wet and are used in road sweeping and farm brushes. **Calabar bass** from Nigeria is currently unobtainable due to the local political situation; (3) **Gumati** is from the leaf sheaths of *Arenga pinnata*, syn. *A. saccharifera*. The fibres are softer and finer than Bahia bass but have similar excellent wearing and sweeping qualities. They are used for sweeping dry concrete floors; and (4) **Bassine** or **palmyra** from the leaf bases of *Borassus flabellifera* (palmyra palm). The fibres are less resilient than the others and are used in the cheaper warehouse brooms and household brushes, and for scrubbing brushes. The split midrib of *B. flabellifera*, known as **split palmyra** or **split cane**, is always used with other fibres, such as Sherbro bass, to enhance the stiffness and add a decorative colour contrast (Coward, 1997).

Coir or **coir fibre** is the name given to the twisted fibres obtained from the mesocarp (husk) of the coconut and is the only seed fibre used in brush making. It is cheap and abundant but is liable to crush and distort; it is therefore used for the cheaper brushes. The coconut husk was traditionally retted for 10 months and then beaten with sticks to free the fibres, which were then **hackled**, i.e. combed, with a steel comb. Now the unretted husks are passed through a defibering machine and then retted for only 3 days. One tonne of nuts will yield *ca.* 150 kg coir. Coir is unusual in that it does not retain smells. Its use, however, is declining as synthetic substitutes become cheaper and more readily available. Coir is also widely used for matting, especially door mats, ropes, and traditionally for wrapping around bedsprings in mattresses. The short fibres and dust are now widely used in horticulture as a peat substitute (Robbins, 1995; Coward, 1997).

The peeled and bleached rhizomes of the Mexican grass *Muhlenbergia macroura*, syn. *Epicampes macoura* (zakaton) are known as **Mexican whisk**, **broom** or **rice root**; it is regarded as the best material for brushes used for grooming animals, otherwise it is not very widely used. The culms of *Miscanthus sorghum*, syn. *Miscanthidium sorghum* (broom grass) from Lesotho mixed with other fibres are used for the lighter domestic and yard brooms. A native of eastern South Africa the upland stream banks of Lesotho provide the environment that gives the fibres their resilience (Coward 1997).

During brush making the fibres are doubled over and retained in the wooden or plastic brush back with a wire staple, an operation that is now carried out at speeds up to 300 tufts per minute using computer controlled production machines. It is important, therefore, that the fibres do not break when doubled over at speed. Some fibres, such as Bahia piassava and Sherbo piassava are first soaked in hot water in order to make them more supple but most fibres, such as coir and gumati, are punched dry (P.W.Coward, 1997 pers. comm.).

Broom-corn from a cultivar of *Sorghum bicolor* is grown for its long, fibrous panicle branches, which are made into brooms and brushes in many parts of the world. A native of Manchuria and originally developed by the Chinese as broom kaoling, broom-corn was cultivated in Europe during the mid 17th century and introduced to America in 1797, and has since spread to Australia and South Africa. The seed branches are long and straight, but with the rachis much shorter and telescoped than in the grain sorghums. The inflorescences are harvested by cutting the stem 15 cm below the node. The leaf sheath is then removed and the heads threshed and spread out to dry in the open or in curing sheds, or they may be ricked and then threshed, although the former method causes less damage to the fine fibres. The heads are then bundled together into brooms (Hill, 1952; de Wet, 1990; Doggett, 1970; Rooney, 1996).

9.   **PLAITING AND BASKETRY**

The range of raw materials used in the domestic and local economy for plaiting and basketry include the stems of reeds, rushes, grasses, bamboo, rattan, willows, etc. as well as leaves and roots used either entire or split. They are either woven or twisted together into hats, sandals, mats and matting screens, chair seats, baskets, etc. Despite the numerous plant materials available relatively few are of any commercial importance.(Hill, 1952).

Of these, the Panama hats are of interest since there has been a revival in their use in recent years. The hats, despite their name, are made in Ecuador, with over 1 million hats being exported annually The hats are characterised by their uniformity and fineness of texture, their strength, durability, elasticity and water resistance. They are manufactured from the young leaves of *Carludovica palmata* (Panama hat palm, toquilla), an almost stemless, palm-like shrub of the Cyclanthaceae from the forests of Central and South America. Six leaves are required per hat; they are cut before they unfold (Hill, 1952; Kirby, 1963; Mabberley, 1987).

Chapter 15

Phytochemicals

Phytochemicals are the plant's chemical constituents, their type and quantity being affected by environmental factors, e.g. climate and soil, also by the karyotype and stage of development, especially the former. *Acorus calamus* (sweet flag), for example, contains *β-asarone*, which has halucinogenic and carcenogenic properties. The Asian tetraploid contains 70-96% β-asarone, the European triploid contains less than 15% while it is absent in the North American diploid (Schultes and Hofman, 1992; Motley, 1994).

The chemical products that can be obtained from plants include gums and resins, tannins, dyestuffs, latexes and rubbers, lipids, essential oils, waxes, alcohols and other chemicals (Cook, 1995). However, there are some substances that cannot be readily classified chemically and can be variously considered within several different groups. The toxic substance *solanine*, for example, present in the potato and other members of the Solanaceae, can be considered for inclusion in four different groups. It has the physical properties of a saponin because it forms a semi-permanent froth when shaken with water. Chemically it is a glycoside because it consists of the sugar *solanose* bound to the aglycone *solanidine*, the non-sugar part of the glycoside molecule. During digestion the solanine molecule is broken up by enzymes and the sugar is split off, leaving the solanidine. Because solanidine is a nitrogenous organic substance it conforms to the definition of an alkaloid. Furthermore, because part of the molecule of the alkaloid is a sterol group, it can also be classified as a sterol. As a consequence solanine can be regarded as a steroidal, alkaloidal glycoside with the properties of a saponin (Kingsbury, 1964, cited by Everist, 1972)!

1.   GUMS, MUCILAGES, RESINS AND OLEORESINS

1.1   Gums

The true gums are complex polysaccharides, i.e. carbohydrates derived from monosaccharides, that either dissolve or swell in water to produce very vicious colloidal solutions (sometimes incorrectly referred to as mucilages) that are insoluble in organic solvents. They form the dried plant exudates that are usually obtained when the bark is cut or the plant otherwise injured. Carob bean or locust bean gum, however, is obtained from the pods of *Ceratonia siliqua* and guar gum from the seeds of the cultigen *Cyamopsis tetragonolabus*. The poor water solubility of carob bean gum at temperatures below 85°C make the gum highly suitable for use in processed foods where its thickening and texturising properties do not interrupt the cooking process until high temperatures are reached. By way of contrast, the water absorbing properties of the readily water-soluble guar gum are used to control the viscosity of drilling muds. Mention should also be made of xanthan gum from the bacterial fermentation of waste sugar products (see Chapter 18).

A selection of commercial gums that are widely used as thickening agents in the food and pharmaceutical industries are shown in Table 18).

TABLE 18. A selection of commercial gums and their applications (Brouk, 1976; Anderson, 1985; Sharp, 1990; Robbins, 1995; Mabberley, 1997)

Plant source	Gum	Application
Acacia senegal	gum arabic	see text below
Anogeissus latifolia	gum ghatti	substitute for gum arabic
Astracantha gummifera, syn. *Astragalus gummifera*	gum sarcocolla	stabiliser in food and pharmaceutical industries; formerly used for sizing textiles
Astracantha microcephala, syn. *Astragalus microcephalus*	gum tragacanth	stabiliser in food and pharmaceutical industries; formerly sizing textiles
Ceratonia siliqua	carob gum	thickener and texturiser in processed foods
Cyamopsis tetragonolobus	guar gum	thickening agent in sauces, etc.; in drilling mud, flocculant in ore recovery, filtering and suspension agent in coal mining, sizing agent, also used in pharmaceuticals, cosmetics, printing inks, explosives
Larix occidentalis	larch gum	thickener, stabiliser, emulsifier and binder in food
Sterculia urens	karaya gum	texturiser, stabiliser and binder in food; colostomy bags, dental fixative, cosmetics, dyes, inks

1.1.1   Gum Arabic

The gum from *Acacia senegal* known as **gum arabic** is one of the major gums of commerce. Its history in the world market well illustrates the influence of taxonomy and the tightening of quality control in defining the product. It was originally defined as "the gummy exudate from *Acacia senegal* or its related species", embracing 18 species. The group included the unrelated *A. seyal* (gum talha from East and West Africa), *A. xanthophloea* (from East Africa) and *A. karroo* (from southern Africa). All were traded on the international markets. This was despite the fact that the Test Article, evaluated as toxicologically safe as a food additive, refers solely to gum from *A. senegal*. The increasing international pressure towards tighter trade specifications and labelling regulations, identity and purity, has led to the Revised Specification (WHO. 1990a, b; FAO, 1990) where gum arabic is defined as originating from *A. senegal* or closely related species, with a specific optical rotation range of -26° to -34° and a Kjeldahl nitrogen content of 0.27-0.39%. This has limited the gums permitted for the food trade and designated gum arabic to members of *Acacia* subgenus *Aculeiferum* which, in addition to *A. senegal*, include *A. laeta*, *A. mellifera*, and *A. polyacantha*. Since the exudates from the other authorised species occur as small tears and driblets, their collection is consequently extremely time-consuming, so much so that the marketing of these gums is not commercially viable.

The other gums are now restricted to industrial use only. Nevertheless, acacia gum 'from *A. senegal* and other African species' is still listed as official by the British Pharmacopoeia Commission (1993) for use as a bulk-forming laxative and pharmaceutical aid. While the US specification for the use of acacia gums in the food trade is restricted to gum arabic, the pharmaceutical specification permits the use of gum talha. The rational for the less stringent pharmaceutical specifications is because the small quantities of gum talha used are given under medical supervision (Anderson, 1993).

In the Sudan, which is the largest exporter of gum arabic, the best grades are known as Kordofan gum. The marketing of gum arabic is further discussed in Chapter 7. High quality gum droplets should be of a light yellowish or rosy pink colour, globular and 2-4 cm in diameter. Gum arabic is a polysaccharide, consisting of D-**galactose**, D-**glucuronic acid**, **arabinose** and **rhamnose**.

Its use as the food additive E414 within the European Union is subject to a rigorous specification regarding identity and purity. The gum is used in the food industry to fix flavours and as an emulsifier. It is also used to prevent the crystallisation of sugar in confectionery products, as a stabiliser in frozen dairy products and as a foam stabiliser and clarifying agent in beer, while its viscosity and adhesive properties find use in bakery products. In the pharmaceutical industry gum arabic is used as a stabiliser for emulsions, a binder and coating for tablets, and as an ingredient of cough drops and syrups. Gum arabic is also extensively used in folk medicines as a soothing and softening agent. In cosmetics it finds use as an adhesive for facial masks and powders, and to give a smooth feel to lotions. The poorer grades of gum arabic are used industrially in foundry sands to give the sand moulds cohesive strength, applied as an adhesive, as a protective colloid and safeguarding agent for inks, as a coatings for special papers, a sizing agent to give body to certain fabrics, as a sensitiser for lithographic plates and as an anti-corrosive coating for metals. It is also used in the manufacture of matches, paints, boot polish, ceramic pottery, etc. (Anderson, 1985; Sharp, 1990; Cossalter, 1991; Robbins, 1995).

The tannin-containing gums with a positive rotation, i.e.. from *A. karroo, A. nilotica, A. seyal* and *A. xanthophloea*, are considered to be carcinogenic and, because they are now excluded for use in food, attract a lower price. These factors, together with the low price, currently at US$ 1000 tonne^{-1}, compared to US$ 5000 tonne^{-1} for gum arabic, is likely to kill the export trade in these gums (Anderson, 1993)

Although a prerequisite to selection and management for sustainable high yields, the physiology of gum production, i.e. *gummosis*, is still not fully understood, although the similarity in the structure of the *arabinogalactans* of the cell walls and the gum, suggests an arabinogalactan as a precursor of gum arabic (Anderson and Dea, 1968). Although unconfirmed, gummosis appears to be stress induced, possibly from the effects drought, natural or artificial wounds, parasite attacks, etc. The problem is currently under investigation at the Université P. Sabatier, Toulouse.

While gummosis may be accompanied by some cell wall restructuring in response to cell adaptation, it is now believed to be directly related to starch metabolism within *A. senegal*. Histological investigations have shown that there is a pectic trend in the tissues of both tapped and untapped trees. This tendency was also found in the walls of the phloem parenchyma of seedlings, i.e. before any possibility of gum exudation. Such findings suggest that there is an early preformation of gum in the tissues plus a genetic potential for gummosis. This would suggest that clonal reproduction rather than seed from selected elite trees should be used for increasing plantation yields.

The first conspicuous phase of gummosis is the thickening and chemical modification in the cell walls of the phloem parenchyma, which is eventually followed by a breakdown of the cell walls to form lacunae. An intermediate stage involves the modification of the phloem and cortical parenchyma. Later the sclerenchyma and fibres in the vicinity of the point of exudation are also modified, while the starch grains within the affected tissues appear to be progressively assimilated into the amorphous mass of gum (Joseleau and Ullmann, 1985; Mouret, 1985).

1.2   Mucilages

Of plant origin, *mucilages* are polysaccharides consisting of a mixture of a complex *polyuronide*, proteinaceous matter and cellulose, which swell in water and have glue-like properties. Mild hydrolysis of the polyuronide yields *xylose* and *galactose* residues and a more resistant fraction consisting of *galacturonic acid* and *rhamnose*. Within the plant the mucilages are mainly concerned with water retention, e.g. the pentosan mucilages in succulent xerophytes help by increasing the water-holding capacity of the cells and reduce transpiration losses. The mucilaginous coating of many seeds aids dispersal and water uptake during germination (Tootill, 1984; Sharp, 1990; Walker, 1991; Gutterman, 1993).

Important mucilages include *slippery elm* from the dried inner bark of *Ulmus fulva* (Indian moose, sweet elm), which is used as an invalid food and medicinally as a laxative and emollient of the gastro-intestinal tract. Also *psillium* from the mucilaginous seeds of *Plantago afra*, syn. *P. psyllium* (psyllium plantago), used as a laxative, as an emollient for relieving skin irritations and cosmetically in face masks (Chiej, 1984; Mabey, 1988; Mabberley, 1997). See also Chapter 19 regarding agars from seaweeds.

1.3   Resins

Resins are high molecular weight materials consisting of highly polymerised acids and natural substances mixed with terpene derivatives which soften at high temperatures and are insoluble in water. The natural resins present in plants are known as *rosins* (see also Chapter 12), those from insects are *shellacs* (Chapter 11).

Hard and brittle, more or less translucent, non-volatile, with no particular odour or taste, rosins are readily fusible and burn with a smoky flame. They are very poor conductors of electricity but become negatively charged when friction is applied. Rosin consists of a complex mixture of diterpene resin acids, mainly monocarboxylic acids related to *abietic acid*; they contain only a little, if any, essential oil. They occur naturally in pine oils and are also obtainable from tall oil, a by-product of the kraft paper industry (see Chapter 12). Rosin was formerly used for caulking ships' hulls, and is now widely used in various industrial applications. Rosins, because of their low oil content and ready solubility in alcohol, and are an important source of varnishes; they are also used in paints, inks, plastics, sizing, adhesives, fireworks, etc.

Rosin is generally modified by a combination of chemical reactions to form salts, esters and maleic anhydride adducts, and as hydrogenated, disproportionated and polymerised rosins, including their hydrogenation, esterification and adduct formation, for use in lacquers, plasticisers and floatation agents. *Plasticisers* are high boiling point liquids incorporated into lacquers and various plastics such as PVC in order to preserve workability, flexibility, flow, and impact resistance, while *floatation agents* are used to form a moderately stable foam for the separation of ores in aqueous suspension. Esters obtained by the fusion of rosin with metal oxides to form *resinates* are used as a drier in paints, in plasticisers and floatation agents (Sharp, 1990; Walker, 1991).

They are referred to commercially as *hard resins*. The most commercially important of the hard resins are the copals and damars. *Copals* are those of recent or fossil origin containing very little or no essential oils. The word is Spanish, from the Mexican Nahuati *copali*, for rosin. They yield a hard, elastic varnish, which is much used for exterior work. The true **damars** are distinguished by being insoluble in chloral hydrate but completely soluble in alcohol and turpentine. The term is now commercially applied to the hard rosins obtained by tapping members of the Dipterocarpaceae, especially *Shorea* spp. and a few members of the Burseraceae, also species of the conifer *Agathis*. The word damar or dammar is from the Malayan *damar* for rosin and is applied to a torch made of decayed wood and bark mixed with rosin, wrapped in leaves and bound with rattans. True damars are mainly used in spirit varnishes and in the manufacture of **cellulose nitrate** (nitrocellulose) lacquers. Because of their lustre and light colour they are especially suitable for varnishing paper; they are also used for interior work and histology.

Other hard resins include the Eocene fossil **amber** from now largely extinct conifers, and is used mainly for jewellery and tobacco pipes. Also **lacquers**, chiefly from *Rhus verniciflua* (Chinese or Japanese lacquer tree), which are used as a lacquer and for candles. The **acaroid resins** from *Xanthorrhoea* spp. (blackboy) are used as adhesives, metal lacquers, gold size, a mahogany stain and as a source of **picric acid**. **Sandrac** from *Callitris* spp. (Cypress pines) and *Tetraclinis articulata* (thuya) is used for varnishes, and **mastic** from *Pistacia lentiscus* (mastic) for varnishes, quelling halitosis, as a filler for caries and as an ingredient of ouzo. **Dragon's blood** from American species of *Dracaena* and Asian species of *Daemonorops* are used in varnishes and photo-engraving. Finally, the **gum kinos** from *Eucalyptus* spp. and *Pterocarpus* spp. have astringent properties and are used medicinally.

Not all resins are obtained from woody species. **Jalap**, for example, is from the dried tubers of *Ipomoea purga*, syn. *Exogonium purga* and, as its specific name implies, is used as a purgative. The Aborigines of central Australia also extract a hard rosin from the highly inflammable spinifex grasses *Plectrachne* spp. and *Triodia* spp. for use as an adhesive. The rosin is secreted by the leaf epidermal cells and coats the leaf surface, and helps to reduce water losses (Hill, 1952; Tootill, 1984; Walter and Breckle, 1986; Long, 1994; Mabberley, 1997).

1.4   Oleoresins

In addition to resinous materials the **oleoresins** also contain considerable quantities of essential oils (see Section 6) and are consequently more or less liquid; they are widely used in paints and varnishes. Among the oleoresins are the balsams, elemis and turpentines. It should be noted, however, that the groups are not clearly distinguished and the terms are often confused. For example, the term 'balsam' is often wrongly applied to quite different substances, such as the turpentine known as Canada balsam from *Abies balsamea* (balsam fir), and copaiba balsam obtained from *Copaifera* spp.

A *balsam* is an oily or gummy exudation containing **benzoic** or **cinnamic acids** and their corresponding esters; they are consequently highly aromatic, yielding essential oils on distillation. Balsams are used pharmaceutically as a base for cough mixtures and other medications; they are also employed as fixatives in the perfume industry. Examples include the balsam of Peru from *Myroxylon balsamum* var. *balsamum*, and Tolu balsam from var. *pareirae,* syn. *M. pareirae.*

An *elemi* refers to various oil resins, e.g. Manila elemi from *Canarium luzonicum* (Java almond). They are exuded as clear, pale liquids which tend to harden on exposure to the atmosphere, although some may remain soft while others become quite hard. According to Long (1994) the word is from the Spanish *elimi,* from the Arabic *elemi,* a dialectal variant of *al-lāmi,* the elemi. They are used in making inks and varnishes, and in the pharmaceutical and perfume industries.

The *turpentines* are light, volatile, essential oils obtained almost exclusively as exudates or by distillation from conifers. They are viscous, honey-like liquids or soft, brittle solids, consisting of a mixture of cyclic terpene hydrocarbons, the chief being **α-pinene.** They are used as thinners for paints and varnishes, and as solvents. See also Chapter 12 for further discussion (Hill, 1952; Sharp, 1990; Coppen and Hone, 1995; Mabberley, 1997).

2.   LATEXES AND RUBBERS

A *latex* is the stable aqueous dispersion of a polymer. It is a term formerly applied to natural rubbers obtained from plant exudates, although these have now been largely replaced by synthetic rubbers and polymers such as PVC and polyacrylates. Latexes may be used for the direct manufacture of rubber and plastic goods by dipping, moulding, spreading, electro-depositing and impressions. In green plants the latex is stored in laticifer cells, while in fungi, such as *Lactarius* spp. (milk caps), the latex is produced in a latex duct produced by anastomising hyphae.

A *rubber (caoutchouc)* is defined as a high molecular weight natural or synthetic polymer which exhibits elasticity at room temperature. **Gutta-percha** is defined as a natural occurring polymeric material, isomeric with rubber but having the *trans* configuration. Interestingly, the formation of caoutchouc and gutta-percha appears to be mutually exclusive, with no plant yet found to produce both. Gutta-percha is now chiefly used as a rubber additive. Sources include *Palaquium gutta* and *Payena leerii* (Tootill, 1984; Sharp, 1990; Mabberley, 1997).

Commercial natural rubber, a cis-polyisoprene, is almost exclusively obtained from clonal plantations of *Hevea brasiliensis* (Pará rubber, Indian rubber, caoutchouc), a native of the rain forests of the Amazon basin, but mainly cultivated in SE Asia. Of the nine rain forest *Hevea* species, the only other species yielding acceptable quantities of latex are *H. benthamiana* and *H. guianensis*, the remainder providing a valuable gene bank for a range of physiological attributes. Indian rubber from *Ficus elastica* is a minor rubber source; the species is probably extinct in the wild. Former rubber sources, now superseded by Pará rubber, include *Castilla elastica* (Panama rubber), a species apparently seen by Columbus, and *Manihot glaziovii* (Ceará rubber).

Tapping Pará rubber trees yields a latex containing 30-36% rubber. The latex is then strained, diluted with water and coagulated with methanoic or ethanoic acids to yield a solid rubber. *Vulcanisation* is the process by which wear and tensile strength is increased by rendering the rubber less plastic and sticky. The necessary cross-linking is achieved by heating with a vulcanisation agent, generally sulphur. The process was discovered in 1839 by the US inventor Charles Goodyear while trying to find a method of raising the melting point of rubber when rubber mixed with sulphur accidentally dropped onto a hot stove. The method was patented in 1844 (Brücher, 1989; Sharp, 1990).

3.   TANNINS

The *tannins* constitute a large class of amorphous, bitter and astringent plant metabolites, which are often present in the bark, leaves, fruit, etc. They are either rare or only present in small quantities in the Lower Plants, comparatively rare in the Monocotyledonae except for the Palmae, and common but scattered among the Dicotyledonae. For example, they are absent or rare in the Cruciferae and Labiatae, and invariably present in the Rosaceae and Guttiferae; large quantities are often present among members of, for example, the Combretaceae and Rhizophoraceae. Globally the most important tannin sources are the Anacardiaceae (*Rhus* spp. and *Schinopsis* spp.), Combretaceae (*Terminalia* spp.), Leguminosae (*Acacia* spp. especially *A. mearnsii*) and Rhizophoraceae (several genera).

Tannins are defined as complex polyhydric phenols with a molecular size and shape permitting suitable solubility in water. They represent the condensed products of various phenols, the most important being *pyrogallol* and *catechol*. Depending on whether they can by hydrolysed by acids or enzymes, or whether they condense the components to polymers, tannins may be separated into *hydrolysable tannins* and *condensed tannins* respectively. Such separation roughly corresponds to groups based on *gallic acid* or *flavone-related* components, and are of some importance when considering dietary tannins (see Chapter 8). Although tannins can act as a deterrent to herbivores, their major evolutionary role is believed to deter fungal and bacterial attack.

.Tannins can be extracted from the raw material by leaching with water or other solvents and precipitated with lead ethanoate. The plant extracts will contain both tannin and non-tannin components; clearly the higher the tannin content the more suitable the plant source. The property of tannins to precipitate the gelatine and other proteins present is widely used in the tanning industry in the treatment of hides to make leather. Tannins are also used as mordants in the textile industry, to clarify wine and beer, and as an astringent and styptic. Some tannins are dual purpose and used for both dyeing and tanning, e.g. the dark extract known as *catechu* obtained by boiling heartwood chips of *Acacia catechu* (Lemmens and Wulijarni-Soetjipto, 1991; Lemmens *et al.*, 199i; Sharp, 1990).

3.1   Tanning

The skins used for tanning are composed of two outer epidermal layers and an inner dermis or true skin, and of these it is the inner layer, consisting of *ca.* 98% collagen (fibrous protein), that is important for leather. An essential property for tanning is a stereochemical resemblance between the tannin and the protein, without which some polyphenol molecules are unsatisfactory, even though their molecular weights are between 500 and 3000. If the gram molecular weights of the tannin molecules are <300 there is, for some unknown reason, little or no reaction; if >3000 they physically prevent a complete reaction along the polyphenol molecule.

For tanning purposes, the **gallotannins** and **ellagitannins**, such as those obtained from members of the Fagaceae and *Terminalia* spp., are generally preferred as they produce good quality leather with a pale tan that does not fade in the light. A darker tan can be obtained from **proanthocyanidines**, such as those obtained from *Acacia* spp. (wattles) and members of the Rhizophoraceae. The tendency of such leathers to fade can be lessened according to the tanning techniques used.

Tannage, the art of tanning hides and skins to make them flexible, dates back into prehistory, with archaeological evidence from northern Germany dating from 10,000 BC and Pharaonic Egypt from 5000 BC. Commercial tanning requires the hides to be trimmed and soaked, and for the removal of any remnants of flesh, after which the hides are depilated by placing them in lime for *ca.* 7 days. Following deliming and treatment with detergents, the hides are soaked in ever increasing concentrations of tannin for *ca.* 3 weeks. The shoulder and belly leathers are then trimmed, leaving the thicker and more valuable *butt leather* from the back and sides. The butts are halved into *bends* and receive further tanning for several weeks, after which they are cleaned and bleached. The leather is then treated with various oils and chemicals, rolled with a heavy cylinder, sponged with a wax coating and finally dried. Two notable and possibly linked characteristics of leather are a decrease in hydrophilic properties and a stability against rotting.

The discovery in 1851 of *chrome tanning* by impregnating the hide with chromium salts was later followed by the development of synthetic tannins, such as syntans, resin and aldehyde tannages, and has enabled the tanner to obtain greater control over the tanning process. Chrome tanning, which is mainly used for upper and light leathers, can be completed in a few days. By comparison, vegetable tanning can take 2 months or more and is mainly used for heavy leathers, e.g. for soles, belts, straps and mechanical leathers. The advantages of vegetable tannins is that they impart greater mouldability to the sole leathers, and also provide greater weight and better durability, whereas chrome leather has the advantage of being more heat-proof, stronger, more supple, elastic and water repellent, and easier to dye.

The use of synthetic tanning has increased rapidly since 1950, so much so that world shipments of the three major tannins, from *Acacia mearnsii* (mimosa), *Castanea sativa* (chestnut) and *Schinopsis* spp. (quebracho), fell from 440.8×10^{-3} tons to 179.6×10^{-3} tons between 1950 and 1988. Although vegetable tannins still remain an important commodity, the shift from vegetable to synthetic tannin materials is likely to continue. Such a shift could also be environmentally detrimental since the waste products from chrome, aluminium and titanate tannins are heavy pollutants, whereas the effluent from vegetable tannins are relatively readily biodegradable and, provided the waste products are not dumped in excessive quantities, less detrimental to the environment. Vegetable tannins also have the advantage of being a renewable resource, although there is an unfortunate tendency for the mangrove sources to be over-exploited. Obviously, there is an urgent need for their conservation for sustainable development (Lemmens and Wulijarni-Soetjipto, 1991).

4.   DYESTUFFS

Dyestuffs are intensely coloured compounds, i.e. dyes, pigments, inks and stains, that are applied to a substrate such as fibre, paper, cosmetics, hair, etc. in order to give colour. Plant dyestuffs are extracted by fermentation, boiling, or chemical treatment of plant tissue. A definitive listing of dyestuffs and pigments is given by *The Colour Index* (Anonymous, 1971) and later supplements, providing information on their chemical nature, commercial names, method of application, etc. The method of application may be acid, basic, direct, disperse, azo, sulphur, vat, or fibre reactive. Gums, resins, such as **gum damars** and **karaya gum** and latexes (Section 1) are often used as thickening agents for solutions of dyes used in paints and inks, while the golden-yellow dye gum resin known as **gamboge** from *Garcinia hanburyi*, is used in paints, varnishes, lacquers and inks (Lemmens *et al.* (1991).

4.1   Natural Dyes and Pigments

The terms natural dyes, colorants, and pigments are used indiscriminately in both commerce and the literature. They can be defined as follows: (1) **Natural dyes** or **dyestuffs**, as distinct from natural colorants, are the natural plant (or animal) products used to impart a desired colour to non-food materials such as textiles, wood, leather, etc. by a process known as dyeing; (2) **Natural colorants** are natural products which are incorporated into foodstuffs to provide an attractive colour to the final product (see Chapter 9); and (3) **Natural pigments** are specific chemical compounds that are responsible for the colour in living plant organs, e.g. the yellow pigment *crocein* present in the stigmas (*saffron*) of *Crocus sativus*, and the green, photosynthetic pigment *chlorophyll* in plant tissues, which are sometimes used for colouring food. Pigments are used to impart colour to surfaces, plastics, inks, etc.; they may incidentally affect other properties of the substrate. Unlike dyestuffs, which operate at molecular level, pigments tend to be particulate and insoluble, and to more closely retain their identity on bonding to the substrate. Many insoluble organic dyestuffs and inorganic metal compounds are also used as pigments (Sharp, 1990; Green, 1995).

Some vegetable dyes, such as *indigo* from the leaves of *Indigofera* spp., *madder* from the roots of *Rubia tinctoria*, and *woad* from the leaves of the herb *Isatis tinctoria*, have been used for dyeing fabrics, implements and utensils for thousands of years. Indigo was recorded as being used in China as long ago as 4000 BC and in the Sanskrit writings of 2000 BC. Indigo dyes from India are currently being used to produce the 'faded look' to denim fabrics.

Woad was reputed to have been used by the Ancient Britons to daub their bodies. Cloth first dyed blue with woad and then yellow with *Reseda luteola* (dyer's rocket, weld) yields *Saxon green*, a colour associated with Robin Hood and the Sherwood Forest. Woad was cultivated in the English Fens until the beginning of the 20th century, the last factory at Wisbech in Cambridgeshire being closed in 1914. The dye was allegedly used to dye policemen's uniforms blue before being replaced by cheaper imports of indigo. Indigo gave a stronger and faster blue and was, in turn, replaced by cheaper synthetic dyes. The red and yellow dyes from *Lawsonia inermis* (henna) and *Punica granatum* (pomegranate) respectively, were being used in Ancient Egypt to dye leather in 2000 BC.

The first synthetic aniline dyes, 'mauve', was discovered in 1856 by the English chemist W.H. Perkin while attempting to produce synthetic quinine from coal tar. *Indigotin*, the active principal of indigo (see below) was first synthesised in Germany by Adolf von Bayer in 1880, but it was not until the end of the 19th century that synthetic indigotin could be produced more cheaply than the natural product. This led to a rapid replacement of many vegetable dyes by the beginning of the 20th century, although there has been a revival in recent years in the use of woad and other natural dyes by the traditional craft industries. For example, following a study at the School of Plant Sciences, University of Reading into the mediaeval techniques used for indigo extraction and dyeing, a methodology is now being developed to enable indigotin to be purified from future woad crops grown in the UK (Baker, 1964; Green, 1995; Robbins, 1995; Mabberley, 1997; Wigmore, 2000).

The major classes of plant dyes and pigments recognised by Lemmens *et al.*, 1991) are: (1) *Chlorophyll*, a generic term embracing a number of closely related green pigments, e.g. *chlorophyll a* present in all autotrophic plants and algae, *chlorophyll b* in the Chlorophyta and in land plants, chlorophyll a and b in the Cyanobacteria and *chlorophyll c* and *d* present in certain algae. They are sometimes used for colouring foods and beverages; (2) *Carotenoids* embrace a tremendous variety of chemical structures exhibiting a range of yellow, orange, red and purple colours. Examples of carotenoid pigments are the red colorant of *annatto dye* of commerce from *bixin* present in the seed coats of *Bixa orellana*, yellow *crocein* from the stigmas of *Crocus sativus*, the petals of *Nyctanthes arbor-tristis* and the fruits of *Gardenia jasminoides*. They also used for colouring foods and beverages. The non-toxic annatto is insoluble in water but soluble in fats. Formerly used by the Caribbean Indians for anointing their bodies, it is now finds similar use as one of the colouring agents of lipstick; (3) *Flavonoid pigments* such as *morin* from several species in the Moraceae, including the yellow-khaki *fustic* from the heartwood of *Maclura tinctoria*, syn. *Chlorophora tinctoria*, and *rutin* from the flowers of the legume *Sophora japonica* (pagoda tree). During tanning the derivatives of flavonoid tannins present often impart a particular colour to the leather; and (4) *Quinones*, which usually produce a yellow to red colour. They include the black to red to blonde naphthaquinone dye *lawsone* from the dried leaves of *Lawsonia inermis*, and the anthraquinones present in members of the Rubiaceae, e.g. the orange-red *alizarin* and *purpurin* from *Rubia tinctoria* (madder), and the Turkey red *morindin* from the root bark of *Morinda citrifolia* (Indian mulberry). The glandular young shoots and leaves of *Cordeauxia edulis* (jicid, ye'eb) contain *cordeauxiaquinone*, an orange to magenta naphthaquinone-derived pigment which forms a fast and insoluble combination with many metals. The dye is also noteworthy in being the only known naphthaquinone found in the Leguminosae (Lister *et al.*, 1955; Baker, 1964; Purseglove, 1987; Booth and Wickens, 1988; Sharp, 1990; Lemmens *et al.*, 1991; Green, 1995).

There are also other important dyes that do not conform to the above groupings, including the dark blue *indigo dyes* extracted from the leaves of *Indigofera* spp. by the hydrolysis of the colourless *indican* glucoside to *indoxyl* and its subsequent oxidation to *indigotin*, which is also present in woad. Indigo is unusual in being one of the few natural dyestuffs whose fastness is not improved by a mordant process.

The fiery red crystalline brazilwood dye *brazilein* is produced by the oxidation of the whitish, water-soluble phenolic compound *brazilin* present in the heartwoods of *Caesalpinia echinata* (Bahia wood, Brazilian redwood), *C. violacea*, syn. *C. brasiliensis*, *Haematoxylum* spp. (logwood, campeachy wood, campeche), etc. from South America. The heartwood of *H. campechianum*, for example, contains the colourless *haematoxylin*, which is then rapidly oxidised to produce the violet-blue *haematoxein*, known commercially as *haematein*. The brazilwood dye was originally obtained from the Asian *Caesalpinia sappan*. Following the discovery of the Americas, the name was transferred to brazilwood dyes from the more productive South American species. It was because of the large quantities of the dye (*pau brasil*) then being exported to Portugal that Brazil owes its name. The dye derives its name from the Middle English *brasil*, from the Old French *bresil*, 'red-dye wood', probably from the German *brese*, 'burning coals' (Long, 1994; Green, 1995; Mabberley, 1997).

4.1.1   Dyeing Textiles

Several basic types of vegetable dyes are used for dyeing textiles. These are: (1) *Direct dyes* forming hydrogen bonds with the hydroxyl groups of the fibres. Dyeing is direct from an aqueous solution of the dyestuff. Such dyes are not fast, e.g. the yellow *curcumin* colorant and former dyestuff from the tubers of *Curcuma longa*, syn. *C. domestica* (curcumin, tumeric), and used as a food colorant. Synthetic direct dyes are now available, of which the *azo dyes* are probably the most important; (2) *Acid dyes* are dyestuffs containing an aromatic chromophoric group and a group conferring solubility in water, generally with the SO_3H group as its sodium salt. They are relatively simple in application. Examples are to be found among the flavonoid pigments; (3) *Basic* or *catonic dyes* are dyestuffs containing ionic species. Their chlorides are generally water soluble organic salts, oleates or stearates soluble in organic solvents. They are used in printing inks, forming insoluble salts with heteropolyanions; they have a high fastness and brilliant shades. They are also used in paints and wallpaper pigment. A mordant is usually required when used with natural fibres; (4) *Vat dyes* are regenerated in the fibres by a redox process, i.e. oxidation reduction. The water-insoluble dyestuff becomes water-soluble on reduction in an alkaline solution. The insoluble dyestuff is precipitated within the fibre on re-oxidation, generally in the atmosphere. Such dyes, e.g. *indigo*, often display excellent fastness to light and washing; (5) *Disperse dyes* form a group of water-insoluble dyes which are generally used from an aqueous suspension, the dyestuffs having a high affinity for the fibre, especially nylon and other synthetic fibres. The main types are anthraquinone, e.g. *alizarin*, and synthetic aminomonoazo compounds; and (6) *Mordant dyes* are used to dye textiles that have been treated with a mordant. Such dyes can be very fast, e.g. *alizarin* and *morindin*. Vegetable tannins are sometimes used in alizarin stain complexes in order to prevent white-coloured portions of a textile from being dyed differently by binding the stain that is lost from the coloured parts.

The majority of plant dyestuffs fade rapidly when exposed to sunlight or detergents, consequently the importance of a dye is judged by the fastness of the colour. A *mordant* is used to increase the adherence of the dye to the fabric. They are usually salts of aluminium, iron, tin, or chromium, and form a chemical link between the dye and fibre molecules; some will affect the colour of the dye. There are also some plant products that can also be used as aluminium mordants, e.g. the leaves and bark of *Symplocos* spp. (Hill, 1952; Sharp, 1990; Lemmens *et al.*, 1991; Walker, 1991; Green, 1995; Mabberley, 1997).

4.2   Inks

In inks the colouring matter is dissolved or dispersed in a solvent or carrier, and on drying the colouring matter is bonded with the substrate. *Writing ink* usually consists of a fluid tannin extract with the addition of solutions of iron salts, with which it reacts to form dark blue or greenish-black compounds. The galls from the twigs of *Quercus pubescens*, syn. *Q. infectoria* (Aleppo oak), contain 36-58% tannin and were an early and important source of tannin inks. The inks used in ball-point pens are highly concentrated dyes in a non-volatile solvent. Interestingly, the arils of the seeds of *Acacia cowleana* contain a powerful solvent of ball-point ink and may have a future industrial or domestic application. *Coloured inks* are prepared from natural dyestuffs or analine dyes in combination with alum, water and a gum, e.g. gum arabic. Examples include the *betalain* dye from the berries of *Phytolacca americana* (inkberry) and brazilwood, the red ink of the latter being especially noteworthy for the presence of both tannin and a colouring agent. The *carbon inks*, prepared from charcoal, gums and varnish, are known from Chinese writings as early as least 2600 BC, and from the Egyptian papyruses of 2400 BC. Like paint, they differ from other inks in remaining on the paper surface, unlike the tannin inks which combine chemically with the paper fibres. *Chinese* or *India ink* is a virtually permanent ink made from carbon black, lamp black, or soot from burning pine wood or vegetable oils, such as sesame or tung oils, mixed with a sizing agent, such as glue or gum arabic. *Printing inks* consist of carbon in combination with rosin, gum arabic, a drying oil, such as linseed and tung oils, fractionated palm and coconut oils, a chemical drier and a soap. Large quantities of alginates (see Chapter 19) are now used for sizing and the thickening of printing inks. For lithographic printing the application of gum to relevant parts of the printing plate makes the parts more receptive to the printing ink, while areas free of gum will repel the ink (Hill, 1952; Lewington, 1990; Sharp, 1990; Walker, 1991; Thomson, 1992; Mabberley, 1997). The use of woad is currently under investigation at Bristol University for use in inkjet printers.

4.3   Stains

Stains are specific dyes that are used to treat specific organs or chemical substances in biological specimens so that their identity may be more clearly visible, as in microscopic examinations, e.g. *haematoxylin*, which continues to be successfully used for histological staining. Other well-known examples are *iodine*, a product of the seaweed industry and used to highlight the presence of starch, also *litmus* obtained from *Roccella* spp. following the oxidation of the lichens in the presence of NH_3 and is used as an indicator of acidity (Sharp, 1990, Green, 1995; Mabberley, 1997).

In Somalia the teeth and bones of goats browsing the glandular shoots of *Cordeauxia edulis* are stained an orange to red by the *cordeauxiaquinone* present, suggesting the possibility of a histological use. It has been suggested that the pigment may directly or indirectly act in stimulating the haemopoetic tissue to produce erythrocytes (Gutale and Ahmed, 1984; Booth and Wickens, 1988).

5.   LIPIDS

Natural substances of a fat-like nature are known as *lipids*, although the exact definition is somewhat variable. Strictly they fatty acids or their derivatives, that are insoluble in water but soluble in organic solvents. They include simple fats and waxes, also the phospholipids and cerebrisides (largely present in nerve sheaths). Many would also consider such compounds as sterols (phytosterols) and squalenes to be lipids. The seeds of the grain amaranths (*Amaranthus* spp.), for example, contain 7-8% fat (oil), of which 4-11% of the total oil fraction is the acrylic triterpene *squalene*. Both squalene and the more stable saturated hydrocarbon *squalane* act as non-toxic vehicles for cosmetics, for promoting the absorption of drugs applied to the skin, and as a lubricant for the computer industry (Sharp, 1990; Jain and Sutarno, 1996).

Fatty oils are produced in many plants; they are usually stored in the seeds but may also be found in other organs. Most oils function as energy storage compounds and are especially useful during seed germination. In the past vegetable oils and fats or tallows have mainly been utilised for food purposes, with many also finding applications as lubricants and greases, illuminants and candles, and in soaps and paints. Even in this age of petrochemicals fats and oils still have a major role in non-food applications. Although water-based latex systems have now replaced much of the traditional vegetable oil-based paint market, approximately one third of the binders used are still based on vegetable oils or their derivatives. Large quantities of fatty acids derived from vegetable sources are also required as *surfactants*, i.e. soluble, surface-active agents, such as detergents and soaps, capable of altering the interfacial tension of water and other liquids or solids (Princen, 1983; Walker, 1991).

The term '*oil*' is applied to those glycerides that are liquid at 20°C. Four classes of oils are recognised: (1) *Drying oils* which, on exposure to the atmosphere, are oxidised to form thin, elastic films of dry resin. They are of great importance to the paint and varnish industries. The oils consist mainly of unsaturated triglycerides and unsaturated hydrocarbon polymers; (2) *Semi-drying oils* can only slowly absorb a limited amount of oxygen and will only form a soft film after long exposure to the atmosphere. Some of these oils are edible, others are used as illuminants, or in the manufacture of soap and candles; (3) *Non-drying oils* remain liquid at ordinary temperatures and do not form a film. They are edible and, in addition to the food industry, can be used in soaps and lubricants; and (4) *Fats and tallows* are solid (tallows) or semisolid at 20°C. They are edible and also used in the manufacture of soaps and candles. Examples of these four classes are shown in Table 19.

Fatty oils (*fixed oils*), as distinct from volatile oils (see Section 6), do not evaporate or become volatile, neither can they be distilled without being decomposed. They are also bland, insoluble in water but soluble in various organic solvents, unlike 'fats' which are solid at 20°C. Chemically they are close to animal fats, consisting of triacylglycerols, esters of glycerol (glycerin) and three long-chain carboxylic acids or fatty acids, with oleic, palmitic and stearic acids predominating. They may be triesters of either the same fatty acid, such as palmitic, stearic, oleic, and linolenic acids.

TABLE 19. Examples of the four classes of vegetable oils and their sources'

Non-drying oils	Semi-drying oils	Non-drying oils	Fats and tallows
tung (*Aleurites* spp.); saffseed (*Carthamus tinctoria*); soya bean (*Glycine max*); Niger seed (*Guizotia abyssinica*); oiticica (*Licania sclerophylla*); linseed (*Linum usitatissimum*); perilla (*Perilla frutescens*)	rape or colza (*Brassica napus*); cottonseed (*Gossypium* spp.); sunflower (*Helianthus annuus*); sesame (*Sesamum orientale*); corn (*Zea mays*)	groundnut *(Arachis hypogea*); olive (*Olea europaea* subsp. *europaea*); castor (*Ricinus communis*)	murumura (*Astrocaryum murumuru*); coconut (*Cocos nucifera*); palm and palm kernel (*Elaeis guineensis*); cohune (*Orbignya cohune*); babassu (*O. oleifera* and *O. phalerata*); Chinese vegetable tallow *Sapium sebiferum*); nicuri or macauba (*Syagrus coronata*); cocoa butter (*Theobroma cacao*); shea butter (*Vitellaria paradoxa*)

Two classes of fatty acids are recognised: (1) **Saturated fatty acids**, i.e. with molecules to which no further atoms may be added. They have the general formula $C_nH_{2n}O_2$, e.g. palmitic and stearic acids with methanoic acid (formic acid) from *Urtica* spp. as the lowest members of the series; and (2) **Unsaturated fatty acids** include the **monounsaturated fatty acids**, or oleic acid series, with one double bond and the general formula $C_nH_{2n-2}O_2$, and the **diunsaturated fatty acids**, or linolenic acid series, with two double bonds and the general formula $C_nH_{2n-4}O_2$, etc. The lower members of the series are liquids, soluble in water and volatile in steam. As the number of carbon atoms increases, the melting points rise and the acids form solids, insoluble in water and soluble in organic solvents. They occur mainly as oils. The only exceptions in major commercial seed oils are linseed oil (methyl-interrupted triene), tung oil (conjugated triene), castor oil (hydroxy fatty acid) and high-erucic rapeseed oil (long-chain fatty acid). High-erucic rapeseed is now relatively scarce due to the increasing requirement by the food industry for low-erucic acid rapeseed. The major fatty acids available commercially are shown in Table 20, and their application in Table 21.

TABLE 20. Major commercial fatty acids from vegetable fats and oils (Princen, 1983; Appleqvist (1989); MAFF, 1994a)

Class	C_n content	Name	Plant source
saturated	8.0	octanoic (caprylic)	*Cocos nucifera* (coconut), *Cuphea* spp., *Elaeis guineensis* (oil palm)
saturated	10.0	capric	*Cocos nucifera*, *Cuphea* spp., *Elaeis guineensis*
saturated	12.0	lauric	*Cocos nucifera*, *Coriandrum sativum* (coriander), *Cuphea* spp., *Elaeis guineensis*
saturated	14.0	myristic	*Cocos nucifera*, *Cuphea* spp., *Gossypium* spp. (cottonseed)
saturated	16.0	palmitic	*Elaeis guineensis, Gossypium* spp.
monounsaturated	16.1	palmitoleic	*Gossypium* spp.
saturated	18.0	stearic	*Theobroma cacao* (cocoa)
monounsaturated	18.1	oleic	*Brassica napus* (rapeseed), *Euphorbia* spp., *Helianthus annuus* (sunflower), *Olea europaea* subsp. *europea* (olive)
	18.1	petroselenic	*Coriandrum sativum* (coriander), *Daucus carota* (carrot), *Petroselenium crispum* (parsley)
hydroxy	18.1	ricinoleic	*Lesquerella* spp., *Ricinus communis* (castor)
diunsaturated	18.2	linoleic	*Glycine max* (soya), *Guizotia abyssinica* (Niger seed), *Helianthus annuus* (sunflower), *Zea mays* (maize)
multiunsaturated	18.3	α-linolenic	*Aleurites* spp. (tung), *Camelina sativa* (cameline), *Linum usitatissimum* (linseed)
multiunsaturated	18.3-1	γ-linolenic	*Borago officinalis* (borage), *Oenothera biennis* (evening primrose)
saturated	20.0	arachidic	*Arachis hypogea* (groundnut), *Limnanthes alba* (meadowfoam)
monounsaturated	20.1	gadoleic	*Brassica napus* (rapeseed)
saturated	22.0	behenic	*Moringa oleifera* (moringa)
diunsaturated	22.2	erucic	*Brassica hirta* (mustard), *B. napus* (rapeseed), *Crambe hispanica* (crambe), *Lunaria annua* (honesty)
saturated	24.0	lignoceric	*Arachis hypogea*
monounsaturated	24.1	nervonic	*Lunaria annua*

TABLE 21. Examples of characteristic fatty acids and their use and potential use (compiled from Rexen and Munck, 1984; MAFF, 1994a; Mabberley, 1997)

Characteristic acid	Use and potential use
octanoic (caprylic)	fuel, detergents, soaps, potential synthetic resins and C_{12} compounds in surfactants
capric	fuel, detergents, soaps, potential synthetic resins and C_{12} compounds in surfactants
lauric	fuel, detergents, soaps, potential synthetic resins and C_{12} compounds in surfactants
petroselenic	detergents, plastics
myristic	fuel, detergents, soaps, potential synthetic resins and C_{12} compounds in surfactants
palmitic	food
palmitoleic	food
stearic	food
oleic	food. lubricants
linoleic	food, alkyd paints, varnishes, linoleic acid
α-linolenic	linoleum, paints, varnishes
γ-linolenic	pharmaceuticals
ricinoleic	grease, lubricants, paints, varnishes, plasticisers, dyes
arachidic	food, lubricants, cosmetics
gadoleic	food
behenic	salad oils, artist's paints, soaps
erucic	nylons, erucamide, perfumes, cosmetics
lignoceric	soaps, detergents
nervonic	lubricants

The commonest fatty acids found in plants are palmitic and oleic acids, with linolenic acid predominant in specialised tissues, e.g. chloroplasts. Other important fatty acids are lauric, linoleic, myristic, palmitoleic and stearic acids. Both the linolenic and linoleic acids can be synthesised by plants but not by animals. They are the precursors of prostaglandins and are an essential requirement for animal diets; they are consequently known as *essential fatty acids*.

The fatty acids are loosely grouped according to their carbon content as being *short-chain*, e.g. octanoic acid (caprylic acid), *medium-chain*, e.g. palmitic acid, and *long-chain*, e.g. erucic acid. The short- to medium-chain fatty acids are required for the production of soaps, detergents and emulsifiers; they are also used for lubricants and other materials. The soap and detergent industries in particular are heavily dependent upon coconut oil for their lauric acid requirements. *Medium-chain fatty acids* (C_{16-20}) are used in the manufacture of plastics, fabric softeners, adhesives, and coatings, i.e. films forming the plates of a capacitor in the electrical industry.

Long-chain fatty acids (C_{22}) are largely responsible for the physical properties of complex lipids, being able to withstand considerable heat before breaking down. They are consequently suitable as high temperature, non-foaming lubricants for jet engines, etc. Until recently sperm whale oil from *Physetes catodon* (sperm whale) was the sole commercial source of liquid wax esters for high-performance lubricants, cosmetics, etc. Jojoba oil from the seeds of *Simmondsia chinensis* is now a potential alternative, containing virtually 100% liquid wax ester without any triglycerides, whereas sperm whale oil contains up to 25%. Despite the need to conserve the sperm whale, jojoba oil is only commercially viable for use in cosmetics and other high value products.

The major commercial source of commercial **hydroxy fatty acids** (the prefix hydroxy denotes the presence of a hydroxyl (-OH) group) is ricinoleic acid from castor oil seed, with *Lesquerella* spp. (bladder pod) from southern USA and northern Mexico as potential alternatives. **Epoxy fatty acids** (the prefix 'epoxy' indicates the presence of an oxygen bridge across an alkene bond) are used extensively in the coatings and adhesive industries. They are largely derived from petrochemicals, although some $45\text{-}90 \times 10^{-6}$ kg of linseed and soya bean oils are also converted annually. Other potential sources are *Vernonia anthelmintica* (ironweed), *V. galamensis* and *Stokesia laevis* (Stokes' aster) (Hill, 1952; Princen, 1983; Tootill, 1984; Perdue *et al.*, 1986; Horrobin, 1990b; Sharp, 1990).

6.   ESSENTIAL OILS

Essential oils are more or less volatile oils which are mainly formed in specialised glands, rarely in ducts, and are extracted from plants. They occur throughout the plant kingdom, among both higher and lower plants. Among the natural exudates (see Section 1.4) are the **balsams**, **elemis**, and **gum resins**, including **oleogum resins** and **oleoresins**. They are called 'essential' because the oils are believed to possess the very essence of colour and flavour. They were certainly used for aromatics and perfumes by the early Egyptians and Hebrews, amongst others, and were recorded by Theophrastus *'On Odours'* in *ca.* 288-287 BC.

Essential oils are secondary metabolites consisting mostly of terpenoids, also as aliphatic and aromatic esters, phenolics and substituted benzene hydrocarbons. They are usually liquid but can also be solid (*orris*) or semisolid (*rose*) depending on the temperature. Essential oils are solid in pure alcohol, fats and oils but insoluble in water and, on evaporation following exposure to the atmosphere, they leave no oily residue (Hill, 1952; Walker, 1991; Lawrence, 1995; Robbins, 1995; Scarborough, 1996).

The function of essential oils is either to attract pollinating insects or to repel hostile insects and animals; sometimes their function is allelopathic. A number have antiseptic, insecticidal, fungicidal and bactericidal properties. Insecticidal activities have been found in the steam volatile fraction of cedarwood oil obtained from the heartwood of *Juniperus recurva* (Himalayan weeping juniper) of Nepal, of which the insecticidal components are **thujopene** and **8-cedren-13-ol**. It has also been found that *Reticulitermes flavipes* (termites) are unable to survive on sawdust from *J. virginiana* (pencil or eastern red cedar), nor on filter paper treated with a pentane extract of cedarwood oil. Twelve other US juniper species used for post timbers are also known to contain similar natural wood preservatives. Essential oils with fungicidal properties from *Cymbopogon flexuosus* (Malabar oil grass), *Santalum album* (sandalwood), *Vetivera zizanioides* (vetiver grass) and in particular *Trachyspermum ammi* (ajowan), have been shown to inhibit growth of *Microsporum gypseum*, *Trichophyton equinum* and *T. rubrum*, which are among the pathogenic fungi responsible for ringworm (Dikshit and Husain, 1984; Adams, 1991).

Essential oils may also be used as food and drink additives, in cosmetics, perfumes, incense, soaps, toothpastes, shampoos, deodorants, detergents, cleaning agents, pharmaceuticals, aromatherapy and insecticides, e.g. *camphor* from *Cinnamomum camphora*, which is also used in the manufacture of celluloid and explosives, and is now largely obtained from synthetic sources. Perhaps rather surprising to the western world where cloves, from *Syzygium aromaticum*, are used for flavouring food, in Indonesia they are used in the manufacture of a clove-flavoured cigarette known as 'kretek' (Hill, 1952; Sharp, 1990; Linskens and Jackson, 1991b; Walker, 1991; Lawless, 1992; Coppen, 1995; Robbins, 1995).

Other examples of plants yielding essential oils that are used as sources of chemical isolates for derivative manufacture are Chinese sassafras oil from *Cinnamomum camphora* (also a source of camphor) and Brazilian sassafras oil from *Aniba pretiosa*, syn. *Ocotea pretiosa*, the oils yielding **safrole**, which is used to manufacture **heliotropin**, a valuable flavour and fragrance compound; they also yield **piperonal butoxide**, an important ingredient of pyrethoid insecticides. Previously unexploited *Piper* spp. are also a potential source of safrole. Originally flavours and fragrances were obtained almost entirely from the plant kingdom, very few being of animal origin. While many are now products of today's petrochemical industry, the building blocks for many of these synthetic compounds are of plant origin. For example, the **α-** and **β-pinenes** from turpentine lack flavour and fragrance but can be converted into more desirable derivatives (Coppen, 1995).

There are approximately 100 commercially marketable essential oils that are derived from plants, of which those from *Citrus* spp. account for about one third of the world's production, the remaining two thirds are almost entirely from cultivated trees and herbs. World production is in the region of 45,000 tonnes and worth *ca.* US$700 million and, although the number of uses and the volume of trade has increased considerably, the development of synthetic substitutes has also grown. For example, in the perfume industry **rosewood oil** from *Aniba rosaeodora* and *A. duckei* was formerly an important source of **linalool**, which was used either alone or as a precursor of other fragrance compounds. Cheaper sources of synthetic linalool are now available, although rosewood oil (***bois de rose***) persists as the established ingredient of the more expensive perfumes. Similarly, the much prized **sandalwood oil** from *Santalum album* remains unsurpassed by synthetic substitutes as a perfume.

Probably the most widely known of the essential oils is eucalyptus oil, or *Oleum Eucalypti*, which is defined in the *British Pharmacopoeia* (1985) as 'the oil distilled from the fresh leaves of *Eucalyptus globulus*, *E. amygdalina*, and probably other species of eucalyptus'.

6.1   Essential Oil Extraction

The mode of extraction of essential oils is dependant on the quality and stability of the compound. The main methods used are: (1) *Steam, water* or *dry distillation*, and is used for the majority of neat essential oils, e.g. cinnamon oil from *Cinnamomum verum*, syn. *C. zeylanicum*, myrrh from *Commiphora myrrha*, eucalyptus oil from *Eucalyptus* spp., lavender oil from *Lavandula* spp., and sandalwood oil from *Santalum album*; (2) *Expression*, i.e. extraction under pressure, is used for most citrus oils; the method also yields neat essential oils; (3) *Solvent extraction* using a hydrocarbon solvent is used where distillation would adversely affect the final product, e.g. jasmine oil from *Jasminum officinale*. The process also produces a more true-to-nature fragrance. The raw plant material, e.g. bark, leaves, flowers and roots, is subjected to solvents to produce a *concrete* rich in soluble material, with very low alcohol levels and devoid of any water-soluble components. The majority of the concretes are solid and of a waxy, non-crystalline consistency and consisting of *ca.* 50% wax and 50% essential oil. Ylang ylang concrete from *Cananga odorata* is a rare example of a liquid concrete with *ca.* 80% essential oil and 20% wax. Concretes have the great advantage of being more stable and concentrated than pure essential oils, and are used in perfumery as fixatives to prolong the effect of the fragrance. Because concretes are not very soluble in perfume bases, they need further conversion into an absolute should the material be required for use in a fine fragrance. Examples of plants that may be either steam distilled or solvent extracted to produce a concrete include *Cananga odorata*, *Lavandula* spp. and *Salvia sclarea* (clary sage). The vernacular name of the latter is based on a corruption of the Medieval Latin '*sclarea*', dry or stiff; the plant is also known as 'hot housemaid' on account of the pungent scent of sweaty armpits emitted when brushed against. It is not unusual for scents to have an objectionable smell when in high concentrations and a more desirable fragrance when diluted. An *absolute* is obtained from the concrete by a second process of solvent extraction using ethanol, in which the unwanted wax is only slightly soluble. The result is usually a highly concentrated viscous liquid, although in some cases, such as clary sage absolute, they may be solid or semisolid. They are normally subjected to repeated treatments with ethanol, even so, as is the case of orange flower absolute from *Citrus aurantium*, syn. *C. aurantium* var. *amara*, a small portion of the wax remains. Absolutes can be further processed by molecular distillation to remove every last trace of non-volatile material. The alcohol is recovered at the end of the process by evaporation, although some absolutes will retain *ca.* 2% or less of ethanol and are consequently not recommended for therapeutic use; (4) *Enfleurage extraction* is primarily used for extracting aromatic materials from flowers that contain minute quantities of delicate aromatic substances which would otherwise by hydrolysed by moisture or decomposed by heat when other extraction methods are used. The process also takes advantage of the phenomenon where certain flowers continue to give off an aroma after they have been picked due to an enzymatic breakdown of bound glycosides, etc. Freshly cut flowers, such as those of *Jasminum officinale* (jasmine) or *Polianthes tuberosa* (tuberose), are placed on a plate covered with a thin coating of specially prepared and odourless fat known as a *chassis*; the chassis is repeatedly renewed. After 12-30 hours for jasmine and 24-100 hours for tuberose, the fat becomes saturated with the volatile oils. The fragrance-saturated fat, known as a *pomade* or *corps gras*, is then extracted with ethanol to yield the pure absolute or perfume. A tincture of a pomade is known as an *extrait*. The method is expensive and is now rarely used. Pomades can also be produced by *hot fat extraction*, also known as *maceration*, but the technique is now obsolete; and finally (5) *CO₂ extraction* is a recent development where liquid CO_2 under pressure is used to extract the essential oils and other aromatic substances from dry plant materials. Such extracts have excellent odour quality and purity, being entirely free of unwanted solvent residues and non-volatile matter. The method does have the great disadvantage of very high installation costs (Lawless, 1992; Lawrence, 1995; Macdonald, 1995).

Solvent extraction can also be used to produce: (1) *Resinoids* are the solvent extracts from natural exudates or dead, natural resin exudates such as balsams, gum resins or oleoresins using a hydrocarbon solvent such as petroleum ether or hexanes. The resinoids may therefore be regarded as concretes of dried organic materials. They usual form a homogeneous mass of non-crystalline character, but may occur as viscous liquids, semisolids or solids. Resinoids, like concretes, are also used in perfumery as fixatives to prolong the effect of the fragrance. Some resinous materials, such as frankincense from *Boswellia* spp., especially *B. sacra*, and myrrh, are used to either produce an essential oil by steam distillation, or a resin absolute by ethanol extraction directly from the crude oleo-gum. *Benzoin* from *Styrax benzoin*, however, is insufficiently volatile to produce an essential oil by distillation, and liquid benzoin is often just a benzoin resinoid dissolved in a suitable solvent or plasticising dilutent. It is used a fixative for perfume, medicinally as an inhalant for respiratory infections, and locally for incense; (2) *Extracts* are. concentrates obtained from resin-free dried aromatic plant material using a polar solvent, by CO_2 under pressure, or extraction using microwaves and microwave transparent solvents; (3) *Infusions* are the hot extraction of plant materials or exudates with water or an organic solvent. The technique is not popular due to the difficulties in controlling the chemical composition of the infusions; and (4) *Tinctures* are the alcoholic or aqueous alcoholic extracts of natural raw materials where the eventual alcohol content is usually adjusted to 20-60%, although some 95% aqueous alcoholic tinctures are known (Lawrence, 1995).

There are approximately 100 commercially marketable essential oils that are derived from plants, of which those from *Citrus* spp. account for about one third of the world's production, the remaining two thirds are almost entirely from cultivated trees and herbs. World production is in the region of 45,000 tonnes and worth *ca.* US$700 million and, although the number of uses and the volume of trade has increased considerably, the development of synthetic substitutes has also grown. For example, in the perfume industry **rosewood oil** from *Aniba rosaeodora* and *A. duckei* was formerly an important source of **linalool**, which was used either alone or as a precursor of other fragrance compounds. Cheaper sources of synthetic linalool are now available, although rosewood oil (**bois de rose**) persists as the established ingredient of the more expensive perfumes. Similarly, the much prized **sandalwood oil** from *Santalum album* remains unsurpassed by synthetic substitutes as a perfume.

Some 650 tonnes of eucalyptus oil are produced annually from more than 15,000 tonnes dry weight of leaves. harvested from coppiced trees; oil production can be a year-round activity. China by far the largest producer, although it is suspected that some of the eucalyptus oil from China is Chinese sassafras oil, which is not readily distinguishable from that of *Eucalyptus.*

The grades of eucalyptus oils vary according to the species. Those from *E. camaldulensis* (Red River gum), *E. dives* (broad-leaved peppermint), *E. exserta* (Queensland peppermint), *E. globulus* (Gippsland blue gum), *E. polybractea* (blue mallee), *E. radiata* (candle bark), *E. smithii* (gully gum), and *E. viridis* (green mallee) are characterised by their high **cineole** content, below 70% being considered undesirable for pharmaceutical purposes. These oils are also used in confectionery, as a general disinfectant, cleaner and deodoriser, as a fuel additive and solvent, and in formulations with other oils. The oil from *E. dives* is also rich in **pipertone** and **phellandrene**, and is used industrially in a small and diminishing market as a substitute for natural and synthetic methanol. While the oils from *E. citriodora* (lemon-scented gum) and *E. staigeriana* (lemon-scented ironbark) are rich in **citronellal**, and are used either in the cheaper soaps, perfumes and disinfectants, or for the production of citronellal by the fractionation of the crude oil for use in the aroma and chemical industries (Inman *et al.*, 1991; Coppen, 1995; Robbins, 1995).

7.   **WAXES**

The term **wax** was formerly limited to fatty acid esters with monohydric fatty alcohols having plastic and water-repelling wax-like properties. The term is now arbitrarily used for any organic substance having such properties. The waxes are important components of the cuticle covering the stems, leaves, flowers and fruits of most plants. They originate in the epidermal cells as oily droplets and migrate via tiny canaliculi to the cell surface. Waxes are used in paper coating, polishes, electrical insulation, textiles, leathers, cosmetics and pharmaceuticals. Important commercial sources include carnauba or Ceará wax from *Copernicia prunifera* (carnauba wax palm), candelilla wax from *Euphorbia antisyphilitica* (candelilla), and ouricuri or licuri wax from *Syagrus coronata* (ouricuri, nicuri, palm nut), with cauassú wax from *Calathea lutea* (balasier, cachibou) as a potential source. Extraction usually involves beating the harvested leaves to free the wax (Tootill, 1984; Sharp, 1990; Robbins, 1995; Mabberley, 1997).

8.   ALCOHOLS

Alcohol is the general term for compounds containing hydroxyl groups attached to carbon atoms in place of hydrogen atoms, and having the general formula R.OH, where R represents the aliphatic radical (Walker, 1991).

In the past *ethanol (ethyl alcohol, industrial alcohol, spirits of wine)* was almost exclusively manufactured from the fermentation of materials containing starch and sugar by yeasts, and to a lesser extent by other moulds and bacteria (see Chapter 19). Ethanol is now largely produced by the direct hydration of ethene (ethylene) obtained as a by-product of the petroleum industry. Ethanol is the starting point in the manufacture of other chemicals, principally *ethanal (acetaldehyde)* as a solvent and chemical intermediate for ethanoic (acetic) acid and other chemicals, perfumes and pharmaceuticals. The recent general trend for increasing oil prices has stimulated an interest in using alternative sources to the petrochemical industry for ethanol from the fermentation of maize and other cereals, sugar cane, molasses, potatoes, sugar beet, etc., even though the production costs are greater than for gasoline. The ethanol production from one ton of sorghum, for example, yielding 372 l of 182 proof ethanol has a potential comparable to the 367 l of ethanol obtainable from maize. In terms of work output ethanol used as a fuel contains 60% of the energy value of petrol, while *gasohol*, i.e. petrol containing 10% ethanol, is comparable to that of petrol plus a lead additive. The technology is particularly well developed in Brazil (Cross, 1984; Rexen and Munck, 1984; Sharp, 1990; Eckhoff and Paulsen, 1996; Rooney, 1996).

Methanol (wood alcohol) is produced by the distillation of hardwood lumber and sawmill waste - softwood waste contains very little methanol. It is used to produce *methanal (formaldehyde)*, which is used as a powerful germicide and in the manufacture of polymethanal resins and other products, *methanoic acid* for use in the dyeing and finishing of textiles, leather tanning, and as an intermediate for other chemicals, including *chloromethane (methyl chloride)* used in the production of silicones and the manufacture of anti-knock additives, butyl rubber, methyl cellulose, and for quaternising organic bases, such as Paraquat, and numerous other organic compounds (Hill, 1952; Sharp, 1990).

9.   CARBOHYDRATES

Carbohydrates are one of the principal classes of natural organic compounds. They have the approximate formula $(CH_2O)_x$ and include sugars, starches and cellulose. They are produced by plants as a result of photosynthesis and are stored mainly as starches, fructoses, mannans and galactomannans, primarily in parenchymatous storage tissues of the roots, tubers and piths, where they function as energy storage molecules and as structural elements.

The simple carbohydrates are mono-, di- or polysaccharides, having repeating units usually of five or six carbon atoms joined through oxygen linkages. It is this basic sugar skeleton of the carbohydrates, involving hydroxyl groups, that gives them their properties, such as water solubility and sweetness (Whistler and Corbett, 1957; Tootill, 1984; Sharp, 1990).

9.1   Sugars

The sugars include any of the lower molecular weight carbohydrates, namely monosaccharides, the smaller oligosaccharides, and their derivatives. Any monosaccharide with the carbonyl (CO) group on the terminal carbon and forming an aldehyde (CHO) group is known as an *aldose sugar*. When the carbonyl group is positioned elsewhere, it is referred to as a *ketose sugar*. The majority of the natural sugars contain 6 or 12 carbon atoms in the molecules. They are crystalline, extremely soluble in water, and generally have a sweet taste, with sucrose as the standard (1) against which sweetness is measured. Those sugars possessing a potentially active aldehyde or ketone group, and therefore capable of reducing an oxidising agent, are known as *reducing sugars*, and can be detected by using Benedict's and Fehling's solutions. All monosaccharides have this capacity, but for a disaccharide to be a reducing sugar one of the reducing groups of either of the two component monosaccharides must be left intact. Thus, *maltose*, which consists of two glucose units linked by an $\alpha(1\text{-}4)$ glycoside unit, is a reducing sugar because the second glucose unit can undergo oxidation, having an aldehyde at carbon 5, while *sucrose* is a *nonreducing sugar* because the component glucose and fructose units are linked by the aldehyde and ketone groups. The enzyme *invertase* hydrolyses sucrose to *D-fructose* to produce *invert sugar*, the name indicating the change of optical rotation from the dextrorotatory sucrose solution to the laevorotatory mixture.

The major sugar of commerce is the disaccharide *sucrose*, and is obtained from either sugar cane or sugar beet. Other natural sugars include the monosaccharide *glucose* (*dextrose* or *grape sugar*), commercially produced from starch fermentation. The monosaccharide *fructose* (*laevulose* or *fruit sugar*) is also present with glucose in many fruits. Commercial production is from *inulin*, present in the tubers of *Helianthus tuberosus* (Jerusalem artichoke) and some other species. Inulin is a chain-like polysaccharide of fructose units, each chain with a terminal glucose unit. It cannot be digested by man until broken down by micro-organisms in the colon, hence its usefulness as a sweetener for diabetics and is, by association, sometimes confused with the pancreatic hormone *insulin*. The disaccharide *maltose* occurs free in barley grains and a few other plants. It is readily produced from starch or glycogen by the action of the enzyme *amylase (diastase)*, and is of considerable importance to the brewing and soft drinks industries (see also Chapter 9).

The aldohexose sugar *mannose* does not occur free in nature but, instead of glucose, chains of mannose form the reserve polysaccharide *mannan* present in many of the Leguminosae; it is also a frequent component of some hemicelluloses. The extremely hard endosperm known as *vegetable ivory* from *Phytelephas* spp. (ivory palm) is also composed of mannans, and is used for billiard balls, chessmen, dice, buttons, etc. The reduction product of mannose is the sugar alcohol *mannitol*, which is the principal soluble sugar in fungi and lichens, as well as being a major product of photosynthesis in the brown algae, lichens, and some higher plants.

Neither does the aldohexose sugar *galactose*, an isomer of glucose, normally exist in the free state in plants, but as polymers such as *galactomannans*, with glucose-forming *glucomannans*. The tubers of *Amorphophallus* spp. (elephant foot yams), for example, store their carbohydrates in the form of large amorphous granules of glucomannan. That obtained from the tubers of *A. konjac* has been used industrially in China and Japan as emulsifiers and stabilisers in the food, drinks and cosmetic industries, and in drilling fluids (Tootill, 1984; Sharp, 1990; Jansen *et al.*, 1996).

9.2 Starch and Starch Products

Starch is the an early end product of photosynthesis which at night is rapidly broken down into sucrose and transported to other organs. It is the most abundant and important of the plant reserve polysaccharides. Natural starch is also unique among the carbohydrates in occurring as discrete granules whose characteristics vary according to the species (Table 22).

Most commercial starches are in the form of single granules. Accidental overheating during flash drying can cause the granule surface to gelatinise and for individual grains to stick together. They can usually be readily separated, although severe overheating can result in large aggregates which do not readily disperse and have an undesirable proportion of either gelatinised or cold water swelling (Whistler and Corbett, 1957; Snyder, 1990).

Starch consists of two structurally different fractions, *amylose* and *amylopectin*, the relative proportions varying according to the species. Amylose is composed of long, straight chains of glucose units, soluble in water and tending to set to a stiff gel. It forms strong, flexible fibres that can be used industrially as a coating agent and for making edible films

TABLE 22. Examples of plant starch grains and their characteristics and uses (Doggett, 1970; Göhl, 1981; Rexen and Munck, 1984; Snyder, 1984; Purseglove, 1985, 1987; Ensminger *et al.*, 1994; Flach and Rumawas. 1996)

Species	Grain size (μm)	Starch (% DW)	Amylose (%)	Amylopectin (%)	Use
CEREAL GRAINS					
Hordeum sp. (barley)	2-3[1]	68-78	35-40	60--65	food, alcohol, fodder, adhesives
Sorghum bicolor (sorghum)	2-25	68-85	14-25	75-86	food, alcohol, fodder, industrial starch
Oryza sativa (rice)	2-10	69-88	14-32	68-86	food, alcohol
Zea mays (maize)	5-30	72-84	23-28	77-77	food, alcohol,
high amylose	5-35		50-85	15-50	fodder, industrial
waxy maize	5-25		1-5	95-99	starch
Triticum aestivum (wheat)	5-40	60-85	19-28	72-81	food.alcohol, fodder, ethanol
CORMS, RHIZOMES, ROOTS AND TUBERS					
Dioscorea esculenta (lesser yam)	0.5-2.0	83-88			mainly food
Colocasia esculenta (dasheen, taro)	1-6.5	85-89			food
Ipomoea batatas (sweet potato)	5-50	88-91			food, fodder, industrial starch
Pachyrhizus erosus (yam bean)	8-35	84			food, fodder
Maranta arundinacea (W. Indian arrowroot)	10-60	19-22	20	80	food, fodder, glues, soap
Curcuma zedoaria (shoti, zedoary)	10-60[2]	23			food
Solanum tuberosum (potato)	10-185	75-82	20-24	76-80	food, fodder, adhesives
Xanthosoma spp. (tannia)	17-20	86-88			food
Canna indica (Australian or Queensland arrowroot)	30-130	85-86			food, feed, industrial starch
STEM PITH					
Metroxylon sagu (sago palm)	81-88	81-88	27[3]	73[3]	food, industrial starch

[1]sharply bimodal; [2]non-spherical; [3]purified sago starch

Amylopectin differs in that it is a branch-chained polysaccharide which tends not to gel readily in aqueous solution but will gelatinise in hot water at 60-80°C. The higher the amylopectin content the higher the glutinosity, the viscosity retrograding with time. It has poor film-forming properties and is used in the textile industry for sizing and finishing, and in the food industry as a thickening agent. The ratio between amylose and amylopectin strongly affects the palatability and industrial application of the starch; their industrial separation is based on the solubility of amylose in water. In general the cereal starches contain 15-30% amylose and 70-85% amylopectin. Amylose is, however, virtually lacking in the endosperms of the waxy cultivars of barley, rice, sorghum and maize (Sharp, 1990; Grubben *et al.*, 1996).

Both morphologically and chemically starch can be regarded as an inhomogenous polymer system. The granular structure can be broken down and the hydrophilic character changed by various mechanical/thermal and/or chemical treatments, giving rise to considerably increased possibilities for combinations of starch with different synthetic polymers to form fibres, adhesives, absorbents, surfactants, etc.

Starch, like cellulose, possess one primary and two secondary hydroxyl groups per glucose unit, to which alkyl or acid residues may be coupled to produce ethers and esters respectively. These find use in the food, textile and paper making industries, both *hydroxyethyl* and *hydroxypropyl starches*, for example, are used in the food industry as coatings, the former also as a binder.

The textile industry too is an important outlet for natural starch, slightly modified starch, i.e. oxidised amylopectin starch, as well as starch derivatives such as starch phosphates and acetates, hydroxyethyl and hydroxypropyl starches. Their use include: (1) As a size to strengthen warp yarns, and to improve resistance to abrasion during weaving. The use of *cationic starches*, which in aqueous solution form positively charged surface-active ions, are becoming increasingly popular for sizing; (2) To change the appearance of the fabric after bleaching, dyeing, or printing; (3) To confine dyes and other chemicals within given areas during textile printing by preventing their diffusion into surrounding areas; and (4) As a component in finishes to give glaze and polish to sewing threads.

In the paper and board industry *starch phosphate* may be used as a binder and emulsifier, and *starch sulphate* as a hydrophilic sol. Their use helps to compensate for the loss of the natural binding ability in recycled fibres. The chemical similarity of starch and cellulose polymers and their compatibility leads to their use for (1) *Beater sizing* using *ionogenic starch*, i.e. forming ions, thus saving milling costs, improving starch retention, and improving optimal strength values. While 1-3 kg starch dry matter 100 kg^{-1} fibre is normally used and although additional starch would further improve paper quality, any additional starch would also increase pollution problems. Whereas cationic starch or starch and xanthates would both improve starch retention and allow further additions, thereby improving the ultimate strength of the fibre; (2) *Surface sizing* using modified starch products of low viscosity, such as oxidised starch, starch esters or slightly hydrolysed starch, with protein wax emulsions, carbonomethyl cellulose, etc. will improve the surface for printing as well as improving burst factors, folding number and breaking length; and (3) *Paper coating* for smoothing out any unevenness in the paper surface by using pigmented suspensions containing special starches as binding agents, or more often, to coat the printing paper with organic pigments, such as white clay fixed in organic adhesives such as starch, proteins and synthetics.

Starch is usually used as an adhesive in the manufacture of corrugated paper, paper bags, laminated paperboards, wallpaper, etc. The increased addition of starch facilitates the production of a thinner paper of satisfactory strength, thereby saving in cellulose fibres. Excessive starch, however, will make the paper transparent and brittle. The new starch derivatives now available can also improve the strength of short fibre pulp, e.g. straw pulp. Potato starch is preferred for some application, although barley starch, which has similar viscosity properties, can also be used but has the disadvantage of being greyer. The use of starch as a substitute for methanal (formaldehyde) and phenol in phenol-methanal polymers, reduces the emission of methanal from phenolic glued wood boards, which is of significant benefit when such boards are used indoors.

In the rubber industry a cross-linked *starch xanthate* is used to replace carbon black derived from petroleum. There is also a promising use of starch xanthates in encapsulating pesticides for their safe handling, the pesticides later being released from the enveloping capsule by wetting. In the future it is possible that up to 25% of the structural chemicals in paints could be replaced with purified vegetable starch. The standard household paints use petroleum derivatives such as acrylic and vinyl polymers that harden on exposure to air by linking together into long chains. Amending the catalysts used to speed up the hardening process has enabled the petroleum derivatives to be replaced by starch. At present such paints have the disadvantage of being less watertight due to some of the starch breaking down into their component sugars (Rexen and Munck, 1984; Sharp, 1990; Grubben *et al.*, 1996; Knight, 1997).

Polyhydryl compounds known as *polyols* are an important group of chemicals used for the production of polyesters, polyurethanes, surfactants, emulsifiers, etc. The polyols are largely derived from the petrochemicals ethene (ethylene), propylene, etc. However, starch hydrolysates may also be used for producing a number of compounds, including: (1) *Arabitol*, a sweet tasting penthydric alcohol present in lichens; (2) *Erythritol*, which also occurs naturally in lichens and some algae, and has twice the sweetness of sugar; (3) *1,2-dihydroxyethane (ethylene glycol)*, used in anti-freeze and coolants, also in the manufacture of polyester fibres such as Terylene, and of various esters used as plasticisers; (4) *1,2-dihydroxypropane (propylene glycol)* used as an anti-freeze agent, in the manufacture of perfumes and flavouring extracts, as a solvent, and to inhibit mould growth; (5) *Glycerol (glycerin)*, used in the manufacture of synthetic resins, esters, gums, explosives, and cellulose films, as a moistening agent for tobacco, etc.; and (6) *Sorbitol*, which is naturally present in fruits of the Rosaceae, e.g. *Sorbus aucuparia* (rowan) and is used in the production of vitamin C. It is also extensively used for making surfactants and emulsifiers, especially for the food industry, as well as in the chemical, pharmaceutical, cosmetic, plastic, paper and textile industries.

Depending on oil prices, starch polymers can be a cheaper substitute raw material than the petroleum derived, synthetic polymers polypropylene and polyethylene. They can be used as a *filler* for polyvinyl acetate (PVA), polyvinyl chloride (PVC), and polyethylenes, thereby reducing the cost of raw materials and making the product more biodegradable, e.g. PVC will degrade in 30-120 days. Starch-PVA plastics are water-soluble and are used in the USA for hospital laundry bags that dissolve in the washing machine. Like starch xanthates, there is a similar potential for packaging agrochemicals to improve safety during handling. Starch polymers are also used in PVC for vinyl-coated paper, such as washable wallpaper. Also for low density polyethylene (LDPE) bags, e.g. carrier bags, envelopes, air-sickness bags, which contain 6% maize starch treated with silicone to create an oleophilic surface, thereby providing an attractive, satin-like appearance with improved machine-printing qualities. Such starch polymer films can also be used as an agricultural mulch to control soil moisture and temperature, reduce nutrient leaching, prevent weed growth, and thereby increase crop yields by 50-350%.

An alkali-finished starch produces strong bonds with polyester castings and laminating resins. The presence of starch gives fire-retarding properties and considerably reduces the inflammability and smoke generation of the composites. There are also wide uses for polyester/glass fibre laminates in the vehicle and construction industries. For example, in the vehicle industry polyurethane foam accounts for *ca.* 70% of the organic polymers used. For use as a filler the starch is mixed with the polyol component to provide improved smoke reducing properties as well as reducing the coefficient of thermal expansion.

Other uses of starch and starch-derived products are as active components of synthetic polymers, including starch graft co-polymers in the production of natural film, moulded products, etc. The graft polymerising of acrylonitrile into gelatinised starch and then subjecting the co-polymers to alkaline saponification results in a saponified starch-polyacrylonitrile. Such graft polymers act as suprabsorbents with remarkable high water absorption properties. The polymer surface acts as a semipermeable membrane, thus providing a capability of cyclic absorption and desorption over long periods of time. Solid polymers, known as *super-slurpers*, are capable of absorbing many hundreds of their own weight of water without being dissolved. They are used in horticulture and forestry, especially in the arid and semi-arid regions, as soil conditioners to increase the water absorbing properties of porous soils and decrease evaporation losses, thus making the soil moisture more readily utilisable for plant growth. They are similarly used in hanging baskets and potted house plants to reduce the need for watering; the polymers can also be used for coating seeds, cuttings and roots to aid establishment. Super-slurpers may also be used medicinally as an absorbent for disposable bandages, bed pads, baby powder, nappies, sanitary towels, etc. (Johnson, 1984; Rexen and Munck, 1984; Callaghan *et al.*, 1988; Aronson *et al.*, 1990; Sharp, 1990).

9.2.1 Starch Fermentation Products

Starch fermentation is the source of the extremely important intermediary product *glucose* which is, in turn. used for conversion to various cyclic and acyclic polymers, aldehydes, ketones, acids, esters and ethers. Starch fermentation is also used in the synthesis of starch glycosides with *1,2-dihydroxyethane* (*ethylene glycol*) or *glycerol* (*1,2,3-trihydroxypropane*); the resultant *glycol glycoside* being used in the production of rigid urethane foams, biodegradable surfactants and alkyds. Biopols using microbial polysaccharides produced from starch and starch hydrolytes are also finding increasing application. Molasses may also be used as the raw material for fermentation and while it has the great advantage of being cheaper than other raw materials, it has the distinct disadvantage of containing secondary carbohydrates, nitrogenous compounds and salts. The process consequently requires more rectification and purification and consequently produces more pollution (Rexen and Munck, 1984). See also Chapter 18 for further information on fungal and bacterial fermentation.

The starch fermentation products include (1) *Ethanol*, used as a solvent and feed for other chemicals, principally ethanal, glycol ethers and amines, acrylic and ethanoic esters (acetic esters); (2) *Glycerol*, which is mainly produced from propylene derived from petrol and natural fats. although there is the alternative possibility of carbohydrate fermentation, the process is complicated by the production of ethanol and ethanal as by-products. Consequently, it is more economical to produce glycerol by chemical synthesis than by fermentation. Glycerol is used in the manufacture of synthetic resins and ester gums, explosives and cellulose films, and as a moistener agent for tobacco; (3) *Fumaric acid*, produced by the fermentation of *malic acid* and *maleic anhydride*, is used as a food acid; (4) *Itaconic acid*, which is produced by the fermentation of sugars, is used as a commoner in plastics, and its esters polymerised to lubricating oils and plasticisers; (5) *Industrial ethanoic* (*acetic*) *acid* is an important feedstock for the chemical industry. It is now entirely produced by chemical processes. However, food grade aqueous ethanoic acid, i.e. *vinegar*, is exclusively manufactured by the oxidation of ethanol with *Acetobacter aceti*; (6) *Propanoic acid* (*propionic acid*) occurs as a product of wood distillation and as a major end product of glucose fermentation by *Propionibacterium* spp., with ethanoic acid and CO_2 as by-products. The bioconversion of propanoic acid to the highly volatile chemical *propenoic acid* (*acrylic acid*) is by the action of *Clostridium propionicum*. The propenoic acid can then be polymerised to important polymers used as thickeners in textile treatment, as drilling mud additives, flocculating agents, in paper making and, if co-polymerised with, for example, divinylebenzene, as ion-exchange resins; (6) *Citric acid*, obtained commercially by the fermentation of glucose with *Aspergillus niger*, is used in the soft drinks and food industry; and (7) *Lactic acid*, which is potentially a key biomass intermediary and is produced synthetically from ethanol. It can also be produced from the fermentation of hexoses such as starch hydrolysates, although the process is not yet commercially viable. Lactic acid forms *lactide* through internal esterification, which are able to form polymers with many hydroxyacids, and are then used in the manufacture of transparent films and strong, biodegradable fibres; it is also particularly effective in preventing putrefaction (Montagné, 1977; Rexen and Munck, 1984; Sharp, 1990; Postgate, 1992).

9.3   Cellulose and Cellulose Products

Cellulose is the major constituent of the cell walls of all plants as well as being the most abundant organic matter found in nature. It is a glucose polymer, with a chain of over 3500 repeat units. It is completely hydrolysed by strong acids to *glucose*, while mild hydrolysis produces *hydrocellulose* with shorter chains, lower viscosity and lower tensile strength. Wood pulp and cotton linters are the most important commercial sources of cellulose, with its largest use in the rayon industry. The *hemicelluloses* are rather ill-defined and ill-differentiated from cellulose and represent the more easily hydrolysed portion of cellulose. *Pentosans* are those hemicelluloses, such as *araban* and x*ylan*, that yield pentoses on hydrolysis (Tootill, 1984; Sharp, 1990).

There is an increasing interest in Europe in the use of cereal straw as a source of cellulose, especially now that such countries as the UK have banned the burning of straw after harvesting. Cereal straw contains *ca.* 70% carbohydrates, principally as cellulose and pentosans, and can be hydrolysed to yield low molecular weight sugars, mostly as *glucose* and *xylose*, for use as a raw material for the food, feed and chemical industries. Unlike starch, cellulose is highly resistant to hydrolysis due to its highly ordered crystalline structure and the physical barrier imposed by the lignin enveloping the cellulose fibres. There is an urgent need to develop an economically feasible hydrolysis process, such as that which already exists for wood cellulose. See also Chapter 14 regarding the use of straw pulp for the paper industry.

There are three basic processes that can be used to obtain glucose from cellulose: (1) *Dilute acid hydrolysis*, yielding only *ca.* 50% of the potential glucose as well as seriously degrading the lignin; (2) *Concentrated acid* using H_2SO_4 or HCl and giving a 85-90% conversion. The process has the serious disadvantages of high material costs and in the lignin molecules being seriously affected. Alternatively, HF offers prospects of minimal feed-stock pre-treatment, with more than 90% conversion and undamaged lignin. This has the disadvantage that HF is expensive and highly toxic; and (3) *Enzyme treatment*, yielding pure glucose and undamaged lignin, but with only 50-60% conversion and with the feed-stock usually requiring expensive pre-treatment. Alternatively, treating the straw with phenol as a delignification solvent yields a relatively pure cellulose and, consequently, a pure glucose syrup with the lignin and semicellulose components undamaged and usable for the production of *vanillin*, *xylitol* and *furfural*. Vanillin is an extremely important flavouring and perfumery material, with large quantities being used in the food, pharmaceutical, and toiletry industries (Rexen and Munck, 1984; Sharp, 1990).

The cellulose ester *cellulose ethanoate* (*cellulose acetate*) formed by treating cotton and purified wood pulp with ethanoic anhydride, is used to make wrapping and photographic film, and in the production of a wide range of injection-moulded, sheet fabricated and extruded plastics, including the cellulose ethanoate fibre *ethanoate* (*acetate*) *rayon*, which is superior to *viscose rayon* since it absorbs much less moisture, it is stronger when moist and also less susceptible to wrinkling. *Cellulose nitrate* (*nitro-cellulose*) is prepared by treating cotton or wood pulp with HNO_3 plus H_2SO_4 or H_3PO_4. When treated with camphor, cellulose nitrate provided the first successful synthetic plastic 'Cellulose' and is still widely used for moulded articles, e.g. piano keys and table tennis balls, and as a surface coating. Highly nitrated cellulose containing more than 13% N is used in the manufacture of guncotton and cordite. Both cellulose ethanoate and cellulose nitrate can also be used in solution as adhesives and lacquers. *Cellulose esters*, formed by alkyd- and aryl-halides on cellulose in alkali solution, are used as thickening and emulsifying agents in foodstuffs and paints, and as sizes and adhesives in the paper and textile industries. The most widely used ester being *carboxymethyl cellulose* as a stabiliser in drilling muds, as an adhesive and pigment binder, as a strengthener in unfired ceramics, and in laundering to aid dirt removal (Bramwell, 1982; Sharp, 1990).

10.   ALKALOIDS

The term *alkaloid* encompasses an extremely diverse group of chemical structures for which it is difficult to provide a succinct definition. Molyneux and Ralphs (1992) cite that of Pelletier (1983): *'An alkaloid is a cyclic-organic compound containing nitrogen in a negative oxidation state, which is of limited distribution among living organisms.'* All alkaloids are basic and combine with acids to form crystalline salts that are usually soluble in water and usually less soluble in organic solvents, e.g. alcohol, chloroform, and ether.

Alkaloids occur widely throughout the plant kingdom, especially among the angiosperms. Alkaloids have been isolated from the roots, seeds, leaves, or bark of some members of at least 40% of the plant families, with the Amyrilidaceae, Buxaceae, Compositae, Euphorbiaceae, Leguminosae, Liliaceae, Papaveraceae, Ranunculaceae and Solanaceae being particularly rich sources.

TABLE 23. Commercial uses of some alkaloids (Petterson *et al.*, 1991, reproduced by kind permission of the Royal Society of Chemistry; [1]fide Skorupa and Assis, 1998)

Plant source	Alkaloid	Use
Apocynaceae		
Catharanthus roseus	vincristine	leukaemia treatment
(Madagascar periwinkle)	vinblastine	
Colchicaceae		
Colchicum autumnale (autumn crocus)	colchicine	agricultural and medical genetics
Erythroxylaceae		
Erythroxylum coca (coca)	cocaine	local anaesthetic
Leguminosae subfam. Papilionoideae		
Lupinus luteus (yellow lupin)	sparteine	oxytocic agent in obstetrics
Loganiaceae		
Strychnos nux-vomica (nux-vomica)	strychnine	rodenticide
Papaveraceae		
Papaver somniferum (opium poppy)	codeine	analgesic, anti-tussive
	morphine	analgesic
Rubiaceae		
Psychotria ipecacuanha, syn. *Cephaelis*	emetine	amoebiasis emetic
ipecacuanha (Brazilian ipecacuanha)	cephaeline[1]	emetic
	psychotrine[1]	emetic
Cinchona officinalis (cv = *C. ledgeriana*)	quinine	anti-malarial
(quinine tree)	quinidine	cardiac stimulant
Solanaceae		
Atropa belladonna (deadly nightshade)	atropine	anti-cholinergic
Lycopersicon esculentum (tomato)	tomatine	fungicide
Nicotiana tabacum (tobacco)	nicotine	insecticide

Many alkaloids are toxic, and even minute quantities produce characteristic physiological effects (see Chapters 8 and 17), a number of which are important sources of medicinal drugs (Table 23). Alkaloids have also been found among some members of most groups of the lower plants apart from the algae. Their precise function is not fully understood. Some may protect plants from animal predation. Others are believed to be involved in nitrogen metabolism, while others may be the stored end products of nitrogen metabolism (Tootill, 1984; Sharp, 1990; Duffus and Duffus, 1991).

11.   MISCELLANEOUS PRODUCTS

A high grade *silica* for the glass and ceramics industries can be obtained from the ash of rice hulls. The ash is also used in a cement that is more acid-resistant than Portland cement. When the hulls are heated to 700°C an amorphous silica is obtained that is suitable for reduction to solar grade silicon for solar cells (Grubben *et al.*, 1996).

Chapter 16

Human and Veterinary Medicinal Plants

Medicinal plants are defined as those used for human and veterinary application in traditional medicines, galenicals and herbal tisanes, phyto-pharmaceuticals, new drugs, intermediates for drug manufacture, industrial and pharmaceutical auxiliary products, and for health foods. The efficacy of many plants currently used in traditional herbal medicine are often lacking in reliable clinical evidence. Other plants formerly regarded as official, i.e. recognised and listed in national pharmacopoeia, have either been superseded by other products or, in the light of research, found wanting and discarded. Finally, there are those plants or their chemical analogues that are currently under investigation to provide new medically proven pharmaceutical products.

The following terms refer to some of the uses and therapeutic applications of medicinal plants: (1) *Pharmacy* (from the Ancient Egyptian *pharmaki* and Ancient Greek *pharmagia*) is the term applied to the art of preparing and compounding medicines, or to the place where medicines are dispensed; (2) *Pharmacology* is the science of drugs, including their composition, uses and effects; (3) *Pharmacognosy* is that branch of pharmacology dealing with crude, natural drugs, including medicinal plants; (4) *Aromatherapy* is a form of therapy in which body disorders are treated by massage with aromatic oils which, apart from their perfume, have strong anti-bacterial properties, often with antispasmodic or spasmolytic, stimulatory, cicatrizant, antifermentative and hormonal properties; (5) *Homeopathy* is a system of medical treatment based on the use of minute quantities of remedies that in large quantities produce effects similar to the disease being treated; and (6) *Naturopathy* is a system of therapy that relies exclusively on natural remedies, such as sunlight, clean fresh air, organically-grown foods and massage (see Chiej, 1984 for further discussion).

The histories of the development of both human and veterinary herbal medicines is dealt with in some detail. It is by understanding the past history that the present can be better understood, even though the therapeutic value of many early remedies have still to be thoroughly investigated and their efficacy evaluated. It is, for example,

317

sobering to realise that if scientists had appreciated the practices outlined in the scrolls of the Pharaohs or the use of mouldy bread in medieval eastern Europe, the world would not have had to wait until 1928 for Alexander Flemming's chance discovery of *penicillin*. Also, it is only in recent years that the Western World has begun to appreciated the therapeutic value of some of the more than 2,000 traditional plant remedies being used in China. Today over 125 pharmaceutical products in current use in the West are plant-derived, of which *ca.* 75% were discovered by investigating traditional medicines (Böttcher, 1959; Carlson *et al.*, 1977a; Mabey, 1988).

The traditional Western approach to finding novel pharmaceutical compounds has been largely by the high volume, random screening of plants. A more successful collaborative approach has been pioneered by Sharman Pharmaceutical Inc., a leading US pharmaceutical company. It involves examining the medicinal lore of indigenous healers by Western-trained physicians to evaluate the clinical diagnosis; they are accompanied by ethnobotanists to identify the plants. They found the quality of the research to be greatly enhanced, especially if the field researchers possess interdisciplinary training. Using such an approach to isolate pure antiviral lead compounds and depending on the virus, the results were found to be 125 to 630 times more efficient than random *in vitro* assays. Similarly, by interpreting the diagnoses in the medical literature of ancient Egypt, India and China, they were able to identify descriptions of diabetes and its treatment dating back for at least 2000 years. Some 800 plant-derived active principles with antidiabetic activity have now been discovered. Some, although useful, are not safe to use. To date the biguanidine *metformin* is the only approved ethical drug derived from a medicinal plant, i.e. *Galega officinalis* (goat's rue), that was historically used in the treatment of non-insulin-dependent diabetes. Derived from a prototypic molecule in a plant with a long pedigree in the treatment of diabetes, metformin well illustrates the development of an efficacious drug based on traditional plant use (Carlson *et al.*, 1997a; Oubré *et al.*, 1997)

1.   HISTORICAL BACKGROUND TO HUMAN MEDICINE

Archaeological evidence for the actual use of plants for healing purposes before the advent of writing must be largely conjectural. The earliest written evidence of medicinal plants is believed to be the Samerian cuneiform writings on clay tablets from Mesopotamia and dated *ca.* 3400 BC. However, the evidence of successful signs of healing following trepan surgery carried out during the Neolithic and dated between 5100 and 4900 BC suggests an earlier knowledge of anaesthetics (or intoxicants?) and antibiotics (Schultes, 1960; Duin and Sutcliffe, 1992; Rudgley, 1998).

From Ancient Egypt there are, among others, the three medical papyri of Georg Ebers and P. Hearst dated *ca.* 1550 BC and P. Edwin Smith of *ca.* 1500 BC. The Ebers Papyrus alone provides 842 prescriptions containing 328 different ingredients that are not obviously founded on sorcery. These papyri are of particular interest, especially in the light of the comment by Manniche (1989) that no complete ancient Egyptian herbal (as distinct from medical lore) exists today, although some fragments dating from the 2nd century AD still survive. It was such papyri that showed the Early Egyptians' appreciation of the mould *Penicillium*. The stature of this ancient Egyptian medical lore is evident from the fact that it was the acclaimed medical skills of Imhotep (died 2648 BC), vizier to the Pharaoh Zoser, that in 535 BC led to Imhotep being granted the status of a god. He was later adopted by the Greeks as the god Asklepios and identified with the Roman god of medicine Aesculapius (Böttcher, 1959; Duin and Sutcliffe, 1992; Batanouny, 1999).

The 13th century BC Mycenaean Linear B tablets, a form of early Greek, recorded the use of such spices as *Crocus sativus* (saffron), *Cuminum cyminum* (cumin), etc. and provide a link with the plant lore of classical Greek. It was from this knowledge of Western Asia and the pre-Hellenic civilisations that the herbals of ancient Greece evolved. Thus, in *ca.* 750 BC, Homer is reported as being the first to distinguish between good and bad drugs, i.e. poisons. In his *De historia plantarum* (Enquiry into Plants), Theophrastus (*ca.* 370-287 BC) organised the current botanical lore according to habit, recognising some 480 plants, including *Conium maculatum*, the poisonous hemlock that was reputed responsible for killing Socrates in 399 BC. Some 700 species and slightly more than 1000 drugs were recognised by the Greek hebalist Pedanus Dioscorides in his *De materia medica* of 77 AD, who classified the plants as to whether they were pot roots, pot herbs, fruits, trees and shrubs, and arranged the drugs wherein according to their physiological reaction in the body. Riddle (1985) examined the herbal lore of Dioscorides in relation to our present-day pharmacognostic knowledge and noted that a number were of medicinal value. The historic importance of Dioscorides work cannot be ignored because, until the 16th century, it was the most widely used, copied, extended and translated work on medicinal plants. Not only was his work translated and absorbed into Arabic traditional medicine, later Latin renditions from both Arabic and Greek sources strongly influenced European Renaissance medical traditions and thereby set the scene for modern pharmacognosy and, less obviously, botany. Although a gifted physician, the often cited contribution of Galen of Pergamum (129-post 210 AD) to pharmacognosy and dietics was largely confined to earlier works with some personal contributions. During the 5th century AD missionary expeditions of Buddhist monks from China to India and of Indian monks to China, together with commercial relations with Arabia, all contributed to a wider distribution of medical and herbal lore throughout the known world (Davis and Heywood, 1963; Stearn, 1976; Riddle, 1985, 1996; Hoizey and Hoizey, 1993; Scarborough, 1996).

The translation of herbal lore into Anglo-Saxon during the 10th to 12th centuries from 4th and 5th century Latin compilations can be traced back through Gaius Plinius Secondus (23-79 AD) and his encyclopaedic *Naturalis Historia*, especially books xx-xxxii, to the 4th century BC Greek rhizotomist Diokles of Karystos, regarded as the first author of a systematic herbal. Following the Norman Conquest the more erudite Anglo-Saxon herbals were ignored and replaced by Latin translations of earlier works. Indeed, for much of the Middle Ages (*ca.* 1000-1400 AD), European scholars contributed little beyond copying or attempting to decode the herbals of the Classical Period and in the process perpetuated a web of errors, misinterpretations and superstitions. From the Middle Ages up to the last century little encouragement was given to observation or research and any outspoken criticism was regarded as heresy and exposed the author to the wrath of the church.

During the early 11th century it was the noted herbalist of the Arab world, the Persian Avicenna (Ibn Sina) and his voluminous *Canon of Medicine* that was to became the authoritative work during the Middle Ages throughout both the Moslem and Christian worlds. The only original English work during the Middle Ages was written in the mid-13th century by the cleric Bartholomus Anglicus. His 19 books of *De Proprietatibus Verum* ran to at least 14 editions before 1500 AD, and were translated into English, French, Spanish and Dutch. The first printed English herbal was the small, anonymous quarto volume printed by Richard Banckes in 1525 and known as *Banckes's Herbal,* also running into a number of editions (Rohde, 1922; Vriend, 1984; Blackwell, 1990; Mabey, 1991).

The evolution of the printed herbals in Europe during the late 15th to 17th centuries is discussed by Arber (1912), who defined a herbal "*as a book containing the names and descriptions of herbs, or plants in general, with their properties and virtues. The word is believed to have been derived from a mediaeval Latin adjective 'herbalis', the subjunctive 'liber' being understood. It is thus exactly comparable in origin with the word 'manual' in the sense of a handbook.*"

The most important English herbal is John Gerard's *Historie of Plants*, first published in 1597 but seldom quoted, being replaced by Thomas Johnson's enlarged and amended edition of 1633, of which a second and much superior edition was published in 1636. Better known as Gerard's *Herbal*, it was this latter version that long continued to be the standard work for English students. Other works of this period include John Parkinson's *Theatrum Botanicum* of 1640 and a rather fanciful work by the apothecary and astrologer Nicholas Culpeper published in 1652, which he later repudiated in subsequent editions as incorrect and unauthorised. Some doubt must be assumed concerning Culpeper's credibility as an author for he had earlier incurred the wrath of the College of Physicians with the publication in 1649 of his *Physical Directory* in which he linked herbal lore with astrology, and was an unauthorised translation of their *London Dispensary.*

It was a new breed of commercial herbalists during the 17th and 18th centuries who, willy-nilly, attributed a plant's physical manifestation with its medicinal qualities, in what was to become known as the *'Doctrine of Signatures'*. The eventual division between herbal and orthodox medicine began in 1785, when the English physician William Withering discovered that the unpredictable and often fatal effects of foxglove leaves used in the treatment for dropsy was due the heart stimulating the kidneys to clear the body of fluids, and that small and accurately measured doses of digitalis from the leaf was an invaluable treatment for heart failure (Rohde, 1922; Hudson, 1954; Chiej, 1984; Woodward, 1985; Mabey, 1991; Cotton 1996).

The history of botanical literature includes numerous references and, perforce, medicinal plants in China is eruditely discussed by Needham (1986) while the history of Chinese medicine is dealt with by Hoizey and Hoizey (1993). The former includes what he describes as the pandects of pharmaceutical natural history. The earliest such work devoted to the study of natural history and of plants in particular are the no longer extant *Tzu-I Pên Tshao Ching* (The Classical Pharmacopoeia of Tzu-I) written in the 6th century BC and *Shen'nong Bencaojing* (Shen'nong's Classic of Herbal Medicine) in the 1st century BC. These works, unlike the lapidaries, herbals and bestiaries of their European counterparts, were devoid of magical and fanciful ideas. Indeed, it is from a study of the latter work that led to the use of alkaloid extracts from the fruits of *Camptotheca acuminata* in the treatment of liver cancer. The *Pên Tshao* (natural history) works date from the 5th century BC, the most important being the *Pên Tshao Kang Mu* (Great Pharmacopoeia) of *ca*. 1547 AD by the most renowned of Chinese naturalists Li Shih-Chen (see Read and Yü-Thien, 1931) and culminating in the bibliographic survey of the *Pên Tshao* by Lung Po-Chien (1957).

The early medical publications were presented either as encyclopaedias or as dictionaries, of which the earliest encyclopaedia was the *Ehr Ya* (Literary Expositior) dating mainly from somewhere between the 4th and 2nd centuries BC, but with some contributions possibly from the 6th century and some additions from the 1st century BC. Numerous publications based on the *Ehr Ya* then followed. An important and neglected early work on pharmaceutical plants was the *Chi Chiu (Phien)* (Handy Primer) compiled by Shih Yu between 48-33 BC. The oldest medicinal dictionary was the *Shuo Wên* of 121 AD.

According to a nation-wide survey, the China National Corporation of Traditional and Herbal Medicine (1994-1995), there are 11,118 plant taxa being used in Chinese medicine today, of which 10,027 taxa represent a third of the angiosperm flora (Xia and Peng, 1998). Despite this enormous wealth of pharmacognostic knowledge the important contribution made by the Chinese to medicine has only recently begun to be properly appreciated in the West, the language barrier being a particular handicap to overcome. A good introduction to Chinese herbal medicine is to be found in Duke and Ayensu (1985), who discuss some 1200 species.

In India there are three schools of medicine: (1) The *Ayurveda* (i.e. science of life) school of traditional medicine as passed down by the Lord Brahma, dates back to the Vedic period, *ca*. 1500-600 BC. It is based on the premise that the five basic elements of earth, water, fire, air and ether constitute the body organs, mind and soul. It was largely an oral tradition with a strong religious orientation. There are few written records and these are virtually impossible to date with an even remote degree of accuracy. Four Sanskrit books, the sacred *Vedas* (Sanskrit *verda* = knowledge), are the sole survivors. The Ayurvedic medicinal lore is based on two works supplementing the sacred *Vedas*, the *Charaka Samhita* of 600 BC (Handa (1998) suggests 3000 BC). These deal with medical and pharmacognostic matters, and the *Sushruta Samhita*, concerned with surgery; the first two works are mentioned in the early Vedic scriptures, such as the *Atharvaveda*, *Rugveda* and *Yajurveda*; (2) The *Siddha* school of medicine is practised in the Tamil Nadu region of southern India and is variously considered as being either earlier or contemporary with Ayurveda; and (3) The *Unani-Tibb* school is based on the teachings of the Grecian physicians Hippocrates (*ca.* 460-370 BC) and Galen (*ca.* 130-200 AD) - '*Unani*' is the Arabic corruption of '*Ionian*'. Their teachings were further developed by the Arabs who, by the early Middle Ages, had become a world centre for medical and veterinary knowledge. For example, by the 12th century 12 general treatises on camel diseases had already been written. It was the Arabs who were responsible for bringing their medicinal lore to both India and Africa with the introduction of Islam (Duin and Sutcliffe, 1992; Anjaria, 1996; Schillhorn van Veen, 1996).

The early New World ethnographic records are largely associated with the Maya, Aztec, Inca and other advanced civilisations of Mesoamerica, and are relatively recent. The Mayan pre-Hispanic codices or screenfold books, for example, date from *ca.* 1250-1521 AD. The Aztec cultures are largely recorded by their Spanish conquerors, noted chroniclers being the Franciscan priest Fray Bernardino de Sahagún with his comprehensive *Historia General de las Cosas de Nueva España*, and the physician to the King of Spain, Dr Francisco Hernandez who, between 1570-1575, published *his Nova plantarum, animalium et mineralium mexicanorum historia*. The earliest New World herbal to be translated from Latin was that of the Seville physician Nicholas Monardes (1568), who wrote *Joyfull Newes out of the New-Found Worlde* (Rohde, 1922; Kreig, 1965; Cotton, 1996).

Information on medicinal plants can now be found in documents and databases from a wide range of disciplines, much of which is now being made available in electronic form. In a recent survey of primary publications by Bhat (1997) *ca.* 25% of the total volume of literature currently being generated on medically related subjects appeared in less than 10 periodicals. Approximately 50% of the total volume were spread over some 50 titles and the remaining 50% were scattered through 2500 periodicals belonging to a wide range of disciplines which, in addition to purely medical and related subjects, included botany, anthropology, agriculture, horticulture, chemistry and phytochemistry, aromatic plants, etc. Details of useful databases are also provided by Bhat (1997). Examples of taxonomic publications providing useful information on medicinal plants are to be found in Heywood (1971, 1977) for the

Umbelliferae and Compositae respectively, also regional surveys, e.g. Oliver (1960) for Nigeria and Oliver-Bever (1986) for West Africa.

2.   MEDICINAL PLANTS FOR HUMANS

Of the 250,000 flowering plants in the world only *ca.* 5-10% have been studied either chemically or for their physiological activity. Mendelsohn and Balick (1995, 1997) have estimated that the flora of the world's forests, which represent approximately one half of the world's flowering plants, contain *ca.* 355 drugs, of which a mere 12.5% have so far been discovered.

For centuries there were only three major drugs, opium, digitalis and quinine, with alcohol sometimes considered as a fourth. Quinine, from the bark of *Cinchona officinalis*, is reputed to have been used in 1638 by the Countess of Chinchon (hence the generic name), second wife of the Viceroy of Peru, the bark appearing in Europe in *ca.* 1645 and cited in the *London Pharmacopaeia* of 1677.

A century ago modern pharmacology began to emerge from the accumulation of unusual concoctions which, in addition to fresh and dried plant material, often contained strange and rather revolting ingredients, such as sundry dead creatures and their organs, excrement, ground medicinal earths such as terra lemnia, etc. The new pharmaceuticals were very much influenced by developments in chemistry, and while synthetic drugs may now appear to dominate the pharmaceutical prescriptions of the developed countries in recent years, the percentage of prescriptions based primarily on natural plant products, including antibiotics, has remained fairly constant at *ca.* 25% over a number of years. Furthermore, according to Duke (1992) 80 to 90% of the world's population still rely mainly on traditional ***materia medica*** and their practitioners; similar figures apply to livestock and ethnoveterinary care, with many of the plants being used to treat both humans and livestock. Despite this, the developing countries still largely rely on traditional medicines, of which 85% contain materials of plant origin. Even a developed country such as Chile, with a flora of *ca.* 5215 species, over 570 (11%) are used in traditional medicines and the tally is by no means exhaustive.

In West Africa some 7349 species were recorded by Keay (1954, 1958) and Hepper (1963, 1972). Many of these plants are used in traditional medicine and the active phytochemical compounds for the majority of them were examined by Oliver-Bever (1986), with only 9.6% of unknown quality. Since most medicinal plants are reputed to cure several, often unrelated disorders, their efficacy in most cases must be considered doubtful. Burkill (1985, 1994), for example, recorded over 200 species being used against venereal diseases, yet West Africa certainly doesn't warrant a clean bill of health for those diseases. This does not imply that all herbal medicines should be disregarded, but until clinical trials can be carried out their efficacy remains unconfirmed (Kreig, 1965; WHO, 1976; Zin and Weiss, 1980; Marticorema and Quezada, 1985; Kochhar and Singh, 1989; Waterman, 1989; Lewington, 1990; Cox and Ballick, 1994; Ibrahim, 1996; McCorkle *et al.*, 1996; Sheldon *et al.*, 1997).

2.1 Wild and Cultivated Plant Sources of Pharmaceuticals

Some pharmaceuticals are obtained exclusively from wild sources, including such widely used drugs as *cascara sagrada* from the bark of the North American *Rhamnus purshiana*, *uva ursi* from the circumpolar *Arctostaphylos uva-ursi* (bearberry, uva-ursi) and used in the UK since the 13th century, *physostigmine* from *Physostigma venenosum*, and *gentian violet* from *Gentiana* spp. Others are obtained from both wild and cultivated sources, the better known include *ginseng* from *Panax ginseng* and *P. quinquefolius*, *senna* from *Senna alexandrina* (syn. *Cassia senna, C. acutifolia* and *Senna acutifolia*), *reserpine* from *Rauvolfia vomitoria, R. serpentina* and *R. tetraphylla*, and *nux vomica* from *Strychnos nux-vomica*.

Panax ginseng is highly regarded as a 'dose of immortality' in China and used as a panacea for all bodily ills and an aphrodisiac. Similarly, in North America the Cherokee Indians consider the related *P. quinquefolius* as the 'plant of life' with magical powers of healing and aphrodisiac properties. In western medicine ginseng is regarded more as a stimulant or tonic, although in formal Chinese and Russian medicine it is much more highly regarded (Martin, 1983; Prescott-Allen and Prescott-Allen, 1986; Blackwell, 1990; Rasoanaivo, 1990).

A considerable number of widely used pharmaceutical products are obtained from cultivated plants, including *morphine, codeine, thebaine, noscapine, narceine, papaverine*, etc. from *Papaver somniferum* subsp. *somniferum* (opium poppy), *bromelin* from *Ananas comosus*, syn. *A. sativus* (pineapple), *colchicine* from *Colchicum autumnale* (autumn crocus), *digitoxin* from *Digitalis purpurea* (foxglove), *digoxin* from *D. lanata* (Greek foxglove), *cocaine* from *Erythroxylum coca* (coca), and *leurocristine, vinblastine, vincristine, vincaleucoblastine*, etc. obtained from *Catharanthus roseus* (Madagascar periwinkle). The leaves of the latter were formerly harvested from the wild, they are now obtained from cultivated plants. This is hardly surprising when 2 tons of leaves are required to produce the 1 g of alkaloid needed to treat leukaemia for 6 weeks.

The milky juice from the unripe fruit of *Carica papaya* (pawpaw, papaya) is the source of the enzymes *papain* and *chymopapain* and the polysaccharide *pectin*. Papain is widely used as a meat tenderiser, the usual food grade papain being capable of dissolving about 35 times its own weight of lean meat; it is deactivated by cooking. Its action is similar to the ferments of the gastric and pancreatic juices, hence its use in invalid diets. Papain is also used medically to prevent adhesions, and in very low concentrations for digestion remedies; it can also be used in cleansing fluids for soft contact lenses. Chymopapain is used to dissolve discs in the treatment of prolapsed intervertebral discs. Pectin has been used in the place of blood as a transfusion fluid in cases of haemorrhage and shock (Prescott-Allen and Prescott-Allen, 1986; Robbins, 1995; Macpherson, 1996).

The lower plants must not be ignored. The dried sclerotia (ergot) of the fungus *Claviceps purpurea* (ergot), provides the phamaceutical alkaloids **ergometrinine**, **ergotoxine** and **ergotamine**. The antibiotic **penicillin** is obtained from the mould *Penicillium notatum*, **streptomycin** from the soil bacterium *Streptomyces griseus*, with **aureomycin** and other **tetracyclines** also from soil organisms. The lichen *Usnea barbata* is commercially utilised in Scandinavia as a source of the antibiotic **usnic acid**. Among the marine red algae *Digenia simplex* and *Chondria* spp. have also been successfully exploited in Asia for their anthelmintic properties, and extracts of *Constantinea simplex* used against the herpes simplex virus (Kreig, 1965; Brown, 1977; Launert, 1981; Fenical, 1983).

2.1.1   The Story of the Aspirin

The history of the common aspirin well illustrates the struggle to purify a somewhat hazardous herbal remedy and the apparently poor communications between scientists, while at the same time correcting some misconceptions regarding the origin of aspirin. The of the analgesic and antipyretic properties of the powdered bark of *Salix alba* (white willow) was known to Hippocrates in the 5th century BC and recorded by Dioscorides in his *De materia medica* of 77 AD. It was 'rediscovered' in 1758 by the Rev. Edward Stone and initially ignored. It was in the 1820s that the Swiss pharmacist Johann S.F.Pagenstecher first extracted **salicylic acid** (then known as spirsäure; 'spir' for *Spirea*, 'säure' from the German for acid) from *Filipendula ulmaria*, syn. *Spirea ulmaria* (meadowsweet), a drug with pain-killing properties but having unfortunate severe side-effects on the stomach lining. In 1895 the German chemist Felix Hoffman managed to eliminate these side-effects by converting the salicylic acid to **2-O-acetylsalicylic acid**, a substance that not only acted as an analgesic but also as an antipyretic and anti-inflammatory. Four years later Hoffman and his colleague Heinrich Dreser developed **acetylsalicyate**, better known as **aspirin** ('a' for '*acetyl*', 'spir' for '*Spirea*'), which was then patented by the drug company Bayer in the following year.

It was in 1826 that two Italians discovered that the active ingredient of the bark of *Salix alba* was **salicin** which a French chemist purified 3 years later. In 1839, another Italian chemist prepared salicylic acid from salicin. Thus the active ingredient in aspirin could from henceforth be obtained from both the willow and meadowsweet, although it is not clear whether Hoffman was aware of this alternative source. Today it is obtained synthetically by the action of ethanoic anhydride on salicylic acid; US production in 1981 amounted to 18 000 tonnes. Salicylic acid is also used in the manufacture of dyes, liniments and rust-resisting fluids.

Salicylic acid also occurs naturally as **methyl salicylate** (oil of wintergreen) in *Gaultheria procumbens* (checkerberry), from which it was originally obtained; it is now extracted from the distilled bark of *Betula lenta* (sweet birch) (Hill, 1952; Sharp, 1990; Duin and Sutcliffe, 1992; Mabberley, 1997).

2.2   Developing New Pharmaceuticals

Despite the enormous wealth of information available about reputed medicinal plants in the United States, very few new and marketable drugs have been discovered. After 25 years of testing some 40,000 plants, the National Cancer Institute have failed to identify a single agent of general use in the treatment of human cancer. The single outstanding exception was the more or less accidental discovery of the anti-cancer properties of *leurocristine* and *vincaleucoblastine* in *Catharanthus roseus*. It was an accidental discovery since the researchers of the Eli Lilly and Company were only examining the plant because of its reputed hypoglycaemic properties for the treatment of diabetes! A further screening of some 200 plants with interesting traditional usage yielded a further half dozen plants with these anticancer substances. Similarly, after over a decade of screening of alkaloid-containing plants, the Smith Kline and French Company failed to produce a single new product. This failure is considered to be due to an almost total lack of interaction between chemists, botanists, biologists and physicians. Neither should collaboration between classical scholars and scientists be neglected (Farnsworth, 1984; Tyler, 1986).

By way of contrast and arising from a combination of a national interest in herbal remedies and less rigorous regulations Germany has produced a number of new drugs in recent years, although should any evidence of toxicity be found, the preparation would be promptly withdrawn. *Symphytum officinale* (comfrey), for example, which is widely used in allopathic medicine, was found to contain carcinogenic pyrrolizidine alkaloids. Surprisingly, although no longer available in Germany, it is still available in the USA, despite that country's reputation for stricter drug laws, especially with regard to carcinogens. Interestingly, comfrey contains the glycoside *allantoin*, a useful plant substitute for the maggots of *Lucilia sericata* (sheep blowfly), which were formerly used in the healing of suppurating wounds (Launert, 1981; Tyler, 1986).

Placing a new drug on the market is an expensive process following the initial collection it will take 9-12 years of preclinical and clinical trials before acceptance, and at a cost of *ca.* $125 million per successful drug. Even before any investigations can begin, it is now essential to establish who can purchase and who can sell the genetic plant material. Initially drug companies will compete in identifying potential drugs, since the first to succeed will thereby obtain rights over other competitors. Yet by so doing, the allocation of exclusive rights to a drug company to screen a forest will limit the number of screens that the company can employ and consequently the number of active ingredients likely to be discovered. It therefore follows that open competition will encourage a wider screening and thus increase the number of active ingredients discovered (Baerheim Svendsen, 1984 cited by Tyler, 1986; Heywood, 1990).

Fourie *et al.* (1992) investigated a number of traditional medicinal plants in South Africa recorded on a pharmaceutical data base. They used a number of criteria for selecting potential plants for further investigation, including the convergence of a series of indications on their medical use, the coincidence of information from various sources, reports from southern Africa, and whether their chemistry and pharmacology has been extensively investigated. A much better chance of finding pharmaceutically active ingredients in a plant that has been widely used for treating ailments throughout all or much of its distribution range (endemics are the exception) can be generally be expected from such studies. However, it is not unknown for some traditional compound medicines to contain placebo plant materials which have been deliberately introduced in an attempt to disguise the source of the active ingredients. The presence of toxic secondary metabolites in herbal remedies is an additional hazard. For example, the genus *Heliotropium* has a number of species, including *H. bacciferum* and *H. ramosissimum*, that are still widely used as herbal remedies yet are known to possess carcinogenic properties.

2.2.1   Screening Processes

The screening of plants with known active ingredients can be expected to have a greater chance of yielding useful drugs than a blanket screening process. Although there are *ca.* 500 screening processes that can be used to test new drugs, a major drug company is only likely to use 50-75 screens (Macksad *et al.*, 1970; Sims, 1981; Fourie *et al.*, 1992).

Fourie *et al.* (1992) describe how, after a potential plant has been identified, collected, dried and milled, it is exhaustively extracted with dichloro-methane-methanol to ensure maximum removal of all the soluble compounds regardless of any conventions regarding the use of the plant in traditional medicine. The removal of unwanted tars and macromolecules prior to the initial pharmacological screening is achieved using an open gelica gel column, through which they are unable to pass. Any pharmaceutical activity identified in the crude extract is then likely to show enhanced activity on testing, while the fractions obtained after careful combinations are well suited to further testing. A number of basic pharmaceutical screening tests are available in the pharmacological programme. These are: (1) General toxicological and central nervous system effects (Irwin screen); (2) Analgesic, anti-inflammatory, anti-hypersensitive, anti-ulcer, narcotic analgesic, anti-metrazol (anti-convulsant), anti-depressant, anti-arrhythmic and diuretic tests; and (3) Anti-microbial/ anti-fungal activity tests. In addition to the above the structural determinations of purified compounds are carried out.

In the past the lack of simple bioassay procedures have inhibited the screening for physiological activities and in the identification of active constituents by fractionation. For bioassays exposing brine shrimp (*Artemia salina*) to varying concentrations of the test material to obtain an LD_{50} value (median lethal dose - the dose at which a toxic substance will kill 50% of the brine shrimps) is a convenient, rapid, reliable and inexpensive method. Preliminary development trials with the potato-disk assay for observing the inhibition of crown gall tumours induced by *Agrobacterium tumefaciens* appears promising for testing certain kinds of anti-tumour activity.

The identification and quantification of nanogram quantities of constituents in *ca.* 1 mg of a tissue sample is now possible using multistage or tandem mass spectrometry (ms/ms) without the need for prior extraction or purification. While the structure of complex plant constituents can now be determined using X-ray crystall-ography and the various variations of nuclear magnetic resonance spectroscopy.

Providing adequate supplies of raw material for drug production can be a problem Plant-cell-culture techniques can now be used to bulk up sparse raw material and produce large quantities of undifferentiated plant cells. However, the apparently homogeneous populations of cells are, in fact, quite heterogeneous regarding their ability to biosynthesise and accumulate desired secondary constituents. (Barz and Ellis, 1981 cited by Tyler, 1986). Selection procedures are necessary to obtain high-yielding subpopulations. At present the method is only economically feasible for costly and unique plant products such as **diosgenin** from *Dioscorea deltoidea*, **serpentine** from *Catharanthus roseus*, and **ubiquinone-10** from *Nicotiana tabacum* (Tyler, 1986).

Genetic engineered micro-organisms are now being successfully used for the commercial production of various enzymes, hormones, antibodies, vaccines, etc. Recently Dr Miuguel Goez of Cinvestav, Irapuato, Mexico developed the world's first vaccine-carrying banana for use against cholera, enabling the vaccine to be used in remote areas and thereby dramatically reducing the cost of traditional jabs and the need for refrigerated vaccine (BBC Tomorrow's World, 3 May 1966). Bananas, however, do have the disadvantage of a short shelf life. Hopefully research in Holland may be able to overcome this problem by substituting honey from GM vaccine-carrying crops. Another recent technique developed by Canadian researchers is to insert GAD protein from the pancreas into the DNA of the potato to provide protection against type 1 diabetes in mice since they have the same pancreas structure and immune response as humans. It is hoped the technique can be adapted for human use. There is also the possibility of producing other GM plants for treating a variety of auto-immune diseases such as multiple sclerosis and rheumatoid arthritis (Pincock, 1997).

2.3   Marketing Pharmaceuticals

The commercial marketing of approved plant pharmaceutical products, whether whole plant products such as senna pods, or their extracts, have a number of disadvantages. These are: (1) The enormous bulk of material that has to be processed in order to extract a particular drug. An extreme example being the 0.0003%

vincristine yielded by *Catharanthus roseus*; (2) Variation in bioactive phytochemical content due to the season, environment and the genetics of the individual; (3) Plant availability and the cost of harvesting; (4) Storage losses coupled with problems in marketing and their affect on growers/harvesters, middlemen and the pharmaceutical companies; (5) The commercial necessity of obtaining a steady supply of consistent quality; and (6) The high cost of producing and marketing the product to approved standards. When the cost of research and development by the top 20 drug companies in the world over the past 15 years is divided by the number of genuinely new compounds discovered and marketed, the cost of developing each new compound is between $750 million and $2000 million (Office of Technology Assessment, 1983; Horrobin and Lapinskas, 1998).

In the United States the development costs for a new drug product is almost prohibitively high, in the region of $50-100 million. It is lower in Germany where a doctrine of reasonable certainty based on the clinical experience of general practitioners supplemented by literature evidence and manufacturer's data, is substituted for strict clinical trials. These extreme examples among the developed countries illustrate the inhibiting influence of a lack of interest and strict federal regulations in the United States and the more realistic government regulations prevailing in Germany. Problems with the introduction of oraflex and thalidomide have demonstrated that neither regime is foolproof.

In order to overcome such difficulties many pharmaceutical drugs are now synthesised, the chemical analogues either resembling the natural active principle or are an improved variation thereon. They have the definite practical advantage of assured quality, thus enabling accurate dosages to be administered. For example, *Morphine* has provided the prototype of *hydromorphine*; *lysergic acid* from the hydrolysis of ergot alkaloids has been converted to *methysergide*; *cocaine* has produced *procaine*, while *physostigmine* has been metamorphosed to *neostigmine* and *salicin* improved to provide *acetylsalicylic acid* (Tyler, 1986).

3.   HISTORICAL BACKGROUND TO VETERINARY MEDICINE

Physicians for animals were officially recognised in China according to the *Zhou*li (Rites of Zhou) during the 11th century BC and their ancient tradition of herbal veterinary medicine continues to be practised to the present day. In Ancient Egypt the priests of the lion goddess Sekhmet practised both human and veterinary medicine, the latter with special regard to the sacrificial bulls. Indeed, the Kahun veterinary papyrus written in *ca*. 185 BC, and discovered by Flinders Petrie in 1889, is unequivocally a religious publication. Many of the beliefs and practices of Sekhmet's priestly healers even appear to have been handed down to the Dinka pastoralists now living in the Nile Valley of southern Sudan (Smith, 1976; Hoizey and Hoizey, 1993; Lin and Panzer 1994; Schwabe, 1996).

Veterinary medicine was certainly practised in Mesopotamia during the 2nd millennium BC and is cited in the Code of Hammurabi, who reigned 1792-1750 BC, and an honorific inscription for a *hippiatroi* (horse doctor) was noted *ca.* 130 BC. The earliest major Greek work on veterinary medicine is a collaborative compilation known as the *Corpus Hippiatricorum Graecorum*. It contains textual and nontextural excepts from lost writings on the treatment of horses and other equines compiled during Byzantine times by various unknown or little-known 4th century AD authors, and later revised for the emperor Constantine VII as *Porphyrogenitus* (905-959 AD).

In Rome an *equarius medicus* (horse doctor) was recognised by the end of the 1st century BC; certainly a *mulomedicus* (mule doctor) and a *medicus veterinarius*, *medicvus iumentarius* or *medicus pecuarius* (livestock doctor) are attested for the late-Roman empire. Although emphasis was placed on the care horses and other Equidae, other domestic animals were also included. This is evident from the three books of *De re rustica* by Marcus Terentius Varro, who lived from 116-27 BC, book 5 of Virgil's *Georgics* of *ca.* 29 BC and, in the 1st century AD, of Renatus Flavius Vegetius's *Digestorum artis mulomedicine libri*, and books 6-9 of Lucius Iunes Moderatus Columella's *De re rustica* of *ca.* 60-65 AD. Veterinary medicine was and remained throughout ancient antiquity primarily a matter of empirical practice.

There were prejudices against the action and status of ancient veterinarians, stemming from the lack of any theoretical or philosophical development of animal medicine and other factors, such as the treatment of animals being considered undignified, while some philosophers considered the subject trivial or even disagreeable. A similar view was expressed by Publius Vegetius Renatus (*ff.* 450-500 AD) in his *Artis veternariae, sive mulomedicinae libri quatuor*, first printed in Basle in 1528 (Smith, 1976; Bodson, 1996). In the UK a similar professional disdain for veterinary medicine prevailed. The only early English herbal to devote a chapter to herbs useful for animals was that of W. Coles, *The Art of Simpling* and published in 1656 (Rohde, 1922).

Early veterinarians, like surgeons, served an apprenticeship, unlike the physicians who, as early as the 10th and 11th centuries could receive a professional education at Salerno's flourishing medical school and during the 12th century at the universities of Bologna and Paris. It was not until 1762 that the first veterinary college was established, the École Nationale Vétérinaire at Lyons. In the UK the Veterinary College, Camden Town was founded in 1792, later to become the Royal Veterinary College. Such was the low status of veterinary practice in the UK that it was not until the passing of the Veterinary Surgeons Act of 1948 that veterinary practice by unqualified persons became illegal (Miller and West, 1956; Duin and Sutcliffe, 1992; Robertson, 1994).

In India the three major traditional schools of human medicine, Ayurveda, Siddha and Unani-Tibb, were also involved in veterinary medicine. Although the Vedic scriptures made reference to veterinary medicine, the *Shalihotra*, in the late Vedic period was the first Sanskrit book entirely devoted to the subject. The first Ayurveda veterinary clinics were established during the 3rd century BC. Among the Sanskrit scripts associated with the period are the *Aswa Chikitsa* and the *Gaja Chikitsa* on equine and bovine medicine, respectively (Anjaria, 1996).

4.   MEDICINAL PLANTS FOR ANIMALS

Ethnoveterinary medicine, or traditional veterinary remedies, is defined as dealing with the folk beliefs, knowledge, skills, methods and practices pertaining to the health care of animals and appears to have a practical beneficial use for livestock, especially in the developing countries. The advantages are: (1) Stockmen are familiar with their use; (2) A significant proportion of the remedies appear to work; (3) The remedies are freely available or at a reasonable cost; and (4) They are usually easily administered, either topically or orally. The disadvantages are (1) Because particular methods are often very localised, scope for further dissemination is limited; (2) Efficacy is variable, depending on the season, method of preparation, etc. and few remedies have been clinically validated; (3) Some remedies are totally ineffective; (4) Remedies are ineffective against acute viral diseases of animals; and (5) Large scale application of ethnoveterinary medicine often impractical due to the bulk of raw material required for treatment (Mathias-Mundy and McCorkle, 1989; Fielding 1998).

A number of regional and national studies have already been made, including India (FAO, 1984a), Thailand (FAO, 1984b), Nepal (FAO, 1984c), Pakistan (FAO, 1986b), Sri Lanka (FAO, 1991a), Indonesia (FAO, 1991b), Philippines (FAO, 1992), and Africa (McCorkle and Mathias-Mundy, 1992). However, traditional veterinary herbal medicine is not the prerogative of the developing countries. The ancient custom of the shepherds of the Cévennes region of France hanging a bouquet of herbs from the roof of sheep pens persists to this day, some 26 plant species are currently or traditionally used in the bouquets to combat dermatological problems in livestock and people (Brisebarre, 1996). Yet, despite the wide-spread prevalence of animal herbal lore I am unaware of any popular modern herbals on the subject, apart from publications intended for organic farmers.

Chapter 17

Plant Toxins and their Applications

Throughout the plant kingdom plants are subjected to mammalian herbivory, and insect and fungal infestations. During the course of evolution, they have produced a number of secondary compounds that are not directly involved with the basic metabolism but, by chance, serve with varying degrees of efficacy to enhance, reduce or destroy the plant's palatability to phytophagous members of the animal kingdom. All angiosperms tend to accumulate concentrations of at least one type of secondary compound, whether it be alkaloid, flavonoid or terpenoid, but rarely concentrations of different classes of secondary compounds. Such defensive phytochemicals offer an explanation of what Feeny (1975) cited by Harborne (1988) considered a conspicuous non-event as to how the plant kingdom had managed to survive and dominate the earth against the predations of the numerically greatly superior phytophagous insects. Cruciferae members, for example, commonly contain a glucosinolate-myriosinase system that principally acts as a defence against herbivore insects (Blau et al., 1978; Chew, 1988). Yet another group of secondary compounds are responsible for disease resistance (see Section 4).

The effect of plant toxins are governed by a number of factors. The quantity of toxic principle present or absent can depend on climatic or soil conditions, stage of plant growth and genetic constitution. Even with such variables, the effect of the toxin may vary with the breed, health and susceptibility of the recipient, and quantity ingested. See also Chapters 8 and 10 for further discussion on human and livestock toxins.

1.   VERTEBRATE TOXINS

Vertebrate plant toxins can be presumed to have evolved as a defensive mechanism against herbivory. However, Molyneux and Ralphs (1992) argue that insect herbivory subject plants to far greater stress than mammals, and since the rangeland floras of western North America and Australia evolved before the arrival of large herbivores, it follows that the toxins must have developed primarily in response to insect herbivory. For instance, the concentrations of pyrrolizidine alkaloids present in *Senecio* spp. are at their maximum at the bud stage of growth; in *S. riddellii* (Riddell's groundsel) it is by as much as 18% of the dry weight, thereby ensuring seed production will be uninterrupted by insect predation. The toxins offer no deterrent to grazing livestock early in the growing season, although the consumption of large quantities could eventually lead to fatal liver failure. The authors therefore regard livestock poisoning to be an unfortunate accident and, paradoxically, an indicator of the potency of the toxin. Of course, the reverse can also occur, as in *Conium maculatum* (hemlock) where the alkaloid *coniine* concentration decreases with age.

It is fortunate that while many toxins offer protection against insect predators, they are not all toxic to mammals, otherwise there would be no grazing livestock! Molyneux and Ralphs (1992) conclude: (1) Plant toxins confer a competitive advantage to the plant by providing protection against insect predation or plant pathogens; livestock poisoning is coincidental; (2) Increased or decreased palatability to livestock is poorly correlated with the presence of toxins; and (3) Because insects are specifically targeted, either by concentrations of toxins in a particular location in the plant or by production at concentrations sufficient to intoxicate the insects, livestock poisoning can be avoided by careful management. Since most plant communities will include some toxic species it is inevitable that livestock will be exposed to such plants. Toxicity follows ingestion and its effect will depend upon the quantity and rapidity with which it is eaten, the risk being greatly increased during times of dearth. Local immunity to toxicity can sometimes be acquired, alternatively the animals learn which plants to avoid. In southern Africa, for example, locally bred cattle are reputed to be unaffected by *Moraea* spp. (tulps) although the plants are definitely lethal to introduced and hungry cattle.

The ingestion by grazing livestock of toxic plants are a source of obvious direct economic loss to the grazier through death, emaciation, poor growth, abortion, impaired reproductive efficiency or birth defects. An example of the latter are ewes grazing *Veratrum album* (white false hellebore) in the upland pastures of Eurasia giving birth to lambs with a single, central eye. This is possibly the inspiration for the one-eyed god Polyphemus in Homer's *Odyssey*, who captured and was later blinded by the escaping Ulysses. The plant has been used in Europe since the 1st century to control rodents and plant pests. The active principles are the extremely toxic *veratrum alkaloids*. *Veratrine* and other alkaloids act as a sedative and depressant of the heart and nervous system, and are used for treating high blood pressure (Blackwell, 1990; James *et al.*, 1992; Macpherson, 1995; Mabberley, 1997).

The intensity of toxicity may also vary within a species. For instance, in some areas *Acacia georginae* (Georgina gidya) are toxic while in other areas they are non-toxic and may be safely browsed. The toxicity is attributed by some to the amount of fluorine in the soil and the differing abilities of the plants to take up the fluorine and convert it into fluoroacetate. The toxic principle is probably *monofluoracetic acid*, which occurs in diminishing concentrations in the seeds, pods and leaves (Askew and Mitchell, 1978, Everist, 1981; James *et al.*, 1992). There is a further complication in northern Australia where Georgina gidgee and *Eremophila maculata* (fuschia bush, spotted berrigan) may be grazed separately with impunity but are highly toxic when grazed together. The acacia pods are harmless but contain an enzyme that liberates prussic acid from the cyanogenic glycoside *prunasine* present in the fuschia bush (Jackson and Jacobs, 1985). To add to the confusion, Dowling and McKenzie (1993) contend that the leaves of *E. maculata* are potentially toxic at all times but sufficient enzymes have to be produced to release toxic amounts of HCN.

Although it is often assumed that ruminants are resistant to the toxic activity of many substances commonly found in plants, what is non-toxic to one ruminant is not necessarily non-toxic to others. Verdcourt and Trump (1969) cite the example of the giraffe in the Nairobi National Park, Kenya habitually eating the foliage of *Elaeodendron buchananii* with impunity, foliage that is known to be fatal to other ruminants, especially sheep. While there is no doubt that rumen fermentation uniquely confers immunity to the grazing animal, in at least one compound, *S-methylcysteine sulphoxide*, the rumen fermentation actually enhances the toxicity through the synthesis of a highly active metabolism (Duffus and Duffus, 1991).

Examples of plants containing *pyrrolizidine alkaloids* such as species of *Amsinckia, Astragalus, Crotalaria, Echium, Heliotropium, Senecio* and *Trichodesma* are known to be toxic, even the nectar of *S. jacobaea* (ragwort) species is toxic but fortunately any honey produced is bitter and off-colour. However, under certain controlled conditions based on a knowledge of safe levels of toxin concentrations, etc., *Lupinus* spp. and *Delphinium* spp. (larkspur), can be extensively grazed without any adverse effects from the toxic *anagyrine* (Dayton, 1948; Everist, 1972; Harborne, 1988; James *et al.*, 1992).

Contact poisons, such as *Dieffenbachia* spp. (dumb cane) and *Euphorbia* spp. contain an irritant sap, the toxic latex of the latter causing skin allergies, it is also used to stupefy fish. Contact dermatitis can be caused by the transfer of an allergenic substances from the plant to the skin. The classic example is *Rhus radicans* (poison ivy), where the 3-*n* pentadecycatechnol *urishiol* promotes dermatitis in over 350,000 cases annually in the US alone. Similarly, the furanocoumarins present in *Heracleum montegazzianum* (giant hogweed) and *Ruta graveolens* (rue) cause photodermatitis, sensitising the skin to ultraviolet radiation and giving rise to severe blistering in areas exposed to sunlight (Blackwell, 1990; Mabberley, 1997; Cooper and Johnson, 1998).

1.1   Homicide Poisons

Phytochemicals acting as homicide poisons may be administered deliberately with murderous intent, used as an ordeal poison, or ingested accidentally. Murder is outside the scope of this book, although the distinction between murder and ordeal poisons can be a very fine one. Trial by ordeal poison subjects the accused to doses of toxic substances in order to determine guilt or innocence, and it often depends on the skill of administering witch doctor, shaman, etc. in controlling the dose, as well as whether guilt has been prejudged by the administer, as to whether the defendant is deemed not guilty and lives or is guilty and dies. The axiom of the Renaissance physician, the Swiss Phillipus Paracelsus *ca.* 1493-1541, *'Sola dosis facit venenum'* (Only the dose makes the poison) is particularly apposite when referring to ordeal and accidental poisoning.

Physostigma venenosum (Calabar or ordeal bean) was formerly widely used as an prdeal poison in West Africa, the active ingredient being the alkaloid *physostigmine*, also known as *eserine*. The action of physostigmine produces an identical effect to stimulating the parasympathetic nervous system, i.e. it constricts the pupil, stimulates the gut, increases saliva secretion, and increases the irritability of voluntary muscle, and in excess causes general paralysis. It is now used in ophthalmic medicine for protracted pupillary contraction and in the treatment of glaucoma. It also counteracts the action of curare, as well as being used as an antidote to atropine poisoning (Blackwell, 1990; Macpherson, 1995; Mabberley, 1997).

1.2   Arrow Poisons

The earliest records of the use of plant toxins for arrow and spear poisons date from the *Rig Veda* of *ca.* 1200 BC, although doubtless they date back into pre-history. They continue to be used by traditional hunters today. Some of the phytochemicals involved are shown in Table 24.

TABLE 24. Plant toxins used as arrow poisons (Cotton, 1995, reproduced by kind permission of John Wiley & Sons)

Chemical type	Biological activity	Plant families (not exhaustive)
Toxic alkaloids	Acetylcholine inhibitors, leading to muscle paralysis, e.g. *d*-tubocurarine chloride	Loganiaceae, Solanaceae, Umbelliferae
Cardiac glycosides	Inhibit sodium/potassium ion pumps in heart muscle cells, leading to abnormal heart activity	Common among Apocynaceae, Asclepiadaceae and Moraceae

Probably the best documented is *curare*, the crystalline alkaloid *d-tubocurarine chloride*, obtained from the bark of the vine *Chondrodendron tomentosum* of the Amazonian rain forest. The curare acts by paralysing the muscle nerve endings Fortunately it is only toxic when it enters the blood stream, consequently game can be eaten without any ill effects provided there are no open sores in the mouth or throat. The action of curare as an arrow poison has been reapplied for medical use, with pure curare now being used in anaesthesia as a muscle relaxer. Interestingly, its action is antagonised by *neostigmine*, a synthetic analogue of physostigmine (Kreig, 1965; Blackwell, 1990; Macpherson, 1995; Cotton, 1996).

1.3   Fish Poisons

A wide range of plant-derived fish poisons, piscicides or ichythyotoxins have been used world-wide by traditional fishermen to kill or stupefy fish, especially in the tropics and subtropics. Many of the toxins interfere with the respiration processes, while others act upon the central nervous system, resulting in a range of effects, including heart or respiratory failure. The phytochemicals involved are shown in Table 25.

TABLE 25. Plant toxins used as fish poisons (Cotton, 1996, reproduced by kind permission of John Wiley & Sons)

Chemical type	Biological activity	Plant families (not exhaustive)
Isoflavonoids	Interferes with mitochondrial activity, leading to asphyxiation, e.g. rotenone, tephrosine, lonchocarpin	Restricted to Leguminosae subfamily Papilionoideae
Saponins	modify water tension, blocking respiration at gills	
Toxic alkaloids	Acetylcholine inhibitors, leading to muscle paralysis, e.g. *d*-tubocurarine chloride	Particularly common among Loganiaceae, Solanaceae and Umbelliferae
Cardiac glycosides	Inhibit sodium/potassium ion pumps in heart muscle cells, leading to abnormal heart activity	Common among Apocynaceae, Asclepiadaceae and Moraceae
Tannins	May act through cross-linking with gill proteins, leading to asphyxiation	
Cyanogenic glycosides	Release hydrogen cyanide to inhibit cytochrome oxidase	Common among Euphorbiaceae. Flacourtiaceae and Rosaceae
Ichthyoctherol	Polyacetylenic alcohol interfering with mitochondria, leading to asphyxiation	Restricted to members of the Compositae

Needham (1984) reports that in China the use of *Buddleja lindleyana* to stupefy fish was recorded as early as the 6th century BC in the *Tzu-I Pên Tshao Ching* (The Classical Pharmacopoeia of Tzu-I), and cited in the *Pên Tshao Kang Mu* (The Great Pharmacopoeia) of 1596 AD, which is a compendium of plants recorded in the Chinese literature since antiquity. The seeds of *Illicium anisatum* (Japanese anise) were similarly used to stupefy fish. Slow-moving small rivers or pools are usually required in order that the toxins remain sufficiently concentrated to function properly.

It should be noted that the indiscriminate use of piscicides may seriously deplete the reserves of fish in the area since they will destroy all age classes.

The only plant piscicide recorded for the British Isles is *Euphorbia hybernica* (Irish spurge); the local fishermen in Kerry and West Cork throw the crushed herb into the water. The saponins present in the milky latex destroy the gill tissue and the dead salmon and trout are then caught floating on the surface (Vickery, 1995; Mabberley, 1997).

2.   INVERTEBRATE TOXINS

In Section 1 it was pointed out that toxic secondary metabolites probably developed as protection against insect rather than vertebrate hebivory. In the wild such protection may be adequate but under cultivation plants will be exposed to far higher concentrations of insect pests and will consequently require greater protection.

While it is possible to breed for insect resistance, it is conceivable that a species that has to use its energy on chemical, physiological or anatomical characteristics against insect predation is less likely to devote all its energy into producing maximum yields. Indeed, for the subsistence farmer low yielding, resistant varieties will out-yield the non-resistant but potentially higher yielding varieties/and may also be better suited to their needs because they do not require intensive inputs. However, in breeding for disease resistance using genetic engineering techniques there is a danger that it could encourage resistance to the toxin if the major gene being introduced does not originate from the general genetic diversity of the plant material. The introduction of the BT toxin in tobacco, for example, can be a cause for concern because it exposes the insects to higher levels of BT toxin than it would normally experience in the field, thereby encouraging a rapid selection for resistance to the BT toxin (van Emden, 1989).

An alternative which is rapidly gaining support among environmentalists is *biological control* which, from an ecological viewpoint, is defined as '*the action of parasites, predators, or pathogens in maintaining another organism's population density at a lower average than would occur in their absence*'. It is the study and utilisation by man of parasites, predators and pathogens for the regulation of host population densities (DeBuch, 1964).

The successful use of pathogens for insect control is dependent upon the biology and characteristics of both the host insects and parasitic micro-organisms and the suitability of environment conditions for infection. The advantages of microbiological control are: (1) Pathogens are harmless and non-toxic to other forms of life; (2) The relatively high degree of specificity of most pathogens tends to protect beneficial insects; (3) Many pathogens are compatible with many insecticides; (4) They are relatively inexpensive; (5) They are highly variable in their application. Some micro-organisms may bring about permanent control, others may be used as sprays or dusts;

(6) The apparent slowness by which susceptible insects develop resistance; and (7) Only doses are often necessary for control.

The disadvantages are: (1) Careful and correct timing of the application in relation to the incubation period of the disease; (2) There is a relatively marked specificity; and (3) A danger of introducing the micro-organisms to new areas where they may become uncontrollable (Hall, 1964).

2.1   Molluscicides

Schistosomiasis, formerly known as bilharziasis, is a common infection of humans in Africa, Middle and Far East and South America. The intermediate host of the fluke responsible are fresh-water snails, mainly belonging to the genera *Biomphalaria*, *Bulinus* and *Oncomelania*, against which a number of plant molluscicides have been evaluated. The active principles involved include monoterpenes, sesquiterpenes, quinones, flavonoids, rotenoids, triterpene saponins, spirostanol saponins, iridoids, coumarins, alkaloids, isobutylamides, and tannins. Unfortunately very few plants satisfy the criteria for large-scale application; many do not have sufficient activity. The LC_{90} (the concentration in water that kills 90% of the target snail population) should be less than 10 mg l^{-1} in order to be competitive with synthetic molluscicides, and not require the use of excessive amounts of plant material. The plants must be readily available, and easy to propagate, maintain, harvest and, if required, process. Leaves and fruits are definitely preferred to root sources, which would be injurious to the plant. Promising candidates include *Swartzia madagascariensis*, and *Phytolacca dodecandra* (endod, pokeweed), their respective pods/seeds and berries possess saponins with activities of a similar order of magnitude to those of synthetic molluscicides. Both also have the disadvantage of being effective piscicides, as too are the fruits of *Balanites aegyptiaca*. Despite the effectiveness of plantings along river banks in controlling schistosomiasis, they have met considerable resistance from local fishermen who prefer to have the disease rather no fish to eat!

The cut herbage of the composite herb *Ambrosia maritima*, syn. *A. senegalensis* (damsissa) has been found effective in Egypt. It has the great advantage of already growing along the muddy banks of the irrigation canals and, at concentrations of 1000 ppm, harmless to fish. The active ingredients are the sesquiterpene lactones **ambrosin** and **damsin** (Verdcourt and Trump, 1969; Hostettmann and Marston, 1987; Kloos and McCullough, 1987; Lugt, 1987; Mott, 1987; Hall and Walker, 1991; Wickens, 1998).

I am unaware of any similar work on mollusicides in the temperate regions for the control of Trematoda (liver fluke) affecting domestic livestock, or for the control of garden snails..

2.2   Insecticides

It has been estimated that approximately one-third of the world's food crops are either damaged or destroyed by insects during crop growth, harvest and storage. In many of the developing countries the losses are even higher. Crop pest control is, therefore, a major economic necessity. The use of plants to destroy or deter insects have a long history. While their early use may be suspected, early documented evidence is sparse. From ancient Egypt the Ebers papyrus of *ca.* 1550 BC refers to the use of *Inula* sp. (fleabane) to repel fleas. The earliest Chinese references to plant-based pesticides are to be found in *Chou Li* (Records of the Rites of Chou), with parts dating from the beginning of the Chou dynasty (1030-221 BC). It provides descriptions of the duties of government officials during the Chou dynasty, and includes herbal remedies for insect pests using *Chrysanthemum indicum, Glycyrrhiza glabra, Illicium* sp. (Chinese anise) and *Melia azedarach*, and for insect fumigants using. *Chrysanthemum indicum* and *Illicium* sp., as well as herbal remedies for the control of human internal parasites, e.g. *Illicium* sp. but whether the latter is that used as an insecticide or a different species is not clear. All the available information on indigenous plant pesticides has now been gathered together in a compendium on Chinese indigenous agricultural drug plants (Needham, 1986; Manniche, 1989).

Jacobson (1982) recognises six groups of plants that are physiologically active to insects. They are: (1) Plants attractive to insects. This ability can be utilised by growing companion crops that are attractive to beneficial insects or attract harmful insects away from the crop. Thus, *Tropaeolum majus* (nasturtium) will attract aphids away from broccoli, while ornamentals such as *Convolvulus tricolor, Fagopyrum esculentum* (buckwheat) and *Limnanthes douglasii* (poached egg plant) will attract such beneficial insects as hoverflies and ladybirds that feed on aphids; (2) Plants repellent to insects. Crop losses from caterpillars in sorghum, wheat and maize have been dramatically reduced from 80% to 5% by interplanting with *Melinis minutiflora* (molasses grass) as a companion crop; it is a native of tropical Africa but has now been introduced throughout the tropics. Molasses grass acts by continuously producing the aromatic compound **dimethyl nonatriene**, which repels female butterflies and moths, thereby reducing the number of eggs laid; it also attracts predatory wasps that feed on any caterpillars; (3) Plants toxic to insects, such as **rotenone** from the roots of *Derris, Lonchocarpus* and *Tephrosia*, **pyrethrin I** and **II**, **cinerin I** and **II**, and *jasmolin I* and **II** from *Tanacetum cinerariifolium*, and **nicotine** from *Nicotiana* spp.; (4) Plant produced insect morphogenetic agents, known as **juvenile hormones** (**JH**), that inhibit or abnormally accelerate normal insect growth and development, e.g. the hormones **cyasterone** and **ecdysterone** isolated from *Ajuga remota* produce abnormal head growth in the larvae of *Spodoptera frugiperda* (fall armyworm) and *Pectinophora gossypiella* (pink bollworm); (5) Plants that sterilise insects. The vapours of calamus oil from *Acorus calamas* (calamas, sweet flag), for example, can cause sterility in male *Musca domestica* (house fly), the female

Callosobruchus chinensis (azuki bean weevil), *Dysdercus koenigii* (red cotton stainer) and *Trogoderma granarium* (khapra beetle); and (6) Plants that deter insect feeding. Examples include recent research at the Rothamsted Experimental Station, UK where such anti-feedant compounds as **ajuganin** from the Labiatae and **polygodial** from species of *Warburgia* and *Polygonum* have been identified. The latter can be extracted in quantity from plants, and since a simple synthesis from **farnesol pyrophosphate** is involved (it is ubiquitous in plants), there is ample scope for production using biotechnology techniques. The terpenoid resin produced by *Grindelia camporum* is also believed to function as an antifeedant, and is even produced in sufficient quantities to be considered a potential source of biocrude (Jacobson, 1982; Timmermann and Hoffmann, 1988; van Emden, 1989; Cole, 1992; Dales, 1992; Coppen, 1995; Flowerdew, 1995; Anonymous, 1997a).

2.2.1 Derris

Commercial **derris** is obtained from the legume lianas *Derris elliptica* (derris root, tuba-root) and *D. ferrugiana*, in which the **rotenone** is present in the roots at concentrations of *ca.* 7%. The insecticide is extracted by grinding the dried root to a powder to produce derris dust, which can then be either applied directly to the plant, or as a liquid extract containing 40% rotenone. Derris is toxic to animals, including fish, as well as insects; the toxic properties disappearing within a few days of application (Purseglove, 1987; Robbins, 1995).

2.2.2 Pyrethrum

Pyrethrum products constitute the world's main biodegradable organic pesticide which, while lethal to a wide range of insects, can be relatively safely ingested or inhaled by humans. The *Tanacetum cineariifolium* crop is mainly grown by small-scale farmers because of the absence of a suitable machine for mechanically harvesting the inflorescence discourages large-scale production. As a consequence the demand for consistent quality and quantity is difficult to control. The dried and ground flowers are solvent extracted, usually with a light petroleum solvent to obtain the pyrethrin. It is traded either as a crude extract containing 1-1.5% pyrethrin or as a refined extract with either 25% or 50% pyrethrin in a liquid form. However, the natural pyrethrum alone would not be an economical insecticide without the addition of **piperonal butoxide** (PBO) as a synergist. PBO is an important derivative of **safrole**, a product isolated from sassafras oil (see Chapter 15), so that the future of the pyrethrum industry is linked with the perfume industry and the production of PBO (Purseglove, 1987; Robbins, 1995).

2.2.3   Azadirachtin

In recent years there has been considerable interest in the triterpenoid *azadirachtin* isolated from the seeds of *Azadirachta indica* (neem). This followed observations that swarms of *Schistocerca gregaria* (desert locust) never attacked the tree. The azadirachtin acts as an antifeedant and insect repellent, as well as an insect growth regulator, controlling insects in all the larval life stages.

Water extraction of the crushed seed has been successfully used in the developing countries. The seed from two mature trees, i.e. 20-30 kg, can normally treat 1 ha. The emulsion from *ca.* 500 gm of kernels steeped in 10 l of water overnight providing the necessary concentration. Commercial formulations are now available for the control of a wide range of insect pests for food and feed crops.

Neem oil from *Azadirachta indica* is a powerful germicide, and can be used topically to control skin-attacking insects such as lice and as a fungicide for athlete's foot and ringworm. It is also used for the control of *aflatoxin* in nuts and seeds, and in the control of soil nematodes, the 'bilharzia snail', and as an insect repellent. Nematode control can also be obtained by scattering chopped neem seeds over the soil surface. Neem oil is also used commercially in cosmetics, including toothpaste and soap. Because of the tremendous interest by buyers in neem for its chemical and pharmaceutical products the price of seed has risen quite dramatically. So much so that many local farmers can no longer afford to purchase seed (National Research Council, 1992; Robbins, 1995).

2.2.4   Insect Control by Fungi and Bacteria

The control of insect pests by using suspensions of micro-organisms and/or their products that are specific and lethal to particular insects is becoming increasingly important, especially in the wake of insect resistance, the emergence of secondary pests, and the outcry about toxic residues from the use of conventional insecticides.

The amino-acid derivative *tricholomic acid* produced by the fungus *Tricholoma muscarium* and *muscazone*, a toxin produced by *Amanita muscaria*, also have insecticidal properties. The fungus *Metarrhizium anisopliae* has proved effective against a wide range of insect pests. It is commercially available in Brazil, where it has been successfully used to control Ceropidae (spittle-bugs) in pastures and sugar cane plantations. It has also been successfully combined with a virus to control *Orcytes* spp. (rhinoceros beetle) on palm trees in the Pacific. In the Russian Federation and USA another fungus, *Beauveria bassiana*, has been used to control *Laspreyresia pomonella* (codling moth) and *Leptinotarsa decemlineata* (Colorado potato beetle). In the USA *Hirsutella thompsoni* has been used in an attempt to control *Phyllocoptruta oleivora* (citrus rust mite) in citrus plantations; unfortunately the environmental conditions do not favour the fungus and it has yet to find a role somewhere in citrus pest control. Again in USA, *Nomuraea rileyi* has proved to be very effective against caterpillars, especially on soya bean, while *Verticillium lecanii* is used commercially to control both Aphididae and Aleyrodidae (whitefly) in the UK (Lisansky, 1985; Subba Rao and Kaushik, 1989).

The bacterial control of caterpillars, mosquitoes and blackflies is now also possible. Commercial strains of *Bacillus thuringiensis* (BT) are available and have been successfully used against many leptidopterous insects, including *Boarmia selenaria* (giant looper) and *Cryptoblabes quidiella* (honeydew moth) in avocado orchids in Israel, *Ostrina nubilalis* (European corn-borer) as well as *Agrotis*, *Euxoa* and *Feltia* species (tobacco cutworms) etc.

Unfortunately, some strains of *B. thuringiensis* produce the soluble toxin **β-exotoxin** which, although highly toxic to house flies, when ingested has a slight mammalian toxicity, consequently no products containing β-exotoxin are now marketed in the developed countries except Finland. *B. sphaericus* is also considered as a potentially highly effective pathogen against mosquito larvae, but is not yet commercially available.

The **avermectins**, which are derivatives of pentacyclic 16-membered lactones, isolated from the bacterium *Streptomyces avermitilis* have proved very effective against a number of arachnids, insects and crustaceans under field conditions, including inhibiting the production of queens in *Solenopsis invicta* (red fire ant). (Lisansky, 1985; Subba Rao and Kaushik, 1989; Arnon, 1992).

3.   HERBICIDES

Herbicides are used for the control of weeds, which are commonly defined as 'a plant out of place'. In the past their control has been either through cultivation and management practices or chemical herbicides. More recently the use of fungal pathogens as bioherbicides has become increasingly popular. See Table 26 for examples.

TABLE 26. Bioherbicides (Subba Rao and Kauskik, 1989; Watson, 1989; Holliday, 1990)

Fungus	Susceptible plant	Crop protected
Cercospora rodmanii	*Eichhornia crassipes* (water hyacinth	potential contol
Ceriosporella riparia	*Ageratina riparia*	weed - Asia
Colletotrichum gloeosporioides	*Aeschynomene virginica* (joint vetch)	rice, soya
Phragmidium violaceum	*Rubus* spp. (blackberries)	range - Australia
Phytophthora citrophthora (Oomycota)	*Morrenia odorata* (milkweed vine)	citrus
Puccinia chondrillina	*Chondrilla juncea* (skeleton weed)	wheat

Such ***bioherbicides*** are described by Watson (1989) as "*a preparation of living inoculum of a plant pathogen, formulated and applied in a manner analogous to that of a chemical herbicide in an effort to control or suppress the growth of a weed species. .... bioherbicides should not be viewed as alternatives to chemical herbicides, but rather as complementary tactics in integrated weed management systems.*"

In another approach, seed from maize genetically engineered for herbicide resistance may now be used against *Orobanche* spp, (broomrape) and *Striga* spp. (witchweed). Once the seed has germinated, the parasite absorbs the herbicide from the maize and dies. The herbicide is reported to disappear from the crop as it ripens and consequently does not affect the cob (Abayo *et al.*, 1998). See Chapter 6 for further information).

3.1   Allelopathy

Plant metabolites produced by higher plants may find their way into other organisms, where they may play a major or minor role in a multiplicity of important physiological processes. They often attract or repel, nourish or poison browsing insects and other herbivores; they may also stimulate or suppress the growth of micro-organisms. Some may reduce competition (and indirectly lessen the fire hazard) by interfering with the regulatory function of other higher plants within their immediate vicinity, sometimes stimulating growth, sometimes stopping growth altogether. The plant products producing the latter effect are termed ***phytotoxins***. The negative effect of one plant on another by means of chemical products released into the environment is termed ***allelopathy***, implicating either those belonging to the same species, i.e. ***auto-allelopathy*** or ***auto-toxicity***, e.g. *Parthenium argentatum* and *Pinus* spp., or for different species, e.g. *Eucalyptus camaldulensis*, *Gutierrezia sarothrae* and *Juglans regia*. Allelopathy may also be indicated if one plant inhibits the growth of a second plant or micro-organism that itself is essential to the growth of a third plant (Muller and Chou, 1972; Tootill, 1984; Harborne, 1988). Allelopathy is clearly not only of interest to ecologists, it is also of concern to the agriculturist, horticulturist, forester and agroforester. See Chapter 5 for further discussion.

The multipurpose *Leucaena leucocephala*, for example, is widely grown in agrosystems for livestock feed, fuel, pole timber, N-fixation, green manure, etc. yet the concentration of the non-protein amino-acid ***mimosine*** in the leaves and seeds is sufficient to inhibit the germination and growth of such crops as *Abelmoschus esculentus* (lady's fingers, okra), *Brassica rapa*, syn. *B. campestris*, *Lactuca sativa* (lettuce), *Oryza sativa*, *Vigna mungo* (black gram) and V. *radiata*, syn. *Phaseolus aureus* (green gram), as well as the germination and radicle growth of *Acacia confusa*, *Alnus formosana*, *Casuarina glauca*, *Liquidambar formosana* and *Pinus taiwanensis* (Rizvi and Rizvi, 1992). See Chapter 5 for further discussion.

Allelopathy has proved successful for the control of the obligate parasites *Striga lutea*, syn. *S. asiatica* and *S. hermontheca* (witchweed), which can so drastically reduce the yields of maize, millets, sorghum, upland rice and sugarcane, and *S. gesnerioides*, which attacks *Vigna unguiculata* (cowpea) and occasionally tobacco. The witchweed obtains its nutrients by means of haustoria attached to the host plant's root system, causing stunting, wilting and even death of the host plant. While no totally resistant cultivars of the cereals have been identified, absolute resistance has been found among the cowpea. Formerly cowpeas were grown as a catch crop to act as host to the witchweed and ploughed in as a green manure before the witchweed could produce seeds. High yielding, resistant cultivars of cowpea are now being bred that can be used as trap crops.

The assessment of resistance in field or crop trials usually take 10-15 weeks but, by using *in vitro* techniques, such tests can now be completed in 2-3 weeks. Two distinct resistance mechanisms to *S. gesnerioides* have been discovered. Either the cowpea root in the vicinity of the invading haustoria become necrotic within 2-3 days and the parasite dies, or the haustoria form but fail to develop, causing minimal damage to the cowpea host, which then grows and yields normally (Lane, 1992).

Strains of *Oryza sativa* (rice) have also been discovered that have an allelopathic action against the aquatic weeds *Ammannia* sp. (purple redstem), *Cyperus difformis*, *Echinochloa crus-galli*, *Heteranthera limosa* (duck salad) and *Trianthema portulacastrum*. One rice cultivar, Taichung native 1, has even been found to have an allelopathic effect against all these weeds except *C. difformis* (Olofsdotter *et al.*, 1997).

4.   FUNGICIDES

Some plant metabolites offer disease resistance at the pre-infectional or post-infectional stage, although the distinction is somewhat arbitrary since pre-infectional compounds can undergo significant post-infectional changes. Their classification is shown in Table 27.

While some existing metabolites prohibit or inhibit pre- and post-infection stages, *phytoalexins* are chemical compounds that are only produced *de novo* or are activated by the host plant when they come into contact with a pathogenic fungus in hypersensitive living tissue. While the physiological, ultrastructural and pathological aspects of phytoalexin synthesis are not yet fully understood, they represent a major development in physiological plant pathology. The first phytoalexin to be identified was the phenolic compound *pisatin* in pod tissue of *Pisum sativum* innoculated with *Monilinia fructicola* (brown rot fungus).

The ability of the fungus to parasitise a species is also partly related to the ability of the fungus to deal with the phytoalexin. Thus, the degree of resistance shown by *Ipomoea batatas* (sweet potato) to attack by the fungus *Ceratostomella* is correlated with the concentration of the terpenoid *ipomeamarone* in the plant tissues. While the phytoalexins are most active in inhibiting pathogenic fungi, they are also able to react to bacterial and viral infections. They can also be formed abiotically under stress conditions, including temperature shock and wounding (Tootill, 1984; Harborne, 1988).

TABLE 27. Classification of disease resistance factors in higher plants (Harborne, 1988)

Class	Description	Plant metabolites
PRE-INFECTIONAL COMPOUNDS		
Prohibitins	Metabolites which reduce or completely halt the *in vivo* development of micro-organisms	Terpenoids and phenolics, especially hydroxystilbenes
Inhibitins	Metabolites which undergo post-infection increase in order to express full toxicity	Coumarins and hydroxycinnamic acids
POST-INFECTIONAL COMPOUNDS		
Post-inhibitins	Metabolites formed by the hydrolysis or oxidation of pre-existing non-toxic substrates	Inactive glycosides stimulated by microbial invasion to release toxin
Phytoalexins	Metabolites formed *de novo* after invasion by gene depression or activation of a latent enzyme system	Flavan, pterocarpan, stilbene, terpenes, etc.

Anti-fungal activity is not confined to the higher plants, some strains of the fungus *Alternaria solani* (early potato blight) produce a highly phytotoxic antibiotic, *alternaric acid*, which may be used as a fungicide (Brian *et al.*, 1952).

Chapter 18

Useful Ferns, Bryophytes, Fungi, Bacteria and Viruses

Economic plants are often considered in terms of angiosperms and gymnosperms, the important contribution made by the ferns, mosses, fungi, algae, etc. tends to be overlooked. Here they are being considered separately from the flowering plants in this and the following chapter in order to stress their economic importance for the food, drink, medicine, biochemical industries, etc.

1. FERNS AND FERN ALLIES

These are an ancient group of plants which, according to the Five Kingdoms classification, include the Filicinophyta (ferns), Sphenophyta (horsetails), Lycophyta (lycopods) and Psilophyta (whisk ferns). They are represented by between 12,000 and 15,000 species, and are most widely distributed in the tropics (Schultes and Hofmann, 1992).

1.1 Ferns

Ferns are used for a wide range of purposes throughout the world, including food (starchy rhizomes and fronds), flavourings, fats, oils, fragrances, dyes, fibres, medicine, and various religious/magic purposes. Ethnobotanical and anthropological studies have shown that some play an essential role among primitive societies. For example, the Chácobo Indians of Amazonian Bolivia utilise 16 species of ferns, a number of which are used in decoctions for treating such ailments as appendicitis, rheumatism and diarrhoea. Similarly, in Sarawak 30 different species are used by two small indigenous communities for food (fronds), medicine, fibre, and in various religious/magic ceremonies and properties (May, 1978; Boom, 1985; Christensen, 1997).

Of particular importance in tropical agriculture is the genus *Azolla*, with six tropical and warm temperate free-floating species. It is unusual in being the only fern known to have symbiotic relationship with a nitrogen-fixing endophyte, the cyanobacterium *Anabaena azollae*, which is unique in using fructose to fix nitrogen to ammonia. The ammonia is transferred to the host, who then returns amino-acids, proteins and ribo-neucleotides to the endophyte.

In rice fields the fern is capable of fixing 50-150 kg N ha^{-1} in 1-4 months, and under ideal conditions more than 10 kg N ha^{1} day^{1}, making it a major green manure and, in economic terms, the most valuable fern in the world; it may also be used as a stock feed. Draining the fallow paddy fields kills the *Azolla* and releases the nitrogen. The technique was first developed in China during the 1960s using the indigenous *A. pinnata* var. *imbricata*, syn. *A. imbricata*. The recent introduction of the more cold-tolerant *A. filiculoides* from South America has allowed the rice-*Azolla* cultivation cropping system to be extended to northern and north-eastern China

Azolla's ability to reproduce itself rapidly has both advantages and disadvantages. Thus, the capability of the tropical Asian *A. pinnata* for doubling its own weight in 7 days would be considered disadvantageous in blocking waterways, yet its ability to smother the water surface helps to control mosquitoes (Bumpkin and Placenta, 1982; Sprint and Sprint, 1990; Mabberley, 1997).

A number of ferns may be used for food. For instance, the young shoots of *Diplazium esculentum*, syn. *Athyrium esculentum*, are widely used for food in SE Asia, and attempts have even been made to bring the fern into cultivation. The fronds of *Ceratopteris pteridoides* from tropical America are also edible, while *C. thalictroides* is much cultivated in the flooded rice-fields of tropical Asia as a spring vegetable. The rhizomes of *Blechnum indicum* (bungwall) were formerly a traditional food of the Aborigines of northern Australia. The pith of the tree-ferns, *Cyathea* spp. are also eaten. In North America the steamed crosier fronds of *Matteuccia struthiopteris* (ostrich fern) are traditionally eaten as a spring vegetable in the Maritime Provinces of Canada and in Maine. Unlike bracken the fronds of *Osmunda cinnamomea* (cinnamon fern, fiddleheads) are not carcinogenic. Popular demand has lead to frozen and canned fronds being commercially marketed. However, attempts at cultivation have proved unsuccessful due to difficulties with large scale propagation; the use of tissue culture for the mass production of sporophytes is now being investigated.

Ferns are also used in the brewing industry, e.g. *Osmunda regalis* (royal fern) in the brewing Celtic heather ale, where the presence of a **thiaminase** in the spores destroys the vitamin B, and consequently the activity of the yeast, thereby stopping the fermentation process (Copeland, 1942; May, 1978; Aderkas, 1984; Mabberley, 1997).

The young shoots of the monotypic, cosmopolitan *Pteridium aquilinum* (bracken) are canned and eaten in soups, especially in China and Japan as **sawarabi**. They were also formerly boiled and eaten on toast in North America. However, their regular consumption is not advisable due to the presence of the carcinogenic *shikimic acid*. *Warabi starch* is extracted from the rhizomes and were formerly eaten by the Maori and North Amerindians, and the liquorice flavour of the rhizomes formerly used for flavouring tobacco.

Bracken also contains the enzyme *thiaminase*, which can cause vitamin B_1 deficiency (thiamine deficiency) and eventually death when the bracken is eaten by horses. In cattle bacterial activity in the rumen apparently neutralises the thiaminase activity and, although vitamin B_1 deficiency is not the culprit, cattle eating bracken can die of an unknown chemical cause, with symptoms resembling aplastic anaemia in humans. Although bracken spores are known to be carcinogenic to rodents, the risk to humans is still under investigation. Bracken fronds were formerly used` for bedding, thatch, packing, padding, fuel, tinder, compost, etc. The young fronds are also the source of an olive-green dye. In addition, both the growing plant and the litter are allelopathic (Hedrick, 1972; Brouk, 1975; Saunders, 1976; Launert, 1981; Caulton *et al.*, 1995; Mabey, 1996; Mabberley, 1997).

The fronds of the *Asplenium acrobryum* complex (New Guinea salt fern) were formerly an important source of a vegetable salt rich in Ca^{++}, K^+ and Cl^- for the inhabitants of the salt-deficient inland areas of Papua New Guinea. The reason why this particular species was selected when other equally salt-rich species of *Asplenium* were ignored is unclear. (Croft and Leach, 1985).

A number of medicinal and cosmetic properties are also recognised. For example, the dried rhizomes of *Dryopteris filix-mas* and *D. dilatata* are among the oldest known vermifuges, especially against tapeworm, although their use requires careful medical supervision since they are highly toxic, containing the phloroglucinol derivatives *kosidin*, *protokosin* and *kosin*; their use has now been superseded by *quinacrine*. The green rhizomes also contain *ca.* 6% by weight of vegetable fat. The fern was also apparently used in Ancient China for silk reeling. The cosmopolitan *Adiantum capillus-veneris* (maidenhair fern) is listed as official in a number of European pharmacopoeias. It is used for flavouring decoctions, infusions, fluid extracts and tinctures, especially the 'cure-all' medicine 'Sirop de Capillaire', the flavour being due to tannic and gallic acids and traces of an essential oil. The dried fronds were used in the Isles of Arran as a tea substitute. The delicate leaf tips of *Dryopteris cristata*, syn. *Lastrea cristata* are a source of a fragrant oil which is used in the Black Forest to perfume the soap 'Fougére' (Hedrick, 1972; May, 1978; Launert, 1981; Mabberley, 1997; Dagne, 1998).

The large, cosmopolitan genus *Asplenium* contains a number of useful ornamentals, including *A. scolopendrium*, syn. *Phyllitis scolopendrium* (hart's tongue), plants with forked fronds or otherwise divided at the apex being particularly prized by gardeners. The fronds are also used in herbal medicine, as are also those of *A. adiantum-nigrum*, while in South Africa *A. flabellifolium* is a source of HCN. Species of *Osmunda* (royal ferns) are also cultivated as ornamentals and locally eaten (Brightman and Nicholson, 1966; May, 1978; Launert, 1981; Vickery, 1995).

The metamorphosis steroid *α-ecdysone* has been isolated from several species of *Osmunda, Polypodium* and *Pteridium aquilinum,* micro quantities of which leads to precocious metamorphosis in insect larvae, resulting in extremely abnormal growth and development, and even death. The extract has a potential as an insecticide (Jacobson, 1982).

The fibres from the cinnamon fern are used as an orchid-growing medium, while in Japan the hairs surrounding the young fronds are mixed with wool to make a textile for raincoats. The fibrous black petioles of *Pityrogramma triangularis* (goldenback fern) are utilised by the northwestern Amerindians in basketry (May, 1978; Mabberley, 1997).

1.2   Horsetails

The Sphenophyta, syn. Equisetophyta, is represented by the monotypic genus *Equisetum* (horsetails), consisting of 15 species which, apart from Australia, have an almost cosmopolitan distribution. Its members have a great affinity for accumulating and concentrating gold in solution, albeit only 0.25 g gold kg^{-1} stems and rhizomes. Although the commercial extraction of the gold is not economically viable, the presence of horsetails is seen by prospectors as an indicator plant for gold ore.

While alkaloids, including *nicotine* are present in horsetails, their toxicity, like that of bracken, is due to the enzyme *thiaminase* breaking down the vitamin *thiamine* in the browsing animal and leading to a vitamin B_1 deficiency in horses. Medical applications include a possible treatment for Alzheimer's disease.

The shoots of both *E. arvense* (common horsetail) and *E. fluviatile* (water horsetail) were reported to be eaten like asparagus by the Romans, who also used the dried stems to make a tisane and as a thickener. Because of its rough stem *E. hyemale* (Dutch rush) was used before the advent of steel wool and nylon pot-scourers for cleaning cooking pots. Horsetail stems were also formerly used by watch-makers and brass-workers as an abrasive for giving an extra finish after filing, and by cabinet-makers, including the Dutch-born British sculptor Grinling Gibbons (1648-1721), for polishing wood-carvings. In addition, the Dutch rush was some-times eaten in times of famine. Although invasive and extremely difficult to eradicate *E. hyemale* var. *affine,* syn. *E. praealtum,* and *E. telmateia,* syn. *E. maximum* (giant horsetail), are sometimes grown as ornamentals in water gardens (Brightman and Nicholson, 1966; Hedrick, 1972; May, 1978; Chiej, 1984; Blackwell, 1990; Mabey, 1996; Mabberley, 1997).

1.3   Lycopods

Although especially abundant in the fossil record of the Carboniferous, the Lycophyta today are represented by only six genera and are of very little economic importance.

The genus *Lycopodium* (club mosses) contains 40 tropical and temperate species, some of which are exceptional among the pteridophytes in that they contain alkaloids. Some are cultivated as ornamentals, and in the Philippines they are grown in hanging baskets; others are used for stuffing upholstery and for making baskets, bags and fishing nets. The spores of *L. alpinum* (alpine clubmoss) are used to dye wool yellow, while the very fine, bright yellow spores of *L. clavatum* (stag's horn moss), known as lycopodium powder, were a former constituent of the 'flash powder' used in fireworks and stage lighting, and by pharmacists as *vegetable sulphur* for coating pills and condoms, although it is reported that the latter use may cause allergic granulosis in some users. The spores of the stag's horn moss are also rich in vegetable fat, containing 50% lycopodium oleic glyceride (Brightman and Nicholson, 1966; May, 1978; Mabberley, 1997).

The subcosmopolitan genus *Huperzia* with *ca.* 300 species, like *Lypopodium*, species, contains lycopodium alkaloids; both *huperzine A* and *huperizine B* being present in *H. serrata*. The two alkaloids possess an anti-acetylcholine (anticholine esterase) activity, and in China the use of huperzine A is approved for the treatment of dementia and memory impairment (Xiao and Peng, 1998).

The genus *Selaginella*, with *ca.* 700 species, is widely distributed through the tropical and subtropics; there are also a few temperate species. *S. kraussiana* and *S. willdenowii* are cultivated as ornamentals, while the tufted and poikilohydrous *S. lepidophylla* (rose of Jericho), distributed from southern USA to Peru, is sold as a curiosity; as well as being used in local medicine (Mabberley, 1997).

1.4   Whisk Ferns

The phylum Psilophyta, although richly represented in the fossil record of the Devonian, is today represented by only two genera and, as far as I am aware. are of little economic importance.

The genus *Psilotum* has two tropical and subtropical species, of which the widely distributed terrestrial or epiphytic *P. nudum* has been cultivated as an ornamental in Japan for 400 years (Mabberley, 1997).

2.   HORNWORTS

The phylum Anthocerophyta (hornworts or horned liverworts) contain some five genera, and were formerly included among the liverworts, from which they are distinguished by their long-lived axial sporophytes. Most botanists now consider the hornworts not to be very closely related to the bryophytes. The group appears to be of very little interest to economic botanists.

Some genera, such as *Anthoceros*, house N-fixing cyanobacteria, enabling them to colonise bare rock surfaces. Widely distributed, especially in stagnant water, species are capable of rapid growth, rapidly filling shallow waters and where they eventually decay and become offensive. While *Nitella* spp. are good oxygenating plants they are not recommended for aquaria due to their rampant growth (Perry, 196; Tootill, 1984; Glime and Saxena, 1991).

3.   MOSSES AND LIVERWORTS

The mosses and liverworts together with the hornworts were formerly placed together in the Bryophyta. Under the Five Kingdoms classification they are placed in separate phyla, Bryophyta (mosses), Hepatophyta (liverworts) and Anthocerophyta (hornworts), the latter is briefly discussed in Section 2 above.

The mosses and liverworts are particularly well-represented in the tropics although largely under-investigated. Such lack of interest is partly attributed to their apparent lack of economic importance. However, a global survey of the bryophytes and their uses have now adequately shown more than 350 taxa to be of minor economic importance, a selection of which are given below. Although many contain interesting pharmaceutical compounds, their exploitation has not proved economically viable, others are of domestic importance, and only the *Sphagnum* species appear to be commercially viable (Glime and Saxena, 1991; Schultes and Hofmann, 1992).

Even so, mosses and liverworts do have an environmental value, including: (1) As bioindicators of air and water pollution, soil pH and nutrients, and climatic conditions; (2) SO_2 sensitive species act as indicators of acid rain, while resistant species act as sponges, intercepting the SO_2 and converting it into harmless sulphates; (3) For erosion control by absorbing water, thereby controlling run-off and river flow. *Sphagnum*, for example, can hold up to 30 times its own weight of water; (4) By trapping minerals supplied by the rain and leachates from the canopy, retain minerals that would otherwise be leached from the soil; (4) In arctic, subarctic and alpine ecosystems the *Sphagnum* species in association with Cyanobacteria are capable of fixing nitrogen; (6) Provide nest material for birds, rodents, etc.; and (7) Provide food for a wide range of insects, birds and mammals.

3.1   Mosses

The Bryophyta are widely distributed and represented by *ca.* 800 genera and *ca.* 13,000 species. They are relatively small plants, their short stature being due to the absence of lignification, with the largest upright forms up to 80 cm tall, although some aquatic species can be more than 1 m long.

The presence of unpalatable phenolic compounds renders most mosses inedible. In China they are regarded as famine food, while Laplanders have used *Sphagnum* as an ingredient in bread. Not surprisingly, Lindley (1849) wryly refers to *Sphagnum obtusifolium* as being '*a wretched food in barbarous countries.*' In cold environments the mosses are eaten by a variety of mammalian and avian herbivores. *Polytrichum* and *Hypnum* have even been identified in the stomach contents of *Mammuthus primigenius* (woolly mammoth). It is the aquatic insects, however, that are better known for their feeding on mosses.

In China gallnuts are produced on the leaves of *Rhus javanica* (Chinese sumac) parasitised by the gall aphid *Schlechtendalia chinensis*. The aphids overwinter cocooned on a number of mosses, especially *Plagiomnium* spp. Rich in tannin, the gallnuts are used in tanning, the dyeing of blue silk, and in medicine. Such is the importance of the gallnuts that aphid production is encouraged by cultivating the mosses.

In prehistoric times *Polytrichum commune* (hair moss) provided a fine fibre that could be used for clothing and other purposes. The large and vigorous aquatic moss *Fontinalis antipyretica* (willow moss) was formerly used in the walls of houses in Lapland as a non-inflammable insulating material; a number of other species were similarly used (Dimbleby, 1978; Brightman and Nicholson, 1966; Glime and Saxena, 1991).

The aquatic mosses *Eurhynchium riparioides*, *Fontinalis antipyretica*, etc. have the ability to accumulate heavy metals, and are consequently a valuable tool for monitoring heavy metal pollution. In Japan, which has poor iron reserves, it has even been suggested that *Polytrichum* and *Sphagnum,* could be cultivated near ferruginous springs for iron ore production, even so, the economics and productivity of such a scheme appear rather dubious. Even so, the mineral-tolerant species, including *Merceya ligulata*, syn. *Scopelophila ligulata* (copper moss), do have a potential use in geobotanical prospecting.

3.1.1 Sphagnum

The *Sphagnum* species are of particular economic importance as a source of peat. The total global peat production for 1984 was 260 million tons, with most being produced in the former USSR. More than 45-50% was used as fuel for domestic and industrial consumption, including generating electricity and conversion through a digester to methane. The smoke from the peat-fired malting kilns plays an important role in flavouring the malts for the Scotch whisky industry.

Peat is widely used in horticulture as a soil additive to improve the water-holding capacity, as a medium for the cultivation of acid-loving plants, in air-layering, wreaths, etc. Peat is also used for the reclaiming of strip-mined land. *Thuidium delicatulum* (common fern moss) and *Hypnum cupressiforme* subsp. *imponens*, syn. *H. imponens*, are similarly used as a growing medium by orchid growers. Although generally regarded as a renewable resource, exploitation often exceeds the ability of the peat bogs to recover. Consequently, in the interests of conservation, there is now a move to find suitable waste products, such as bark, coconut fibre, cocoa pod shell, etc. as acceptable horticultural alternatives.

The ability of peat to absorb minerals has been used as an effective filtering and adsorption agent for the treatment of waste water and factory effluents containing acid and toxic heavy metals. The peat can then be burnt to recover the heavy metal ions, such as Ag, Cu, Cd, Hg, Fe, Sb, and Pb. Peat is also used in the treatment of waste oils, detergents, dyes, micro-organisms, air pollution control, rubber reclamation, cigarette filters, sugar refining and as a source of *active carbon*, which is widely used in the chemical industry as a catalyst.

During World War I the absorbent properties of *Sphagnum* spp. were extensively used in surgical dressings but less so in World War II; they are still so used in China today. Its absorptive powers are put to further use in feeding baby pigs, the milled *Sphagnum* being an ideal binder for the iron and vitamins required by anaemic piglets.

In the construction industry peat slabs may be used in the insulation of housing and refrigerators. Peat mixed with concrete and hydraulically pressed to form *peatcrete* has a potential as a cheap, easily sawn and nailed construction material where mechanical strength is not a criterion. It can be cast and moulded into any shape and is already being used for garden rockeries. *Peatwood* from dried, pressed and heat-moulded *Sphagnum* blended with a phenolic resin can be used provide an attractive, lightweight, readily produced material for the construction industry; the ultra-light *peatfoam* based on peatmoss and foamed resin may also have a potential. A synthetic cork, *peatcork*, can be obtained from the coarse peat fraction. Finally, peat may also be used for the manufacture of wrapping paper and pasteboard (Sharp, 1990; Glime and Saxena, 1991).

3.2   Liverworts

The Hepatophyta, contain *ca.* 400 genera and 5000 species. They are relatively simple land plants, small and leafy (leafy liverworts) to large, lobed and thalloid-like thallose liverworts). As a group they appear to be of little economic interest.

The aquatic *Jungermannia vulcanicola* and *Scapania undulata* have the ability to accumulate heavy metals. They are consequently considered useful for monitoring heavy metal pollution. Like the mosses *Polytrichum* and *Sphagnum*, it has also been suggested that *Jungermannia vulcanicola*, could be cultivated near ferruginous springs for iron ore production! The mineral-tolerant species such as *Solenostoma crenulatum*, etc. have a potential for geobotanical prospecting, while other liverworts have been suggested for use as bioindicators of atmospheric pollution.

The Doctrine of Signatures influenced the early medical use of the liverish-looking *Marchantia polymorpha* for treating liver complaints, and the rosette-forming *Riccia* for treating ringworm, despite both lacking in effectiveness. Observations that herbarium specimens of liverworts are seldom eaten by insects has led to the discovery of sesquiterpenoid insect antifeedants in species of *Plagiochila* (scale moss) and other genera; when powdered the liverworts are used to protect stored grain (Glime and Saxena, 1991).

4.   FUNGI

The Fungi, as defined in the Five Kingdoms classification, are a distinct kingdom of saprobic (organisms using dead organic materials for food and commonly causing its decay, on timber, corals, sea grasses, etc.), symbiotic (in lichens), or parasitic eukaryotic organisms. They are believed to contain between 30,000 and 100,000 species. In view of their abundance in the tropics and the paucity of their collection, the total may even be as high as 2 million (Schultes and Hofmann, 1992).

Lichens, formerly classified separately, are also now included in the Fungi. Certain primitive fungi, such as the chytrids, that possess motile stages in their life cycle are now placed in a separate kingdom, the Protoctista, although some authorities prefer to split the Protoctista into Chromista and Protozoa. The most comprehensive classification of the fungi is that of Hawksworth *et al.* (1995) who recognises the phyla Ascomycota, Basidiomycota, Chytridiomycota and Zygomycota. This predates the latest version of the Five Kingdom classification of Margulis and Schwartz (1998) who refer the Chytridiomycota to the Protoctista. The former Fungi Imperfecti, which have no known sexual state, are referred to an artificial assemblage known as mitosporic fungi. Where such mitosporic fungi can be correlated with the teleomorphs, i.e. sexual state, in the Ascomycota and Basidiomycota, they are termed anamorphs or anomorphic states of those groups. While it is probably that many more teleomorph/anamorph state connections will be established, a permanent residue of mitosporic fungi will remain. However, it is possible that in the future advances in molecular technology may enable the residue to be placed with the groups of teleomorphic fungi from which they were derived (Hawksworth *et al.*, 1995).

4.1   Edible Mushrooms and Toadstools

The name ***mushroom*** refers to the usually umbrella-shaped edible fruiting body of members of the Agaricales. The term may also be used in the wider sense for any macroscopic fungal fruiting body. The designation ***toadstool*** is essentially synonymous with mushroom in both the narrow and broad senses but with the implication of toxicity. No account has been taken in the differentiation between mushrooms and toadstools of the large number of fungi that are too leathery to be edible, even though they are not poisonous. According to *The grete herball* of 1526, an anonymous translation from the French and cited by Ramsbottom (1960) "*Fungi ben mussherons .... There be two maners of them, one maner is deedly and sleeth them that eatheth of them and be called tode stoles, and the other doeth not*". The

Middle English use of *tode*, i.e. toad, is based on the mediaeval belief the toad was poisonous.

The fruiting bodies of quite a large number of mushrooms *sensu lato* are eaten by people world-wide, either as a vegetable or condiment, boiled, fried or pickled, but rarely eaten raw. Despite their abundance, there is the general belief in the UK that very few of the many edible wild fungi available are safe to eat, which has resulted in the terms 'mushroom' and 'toadstool' being used to emphasise edible and poisonous fungi respectively. In addition to *Agaricus* spp. the major edible fungi include *Lepista saeva*, syn. *Tricholoma personatum* (blewit), *Macrolepiota procera* (parasol mushroom), *Morchella* spp. (morel) and *Fistulina hepatica* (ox-tongue, lange de boeuf, poor man's beefsteak), the latter being one of the few bracket fungi eaten by man. Although rarely eaten in the UK, *Lycoperdon* spp. (puffballs) are edible when young, when the gleba is solid, as is also *Langermannia gigantea*, syn. *Lycoperdon giganticum* (giant puffball). *Coprinus atramentarius* (common ink cap) is edible when young but must never be consumed in a meal accompanied by alcohol as it will cause vomiting and palpitations. This is due to the presence of **disulphiram** (tetraethylthiouram disulphide), which is used as the drug Antabuse to treat alcoholism. Disulphiram was originally discovered independently of the fungus and it was only later that it was also found to be present in the fungus (Lange and Hora, 1965; Pegler, 1990; Hawksworth *et al.*, 1995)

In general fungi contain 90% water and are rich in protein but poor in fats. The carbohydrates are mainly in the form of chitin in the cell walls. Vitamin C is present in *Agaricus bisporus*, syn. *Psalliota bispora* (cultivated mushroom), *Boletus edulis* (cep, penny bun boletus or steinpilz) and *Cantharellus cibarius* (chanterelle, horn of plenty); the cep and chanterelle also contain vitamin D, negligible quantities of which are also present in the cultivated mushroom. Vitamin K has been detected in the latter and vitamin E in the cep.

Among the more commercially important edible fruiting bodies of ectomycorrhizal fungi are the penny bun boletus, *Cantharellus cibarius* and *Tricholoma matsutake* (matsutake), plus the truffles, i.e. those members of the Ascomycota whose hypogeous *ascoma* (fructifications) occur *ca.* 10 cm below the soil surface. The ascoma of *Tuber melanosporum* (Périgord, black or winter truffle), *T. blotii*, syn. *T. aestivum* (summer truffle) and *T. magnatum* (Italian white truffle) are highly appreciated by Epicureans for their fine flavour. The truffle hunters normally use trained dogs and pigs to sniff them out, but in Sardinia goats and in Russia bear cubs may be used (Wang *et al.*, 1997). Of local importance are the underground fructifications of species of *Terfezea* and *Tirmania*, which occur in the deserts of the Middle East and only appear above the soil surface when mature.

Few fungal fructifications are cultivated. This is because the mycelia (mycorrhiza) of most fungi occur in a symbiotic association with the roots of living plants. Almost all the commercially cultivated mushrooms are non-mycorrhizal fungi and mainly members of the Basidiomycota. Among the commonly cultivated fungi are *Agaricus bisporus* (cultivated mushroom), which has been grown in France since the 17th century, *A. bitorquis* (pavement mushroom), *Agrocybe aegerita*, syn. *Pholiota aegerita* (southern poplar mushroom), *Coprinus fimentarius*, *Kuehneromyces mutabilis*, syn. *Pholiota mutabilis* (two-tone pholiota), *Pleurotus eryngii* associated with the roots of *Eryngium campestre* (eryngo) and *Stropharia rugoso-annulata* (king stropharia). In the Far East *Lentinus edodes* (Japanese wood mushroom, shii-take) accounts for over 20% by value of the global mushroom production. Other cultivated mushrooms include *Auricularia auricula-judae* (Jew's ear), *Flammulina velutipes* (velvet shank), *Pholiota nameko* (slime mushroom), *Pleurotus ostreatus* (oyster mushroom), *Tremella fuciformis* and *Volvariella* spp. *Volvariella bresadolae* is grown in the Philippines on rice, wheat or sorghum straw, *V. volvacea* (paddy straw mushroom) is similarly cultivated in China, Indochina, Malaysia, Philippines, also in Madagascar and Africa, and *V. volvacea* var. *heimii* in Madagascar.

It is possible to cultivate some of the ectomycorrhizal species (see Chapter 5) such as *Morchella esculenta* (morels) on apple pomace from cider presses and *Lepista nuda*, syn. *Tricholoma nudum* (wood blewit) on beech-leaf compost, while in Italy, inoculums of *Suillus granulatus*, syn. *Boletus granulatus* (granulated boletus) have been successfully grown on *Pinus radiata*. Despite such possibilities, only the Périgord and Italian white truffles are cultivated commercially in host plantations (Ramsbottom, 1960; Brouk, 1975; Rambelli, 1985; Slee, 1991; Hall *et al.*, 1998a, b).

Forming a non-fruiting, resting stage, *sclerotia* are firm and frequently rounded masses of hyphal tissue, with or without host tissue; some are edible. In Australia the underground sclerotium of *Polyporus myllitae*, syn. *Mylitta australis* (blackfellows' bread) is among the world's largest sclerotia. The densely compacted and agglutinated mass is up to 20-30 cm in diameter and weighing 4 kg or more; it is eaten by the Aborigines. Other edible examples eaten in times of food scarcity are the subterranean sclerotia of *P. indigenus* and *P. saporema* from Amazonia; they weigh over 3 kg, half of which consists of carbohydrates. The sclerotium of *Pleurotus tuber-regium* is similarly eaten in Nigeria. Other sclerotia, such as the dark, club-shaped structures of *Claviceps purpurea* (ergot) found in the ears of cereals and grasses are highly toxic, producing the serious physiological disease known as ergotism in both humans and livestock (Brouk, 1975; Prance, 1984; Hawsworth *et al.*, 1995).

4.2   Hallucinogenic Fungi

Some 37 fungal taxa from around the world are recognised as hallucinogenic, 19 of which belong to Central American taxa of the genus *Psilocybe*. The genera *Amanita* and *Psilocybe* are discussed here (Schultes and Hofmann, 1992). See also Chapter 20 for further discussion on hallucinogenic plants and their recreational and symbolic use.

Without doubt the most spectacular hallucinogenic fungus is *Amanita muscaria* (fly agaric), which is widely distributed through Eurasia and North America. The fly agaric owes its vernacular name to the former 13th century use of the sliced cap soaked in milk as a fly trap. The allegedly lethal affect on flies is due to the *ibotenic acid* present which, in humans breaks down into the active hallucinogen *muscimole*, which is biologically interesting in that the active principle is atypically excreted unmetabolised. Reports regarding the neurotoxic alkaloid *muscarine* also being present have proved erroneous.

The fly agaric is probably man's oldest hallucinogen, and is possibly represented in ancient India as the legendary god-narcotic Soma, where the Vedic deity Indra used Soma as the source of his strength; it is believed that the sacred drink of the Soma cult was derived from the fly agaric. The fungus is used by Finno-Ugrian tribes of north-eastern Siberia as a masticatory and shamanistic inebriant. Two small fructifications are dried and chewed until soft, and then swallowed; the intoxication is strong enough to last an entire day. Its habitual use completely shatters the nervous system, so much so that its trade was made a penal offence by Russian law. The magico-religious cult of the fly agaric is believed to have been carried by ancient Asiatic migrants to the New World (Brouk, 1975; Rambelli, 1985; Blackwell, 1990; Schultes and Hofmann, 1992).

The related *A. phalloides* (death cap), *A. verna* (spring amanita) and *A. virosa* (destroying angel) are noteworthy for their extreme cytopathological toxicity, 50 g of fresh mushroom being lethal to adult humans. They rank as the world's major cause of death from eating poisonous mushrooms. *A. phalloides* is the most dangerous fungus known and is responsible for over 90% of deaths due to fungi. The two alkaloids responsible in *A. phalloides*, and probably in the others, are the nerve and gastro-enterological toxin *amanitin* and the liver toxin *phalloidin*. The symptoms are graphically described by Ramsbottom (1949).

Magico-religious ceremonies by the ancient civilisations of Central and South America possibly date back to prehistoric times. Hallucinogenic fungi belonging to the genus *Psilocybe*, of which *P. hoogshagenii* and *P. mexicana* are the two most important and best documented. *Conocybe siligineoides, Panaeolus sphinctrinus* and *Stropharia cubensis* are also used. It should be noted that the latter is a coprophagous species and therefore not indigenous and must post-date the arrival of the Conquistadors and their herbivores. It has been suggested that it was possibly introduced by cattle brought by Spanish traders from the Philippines to Mexico.

The range of fungi used by the shamans is dependent on the season, weather and specific usage. Pairs of fungi are used in the religious rites; the similar Finno-Ugrian tribal use of a pair of *Amanita muscaria* fructifications has already been noted above. The active ingredient is the indole alkaloid *psilocybine*, the phosphoric acid ester of *psilocine*, of which only traces usually occur; both compounds can be synthesised. So far, present evidence indicates that it is only in Mexico that psilocybine-containing mushrooms are used in native ceremonies. Although the archaeological evidence suggests sacred mushroom cults may also have existed as far south as Peru, there is no past or present ethnobotanical evidence to support such claims (Brouk, 1975; Schultes and Hofmann, 1992).

4.3   Other Macro-Fungi

The fruit bodies of the bracket fungus *Fomes fomentarius* (tinder fungus) are the source of the soft, corky material known as *amadou*, formerly used as tinder and now used for drying fishermen's flies. The earliest historical use of amadou appears to be its presence in the belt pouch of the Neolithic 'Iceman' of Hauslabjoch, Austria. The related *F. officinalis* (female, white or purging agaric) was formerly a noted universal panacea (Pegler, 1990; Spindler, 1994; Hawksworth *et al.*, 1995).

Another bracket fungus, *Fistulina hepatica* (beefsteak or oak tongue fungus), infests a number of broad-leaved species, especially chestnut and oak. In the early stages of the infestation of oak the fungal mycelia produce a streaky brown discoloration of the heartwood, before causing any serious decay. Such **brown oak** is much sought after by furniture manufacturers and consequently fetches an enhanced price. Similarly, the mycelia of *Chlorociboria aeruginascens*, syn. *Chlorosplenium aeruginascens* (green wood cup) stain the wood blue-green. The wood of such **green oak** was formerly highly prized for marquetry (see Chapter 12), and especially for Tunbridge ware (Mabberley, 1997).

A number of the macro-fungi are important destructive parasites of trees, shrubs and timber. *Armillaria mellea* (boot lace fungus, honey fungus) especially is a destructive parasite of woods, plantations and garden trees and shrubs. It is readily identified by the resemblance of the blackish-brown rhizomorphs (tough, cord-like and fused mass of parallel aligned hyphae) to leather boot laces. The thin and spreading fruiting bodies of *Coniophora puteana* (wet rot fungus, cellar fungus) is a common species on the trunks of dead conifers; it also attacks wood in buildings, hence its vernacular names. The allied *Serpula lacrymans* (dry rot fungus) occurs inside buildings, especially cellars and cold, damp houses, where it is notorious for the damage it causes to untreated coniferous woodwork. Closely related species occur on wood in the wild (Pegler, 1990).

4.4   Yeasts

The *yeasts* are not a formal taxonomic unit but a growth form exhibited by a range of unrelated unicellular fungi that reproduce asexually by budding and have the ability to ferment carbohydrates. It was Louis Pasteur (1822-1895) who was the first to establish that the yeasts from the grape skin were living, single-celled organisms. Some 590 species of yeasts are recognised by Barnett *et al.* (1990), yet only the appropriate strains of *Saccharomyces cerevisiae* (baker's or brewer's yeasts) are commonly used by the food industry (see Chapter 9), the potential of other and often more versatile yeasts have rarely been exploited.

Yeasts may be used for various purposes in addition to their traditional roles in baking and alcohol fermentation (see Chapter 9), breaking down glucose into *ethanol* (*ethyl alcohol*) and CO_2. For alcohol fermentation the yeasts are sometimes assisted by other saccharifying moulds. In addition to ethanol the yeasts may also be used for: (1) Producing *lactic acid, ethanoic (acetic) acid, gluconic acid, glutamic acid* and many other amino-acids; (2) Producing protein from *alkanes* (paraffins - aliphatic hydrocarbons) and paper-pulp waste; (3) Producing various alditols, such as *glycerol* or *D-glucitol*; and (4) As sources of enzymes such as *β-D-fructofuranosidase* and *lipase*.

Chemists have also used yeasts for producing novel carbon-carbon bonds, optically active compounds such as *methyl-diols* from aldehydes, secondary alcohol derivatives used in chiral building blocks for synthesising natural products, and biologically active molecules such as *prostaglandins* (a group of related unsaturated hydroxylated fatty acids occurring in mammalian organs, tissues and secretions). There is also a potential for the further exploitation of yeasts in synthesising precursors of important natural products as well as for much simpler processes, e.g. removing contaminating compounds such as a sugar of a particular configuration from a racemic acid mixture. Enantio-selectivity, i.e. selectivity of isomers differing in their configuration at a chiral atom, can also be improved by using a mutant that lacks an unwanted enzyme.

Apart from the widely used *Saccharomyces cerevisiae*, chemists have occasionally used *Candida tropicalis* in the production of *dodecanedioic acid* and its derivatives for the plastics industry. They have also used *Rhodotorula glutinus*, syn. *R. rubra, Schizosaccharomyces pombe* or *Zygosaccharmyces bailii* in the processes involved in synthesising prostaglandins. Genetic engineering can enhance such versatility, while further screening could greatly increase the exploitation of more kinds of yeasts. Unfortunately, the *International Code of Botanical Nomenclature* (see Chapter 2) by insisting on the preservation of yeasts for type specimens instead of living material has meant that those species known only from the type specimens, e.g. *Filobasidiella depauperata*, are not available for research (Barnett *et al.*, 1990). The location of resource centres holding culture collections of yeasts are listed in Kirston and Kurtzman (1988).

The use of yeasts for baking and brewing has been practised since early Biblical times where, in the case of leaven bread, the practice is recorded of keeping back a piece of fermented dough to initiate a later fermentation. The commercial production of, for example, bread yeast, is nearly always grown through several stages under laboratory conditions on a solution of molasses and water. A small quantity of a pure yeast culture is first grown in a sterilised and purified molasses solution to which is added ammonium salts and phosphates. Within 24 h the culture is transferred to a larger container and more molasses solution and salts are added, and then aerated in order to minimise alcohol production. The process is repeated four or five times, using larger and larger containers until *ca.* 5 tonnes of a fermenting liquid known as the *mother* or *seed yeast* is obtained. The mother yeast is divided into three separate tanks, which are used to feed fresh ferments until the initial 5 tonnes has increased to 45 tonnes. The yeast growth is then arrested as a suspension of yeast cells in a liquid residue of molasses solution, after which it is washed and separated in high speed centrifugal separators. The white, thick and creamy yeast at the bottom of the separators is then filtered and stored at 2°C until the excess water can be extracted by a rotary vacuum filter, after which it is compressed, extruded and packed in blocks readily for marketing (Moldenke and Moldenke, 1952; David, 1978).

A considerable quantity of *single cell proteins (SCP)* in the form of baker's yeast are produced commercially using starch products; SCP, mostly prepared by using methanol as the feedstock, is also used for animal feed. It is doubtful, however, whether starch-derived SCP can compete economically with, for example, soya proteins. Even if the starch-derived SCP was competitive with soya meal, other more obvious raw materials are available, including waste waters, bran, cellulose waste, molasses, etc.

Candida utilis is commercially multiplied through the fermentation of molasses to yield *food yeast*. During World War II food yeast provided much of the dietary protein requirements. Unfortunately the high vitamin B content was a problem if eaten to excess as it could give rise to hypervitaminosis. The process was later used by the sugar industries in East Africa, India and Malaya with the intention of providing supplementary protein to the diet but the schemes failed due to consumer resistance, even among starving populations. Nevertheless such microbial processes do have the great advantage over traditional crops in being able to be carried out in a controlled sterile environment independent of the climate (Postgate, 1992).

Fermentation processes are also used to improve the flavour of vegetables, spices, beverage materials, etc. Two basic types of fermentation processes are recognised: (1) By the use of the plant's own enzymes, e.g. the enzyme fermentation of the polyphenol derivatives of catechin and gallic acid present in the leaves of *Camellia sinensis* (tea) to produce *o-quinones*, which polymerise to produce coloured astringent condensation products on brewing, and the enzyme fermentation of the glucosides present in the capsules of the South American orchid *Vanilla planifolia* to yield *vanillin*, the source of the fragrance and flavour of *vanilla*, which is widely used in the food and perfume industries. A less commercially desirable vanillin is also produced as a by-product of the wood pulp industry; (2). Fermentation by bacteria and yeasts, as used in the processing of the 'beans' of *Theobroma cacao* (cocoa). That well-known standby of Chinese-American cuisine, soy sauce, is obtained by the fermentation of soya beans, rice and cereal, principally by *Aspergillus oryzae* and *Zygosaccharomyces rouxii*, syn. *Z. soja*. The enormous quantity of *citric acid* required by the soft drinks industry is also obtained by fermentation, by the action of *Aspergillus niger* on sugar (Brouk, 1975; Purseglove, 1985, 1987; Sharp, 1990).

Other industrial uses of yeasts include the development of strains of *Trichoderma reesei* for the fermentation of hemicellulose and lignin to yield *ethanol*. Plastics, which are usually a product of the petrochemical industry, can also be obtained from fermentation processes. The enzymes required for the production of alkene oxides for polymerisation to yield plastics, can be synthesised from alkene by the lichen *Cladonia*, syn. *Cladoniomyces*, the fungus *Oudemansiella mucida* and the bacteria *Flavobacterium* spp. Fermentation of starch or sugar by the fungus *Aureobasidium pullulans*, syn. *Pullularia pullulans* is used in the production of the derived plastic **pullulan**, which resembles styrene in gloss, hardness and transparency but with much greater elasticity. Since such compounds do not require a plasticiser, they are especially safe for food packaging. An alternative source of imported **γ-linolenic acid (GLA)** from *Oenothera biennis* (evening primrose) has been developed in Japan using strains of *Mortierella* sp. in a liquid culture medium for the commercial production of a lipid rich in γ-linolenic acid (Suzuki, 1988; Subba Rao and Kaushik, 1989; Robbins, 1995).

Although in the past the bacteria have been largely favoured by industry for the biodegradation of lignocellulose and other waste products, there is now an increasing interest in the use of fungi, especially the various white rot fungi for treating such waste (Hawksworth *et al.*, 1995).

4.5   Other Micro-Fungi

Fungi provide the largest group of plant pathogens, although only *ca.* 8% of the 6000 genera are responsible. The term **mould** is used to describe those micro-fungi that produce a distinct mycelium or spore mass, i.e. mildew, often resembling a velvety pad on the surface of its host. Among the micro-fungi are members of the Erysiphaceae (powdery mildews), Uredinales (rust fungi) and Ustilaginales (smut fungi). The powdery mildews differ from the downy mildews Peronosporaceae (downy mildews) of the Oomycota, in that they are chiefly superficial and unlike the downy mildews do not penetrate the inner tissue.

The micro-fungi are also responsible for a number of human diseases, e.g. *Candida* spp. for thrush, and *Microsporum* spp. for ringworm. Fungal spores, especially those of *Aspergillus*, *Penicillium*, are among those responsible for cases of allergic alveolitis, i.e. respiratory diseases, examples of which include farmer's lung, cheese-maker's lung and mushroom-worker's lung. Other fungi contain mycotoxins that contaminate food and food products consumed by humans and animals, and render them poisonous and sometimes carcinogenic, e.g. *Penicillium icelandicum* is the causative agent of yellow rice, a known carcinogenic of rodents and possibly of humans (Holliday, 1989; Bailey, 1990; Postgate, 1992; Hawksworth *et al.*, 1995).

The antibiotic properties of micro-fungi such as *Penicillium notatum* and the action of the purified penicillin and its derivatives, **ampicillin, benzylpenicillin, methicillin**, etc. against the bacteria *Neisseria gonorrhoeae, N. meningitidis, N. pneumoniae, Staphylococcus, Streptococcus*, etc. are well documented. Other sources of antibacterial drugs and pharmaceuticals include *Claviceps paspali* and *C.*

purpurea for **ergometrinine, ergotamine, ergotoxine,** etc. The sclerotium (ergot) of the latter is noteworthy in being the only fungus now in the British Pharmaceutical Codex (Subba Rao and Kaushik, 1989; Macpherson, 1995).

In the dairy industry *Penicillium roqueforti* is added to the curd in order to produce blue-veined Gorgonzola, Roquefort and Stilton, and *P. expansum* for such blue-veined cheeses as dolce verde, where the colour of the veins are due to the spores of the inoculum. It is noteworthy that the *Penicillium* spp. used in cheeses are the only fungi that are consumed entire. Many other cheeses are ripened after they have been shaped by smearing the surface of the young rind with the appropriate microbial inoculum, which then penetrates the young rind and spreads throughout the cheese, e.g. *P. camemberti* and *P. caseicolum* for Camembert and Brie respectively. These fungi may also be eaten separately or used for a **mycoprotein.**

In China, Taiwan and Japan the aquatic perennial grass *Zizania latifolia* (Manchurian wild rice) is cultivated for the greatly enlarged and succulent culms infected by the smut fungus *Ustilago esculenta*. The culms are cooked and eaten as a vegetable; they are marketed fresh, frozen or canned (Brouk, 1975; Terrell and Batra, 1982; Postgate, 1992; Hawksworth *et al.*, 1995). See Section 5 for bacterial inoculums, and Chapter 9 for fermented foods.

In the Far East pure cultures or mixtures of moulds as well as yeasts are traditionally used as starters to aid alcoholic fermentation processes by hydrolysing the starch into sugars. For example, the yeast *Candida sake* and a pure culture of *Aspergillus oryzae* are used for fermenting rice for the alcoholic beverage sake, while a mixed starter would be used in the fermentation of soya for tempeh. In China the starters, known as kyoku-shi and in Japan as koji, contain in addition to the yeasts species of *Absidi, Aspergillus, Monascus, Mucor, Penicillium* and *Rhizopus.*

A number of predatory and parasitic fungi that attack and help control nematodes have been found among the Fungi and the fungal-like organs of members of the Chytridiomycota, syn. Chytridiomycetes, and Oomycota, syn. Oomycetes which, in the Five Kingdom classification are both referred to the Protoctista.

The predatory fungi capture the nematodes by adhesive processes through an anastomising network of hypha, some of which are sticky, as in *Arthrobotrys,* especially *A. oligospora,* or by a hyphal network with adhesive loops or projections. Others form mechanical ring traps of constricting or non-constricting rings in which the nematode becomes wedged. More sophisticated trapping devices by non-trap forming endoparasitic fungi involve a germ tube from a sticky spore penetrating the nematode cuticle, while *Haptoglossa* of the Oomycota has evolved a spore so complex that it is capable of injecting a passing nematode. Others retard nematode development by chemical secretions. *Fusarium oxysporum,* for example, inhibits the growth of *Heterodora schactii* (beet cyst nematode) when both are present in a field of sugar beet. However, the converse can also occur, thus the severity of rice stem rot caused by *Leptosphaeria salvinii* decreases when the rice plant is also attacked by

Aphelenchoides besseyi, the nematode responsible for white tip (Drechsler, 1934; Webster, 1972).

The University of Reading and the Institute for Agricultural Crop Research, Harpenden are currently investigating the ability of the soil-borne V*erticillium chlamydosporum* to suppress root-knot nematodes, the principal nematode pests of vegetables and some field and perennial crops throughout the tropical, subtropical and warmer temperate regions. But before such nematophagous fungi can become available for field control use there remain considerable problems regarding their formulation and application (Barron, 1977; Hawksworth *et al.*, 1995; Gowen, 1998).

4.6   Lichens

Lichens, from the Latin *lïchën*, from Greek *leikhën*, licker, from *leikhein*, to lick (presumably referring to their often tongue-like thallus), form a large and successful but curious group of an essentially obligate, stable, self-supporting association of a *mycobiont* (fungus) and a *phycobiant* (green alga or cyanobacteria). Their fungal and photosynthetic parts each have a separate name, but the name by which the lichen is known refers to the fungal partner. They are no longer regarded as members of the Plantae, with most placed in the phylum Ascomycota of the Fungi Kingdom. There are some 450 genera containing from 16,00 to 20,000 species (Perez-Llano, 1944; Long, 1994 Hawksworth *et al.*, 1995).

Three main types of growth habit are recognised: (1) *Crustose lichens* growing closely attached to the substrate; (2) *Foliose lichens*, generally attached loosely to the substrate by tufts of hyphae known as rhizinae, the thallus with lobed, leaf-like extensions; and (3) *Fruticose lichens*, which are either erect and bushy or hanging and tassel-like, attached at only one point, e.g. *Usnea* (Brightman and Nicholson, 1966; Tootill, 1984; Schultes and Hofmann, 1992).

A number of lichens were formerly used, especially in northern Europe and America, for dyeing wool and other fibres. These dyes have managed to resist supplantation by synthetic dyes, especially by Scottish and Irish tweed industries, for far longer than most other natural dyes. Their use meanwhile by the craft textile industries in Europe and North America has been undergoing a revival in recent years, although it is doubtful if there will be any significant global improvement in the present low level of international trade. For architectural and other models of trees and shrubs species of *Cladonia* are dyed green and made soft and pliable with glycerol (glycerine).

Roccella tinctoria (orchil), from the Mediterranean region, was formerly used for colouring wool, silks and wines, and for staining wood. The dye is now used by the food industry for pickled tongue, sauces and spices. It is prepared by the slow aerobic fermentation of the macerated lichen with aqueous ammonia for *ca.* 2 weeks. The blue orchil liquor is then extracted with water and the ammonia is driven off by heating to yield *red orchil*. The red orchil is then evaporated and ground to a fine powder or paste known as *orseille*. A purple dye is similarly extracted from *Ochrolechia tartarea* (cudbear) - the name apparently being derived from the Christian name of a Dr **Cuthbert** Gordon, an 18th century Scottish chemist, who patented the dye. A litmus is extracted from the Malagasy *Roccella montagnei* in a similar process that also involves the addition of potash and lime to the aqueous ammonia. In Peru species of *Parmelia* and *Umbilicaria* are also used to produce brown and reddish dyes respectively (Hill, 1952; Brightman and Nicholson, 1966; Brouk, 1975; Montagné, 1977; Zumbühl, 1979; Hale, 1983; Long, 1994; Green, 1995).

The lichen dyes can often be used without a mordant to produce a range of subtle, muted colours, from yellow, brown, red, purple to violet. These lichen pigments involve a diversity of oxygen ring compounds which are generically, if inaccurately, known as *lichen acids*. The orchil type dyes are direct dyes, which are fugitive to the light, and represent a complex series of orcein derivatives of natural depside pigments, consisting of a mixture of *oxy-* and *amino-phenoxazon* or *phenoxazin* formed from micro-aerophilic oxidation of orcinol-type secondary metabolites in the presence of ammonia. The reactive *Parmelia* dyes are light-fast and are produced by colourless lichen metabolites with an aldehyde group, such as the depsidone *salazinic acid* (Green, 1995; Hawksworth *et al.* 1995).

Lichens are also used in the perfume industry as fixatives of other ingredients. An unbelievable 8000-9300 tons of the oak mosses, chiefly *Evernia prunastri* (mousse de chêne, stag's horn) and *Pseudevernia furfuracea*, syn. *Evernia furfuracea*, are collected annually in Yugoslavia, southern France and Morocco (Brightman and Nicholson, 1966; Hale, 1983).

They also serve as a source of food and fodder. Species of *Lecanora* and *Sphaerothallia* occupy vast tracts of barren plains and mountains of western Asia and northern Africa. Following long periods of drought they curl up and break loose from the soil. Being extremely light the winds may occasionally transport them for considerable distances before depositing them on the ground, where they may sometimes form layers several centimetres in depth. These lichens are believed to be the Biblical manna 'that fell from heaven' (*Numbers*, **11**, 9). Even today the lichens are gathered up by the Bedouin and, with the addition of meal, made into bread. Bread can also be made with fermented *Evernaria. prunastri* and *Pseudevernia furfuracea*, while in Japan the foliose *Umbilicaria esculenta* (iwatake, rock tripe) is regarded as a great delicacy. An edible jelly can be made from the fruticose *Cetraria islandica* (Iceland moss), although the plant must first be thoroughly soaked before boiling to leach out the bitter flavour, while species of *Cladonia* require boiling with soda to remove the bitter and irritating *fumaroprotocetraric acid*, etc. before eating. Lichens may also be fermented for alcoholic beverages. The reindeer lichens *Cladonia*, syn. *Cladina*, and *Cetraria* can form 15 cm or more high carpet-like masses in the arctic and subarctic regions, and provide essential grazing, while corticolous lichens, such as *Usnea* are browsed by members of the Cervidae (reindeer, moose, caribou) and Bovidae (musk ox). The lichens are also harvested for domestic stock (Perez-Llano, 1944; Moldenke and Moldenke, 1952; Brightman and Nicholson, 1966; Hale, 1983).

A number of medicinal uses have also been recorded, some have their medical properties attributed under the Doctrine of Signatures for treating cutaneous afflictions, but without success. As a consequence the term lichen is now applied medically to a group of chronic skin diseases. Some contain antibiotics, e.g. *usnic acid*, a yellow dibenzofuran derivative present in *Cladonia arbuscula* and *Usnea* spp. with anti-Gram +ve bacterial and antifungal properties. The foliose *Peltigera canina* is used to treat liver complaints and formerly to treat rabies, and *Cetraria islandica* is used for lung diseases and diabetes. In lowland Amazonia *Dictyonema sericeum* is regarded as powerful shaman medicine of the Auca tribe; it is also possible that it is hallucinogenic. Hallucinogenic lichens have also been reported from north-western North America, although confirmation of such activity is lacking (Perez-Llano, 1944; Brightman and Nicholson, 1966; Hale, 1983; Schultes and Hofmann, 1992; Macpherson, 1985; Hawksworth *et al.*, 1995)

Squamulose desert species of *Catapyrenium*, *Heppia*, *Peltula* and *Psora*, and the crustose *Diploschistes* grow appressed to the soil and effectively seal and stabilise the surface. I have personally observed such lichens in the Sudan resulting in no rain run off from otherwise loose sand dunes, where run off would otherwise have been expected.

Other lichens can provide sensitive indicators of NO_2, SO_2, Cr, Hg, Ni, Pb and Zn pollution, especially SO_2. This is because the mycobiant or phycobiant is sensitive to the pollutants, which can disrupt membranes and lead to chlorophyll breakdown. For example, in Europe, *Hypogymnia physioides* dies when SO_2 levels exceed 60-70 µg m^3. By using the different sensitivities of the lichens it is possible to provide a very sensitive monitoring of pollution levels. The measurement of metal and radionuclides take up has even been used to map pollution from the 1986 Chernobyl nuclear disaster. In California the colour changes in *Lecanora cascadensis* are used as an indicator of copper, while *Cetraria* species is highly correlated with the presence of marble and limestone deposits (Hawksworth *et al.*, 1995).

5.   BACTERIA

The kingdom Bacteria represents a heterogeneous group of prokaryotic organisms, i.e. organisms in which the nuclear material within the cell is not separated from the cell protoplasm by a nuclear membrane and where cell division is amitotic, i.e. without the appearance of chromosomes; the nuclear membrane does not break down and a spindle is not formed. The Cyanobacteria, formerly known as the Cyanophyta (blue-green algae) are consequently now included in the kingdom. Some 10,000 species of bacteria are known, and these probably represent only a fraction of the total number. There are, for example, over 5000 strains of bacteria, actinomycota, plasmids and bacteriophages that are available in the UK for use in a wide range of industrial and teaching institutes. See Dando and Young (1990) for a list of their uses and relevant literature references. Their taxonomy and nomenclature, documented by Holt (1984-1989) and Holt *et al.* (1994), is difficult and imperfectly understood.

Two physiologically distinct types of bacteria are recognised, depending on their staining properties using a violet dye mordanted with iodine or picric acid and then decolorised in alcohol. Bacteria that retain the stain are said to be **Gram positive**, honouring the staining method devised by H.C.J. Gram for staining bacteria in tissue sections. Such Gram positive bacteria tend to be more exacting in their nutritional requirements, more susceptible to antibiotics, and more resistant to plasmolysis. They include the Corynebacteriaceae, Lactobacillaceae and Neisseriaceae. Those bacteria that are decolorised and take up a counterstain, usually pink, are referred to as **Gram negative**; they include the Cyanobacteria.

In terms of distribution and numbers bacteria are the most successful of the life forms. Many occur saprophytically in the soil and are important as decomposers in the carbon cycle, while others are important in the nitrogen cycle, e.g. the nitrifying bacteria, *Nitrobacter* and *Nitrosomonas* (see Chapter 5), and denitrifying bacteria, e.g. species of *Clostridium, Micrococcus, Pseudomonas* and *Thiobacillus*. Some are toxic, others are pathogenic, causing bacterial diseases of plants and animals, while others are important in fermentation processes for the food and drinks industries (Tootill, 1984; Holliday, 1990; Hawksworth *et al.*, 1995).

Among the toxic bacteria are the blooms in stagnant or slow-flowing waters of the blue-green colonial *Anabaena circinalis* and *Microcystis aeruginosa*, syn. *Anacystis cyanea*, especially in nutrient-rich regions receiving sewage and drainage effluents from settlements and agricultural lands. The bacteria contains toxic **microcystins**, a group of related cyclic heptapeptides which damage the liver and are not inactivated by the usual treatments of drinking water. The filamentous *Oscillatoria* spp., which are usually present as mats in bottom sediments, may become detached and rise to the surface, where it too releases substantial amounts of microcystins, as well as producing potent neurotoxins similar to **saxitoxin** found in marine algae. Other marine Dinomastigota (dinoflagellates) of the Protoctista are responsible for red tides, resulting in the death of marine animals, including sea birds and farmed salmon in UK (May, 1978; Hawksworth *et al.*, 1995; Bailey, 1999).

Examples of bacterial plant pathogens include species of *Arthrobacter, Clavibacter, Curtobacterium, Pseudomonas, Rhodococcus*, etc. Among the bacterial diseases of humans are: brucellosis from *Brucella abortus*, boils, pimples and spots caused by *Staphylococcus aureus*, tuberculosis due to *Mycobacterium tuberculosis*, pneumonia from *Streptococcus carinii* and other bacteria, and the bubonic plague which so decimated the population of Europe during the Middle Ages from *Yersinia pestis*, syn. *Pasteurella pestis*. Other pathogenic bacteria can also be advantageous. The soil-borne *Pasteuria penetrans*, for example, is currently under investigation at the Institute for Agricultural Crop Research, Harpenden and by the University of Reading. Field results have revealed *P. penetrans* able to slowly reduce root-knot nematode populations in continuously cultivated crops (Holliday, 1990; Postgate, 1992; Gowen, 1998).

Other bacteria can be used for the benefit of man as antibiotics. For example, among the non-filamentous bacteria two genera are responsible for *ca.* 500 antibiotics, while *ca.* 3000 antibiotic agents are known from the Actinobacteria (Actinomycetales). However, of the *ca.* 5000 antibiotics known, only *ca.* 100 are marketed. They include: *Streptomyces aureofaciens* (**tetracycline**), *S. erythraeus* (**erythromycin**), *S. griseus* (**streptomycin**), *S. venezuelae* (**chloramphenicol**, which is now synthesised), etc. *Arthrobacter simplex*, syn. *Corynebacterium simplex*, is the source of **prednisolone**, which has five times the activity of **cortisone**, while genetically engineered *Escherichia coli* is now capable of producing **insulin** 100 times faster than was possible using standard animal production processes (Subba Rao and Kaushik, 1989; Macpherson, 1995).

Nostoc commune and *N. ellipsospermum* are cultivated and eaten in China and in central Asia respectively. Their culture and use as food created considerable interest, and in recent years has led to the culture of other highly productive members of the Cyanobacteria, such as species of *Arthrospira, Nematonostoc, Nostoc, Phormidium* and *Spirulina*, especially the latter, as sources of SCPs (single cell proteins) as a feed supplement for poultry, the high **xanthophyll** content giving a good colour to egg yolks. Because of their high cost the SCPs are not fed to ruminants. In the UK SCP is produced by the action of *Methylophilus methylotrophus* on methanol.

Dried cakes made from floating mats of *Arthrospira maxima*, syn. *Spirulina maxima*, from Lake Texcoco in Mexico, and *A. platensis*, syn. *S. platensis*, found growing in shallow lakes near Lake Chad in Africa, have been known and used as a non-toxic, highly nutritious and easily digestible food by the local inhabitants for centuries. *A. maxima* was harvested for food until the Spanish conquest in the 16th century, after which harvesting ceased. During the past two decades commercial firms extracting soda from the lake have resumed harvesting the cyanobacteria as a by-product, mainly for use as a SCP feed for livestock. *A platensis* is of local importance as a food source as well as being the main food of *Phoeniconaias minor* (lesser flamingos) in the saline lakes of the East African Rift Valley. Both contain 62-68% dry weight of protein.

The economics of SCP culture from organisms such as *Arthrospira* and *Spirulina* are interesting. When cultivated in open ponds *Spirulina* consumes 25,000 m^3 ha^{-1} of water, more water per unit area than the 17,000 m^3 required for rice. But in terms of protein production *Spirulina* requires only 1000 m^3 ton^{-1}, considerably less than the 7000 m^3 required by soya. Containing 75% protein, it has been estimated that 1 ha of *Spirulina* could yield 25.4 tonnes of protein (65% of dry weight) as compared to 4.05 tonnes from wheat and 0.4 tonnes from beef.

Spirulina is also a source of various organic compounds, including the essential amino-acid *γ-linolenic acid* (*GLA*), which has a valuable pharmaceutical application. Currently being marketed in health shops as a dietary supplement, the claimed benefits of weight loss, physical fitness and well-being, etc. can probably be attributed to the GLA rather than the protein, vitamin or mineral content. Unfortunately, GLA production from *Spirulina* is more expensive than that from *Oenothera biennis* (evening primrose).

While it is possible to produce cyanobacteria in artificial open ponds, there are problems with contamination by algae and other micro-organisms such as *Chlorella*, *Euglena* and *Scenedesmus*. These problems can be overcome by using a very alkaline growing medium, or by the selection of suitable species, especially thermophilic and halophilic cyanobacteria, for growing in arid or semi-arid environments where the harsh conditions would be lethal to the contaminants. *Spirulina subsalsa* var. *crassior* and *Synechococcus elongatus* var. *vestitus* appear to be suitable candidates, the former with an estimated annual yield of *ca.* 7.5 tonnes ha^{-1} and the latter *ca.* 12.5-25 tonnes. While the *Spirulina* can be effectively harvested by netting the unicellular *Synechococcus* has proved more difficult. The problem has been overcome by feeding directly, without harvesting, to either brine shrimps or *Tilapia* (Léonard and Compère, 1967; Göhl, 1981; Ciferri, 1983; Ciferri and Tiboni, 1985; Roughan, 1989; Shinohara *et al.*, 1989; Horrobin, 1990b).

In the food industry lactic acid fermentation by *Lactobacillus brevis*, *L. plantatrum* and the blue green *Leuconostoc mesenteroides* are used for pickling cucumbers and sauerkraut; they are all naturally present on cabbage leaves. The sour taste of rye bread is also due to lactic acid bacteria. A lactic acid fermentation process dominated by *Streptococcus* sp. with *Pediococcus* sp. is involved in the preparation of the traditional Sudanese food known as sigda from the oilseed cake of *Sesamum orientale* (sesame); other participants in the fermentation processes are the yeasts *Candida* sp. and *Saccharomyces* sp. Also in the Sudan, the bacterium *Bacillus subtilis* and the yeast *Rhizopus* sp. are used in the fermentation of the leaves of *Senna obtusifolia*, syn. *Cassia obtusifolia*, for the production of kowal, a high protein food from what was otherwise an unpalatable plant with toxic glycosides. The crushed seeds of *Hibiscus sabdariffa* (kerkade) are also fermented for furundu. In West Africa *Bacillus* spp., especially *B. subtilis*, are the principal micro-organism used in the preparation of ogili from the seed of *Riccinus communis* (castor oil plant) and ogiri from the seeds of the watermelon *Citrullus lanatus*, syn. *C. vulgaris,* for food (Brouk, 1975; Dirar, 1983; Subba Rao and Kaushik, 1989; Elfaki *et al.*, 1991).

In the dairy industry lactic acid fermentation employing species of *Lactobacillus*, *Leuconostoc* and *Streptococcus* are chiefly used for dairy products in which the glucose is broken down under anaerobic conditions into two molecules of lactic acid. Cultured sour milk and sour cream is obtained by inoculation with *Streptococcus lactis* subsp. *lactis* or subsp. *cremoris*, syn. *S. cremoris,* to produce curdling, and flavoured by the simultaneous us e of *Leuconostoc cremoris*, syn. *L. citrovorum,* and

L. mesenteriodes subsp. *dextranicum*, syn. *L. dextranicum*. The fermentation of milk by *Lactobacillus acidophilus* is used to produce an acidophilic bioyoghurt. The inoculation of milk at 37°C with *Streptococcus thermophilus* produces Bulgarian buttermilk, to which the addition of *L. delbrueckii* subsp. *bulgaricus*, syn. *L. bulgaricus,* also yields a yoghurt. Kefir, from cow's, goat's or sheep's milk and kumiss, usually from mares' milk, also requires fermentation with various species of *Lactobacillus* and *Streptococcus* and lactofermentation yeast species of *Saccharomyces*. Other species of *Lactobacillus* are used industrially in the fermentation of maize and potato sugars.

Cheese from natural milk is soured using a starter. If the milk is being processed at high temperatures the starter is *Streptococcus thermophilus* plus a species of *Lactobacillus* such as *L. delbruekii* subsp. *lactis*; at lower temperatures the starter used is *S. lactis* subsp. *cremoris* or *L. delbruekii* subsp. *lactis*. When the milk reaches the required acidity it is curdled by the addition of rennet (an impure form of the enzyme **rennin**), which curdles the soluble protein **euseinogen** but not the insoluble **casein**, thereby separating the sour milk into curds and whey.

Some cheeses are further ripened using propanoic (propionic) acid bacteria such as *Propionibacterium freudenreichii*, syn. *P. shermanii*, which is responsible for the 'eye' in Emmenthaler and other cheeses, with the formation of propanoic and ethanoic acids and small quantities of succinic acid and CO_2 from the glucose. Inoculating the rind with the short, reddish, rod-shaped bacteria *Brevibacterium erythrogenes* and *B. linens* give the characteristic taste and orange colour to Limberg cheese.

In wine-growing districts ethanoic acid bacteria *Acetobacter* and *Gluconobacter* are used to oxidise wine exposed to air into wine vinegar (from the French *vin aigre* = sour wine), in the UK malt is used for malt vinegar, and in the USA cider is fermented for cider vinegar. The *A. aceti* subsp. *aceti* and subsp. *xylinum* (syn. *A. xylinum*) and *A. pasteurians* (syn. *A. kützingianum*) are capable of breaking down ethanoic acid into CO_2 and water, while *Gluconobacter oxydans* (syn. *A. melanogenus, A. oxydans, A. roseum, A. suboxylans*) is incapable of causing any further breakdown of the ethanoic acid. Since the oxidation by *Acetobacter* of alcohol into ethanoic acid and water occurs under aerobic conditions, it is not a true fermentation process, which is defined as the anaerobic breakdown of glucose and other organic fuels to obtain energy (Brouk, 1975; Postgate, 1992; Bailey, 1999).

In the food industry **coupling sugar** is formed by the action of *Bacillus megaterium* on starch and sucrose. It has 80% of the sweetness of sucrose and has the additional advantage of not causing dental caries. Coupling sugar does not caramelise (discolour) when heated, and as a result has widespread possibilities for use in the food industry..

Bacterial fermentation is also used industrially, including the production of **xanthan gum** by the action *Xanthomonas campestris* on waste sugar products, such as those from the refining of sugar beet and corn starch; other *Xanthomonas* spp. act on a glucose and salts substrate. The gum is used in gel air-fresheners and in pharmaceuticals, and as a thickener and stabiliser in food glazing, ceramics, cleaning, polishing, paint emulsions, oil well drilling, textile printing and dyeing, also in the production of propanone and butanols (butyl alcohols) for industrial use, the latter are used as solvents for resins and lacquers. Other *xanthates* are used in the curing and vulcanisation of rubber, and for the detection of certain metals. *Thiobacillus ferrooxidans* and *T. thiooxidans* are involved in the recovery of copper from low grade ores.

Among other industrial uses is the synthesis of *cyclodextrins* from starch using *Bacillus macerans*; the cyclodextrins are used for the encapsulation of unstable materials. *Alcaligenes eutrophus* is used in the preparation of *polyhydroxybutanoic acid (PHB)*. In many respects the PHB resembles *polypropylene* and is used for surgical sutures and encapsulating materials. The fermentation of sugar and starch products by *Bacterium subtilis*, etc. may be used to produce *butanoic acid (butyric acid)*, the cellulose derivatives of which are used in lacquers and as moulding plastics, and the *butanoates* used in flavouring and as plasticisers. The action of *Clostridium acetobutylicum* on a maize meal feedstock was formerly used for the production of *butyl esters* and *propanone (acetone)* for synthetic flavouring essences and perfumes, solvents, and various other chemicals (Rexen and Munck 1984; Subba Rao and Kaushik, 1989; Sharp, 1990).

6.   VIRUSES AND VIROIDS

Viruses are transmissible, subcellular entities, invisible by light microscopy, and only capable of replication within the living cell by modifying the genetic machinery of the host; they are incapable of multiplication in an inanimate media. They are not accepted as organisms within the Linnaean binomial system, neither do they appear in the Five Kingdoms classification. They are briefly discussed here because of their pathological importance for the economic botanist.

It is in the inert state, outside the host cell, that a hierarchical classification system is currently evolving. The virus name conveys no information as to its taxonomic position in the hierarchy, and the group name follows the virus name, e.g. arabis mosaic nepovirus or carnation etched ring caulimovirus. The five levels recognised are, in descending order: particle, group, sub-group, virus, and virus strain. A particle, for example, may be icosahedral or isometric, a straight rod, a flexuous filament or cylindrical, i.e. bacilliform. The divisions within the isometric particle include the ssRNA Nepoviruses and the dsDNA Caulimoviruses, etc. See Holloway (1990) for further information and literature references.

A virus that occurs in a fungus is known as a *mycovirus*, one that infects members of the Prokaryotes, i.e. a bacterium, is known as a *bacteriophage* or *phage*. The phage particles multiply within the host cell, which eventually bursts, thereby releasing the particles and causing *lysis*, i.e. the dissolution of a bacterium infected with phage. Phage typing is used in identifying bacterial species or types by testing their reactions to selected phages, hence the phagotype, distinguished by sensitivity to a specific phage. Outside the host cell viruses consist of DNA or RNA typically surrounded by a protein coat or *capsid*. It is the serological reaction of the capsid proteins that are used to identify the viruses. Some of the simpler viruses, e.g. tobacco mosaic virus, can even be crystallised. Virologists regard the viruses as subcellular entities with genomes analogous to such molecules as messenger RNA, plasmids or transpons.

Viruses are the causative agent of many important diseases of man, lower animals and plants, e.g. poliomyelitis, foot and mouth disease, tobacco mosaic and tomato mosaic. It should be noted that strains of the tobacco mosaic virus rarely occur in the tomato and contrary to popular opinion they compete poorly with those of the tomato mosaic virus. There are approximately 400 plant viruses, most of which are single-stranded RNA (ssRNA) viruses. Symptoms include mosaic, leaf spots, and deformed growth of certain organs. Some may be of horticultural interest, such as the broken flower colour of certain ornamentals, e.g. the colour breaking in the tepals of Rembrandt tulips with anthocyanins, the tulip varieties so named after a painting by Rembrandt showing characteristic colour breaking on early introductions. Other viruses may not produce any disease symptoms apart from a noticeable reduction in yield. They may be transmitted by vectors, infected seed and pollen. Viruses are generally not found in meristematic tissue, hence the use of tissue culture (see Chapter 6) to obtain virus-free explants.

Virus neutralisation tests are used to identify the antibody response to a virus or, by using a known antibody, to identify a virus. The test depends on a specific antibody neutralising the infectivity of a virus by preventing it from binding to the target cell. The tests may be carried out *in vivo* in susceptible animals or chick embryos or, more usually, in tissue culture.

The *viroids* are extremely small and circular infectious agents consisting solely of RNA with no enveloping coat or capsid. Examples isolated from plants within which they are able to replicate and cause characteristic disease symptoms include the potato spindle tuber viroid, hop stunt viroid, and avocado sunblotch viroid; they have not been found in animals. They are not detectable as particles in infected plants, but when placed in plants they replicate autonomously is susceptible cells and produce a characteristic disease syndrome (Holliday, 1990; Walker, 1991; Bailey, 1999).

Chapter 19

Useful Algae

The algae represent an extremely diverse group of predominantly aquatic plants within the kingdom Protoctista, although the term no longer has any taxonomic significance. They are *eukaryotic organisms*, whose cells contain a distinct nucleus and cell division is usually by mitosis or meiosis. They exhibit relatively little differentiation of tissues and organs, and range from unicellular organisms through colonial and filamentous forms to the parenchymatous seaweeds with lamina over 50 m long.

Depending on their pigmentation, food reserves, cell-wall materials, number and types of flagella and ultrasonic detail, the Protoctista are represented by *ca.* 50 phyla. Among those phyla that contain organisms that at some time or other have been classified with the algae or fungi are: Dinomastigota (syn. Dinoflagellata, Dinophyta), Haptomonada (syn. Haptophyta), Discomitochondria (euglenids, etc.), Cryptomonada (syn. Cryptophyta), Xanthophyta (yellow-green algae). Chrysomonada (syn. Chrysophyta, golden-brown algae), Phaeophyta (brown algae), Rhodophyta (red algae), Gamophyta (including the desmids and conjugating green algae), Chlorophyta (green algae) and Diatoms (syn. Bacillariophyta). Of the former algae groups the Chlorophyta, Phaeophyta and Rhodophyta are of particular economic importance. There was a temporary interest in the genus *Chara* of the Gamophyta when it was suggested that their presence was harmful to mosquito larvae but subsequent investigation failed to support the idea (Allen, 1950; Bailey, 1999).

The economic value, culture and breeding of both marine and terrestrial algae requires a detailed knowledge of their classification and taxonomy in order to better understand their complex life cycles, sexual and asexual reproduction, gametophyte and sporophyte generations, etc., subjects outside this present study, although some of the problems involved can be deduced from the text.

1.   ALGAL BIOCHEMICALS

Many of the macro-algae, especially among the marine algae, present valuable and largely undeveloped sources of important biochemicals. Among the halophilous micro-algae a number are capable of tremendous productivity and provide useful sources of single cell proteins and other products.

1.1   Phycocolloids

A *colloid* is defined as finely divided particles of approximately 1-100 mm in size dispersed in a continuous medium. As such colloids are intermediate between coarse suspensions and molecular or ionic solutions. Their economic importance lies in their surface properties (Sharp, 1990). The colloids obtained from algae are known as *phycocolloids*, consisting of hydrophile colloids of basic polysaccharide structure derived from the cell walls or intercellular matrix of the algae. Their commercial sources are the marine algae (seaweeds), including both the benthic macro-algae of the sea floor and the free-floating planktonic microalgae. In the absence of any reported industrial use as producers of polysaccharides the microalgae need not be considered further here.

The intrinsic properties of the different macromolecular components of a given phycocolloid may differ more or less from those determined from the actual crude phycocolloid, so that a comparison between other gums and polysaccharides may not always be fully justified. *Agar*, for example, contains both *agarose* and *agaropectin*, which differ only slightly in structure but have very different physical properties. Notwithstanding, it is the actual physical properties of agar itself that are described and compared with other polysaccharides having the same industrial application. It is their non-toxicity combined with their unique rheological properties, i.e. elasticity, viscosity and plasticity, that give the phycocolloids certain advantages over other sources of industrial gums, although cheaper and lower quality substitutes may replace phycocolloids for certain uses and regions, especially in times of economic depression. They are used world-wide in the modern food and pharmaceutical industries as colloid stabilisers, emulsifying agents, gelling agents and in immobilising enzymes. Only rarely are they considered to be of therapeutic value.

The four major phycocolloids for industrial application are alginate, laminarin, agar and carrageenans. The alginates and laminarin are mainly from the Phaeophyta (brown algae) and the more or less sulphated galactans such as agar and carrageenans are obtained principally from the Rhodophyta (red algae) and serve the same industrial purposes as the alginates.

1.1.1   Alginates

The principal constituent of the phycocolloid *algin* forming the cell walls of the brown algae and certain bacteria, e.g. *Azotobacter vinelandii*, is known as *alginic acid*, a carbohydrate polymer of *D-mannuronic acid* and *L-glucuronic acid* units. It functions as an ion exchange agent, and is present in both the free state and as *calcium* and *magnesium alginates* in the brown algae. Both the alginic acid and alginates are of commercial importance. The alginic acid is sufficiently acidic to displace CO_2 from a carbonate; it is also insoluble in water but absorbs many times its own weight of water to form a slimy gel, hence its incorporation in tablets as a disintegrating agent.

Alginate production is currently based on *Macrocystis pyrifera* harvested in the USA, *Durvillea* spp. and *Lessonia* sp. from Chile. In the Far East small quantities are obtained from species of *Ecklonia, Eisenia*, and to some extent *Laminaria japonica*. *Laminaria digitata* is the raw material used in France, and *L. hyperborea* and *Ascophyllum nodosum* in the UK and Norway.

The alginates are mostly utilised for *polypropylene glycol alginate* (*PGA*). Stable under acidic conditions, PGA is widely used to suspend the pulp in fruit drinks, and is used in French dressings where an ordinary alginate would be precipitated under acid conditions. It is also used to stabilise beer foam, the addition of 50-100 ppm being sufficient to react with the protein without any haze formation.

The industrial potential of alginates is due to the relatively simple manner by which the viscosity of its solutions can be controlled over a wide range of values. Gel stability is not temperature dependent; whether it is brittle, elastic or soft depends on the type of alginate. Their low-price and thickening, emulsifying, stabilising, suspending and gelling properties find numerous applications in the food, pharmaceutical, cosmetic, textile, paper making, paint, ceramic and rubber industries.

It was formerly used to suspend cocoa in chocolate milk drinks, for which purpose it has now been replaced by carrageenans. It was subsequently used as an ice-cream stabiliser and later extended into food and confectionery products as thickeners, stabilisers and emulsifiers, especially where an oily substance is involved, such as in mayonnaise and sauces. It is also used in preventing water leakage from frozen fish on thawing, and to prevent the degradation of starch. Alginates are being increasingly used for reconstructed foods, such as crab sticks and onion rings, and the pimiento stuffing of olives.

The ability of alginates to immobilise cells and enzymes is utilised in the production of *ethanol* from starch, brewing beer with immobilised yeast, the production of *citric acid*, continuous yoghurt production, *prednisolone* production from *hydrocortisone*, etc. as well as *glycerol* production from the green, unicellular *Dunaliella tertiolecta*. Among the 1583 or more reported industrial uses of glycerol are in the manufacture of solvents, sweeteners, printing ink, antifreeze and shock absorber fluid.

The *Dunaliella* species are unique among aquatic eukaryotic organisms in occupying a range of saline waters ranging from nearly fresh (<0.1 M NaCl) to saturated salt solution (>5 M NaCl). *D. saline* subsp. *saline*, syn. *D. bardawil*, from the Red Sea is currently under investigation as a commercial source of **glycerol**. Free glycerol is the major product of photosynthesis, with the intercellular free glycerol content produced dependent upon the salinity of the growth medium. When grown in outdoors in a 3 M NaCl medium, the halotolerant *D. salina* contains *ca.* 30% glycerol, 29% protein, 18% lipids, 11% carbohydrates, 8% **β-carotene** and 1% chlorophyll, thus providing three valuable commercial products - bulk chemical β-carotene, high protein feed, and bulk chemical glycerol. When grown under conditions of increasing light intensity and light period, or under stress conditions, such as nutrient deficiency or high salt concentrations, it was found that the chlorophyll content per cell decreased and the β-carotene increased. As the ratio of β-carotene to chlorophyll increases from *ca.* 0.4 to 13 g g^{-1} there is an accompanying colour change from green to a deep orange. The β-carotene, the precursor of vitamin A, is used to impart colour and provide vitamin A in animal and human foods, including margarine. It is also used as a sunscreen agent. The algal residue from the extraction process contains no indigestible polysaccharide cell walls and can therefore be used for livestock feed (Ben-Amotz and Avron, 1983a, b; Fenical, 1983; Kochhar and Singh, 1989).

Alginates are used in the pharmaceutical industry as a tablet binder to provide a sustained release. A combination of water-soluble and water-insoluble alginates are also used to act as a binder in dry tablets, with the alginate's water absorption properties promoting tablet disintegration in water. Alginates are also applied as an anti-refluxant, a suspension agent in drugs, e.g. penicillin, and as an X-ray contrast medium thickener, as an adjuvant for haemodialysis, in haemostasis, and wound and burn healing. For treating wounds and burns, threads of calcium/sodium alginate are woven into a fabric or biopaper bandage and gel on contact with the wound. The alginate's selective ion-binding properties also makes it an interesting agent for blocking the toxicity of certain ingested heavy metals. Alginates are also widely used in the cosmetic industry as a base in creams, jellies, hair-sprays, hair-dyes, etc. and as a dispersal agent in shaving soaps and hair shampoos, as well as being used in dust protection creams for use in factories.

In the textile industry alginates are used as a dressing and polisher, as a thickener for colours employed in printing fabrics, and as a hardener and adhesive for joining threads in weaving, either alone or mixed with starch or tragacanth gum. Its cementing and sealing properties are utilised for the glazing and sizing of paper and cardboard, impregnating fertiliser bags; and as an adhesive stabiliser for corrugated boards, brown coal or lignite briquettes where, by stabilising the viscosity of the adhesive, it controls the rate of penetration. By converting soluble salts into insoluble salts alginates enable the insoluble salts to be used to waterproof fabric. For example, **ammoniated aluminium alginate**, which becomes insoluble on drying, is used to waterproof tents and other canvas covers. Because they are pliable when moist, the insoluble salts can be prepared in the manufacture of plastics, linoleum, vulcanite fibre and imitation leather. Alginates may be used as a binding agent for fish foods, in insecticides and fungicides, as well as a binder for printer's ink and in cartridge primers.

Its emulsive ability is utilised in casein emulsion paints, while its suspending properties are used in car polishes and paints. Copper and mercury alginates are a useful component of underwater marine paints. The heavy metal alginates can be dissolved in ammonia which, on evaporation produce a waterproof film that can act as a varnish, while ammoniacal copper alginate is successfully employed for the impregnation and preservation of wood. The reaction between crude algin and metal ions to form insoluble alginates is utilised for descalling boilers, the scale forming metal ions, which react to form flocculent masses which can be blown out of the boiler.

Alginates have also been used in the building industry to fire-proof wood, in fire-retarding compounds formed from chemicals dissolved in sodium ammonium alginate, for the production of non-splinter glass, as a thickener for bitumen, and in the production of a special cement; the waterproofing of bricks and cement has also been suggested. They are also used in can-sealing compounds, for separating the plates in storage batteries, and as a flux for coating welding rods and electricals (Dickinson, 1963; Chapman and Chapman, 1980; McHugh, 1987; Indergaard and Østgaard, 1991).

1.1.2   Laminarin

Laminarin consist of a group of *β-D-glucopyranose* units linked through carbon atoms 1 and 3. They are produced by the brown algae, especially the sporophylls of *Alaria* and *Laminaria* spp. (kelps) and to a lesser extent by members of the Fucaceae (rockweeds), and by the Diatoms. There are both soluble and insoluble forms of laminarin which differ in the size of the colloidal particles and their degree of branching. Soluble laminarin can be obtained from *Laminaria digitata* and the insoluble form from *L. hyperborea*. Laminarin does not form a viscous solution, neither does it gel. It is readily decomposed by micro-organisms; it also stimulates the bovine rumen flora to increased activity.

Now commercially available, there appears to be a potential application for the use of *sodium laminarin sulphate* medicinally as an anticoagulant, where it is reported as being one third as good as heparin (Chapman and Chapman, 1980; Indergaard and Østgaard, 1991; Skjåk-Bræk and Martinsen, 1991; Smidsrød and Christensen, 1991).

1.1.3   Agar

Agar, formerly known as **agar-agar**, is a mixture of two polysaccharides, agarose and agaropectin. The main constituent is **agarose**, containing **3,6-anhydro-L-galactose** and **D-galactose** as the repetitive unit, and **agaropectin** with the carboxyl and sulphate groups of native agar.

It occurs in members of the red algae as a constituent of the cell walls. Agar-like structures are also found in some of the brown algae. Extraction is by hot water. The preferred species for commercial agar and agarose extraction are *Acanthopeltis*, *Gelidiella*, *Gelidium*, especially *G. amansii* and *Pterocladia*. In the UK agar is obtained from *Chondrus crispus* (Irish moss) and *Mastocarpus stellatus*, syn. *Gigartina stellata*, in Ireland from *Gelidium elegans*, syn. *G. pulchellum*, and *G. spinosum*, syn. *G. latifolium*, and in Denmark from *Furcellaria lumbricalis*, syn. *F. fastigata*. From the Arctic region near Archangelsk and in the Far East near Vladivostok agar is obtained from *Ahnfeltia plicata* (landlady's wig), and in the Black Sea from *Phyllophora nervosa*. During World War II the USA produced some agar from *Gigartina versicolor*, syn. *G. cartilagineum*.

Agars have a relatively low sulphate content. They are good gelling agents with water, even at concentrations as low as 0.5%; the relative gel strength of agar being four to five times that of any other phycocolloid. For certain bacteriological and fungal culture media they are regarded as indispensable because, after nutrient materials have been added, even a dilute solution sets to a firm jelly upon which the bacteria or fungi can grow.

Agar is widely used in the food industry as a thickening agent, emulsifier and stabiliser, particularly in confectionery, and in the canned meat and fish industries, especially pet foods. It is used to protect preserved cooked fish against breakage during transport, also to prevent the blackening or detinning of the cans of certain fish, e.g. herrings. It is particularly valuable as a gelling agent in confectionery, marshmallows and candies, and may also replace gelatine in the making of jellies, which set readily as well as being more economical. Agar is also widely used as a thickening agent for soups and sauces, in the manufacture of ice-cream, malted milks, jellies, candies and pasties and as a stabiliser to give smoothness to sherbets, ice-creams and cheeses. However, due to its low whipping capacity, it is necessary to add a gum, for which purpose sodium alginate has now largely replaced agar. Agar has also been used in the manufacture of cream cheeses, to improve the texture of cream, and in the making of custards, mayonnaise and icing. By tying up the free moisture in cake icing, the icing is prevented from adhering to the paper wrapping, doughnut glazes are similarly prevented from cracking. It is also widely used in the protection of bakery products against dehydration, and as a clarifying agent for beer, coffee and wine.

Agar is commonly used in pharmaceutical and cosmetic preparations, including use as a bulk-producing purgative, with its oldest use probably in emulsions with liquid paraffin for treating constipation. It is used in lubrication jellies, suppositories, emulsions and ointments, as a suspending agent for barium sulphate in radiology, a disintegrating agent and excipient in tablets, and in the compounding of slow-release capsules. Agar is also used in prolonged treatments as an anti-rheumatic agent and in the stabilisation of cholesterol solutions.

In industry agar is used as an emulsifying agent and for the sizing of fabrics, with the highest quality agar being used for silks in order not to destroy the sheen. Japanese agar, obtained from *Gelidium* spp. is considered to be definitely superior as a sizing material to agars obtained from *Chondrus* or *Gigartina*. The poorer quality agars are used as a coating for waterproof paper and cloth, as a high class adhesive in the manufacture of plywood, and as a cleaning medium for liquids. It is also used in the hot drawing of tungsten wire for electric lights, and in the photographic industry for making plates and films. Agar is also becoming increasingly important for use as a reagent in molecular sieve chromatography. Other uses include employment in the finishing of leather for imparting both a gloss and stiffness.

Agar is employed by modellers in the making of plaster of Paris moulds and in the moulds for the casting of artificial limbs, and was formerly an ingredient for dental impressions, for which purpose it has largely been replaced by alginates. It is also the raw material used in the manufacture of linoleum, artificial leathers and silks, heat and sound insulations, as an ingredient of water-based paints and in the manufacture of storage batteries for submarines (Brouk, 1975; Chapman and Chapman, 1980; Armisen and Galantas, 1987; Sharp, 1990; Indergaard and Østgaard, 1991).

1.1.4   Carrageenan

Carrageenan is a mucilaginous extract consisting of a complex mixture of at least five polysaccharides, including D-galactose and L-galactose, 3,6-anhydro-D-galactose and sulphate ester groups; they differ from agars in not forming gels with water without the addition of salts. They occur in many of the red algae.

The name carrageenan is popularly reputed to owe its origins to the small coastal town of Carragheen, near Waterford, although no town of that name exists. The name is apparently derived from carraigeen, i.e. moss of the rock, which was apparently the name first given to *Chondrus crispus* at some time between 1821 and 1829 (Mitchell and Guiry, 1983).

The carrageenans are used extensively as an emulsion stabiliser in both the confectionery and pharmaceutical industries. In the food industry it is widely used as a fining agent in the clarification of beer, coffee, honey and wine, and as a thickener for soups, salad dressings, sauces, fruit drinks, etc. They also act as secondary stabilisers in ice-cream, adding creaminess and preventing the spontaneous ejection of liquid from the gel (*syneresis*) and crystal formation under freeze-thaw conditions.

The carrageenans produce a gelling reaction with milk proteins such as casein following the addition of cold milk, an action that is employed in a wide range of dairy products, including jellies, blancmanges and instant milk puddings, and for stabilising the suspension of cocoa in milk chocolate, while in ice-cream carrageenan prevents the migration of colours in multicoloured ices.

Because of their cost, availability and improved properties, the canned-food market formerly dominated by agar has now been replaced by alkali-modified extracts from *Chondrus* and *Eucheuma*. They are also used as thickeners, stabilisers and emulsifiers in beverages and bakery products, diabetic food, dressings and sauces, frozen foods, etc. A semi-refined carrageenan is used almost exclusively for the pet food market.

Carrageenans soaked in whiskey are traditionally used by Irish barkeepers in New York as a cough cure! The inclusion of ι-carrageenan stabilises emulsions or suspensions of insoluble pharmaceutical preparations and also gives body to lotions, while combinations of ι-, κ- and λ-carrageenans enhance the texture of toothpaste. They are also used in the textile industry for producing a soft finish and a surface to which print will adhere. They give smoothness, gloss and stiffness to leather, as well as being extensively used in boot polish, the mucilage holding down and smoothing out any rough projections in the surface of the leather.

In the manufacture of cold water or casein paints, carrageenan is used to hold the film on the surface while the casein dries; it is also used to bind briquettes of charcoal powder (Watt and Breyer-Brandwijk, 1962; Dickinson, 1963; Brouk, 1975; Chapman and Chapman, 1980; Indergaard and Østgaard 1991).

Among the carrageenans recognised are: (1) *Furcellarin*, a carrageenan-like product derived from the mucilage of *Furcellaria lumbricalis*; it consists of 3,6-anhydro-D-galactose and D-galactose sulphate. Originally produced in Denmark during War World II, furcellarin is also known as *Danish agar*; Denmark is still the major source of supply. It is widely used in the food and pharmaceutical industries, especially in Europe, in suspensions, emulsions, foams and tablet disintegration. It is particularly valuable in the commercial manufacture of marmalades, jams and jellies since furcellarin, unlike pectin, does not require prolonged boiling before setting; it is also used in preserved meat and fish, milk puddings, icing bases, and canned food. Its mucous properties makes it useful in the control of stomach ulcers; it is also used in various pharmaceuticals and toothpastes; (2) *Eucheuman*, either as an ι-or κ-carrageenan, is obtained from wild and cultivated species of *Eucheuma* in the Pacific, especially around the Philippines. It is a compound intermediate between agar and carrageenan, It is mainly used in the food, cosmetic and pharmaceutical industries; (3) *Phyllophoran* also has properties intermediate between those of agar and carrageenan, and is considered to be a valuable bacteriological agar as well as yielding a volatile oil, *geraniol*. It is obtained from species of *Phyllophora*, the principal area of production being the Black Sea; (4) *Hypnean* is obtained from species of *Hypnea*. It is essentially a κ-carrageenan, providing a gel, whose strength can be controlled by the addition of up to 1.5% solution of KCl, thus making it a potentially valuable carrageenan source; (5) *Iridophycan* is derived from species of *Iridaea* and *Iridophycus*. It consists of a mixture of κ- and λ-carrageenans. Comparable sulphated polysaccharides are also found among South African species of *Aeodes* and *Pachymenia*. Iridophycan is used in refining beer, as a stabiliser in chocolate drinks, syrups and paint, as well as for the sizing of paper and cloth; and (6) *Funoran*, an agarose-type polysaccharide rather than a carrageenan. It is obtained from species of *Ahnfeltia, Chondrus, Gloiopeltis, Grateloupia, Gymnogongrus* and *Iridaea*. Funoran is employed as an adhesive, known in Japan as funori, for glazing and stiffening fabrics, paper and threads, the cementing of walls and tiles, and in the decorating of porcelain. There are possible pharmaceutical applications in lowering blood cholesterol and anti-tumour activity (Brouk, 1975; Chapman and Chapman, 1980; Prescott-Allen and Prescott-Allen, 1986).

1.2   Minor Chemical Products

The minor chemical products include: (1) *Mannitol*, a sugar alcohol present as a cell sap food reserve in a number of brown algae, including species of *Ascophyllum, Durvillea, Ecklonia, Fucus, Laminaria, Sargassum* and *Turbinaria*. *Laminaria saccharina* (sugar wrack) can even be used as a sweetener. Although costly to extract, mannitol has a number of possible uses in pharmaceuticals for tablets, diabetic foods, etc. It is also given intravenously as an osmotic diuretic. Mannitol is used industrially in paints, lacquers and leather, also in the plastics industry where it is reputed to produce better products than those obtained from glycerol (glycerine). When nitrated to form *nitro-mannitol*, it produces a powerful explosive similar to nitro-glycerine; (2) *Fucoidan*, which is probably the calcium salt of a carbohydrate ethereal sulphate, is present in the intercellular mucilage of such rock weed species as *Ascophyllum, Fucus, Laminaria, Pelvetia,* as well as *Ecklonia radiata*. It has a potential use as a blood anti-coagulant; (3) *Fucosan* obtainable from *Ascophyllum* and *Sargassum* is used as a tanning substance; (4) *Ginnanso* is a Japanese adhesive extracted from *Iridaea cornucopiae* and *Turnerella mertensiana*; (5) *α-kainic acid* is extracted from *Digenia simplex*. It is marketed in Japan as a broad spectrum anthelmintic. Although various compounds have been isolated from the seaweeds and have been shown to be biologically active, α-kainic acid is one of the few exceptions to be utilised; (6) *Lectin*, small but commercially available quantities of lectins are obtained from *Codium fragile* subsp. *atlanticum* and *Ptilota plumosa*; (7) *Iodine* is, with the exception of the Russian Federation, obtained from brown seaweeds of the Fucales and Laminariales. In the Russian Federation the red seaweed *Phyllophora nervosa* is sufficiently abundant in the Black Sea and Sea of Azov for it to be harvested for industrial extraction. Although a number of other red seaweeds contain considerable quantities of iodine in their gland cells, the iodine is present as compounds and not in the free state (Brouk, 1975; Chapman and Chapman, 1980; Tootill, 1984; Sharp, 1990; Guiry and Blunden, 1991; Walker, 1991; Macpherson, 1995).

1.3   Seaweeds for Energy

The large brown algae are readily convertible to methanol which, in turn, can be economically converted to gasoline. Huge kelp farms are necessary to ensure the necessary regular bulk production of raw material for commercial processing. The development of such kelp farms for *Macrocystis pyrifera* and *Pelagophycus porra*, syn. *Pelagophora porra* in Pacific USA, *Durvillea antarctica* in Australia, New Zealand and South America, *Ecklonia maxima*, syn. *E. buccinalis* in South Africa, and *Laminaria japonica* off Priomorye, Russian Federation, testify to the interest now being shown in this source of bioenergy. The possibility of raising kelp bass and oysters in conjunction with the kelp, and the utilisation of kelp by-products for minerals, fertilisers and animal feed, are additional attractions (Chapman and Chapman, 1980).

2.   EDIBLE ALGAE

Seaweeds are a traditional staple item of diet in China and Japan and, to a lesser extent, in the coastal regions of Europe and North America. They may be eaten raw, cooked or pickled, used in sauces and for thickening soups, made into sweetmeats, and less frequently, as a condiment. The green seaweed with which the western world is probably most familiar is *Ulva capensis*, syn. *U. lactuca* (sea lettuce), it too is also eaten in Japan, as is also *U. pertusa*.

Among the red algae *Porphyra tenera* (nori) is the basis of the large Japanese 'nori' industry. It has a protein content of 29-35% dry weight, similar to that for soya beans and three- and six-fold respectively that of wheat and rice; 75% of the protein and carbohydrate are digestible by man. Other important food sources among the red algae include *Palmaria palmata*, syn. *Rhodymenia palmata* (dulse), which is eaten in Scotland, Ireland, Iceland and Kamchatka, and *Porphyra umbilicalis* (laver) cooked and eaten as the renowned Welsh laver bread.

A large number of species from the brown algae, especially *Laminaria japonica* (ma-kombu), *L. cichorioides* (chizi-kombu) and *L. religiosa* (hosome kombu), are eaten in Japan in soups and tisanes or powdered as a spice for soups and sauces. *Alaria esculenta* (murlins) is eaten in Northern Europe and Iceland, and *A. fistulosa* by the Pacific Coast Amerindians.

More than 50% of the dry weight of seaweeds consists of carbohydrates, mainly roughage as only a small fraction is digestible. Also present are, with some notable exceptions, small quantities of protein of mainly undetermined digestibility, fats and salts. They contain adequate quantities of K, Na, Cl and I, although these vary greatly with the habitat and season. Almost certainly the scarcity of goitre in the Orient may be attributed to the high iodine intake from eating seaweeds. Seaweeds are also rich in vitamins A, B_2 and B_{12}, C and E. The vitamin A content of *Porphyra tenera*, for example, ranges from 20,400 to 44,600 IU, the average value being 67 times higher than that for eggs, and the ascorbic acid (vitamin C) content 1.5 times higher than that of oranges. Indeed, for some Eskimo tribes the ascorbic acid accounts for more than 50% of the human requirement (Brouk, 1975; Chapman and Chapman, 1980; Xia and Abbott, 1987).

2.1   Algal Animal Feed

There is a long tradition for the use of seaweeds, especially members of the Fucales and Laminariales, as animal fodder and they are still so used in a number of countries today, either in the fresh state or as a prepared feed. Browsing by sheep, cattle and horses is still practised in Scotland, Norway and Iceland. In Iceland the plants are washed and dried (using geothermic heat), compressed and stored for use in the winter months. On the west coast of Scotland *Pelvetia* is fed to fattening pigs, and also fed either raw or boiled and mixed with oatmeal to calves. *Alaria fistulosa*, *Fucus evanescens* and *Laminaria bongardiana* are also mixed with meal and fed to pigs, especially after farrowing. In the Commander Islands in the Bering Sea seaweeds are fed to arctic foxes as part of their normal diet; mink in Ontario also receive a seaweed ration in their feed. In Cuba *Ulva* spp. are added to meal and are being fed experimentally to poultry. The seaweeds appear to be rarely used in the tropics for animal feed, although dried *Sargassum* spp. are recorded as being fed to pigs in Hong Kong.

Factories have now been established in many parts of the world for the commercial production of a meal from dried and ground seaweeds for inclusion in the rations fed to cattle, pigs and poultry. The increase in fertility and birth-rate in livestock is attributed to the presence of *tocopherol*, the anti-sterility vitamin E, while the improved colour in egg yolks is attributed to the *fucoxanthin* and *iodine* content, although any improvement in milk butterfat due to an increased iodine intake is still inconclusive. The coralline algae known as maërl (see Section 4 below), are rich in calcium and magnesium and may be used as a mineral supplement (Chapman and Chapman, 1980; Briand, 1991).

There is now an increasing interest in the use of fresh water algae as a source of *single cell protein* (*SCP*) supplements for feed. Algal meal from dried fresh water algae such as *Chlorella vulgaris* and *Scenedesmus obliquus*, are used as a feed supplement for poultry, the high xanthophyll content giving a good colour to egg yolks. Because of their high production costs, however, the SCPs are not fed to ruminants (Göhl, 1981).

3.   SEAWEED MANURE

Traditionally seaweeds have long been used for manure in the Orient and somewhat more recently in the west. In Europe the largest users of seaweed manure are the farmers of north-east France, where 30-40 m^{-3} ha^{-1} are regularly applied annually. The autumn manuring of the early potato fields is also practised in the Scilly and Channel Islands. The large, brown algae are mainly used, the others may also be used provided they have been washed up in sufficient quantity, especially *Ulva*, which is rich in nitrogen. The shore driftweed, however, is rarely if ever, composed solely of brown weeds and invariably contains an admixture of red and green algae. Although relatively high in nitrogen, potash and trace elements, the phosphorus content is low and the crops will eventually require the addition of phosphate fertiliser. The presence of growth hormones, auxins, gibberellins and cytokinins in the seaweeds enhance crop growth, especially when used in liquid extracts. Salt contamination can be a problem if the seaweeds are applied directly to crops without giving time for the rain to wash out the salt. On the western and northern coasts of Scotland and Ireland the pounded ashes of seaweeds, containing 2-5% of an impure Na_2CO_3, known as kelp was also formerly applied to the fields.

Dried meal and liquid extracts of seaweeds, or their waste products following phycocolloid extraction, are becoming increasingly popular among horticulturists and agriculturists. The meal, with its delayed bacterial breakdown, is the better soil conditioner and can be advantageously used on such crops as potatoes, asparagus, flowers, fruit and hops, but not the cereals. The seaweed extracts are also credited with improving seed germination, controlling certain fungal infestations and increasing frost hardiness and shelf life of fruit, as well as improving the physical properties of soil. The UK brands of liquid manure, 'Maxicrop' from *Ascophyllum* and 'Alginure' from *Laminaria*, are now widely used in many countries for horticultural and glasshouse crops, while the New Zealand product 'Seagro' from imported Norwegian *Ascophyllum* is primarily used for pastures and orchard crops.

Known as *maërl*, the coralline algae *Lithothamnion* and *Phymatolithon*, form extensive deposits extending from Norway to the Mediterranean basin. Consisting principally of calcium and magnesium carbonates, the maërl is widely used as a manure in north-western France and as conditioner for compost makers (Low, 1847; Chapman and Chapman, 1980; Blunden, 1991; Briand, 1991).

4.   SEAWEED CULTIVATION

The mariculture of seaweeds is widely practised in Asia but is still in its relative infancy in the western world, even though there are insufficient natural resources available to meet demand. This has led to a need in the western world for sustainable harvesting, genetic selection for productivity and efficient, low-cost harvesting techniques. Since the majority of the economically important genera, particularly those that are cultivated for food in Asia, also appear in European waters, there is no compelling reason for the introduction of any alien genera for cultivation and any consequential upset of the ecosystem.

The complex environmental considerations required for the mariculture of seaweeds involves such factors as sea temperatures (which may be modified by water circulation patterns), light, day length, salinity, nutrient supply, etc. as well as a knowledge of the life cycle of the species to be cultivated. This complexity is illustrated by the fact that in the northern hemisphere, European sea temperatures occur at higher latitudes than in Asia, i.e. for the same latitude, the European shore waters are 8-10°C warmer than Asian waters. Since the reduction in winter day length and irradiance act in concert with latitude, midwinter light rather than temperature is the likely factor limiting marine algal growth in Europe.

4.1 Cultivation of Attached Seaweeds

First developed in China and Japan a primitive cultivation technique for cultivating seaweeds consisted of either the seasonal cleansing of shallow rocks or the provision of bundles of brushwood in which the spores from natural populations of *Porphyra* could settle. The more recent method for growing *Porphyra* involves tanks of sea water kept under glass and the artificial release of *carpospores*, i.e. spores formed by members of the Rhodophyta following the division of the zygote encouraged by previous drying, followed by squeezing or pulverisation of the lamina. The carpospores are then poured into sea water tanks containing local mollusc shells and a regular supply of nutrients. In China the tanks are shallow and the shells lie on the bottom, whereas in Japan the shells are suspended in deep tanks. The carpospores germinate, forming 2n vegetative filaments and eventually producing *conchospores*, which eventually develop into the larger haploid thallus. The conchospsores are seeded onto nets, which are then placed on the sea bed or, if space is limited, suspended from floating structures for the haploid thallus to develop and eventually harvested.

In Japan, naturally seeded nets or ropes are also used for the culture of edible species of *Enteromorpha* and *Monostroma* in a similar manner to that used for *Porphyra*. Young sporophytes of *Alaria, Gelidium, Laminaria, Palmaria, Pterocladia* and *Saccorhiza* species may also be successfully grown trapped in twisted rope. In China nets were initially used but more recently the *Gracilaria* are attached to horizontal or vertically suspended ropes. Similarly, in India pieces of *Gracilaria* are inserted into a twisted, not plaited, rope which is then held horizontally between two stakes. In Malaysia fertile plants of *Gracilaria* are suspended over shallow marine nursery tanks containing a substrate of coral, gravel or shells. The seeded substrate are later planted out in a suitable seaweed farm. Somewhat similar methods have been used in Atlantic Canada, the West Indies and Brazil.

In Chile, *Gracilaria lemaneiformis* has been successfully grown from small pieces pushed into the sand, a method also used in Namibia. More recently pieces of algae were held on the bottom by trapping under sand-filled tubes of soft polyethylene. They grew well and had produced a well-developed underground thallus system by the time the plastic had disintegrated (Kain, 1991).

Other attached seaweeds, such as *Ahnfeltia plicata* and *Ascophyllum nodosum*, can be successfully grown on shallow trays kept moist with continuous spraying of sea water. In Scandinavia spray cultivation has been carried out in greenhouses under artificial light and controlled shading, while in Florida such experiments resulted in scalding and heavy epiphytic growth from the high levels of irradiance (Schramm, 1991).

4.2   Cultivation of Unattached Seaweeds

The mariculture of unattached seaweeds includes not only the free-floating species of *Sargassum*, etc. but also the growing of normally attached species free from their substrate. Their cultivation can involve the management of natural populations or growing in land-locked areas such as lagoons, fjords and embayments, offshore cage culture and floating marine farms, cultivation in land-based tanks, ponds or raceways, and spray culture. The economics of such systems is dependent on the maximum exploitation of the plant's physiological characteristics and the optimal use of free sunlight, marine heat sinks and natural nutrient resources.

Land-locked cultivation utilising domestic or industrial waste has been successfully used for increasing production from existing populations of *Enteromorpha*, *Monostroma* and *Ulva*. The low-intensive pond culture in Taiwan of *Gracilaria* species, in combination with milk fish, shrimps or crabs in brackish polyculture systems, has also proved successful, likewise the commercial non-intensive pond culture of *Caulerpa* and *Gracilaria* in the Philippines. However, attempts at growing unattached species of *Gracilaria coronipifolia* in the Philippines and *G. tikvahiae* and *Sargassum* in Florida have proved unsuccessful due to heavy epiphytic growth and fouling, so much so that the initial heavy yields dropped dramatically within 1-2 months.

4.3   Seaweed Harvesting

A wide range of harvesting techniques can be used, depending on whether the seaweeds are attached, free-floating or strand drift, the depth of water, topography and the nature of the substrate. In addition, the maximum productivity of biomass or biochemicals in relation to the life cycle of the seaweed, regeneration cycle and the nature of the resource, have to be considered. Thus, in Europe the strategy has been to develop the natural resources, while in Japan 80% of the production is from mariculture.

Collection by hand or using rudimentary tools such as hand and drag rakes were the traditional methods, the yield being largely regulated by seasonal storms bringing the seaweeds to the strand line. More recently, tractor operated buck rakes and fork-lifts have been used to collect drift weed from the sea shore.

By 1913 the necessity for a constant supply for phycocolloid extraction had stimulated the development of mechanised harvesting techniques. In the intertidal zone, species such as *Ascophyllum nodosum, Chondrus crispus, Fucus serratus, F. vesciculosus, Gigartina acicularis, G. tweedii* and *Mastocarpus stellatus,* may be harvested by hand, using a knife or sickle to sever the stipe just above the holdfast, and the seaweed loaded directly onto a trailer. Where carting is not practical, the air bladders in the thalli of *Ascophyllum nodosum* and *Fucus vesciculatus* enable the weeds to float and to be trapped by nets or ropes and towed to a suitable loading site. The process may be mechanised by the use of floating barges with a cutter bar and elevator or suction cutter to load the barge, although the method does have the disadvantage of poor manoeuvrability and the necessity for calm seas.

Similarly, *Gracilaria verrucosa*, which grow in pools and channels in the lower intertidal and shallow subtidal zones, may be harvested from boats using either grapnels, long rakes and forks to bring the seaweed to the surface, or nets to collect cast weed. A more recent development has been a hooked conveyer belt attached to a specially designed shallow draft boat.

The deep water maërl banks are formed from fragments of members of the Corallinaceae. The coralline algae grow in shallow waters, where wave action and strong currents break off portions of the living plants and transport them to form extensive deposits up to 15 m thick. Formerly, banks exposed by the tide were harvested by bucket and wheel barrow. A more recent development is the use of boats equipped with grabs or pump dredgers.

Species such as *Laminaria digitata* growing in the lower intertidal and shallow subtidal zones require a different harvesting techniques. The earliest method of hand-harvesting involved the use of a long-handled sickle, a method which necessitated the operator being able to see the weed. This was followed by the scoubidou, which had a hook at one end of a long steel shaft and a hand-operated cranking system at the other, and pulled the weed up from the sea bottom. In 1963, Scuba divers were used to cut the seaweeds with sickles, which were then sucked up into the hold of a specially designed trawler. Divers are similarly used to harvest *Gelidium* spp. and *Pterocladia capillacea.* 1967 saw the development of a trawler with an adaptation of the scoubidou at the end of a jointed crane which twisted in a gimlet-like action and pulled the seaweed to the surface.

Slightly different techniques have to be used where only the stipe of the subtidal *Laminaria hyperborea* is required for processing. Various devices have been used, ranging from a grapnel to drag the seaweed to the surface, hooked conveyer belts, reciprocating reaper-type cutters and suction tubes, a cutting dredge operated from a crane and trawl net, cutter and conveyer belt, and finally to a specialised dredger developed to only harvest the mature stipes. Even so, in the north of Scotland the hand collecting of stipes from drift weed still remains a major crofter industry. In the Pacific, the often long-lived perennial *Macrocystis pyrifera* (giant kelp), is easy to harvest since much of its buoyant fronds lie on the surface and need only to be cut close to the surface and removed from the sea.

Not all seaweeds are harvested for their utilisation. In recent years the industrial maritime countries have had to deal with macroalgal blooms, especially species of green seaweeds such as *Enteromorpha* and *Ulva*, which either decompose *in situ* or in large drifts. Such blooms are a result of pollution and consequent eutrophication of coastal ecosystems. Because of their nuisance value they need to be removed from the beaches, using bulldozers, scrapers, ensilage machines, balers or special raking and sifting machines; a boat-mounted conveyor belt may be used for drift weed. The seaweeds thus harvested may be spread on fields as a soil additive, stocked in dumps or dehydrated for use in premixed poultry feed (Briand, 1991).

Chapter 20

Environmental Uses

The prime purpose of plant life is to create a sustainable environment for all living organisms. Without plants life on earth would not be possible. The present human population explosion and the increasing demands for ever higher and higher living standards in both the developed and developing countries has placed an almost impossible burden on the environment. Consequently there is an ever increasing necessity to manage the ecosystems and preserve the existing genetic diversity for future generations.

Within the plant kingdom there are members whose primary or secondary use by man are for improving and maintaining the environment, such as erosion control, shade, windbreaks, hedges and green manures. Many of these plants are multipurpose species whose use and management can not only provide the necessities of life but also help protect the environment.

1.   SOIL EROSION

Geological erosion is a natural process whereby the land surface is degraded by the action of wind, water, frost, earth movements, etc. Erosion *per se* is a natural process whereby, within a geological time scale, mountains are worn down and soils built up in the lowlands. An over-simplification, but it serves to demonstrate the differences between geological erosion and accelerated erosion by man. *Accelerated erosion* is the result of mismanagement of the environment through over-grazing, deforestation, excessive and unprotected cultivation, especially on steep slopes, poor cultivation techniques, such as ploughing across instead of parallel to the contours, etc. Such land mismanagement increases the surface water run-off following heavy rains, as well as increasing surface wind speeds, both assisting in the removal of the more fertile top soil and a lowering of the soil water reserves, etc. and resulting in ever increasing environmental degradation spiral. Soils that have been built up in geological time can be destroyed by a single heavy storm, requiring engineering works to control the gullies and an increased vegetation cover to reduce run-off and favour the penetration of the water into the soil. Erosion is not confined to the tropics, although the effects there can be quite dramatic, it is a universal problem. Poor drainage systems will inevitably cause salinity problems in irrigation schemes. Even good drainage schemes can have surprising effect on some soils. The drainage of the peat soils in the East Anglian fens in 1850 was responsible for a 3.7 m lowering of the soil level by 1956 due to the drying out of the peat, oxidation of the organic matter and wind blow. The latter can destroy seedling horticulture crops grown on the fertile fen soils and necessitate further reseeding. Wind blow problem is further complicated by the high value of the horticultural crops despite one or more replacement reseeding exceeding the return on a reduced acreage protected by wind-breaks (Russell, 1950, 1959; Soil Survey Staff, 1951).

Erosion control can be difficult and expensive but is far preferable to the damage caused by the lack of appropriate control methods. Control of non-cultivated land basically involves the design and maintenance of a permanent tree, shrub and forb cover to reduce the action of wind and water. Rapidly growing species, capable of providing a good surface cover and able to thrive more or less unaided, are obviously desirable. The root system can also be important. In the sand dunes of northern Sudan I can recall seeing stands of the tufted perennial grass *Panicum turgidum* borne on 2 m high columns of sand, each column protected by a fine mesh of roots. A change in the direction of the wind or an increase in strength had deflected the wind around the dense grass stools and gouged out the intervening sand, left unprotected by any ground cover or spreading roots. For cultivated land ploughing parallel to the contour, the construction of graded and non-graded bunds to control run off, grass strips, windbreaks, etc. can be used. See Section 2.1 regarding shelterbelts and windbreaks, and Skerman and Riveros (1990) for suitable tropical grasses.

There is an increasing use by civil engineers of geotextiles in road building, drainage and for the stabilisation of embankments. They are chiefly made from strong, non-biodegradable woven or pierced polypropylene fabrics, although some jute geotextiles and coir matting are also used. When buried in the soil the coir matting will retain its tensile strength for up to 10 years. In addition its ability to retain moisture is an obvious advantage, especially in arid environments (Rankilor, 1986; Robbins, 1995).

1.1   Amenity Plants

The word amenity is from the Middle English *amenite*, from Old French, from Latin *amoenitas*, from *omoenus*, pleasant delightful (Long, 1994). Amenity plants should be both functional and aesthetically pleasing, with the emphasis on function. Such plants may be grown for shade, shelterbelts, windbreaks, hedges, screening for ugly buildings, as street trees, provide recreational facilities, e.g. public parks, golf courses, etc.

1.2 Shade, Shelter, Windbreaks and Hedges

Plants grown for shade, shelter and windbreaks perform distinct environmental functions, although the actual functions are often indiscriminately applied, sometimes as the result of, for example, a shade plant performing a secondary function as shelter and/or windbreak. The desirable characteristics of plants used for windbreaks, shelterbelts and hedges are shown in Table 28.

TABLE 28. Desirable characteristics of plants used for windbreaks/shelterbelts, stock-proof and garden hedges (after Henderson, 1987)

Windbreaks/shelterbelts	Stock-proof hedges	Garden hedges
Woody or succulent shrubs or small trees	Sturdy, woody or succulent shrubs or low-growing trees	Woody or succulent shrubs or small trees
Strong trunks; minimum height usually 5 m	Multi-stemmed from the base, or low-branching spreading crown	Multi-stemmed from the base, or low-branching
Low lateral branches	Branches rigid or entangled, spiny, prickly or thorny	Branches closely arranged; spines, prickles and thorns sparse or absent
Fairly dense, evergreen foliage	Leaves small, sparse, casting little shade	Dense, evergreen and attractive foliage/flowers
Seed or vegetative propagation easy	Seed or vegetative propagation easy	Seed or vegetative propagation easy
Rapid growth and longevity or permanence	Rapid growth and longevity or permanence	Relatively rapid growth and longevity or permanence
Capability of withstanding considerable root competition	Capability of withstanding considerable root competition	Capability of withstanding considerable root competition
Disease and pest resistance	Disease and pest resistance	Disease and pest resistance
	Non-toxic, non-irritant	Non-toxic, non-irritant
Tolerant of wind, and in coastal areas, salt spray	Browse tolerant, capable of regenerating if damaged	Tolerant of wind, and in coastal areas, salt spray
Non-invasive	Non-invasive	Non-invasive
	Impenetrable by livestock	Impenetrable by livestock

Shade plants are used to diminish the intensity of heat or light for those plants or animals sheltering in the vicinity. Permanent or temporary shade trees are often considered desirable for a number of tropical plantation crops, e.g. *Gliricidia sepium* for bananas, coffee and young cocoa. The foliage is also used as a green manure in Sri Lanka. **Windbreaks** are grown to provide protection from the wind and are either fairly narrow strips of trees and shrubs, or even a single row of leafy shrubs. A dense windbreak reduces the wind speed, causing the wind to descend abruptly behind the break and produce turbulent eddying, A more open row of trees allows some of the wind to pass through the break so that the effects of the wind are felt further away. *Shelterbelts* are purposely planted to act as a shield and provide shelter against the weather, particularly the prevailing winds. The belts are generally wider and the trees more closely spaced than for windbreaks. Provided they are sufficiently tall and dense they can protect the downwind areas for distances up to 20 times the height of the shelterbelt. A series of shelterbelts spaced every 250 m, for example, can also provide sheltered grazing. The more permeable shelterbelt causes less turbulence than windbreaks and, although it has less effect on wind speed, the calming effects are apparent for a greater distance downwind. Both windbreaks and shelter belts can play a vital role in controlling wind erosion (Dalal-Clayton 1981; FAO, 1988b; Arnon, 1992).

Hedges are defined as a close row of shrubs or small trees forming a fence or boundary; they are often planted or maintained to be stock-proof. The life of a properly maintained hedge is far longer than that of wire or rail fences but are much more expensive to establish. Hedges may be grown as: (1) Field boundaries to contain the movement of livestock; (2) Decorative hedges to provide privacy or to partition areas within a garden; and (3) To reduce the force of the wind, acting either as shelterbelts to provide shelter to people, livestock and homesteads, or as windbreaks for orchards and crops (Dalal-Clayton, 1981; Henderson, 1983).

Stock-proof hedges will take several years to establish; they also require regular maintenance if they are not to become gappy at the base and allow stock to escape. The traditional rural practice of layering hedges, usually *Crataegus monogyna* (hawthorn), to provide an impenetrable, stock-proof hedge can be seen in many livestock areas of the UK, although rarely seen elsewhere. However, I have seen *Pithecellobium dulce* (Madras thorn) successfully layered in Nigeria; I also believe a number of other tropical species could be similarly layered. Layering certainly provides a more effective and lasting stock-proof hedge than the more common practice of filling any gaps with other living or dead species.

Where hedges are not layered, either due to the unsuitability species or, more usually, from a lack of technical knowledge, they may have to be constructed from mixed plantings in order to provide an effective barrier. Henderson (1983) considers four life forms are necessary for such mixed plantings. They are: (1) Framework plants to form the main structure of the hedge; (2) Short fillers to fill gaps at ground level. Such plants should be tolerant of shade and have numerous, ridged branches; (3) Tall fillers to fill gaps in and add height to the hedge. Such plants should be much-branched and cast little shade; and (4) Entanglers, which are preferably prickly climbers casting little shade, thickening the hedge and making penetration more difficult.

In the tropics very effective, stock-proof, *palisade hedges* can also be made from the close planting of truncheons of such plants as *Fouquieria splendens* (ocotillo), columnar cacti, e.g. *Stenocereus* spp. and columnar *Euphorbia*, e.g. *E. nigrispina*, etc. In Southeast Asia closely planted bamboos, e.g. *Bambusa multiplex* and *Thyrostachys siamensis*, are used for both hedges and windbreaks; the thorny *Bambusa bambos* is also commonly planted to exclude wild animals.

In the Sonoran Desert of south-western North America hedges of closely planted truncheons of *Salix* sp. (willow) and *Populus fremontii* (cottonwood) interwoven with brushwood have been planted by the local Amerindians along the flood-plain margins of their fields to retard channel cutting, prevent erosion and reclaim land by trapping the floodwater sediments (Nabhan and Sheridan, 1977; Dransfield and Widjaja, 1995).

Finally, there are **brushwood hedges** made from branches of trees felled when clearing land. These are widely used in some countries to delimit field boundaries and prevent trespassing by livestock. However they last for only a few years before becoming brittle and ineffective. In the semi-arid areas such hedges are also very effective against wind erosion, with wind-blown sand building up among the branches. While live hedges are preferable the absence of land ownership in many developing countries often inhibits the growing of permanent boundary hedges. .

1.3 Urban Trees

A tree-lined border to a road can provide both shade and pleasure. In rural Zimbabwe I have even seen how an avenue of vigorously transpiring evergreen *Eucalyptus* trees can transform seasonally impassable soggy ground into an all-season road. However, care needs to be taken in the choice of species; trees favoured by bats or flocks of birds can be messy, and their excreta ruin the paintwork of any parked vehicles. A straight, unbranched bole that neither obstructs vision nor interferes with the passage of vehicles is an obvious requirement. Strong branches that do not break readily are also essential. The crown should be reasonably compact, so as not to impede high-sided vehicles or create excessive shade. Thick leathery leaves are easier to sweep up than thinner leaves that stick to the road surface when wet. Roots can be invasive of drainage pipes, etc. and cause heaving of road surfaces and pathways and even cause structural damage to nearby buildings The root system should consequently be deep and compact, not shallow and widespread, neither should they sucker. Root damage to property is fully discussed by Cutler and Richardson (1989).

The ability to withstand air pollution is also important in urban areas, and accounts for the popularity of *Platanus* × *hispanica* (London plane) in the streets of London and other cities. The rain easily washes the leathery leaves clean of pollutants, although the habit of regularly sloughing off flakes of bark is now discounted as a cleansing action. There are disadvantages too. Some people are allergic to the pollen and fruit debris of the London plane and the leaf hairs can also be an irritant; they are sometimes used by children as a form of itching powder!

The female *Ginkgo biloba* (maidenhair tree) is considered objectionable as a street tree because the fallen seeds stink of rancid butter. Similarly, the large quantities of pollen produced by staminate trees of *Populus fremontii* (cottonwood) and the copious quantities of downy seeds produced by the pistillate trees, are likewise considered a disadvantage. Falling coconuts and similar heavy and hard fruits are also unwelcomed (Kearney and Peebles, 1951; Mabey, 1996; Mabberley, 1997).

1.4   Ornamental Plants

Ornamental plants are grown because they are primarily aesthetically pleasing; they are not necessarily functional in the full sense of amenity plants. They are cultivated in public and private gardens, grown indoors in containers to decorate the house, office, restaurant, etc. However, the distinction between the amenity and ornamental plants is blurred. The grass in a public park, for example, is primarily functional, being used for both walking and recreational purposes, i.e. it is an amenity, its aesthetic value secondary. A garden lawn may perform a similar recreational function and/or may be designed as a foreground to beds of ornamental plants, i.e. it serves as either an amenity and/or ornamental. Shade trees, shelter belts, etc. are planted in order to serve a definite purpose but there is no reason why they should not also be aesthetically pleasing. However, what constitutes an ornamental plant is very much in the eye of the beholder. What one nation may regard as a weed can be considered an ornamental by another, e.g. *Calotropis procera* is regarded as a weed of over-cultivated lands throughout Africa, yet it is sometimes cultivated as an ornamental in the Middle East.

The international ornamental plant industry is now largely dominated by companies based in the Netherlands and the USA. Trade, especially among the developed countries, is regulated by phytosanitary regulations and import licences (see Chapter 7), while international trade in prohibited wild species is strictly controlled by the Convention on International Trade in Endangered Species (see Chapter 3).

Ornamentals for the cut flower market in the developed countries is now a growth industry for the developing countries, especially Kenya, Zimbabwe, India and Columbia. This sacrifice of agricultural land for a cash return has its disadvantages. The high quality standards required for export necessitate high inputs of fertilisers, pesticides and water, with applications estimated to be up to 10 tonnes ha^{-1} yr^{-1}, making commercial flower growing the most polluting of all agricultural industries. Rumours of low wages and lax pesticide health regulations abound. Falling water tables are an additional problem (Robbins, 1995). It is a good example of how the living standards of the developed countries impose an unnecessary strain on the resources of developing countries.

2.   GREEN MANURES

Green manures are quick-growing crops and other green vegetation sources that are specifically grown either alone or with other crops for subsequent ploughing-in or surface mulching in order to provide humus to the soil, improve soil structure, conserve soil moisture and, especially in the tropics, assist in reducing soil run-off and wind erosion. If the main crop, for example a cereal, can be undersown with a green cover crop, such as a legume, it may even increase crop yield by improving soil fertility and provide a useful ground cover against soil erosion after the main crop has been harvested. Leguminous crops that are not allowed to seed have also been found more effective as green manure than other green crops under semi-arid conditions. Long-term experiments have shown that green manures are not effective in regions with less than 375 mm rainfall and are only economically beneficial for dryland cropping where the rainfall is more than 450-500 mm.

There are other advantages and disadvantages. For example, legumes can reduce the incidence of *Gaeumannomyces graminis*, syn. *Ophiobolus graminis* (take-all), in cereals by fixing the soil nitrogen and thereby starving the fungus. Similarly, the crop residues from *Melilotus* sp. (sweet clover) have been successfully used to reduced *Phymatotrichopsis omnivora*, syn. *Phymatotrichum omnivorum* (Texas root rot) in the cotton, presumably by stimulating bacteria antagonistic to the fungus. Conversely, ploughing in a green crop before sowing cotton or planting potatoes has increased attack by *Pythium* spp. and *Rhizoctonia solani*, due to the decaying green vegetation favouring the pathogens.

In the temperate regions the green manures are usually herbaceous, such as legumes, mustard and *Lolium multiflorum* (Italian ryegrass). In the tropics, especially where agroforestry is practised, leafy branches from trees or shrubs such as *Leucaena leucocephala* (lead tree), may be used to provide a surface mulch, which can later be incorporated into the soil.

An unusual and widely grown green manure crop in the paddy rice fields of China and Vietnam is the aquatic *Azolla pinnata* (water fern) which, in symbiosis of the blue-green bacterium *Anabaena azollae*, lives on the water surface. The bacterium is capable of fixing atmospheric nitrogen equivalent to 50-80 kg ha^{-1} in a growing season. A secondary benefit from the growing of *Azolla* is that its very rapid growth smothers the water surface and prevents mosquitoes from breeding (Dalal-Clayton 1981; Arnon, 1992; Mabberley, 1997).

3.   AGROFORESTRY

Agroforestry is defined by Maydell *et al.* (1982) as the deliberate planting of perennials on the same land as agricultural crops and/or livestock, either in some form of spatial mixture or in sequence, thereby producing a significant interaction (positive and/or negative) between the woody and non-woody components of the system, either ecological and/or economical.

Agroforestry species have three important functions: (1) Offer protection against the environment. For example, in the lower rainfall areas they act as wind breaks against wind erosion, while in the higher rainfall areas they may protect terraces and contour ridges against water erosion; (2) Contribute to maintaining soil fertility, either by providing foliage for mulching or through nitrogen-fixation; and (3) Provide food, fodder, fuel, etc. They are essentially multipurpose species and need to be carefully selected.

Among the criteria for their selection are: (1) A root system that does not interfere with cultivation or compete with the crops for water and nutrients, preferably extracting them from depth. *Eucalyptus* would be considered unsuitable in such situations due to its high water uptake often depleting the water-table; (2) A canopy that does not produce excessive shade to the detriment of the growing crop. In some cases branch pruning may be necessary; (3) Does not harbour crop pests and diseases; (4) Does not encroach vegetatively onto the growing crops, i.e. does not sucker or produce rhizomes; (5) Easy to establish and maintain; (6) Recovers readily when cut for fuel, fodder, etc.; and (7) Absence of any allelopathic effect on the growing crops.

4.   FIREBREAKS

Fire can be a valuable tool in the management of natural vegetation, provided it can be controlled. Controlled fire can be used to remove old, dead and surplus vegetation to provide easier access by man and grazing animals. Fire may be used in game parks and other recreational areas, including grouse moors, as a conservation measure, or in hunting to create suitable habitats for game. The latter is commonly practised by the Australian Aborigines. Controlled burning may be used to control the growth of undesirable shrubs, to encourage early grass growth, and possibly control livestock parasites. For example, in the Grand Teton National Park, USA, an increase of plague proportions of bark beetles was attributed to the absence of fire and the prevalence of old and sick trees. Control was achieved by fire (Edwards, 1984; Walter and Breckle, 1985).

The traditional method for controlling the spread of fire is by the provision and maintenance of firebreaks devoid of combustible vegetation. However necessary, such breaks can be environmentally unfriendly, expensive and difficult to maintain, especially in remote regions. In some circumstances it is possible to establish firebreaks with plants of low combustibility that will reduce the intensity of the fire and make it easier to control (see Chapter 5).

The suitability of a tree or shrub for firebreaks is a relative concept and is dependent upon a number of factors: (1) The flammability of the leaves and finer branches, and of the bark. A low volatile oil content and a high salt, moisture and ash content will reduce the flammability of the leaves. A low ash residue will also reduce the time during which the leaves will continue glow after burning. Unfortunately the reverse is true of many Australian species of *Callistemon, Eucalyptus, Leptospermum* and *Melaleuca*, hence the severity of their forest fires. A fibrous or flaky bark tends to burn readily and spread the fire to the crown. Obviously plants with a high flammability risk must be avoided since they can lead to spot fires within the firebreak caused by burning debris carried ahead of the main fire; (2) Species that are able to suppress flammable ground cover can sometimes be advantageous, although some carefully selected, low growing vegetation is necessary in order to prevent creating a wind tunnel effect between the trunks of the firebreak; (3) The retention of combustible materials, such as dead leaves, bark and other debris by some plants, can create a fire hazard; such species should also be avoided; and (4) The selection of fire tolerant plants that recover quickly after fire, possibly due to a thick bark and/or the ability to shoot readily from dormant buds (see Phillips, 1993 for further details).

In Australia breaks of selected trees and shrubs for fire protection are being considered. Such plantings, with certain reservations, follow the general principles for shelterbelt design for reducing wind speed, thereby reducing the speed and intensity of the fire, as well as blocking the direct heat of the fire, filtering or deflecting wind-borne burning material and smoke. The effectiveness of the protection depends on the species grown, its density and height, and the structure, ability to suppress ground cover, orientation, location and layout of the firebreak. A moderate to dense cover of the shrub *Eremophila gilesii* (turkey bush), for example, is able to suppress the ground cover to such an extent that fires rarely spread through such areas. The relationship between allelopathy (see Chapter 5), ground cover and fire retardant species requires further study (Hodgkinson and Griffin, 1982).

In North America low growing and readily established shrubs with fire retarding qualities and giving a low heat output when burning are being investigated. Among the promising shrubs for southern California is the low growing *Salvia sonomensis* (creeping sage) which, once established, is able to smother the annual ground flora, especially grasses, thereby eliminating a potential fire hazard. Other promising species are *Atriplex gardneria* (Gardner's saltbush), *A. cuneata* (Castlevalley saltbush) and *A. canescens* (fourwing saltbush). Although the latter will grow into 2 m high, it is apparently less flammable than most chaparral species. The high salt content of these species and their litter being important factors.

Similarly, the abundant salty litter below the canopy of *Tamarix aphylla* (athel, or tamarisk) kills the surrounding vegetation. This absence of any ground vegetation, plus the non-inflammable nature of the athel litter due to its high salt content, makes it a useful tree for growing as a firebreak. The succulent *Furcraea foetida* (Mauritius hemp) is cultivated as a firebreak in Sri Lanka, which suggests other members of the Agavaceae could be similarly used (Nord and Countryman, 1972; National Academy of Sciences, 1980; Booth and Wickens, 1988; Mabberley (1997).

5.   POLLUTION INDICATORS AND CONTROL

Air, water and mineral pollution have always been with us, but has only become an problem of ever increasing severity in the developed and developing countries during the latter half of the 20th century. Poor mining and mineral extraction techniques, increased industrialisation without adequate pollution control, excessive applications of agricultural inputs, etc. have all contributed to the problem. Although there is a greater awareness of what is involved, there is, despite the health hazards, an unfortunate lack of political and financial will to fully get to grips with pollution control.

The increase in the use of herbicides and pesticides in recent years has become an increasing cause for concern, especially when excessively or improperly used, and where tainted crops enter the food chain. The use of lichens in monitoring pollution is discussed in Chapter 18. Mabberley (1997) notes that in Poland the aquatic herb *Wolfffia arhiza* is also used as a qualitative test for herbicide pollution.

5.1   Air Pollution

The distribution of mosses and liverworts can be used as bioindicators of atmospheric pollution (see Chapter 18). The ability of the Bryophyta and Hepatophyta to store pollutants can be subjected for chemical analysis to monitor changes in the atmosphere. The moss *Calymperes delessertii,* for example, appears to be a good indicator of atmospheric lead and to a lesser extent of copper.

Some lichen genera, such as *Usnea*, are more sensitive to pollutants than others, such as *Lecanora*; *L. dispersa*, for example, will grow in cities where other lichens are unable to survive. The absence of lichen growth on trees, walls and buildings can usually be regarded as an indication of atmospheric pollution by SO_2 (Brightman and Nicolson, 1966; Tootill, 1984; Glime and Saxena, 1991).

5.2   Water Pollution

Clean fresh water is essential for health and there are a number of plants can be used as indicators of water quality, a facility that is of particular interest to people who are totally dependent on natural sources of water for themselves and their livestock. In Israel and Sinai, for example, there are species that can be used to indicate salt concentrations ranging from fresh water, e.g. *Typha domingensis*, syn. *T. australis,* and *Mentha* spp., to saline water, e.g. *Zygophyllum* spp. and *Arthrocnemum macrostachyum* (Danin, 1983).

Some plants are also capable of removing impurities from water. Only two plants are discussed here, but see Jahn (1981) for a more comprehensive treatment. *Eichhornia crassipes* (water hyacinth) has been rightly reviled as being the most pernicious aquatic weed of the tropics and subtropics (see Section 10). However, in certain circumstances it has the potential for purifying sewage pollution, and of absorbing and accumulating extremely toxic heavy metals such as Ag, Au, Co, Cd, Hg, Ni, Pb and Sr from waters polluted by industrial and mining operations (Cross, 1984; Kochhar and Singh, 1989).

In some developing countries the pounded seeds of several species of *Moringa*, especially *M. oleifera* are used to purify water. Doses of 50-250 mg l^{-1} of powdered, shelled seeds, equivalent to 1 seed l^{-1} are sufficient to bring about the coagulation and sedimentation of solids in turbid waters within 1-2 hr. The coagulation and sedimentation processes also remove 98-99% of coliform bacteria, although unable to yield completely coliform-free water. Despite the initial removal, the reduction is only temporary. Nevertheless it is a practice that I believe could be more widely used by aid agencies in refugee camps. The seeds also contain the horseradish-like substance *4 α-L-rhamnosyloxy benzyl isothiocyanate*, a glycosidic mustard oil with antibiotic properties (Jahn and Dirar, 1979; Eilert *et al.*, 1980, 1981; Jahn, 1981; Grabow *et al.*, 1985).

5.3 Salt, Mineral and Mine Pollution

Soil salinity is, with deforestation and desertification, a major factor affecting land use world-wide. It is particularly prevalent in the arid and semi-arid regions where precipitation is insufficient to leach sodium and other highly soluble salts from the soil. Similarly, *alkaline soils* are formed where the predominant calcium carbonates and sulphates fail to be leached from the soil. The situation is further aggravated by poorly designed and managed irrigation schemes with inadequate drainage systems. It has been estimated that 3.8 million km^2 out of the *ca.* 49 million km^2 arid and semi-arid lands of the world are saline. The estimated annual loss to agriculture from salinity in the Indian subcontinent alone is *ca.* 400 km^2, and in the USA with its far more sophisticated agriculture 800-1200 km^2. The reduction in crop yields and the often permanent loss of land to agriculture is unacceptable. Halophytic members of the Chenopodiaceae, e.g. *Atriplex, Halogeton, Haloxylon, Salsola, Suaeda*, etc. are very much in evidence. Some halophytic grasses may grow so vigorously that they are being successfully cultivated for land reclamation in some regions, e.g. *Spartina anglia* in the Netherlands, *Distichlis spicata* in Mexico and *Leptochloa fusca* (kalla grass) in India (Aronson, 1985; Bell, 1985). See Chapter 5 for further discussion on salt stress.

The ability of certain plants to adapt to a heavy metal has been accepted by prospectors as a useful technique for identifying new mineral deposits. For example, in Australia the shrub *Hybanthus floribundus* is regarded as an indicator of nickel; indeed, its ash can contain up to 22% nickel. In Montana *Eriogonum ovalifolium* is considered an indicator of silver, and in Colorado certain species of *Astragalus* indicate the presence of uranium. Among the gold accumulating plants of South

Africa is *Phacelia sericea*, whose gold is chelated as a cyanide, a mechanism leading to the accumulation of as much as 21 ppb of gold in the leaves (Harborne, 1988).

In the UK pollen has been found to reflect the presence of copper, manganese, lead and zinc in the soil, but not magnesium. Pollen collected by honeybees in Australia can indicate the presence of *Coelospermum decipiens*, syn. *Morinda reticulata* (mapoon, rotten cheesefruit), a shrub associated with selenium in the soil. Mining companies in North America have even been investigating the possibility of using hives of bees for prospecting purposes. Whether this use of pollen could become commercially viable has yet to be determined (Crane, 1990).

The rehabilitation of unsightly mine tailings by vegetation can be undertaken by selecting strains of plants tolerant of high concentrations of toxic metals in the soils, including the indicator plants discussed above. Strains of the grasses *Agrostis tenuis* (fine bent) and *Festuca ovina* (sheep's fescue), for example, have been identified that have the ability to rapidly colonise the tailings from heavy metal mines. Some strains of the former are even capable of growing successfully on soils containing up to 1% lead. Other useful plants include copper- and zinc-tolerant strains of *Deschampsia caespitosa* (tufted hair grass) and copper-tolerant strains of *Silene vulgaris*, syn. *S. cucubalus* (bladder campion). Among the tolerant selenium-accumulating plants are *Astragalus* spp. from North America, and *Coelospermum decipiens* and *Neptunia amplexicaulis* (selenium weed) from Australia (Harborne, 1988; Jackson *et al.*, 1990).

The actual mechanism by which metal toxicity is overcome is not yet fully understood, although a series of plant peptides known as **phytochelatins** are known to have the ability to chelate the toxic cations. More rarely resistance can be attributed to phenotypic plasticity instead of the evolution of distinct genotypes. *Typha latifolia* (bulrush, reed mace) would appear to fall in this category, although it is possible that *Typha* may have constitutive zinc-resistance (Harborne, 1988; Fitter and Hay, 1989).

6.   WEEDS

A weed is the term loosely applied to any plant growing where it is not wanted by man, and specifically to any unwanted plant growing in cultivated and grazing land competing with crops for light, water and nutrients, thereby reducing crop yields, hindering cultivation and harvesting operations, contaminating the desired product and sometimes harbouring crop pests and diseases. Genetic engineering has now brought the danger of weedicide resistance being conferred from crop plants to weeds, with obvious difficulties for weed control in the future.

Weeds are plants out of place, and it is because of their adverse economic effect that weeds are considered with economic plants! However, such prejudices may be in the eye of the beholder since they are regarded as aesthetically displeasing to the tidy minded.

While such widespread and aggressive weeds as *Elymus repens*, syn. *Agropyron repens* (couch, twitch or witch grass), and *Imperata cylindrica* (lalang, lang-alang) are unloved by both manual and mechanised cultivators, some weeds may even be considered desirable by some subsistence farmers but not necessarily by their more affluent counterparts. Thus, such cosmopolitan tropical weeds as *Cleome gynandra*, syn. *Gynandropsis gynandra* (cat's whiskers), and *Portulaca oleracea* (purslane) may be left among the growing crops by the subsistence farmer and harvested as required for food or medicine.

In grazing lands weeds may replace more palatable species and even be toxic to the herbivores, e.g. *Senecio jacobaea* (ragwort) and *Moraea* spp. When consumed by livestock some weeds. e.g. *Anthemis* spp. (chamomiles) and *Ranunculus* spp. (buttercups), may taint products such as milk. Other weed seeds can contaminate the crop product. In a mustard crop, for example, the seeds of the *Sinapsis arvensis* (charlock, wild mustard) are inseparable from those of *Brassica juncea* (brown or oriental mustard). Similarly, seeds of *Galium aparine* (cleavers, goosegrass) and *Vaccaria hispanica*, syn. *Saponaria vaccaria* (cow cockle), are inseparable from *Sinapsis alba* (white or yellow mustard) and their hard and brittle seed coats can even cause problems during milling (Miller and West, 1956; Hemingway, 1995).

Weeds may also harbour pests and diseases, which can then spread to nearby crops. The following example well illustrates the ramifications of mistletoe infestation in the cocoa plantations of Ghana, especially on old, unshaded trees. The prevalence of cacao swollen shoot virus and other pests and diseases, e.g. the capsids *Distantiella theobroma* and *Sahlbergella singularis*, and the oomycot *Phytophthora palmivora* (Butler's fungus), responsible for black pod disease of cocoa, are associated with mistletoe infestation, especially by *Tapinanthus bangwensis*, and to a lesser extent by other mistletoes. The holes bored by the haustoria are almost invariably inhabited by *Cataenococcus loranthi* (mistletoe mealy-bug), whose abundant honey-dew attracts large numbers of *Crematogaster* ants. These ants tend and possibly transport the mealy-bug vectors of the swollen shoot virus; it is also possible that the spores of Butler's fungus may be spread in the detritus used for the ant tents covering the mealy bugs. Unfortunately the *Crematogaster* ants are antagonistic to the capsids' principal predator, the ant *Oecophylla longinoda*, making biological control difficult. Other mistletoes affect rubber and teak plantations in Cameroon and Nigeria, shea butter trees in Burkina Faso, and citrus and guava in the Sudan. Globally, the greatest damage is caused by species of *Arceuthobium* (dwarf mistletoe), especially in the coniferous plantations of the northern hemisphere (Polhill and Wiens, 1998).

Weeds are opportunists, often invading bare ground, thereby providing protection against erosion and often improving soil fertility during periods when the land is not being cultivated. Provided they can be eradicated during cultivation, such weeds can be considered beneficial. Other weeds can be highly invasive. For example, the South American and now pantropical ornamental climber *Lantana camara* (cherry pie) can, under the right conditions, form impenetrable thickets and rapidly smother and, in extreme cases kill, tree plantations. In the UK the introduced *Fallopia japonica* (Japanese knotweed) is now a major weed of gardens and waste places; its almost indestructible root system makes it extremely difficult to eradicate.

Also from South America, the free-floating aquatic, *Eichhornia crassipes* (water hyacinth) has become a major pantropical weed of waterways. Introduced as an aquatic ornamental for a hotel garden pond in Uganda, it only took two decades from first sighting in 1958 before severely impeding the flow of water and blocking the passage of shipping through the Sudd region of the Nile. The water hyacinth also chokes irrigation channels, blocks hydroelectric installations, and seriously affects fisheries. So far, efforts to control the water hyacinth with herbicides and other methods have failed, primarily because the dead plants sink, decay and recycle their nutrients, thereby enriching the water and encouraging further growth of the weed. Scientists are now trying to make use of this enormous plant resource for animal fodder, paper, insulation board, fertiliser and methane gas generation, and as a source of leaf proteins and hormones. Its potential for water purification are discussed in Section 6.2 (Obeid Mubarak *et al.*, 1982; Kochhar and Singh, 1989). There are two lessons to be learnt from the water hyacinth saga (and other introduced weeds). The first is that plant quarantine arrangements must not be ignored, the second, to look for possible beneficial use!

Chapter 21

Social Uses

Social uses refer to those plants used by people that are not essential for survival but are considered socially desirable, or are of ritual or spiritual significance. In some cases they may even be detrimental to health, yet provide a sense of well-being, e.g. alcoholic beverages and tobacco.

1.   FUMITORIES AND MASTICATORIES

Fumitories and masticatories are plant materials that are smoked and/or chewed, usually for their stimulative and narcotic effects. In some cases the alkaloids present may affect the central nervous system, although not all contain alkaloids, such as the dried herbs smoked as tobacco substitutes, likewise the chewing latex chicle from *Manilkara chicle*.

1.1   Fumitories

Of the fumitories tobacco is by far the best known and widely used. There are two commercially important species of tobacco, *Nicotiana rustica* (Aztec or wild tobacco) and *N. tabacum* (smoking tobacco). Aztec tobacco has been cultivated in Mexico and eastern North America from pre-Colombian times. The species is unknown in the wild, the cultigen possibly originating in Peru as a cross between *N. paniculata* and *N. undulata*. Early observations by Spaniards noted the tobacco being smoked by the Amerindians either through a forked stick or the leaves rolled in the manner of a cigar. It was first introduced into Spain in 1519, and to the UK by Sir John Hawkins in 1573 where it was much popularised by Sir Walter Raleigh. The craze for smoking had started by 1586 and was originally believed to be of medicinal benefit. However, in 1603 King James I issued a pamphlet entitled *A Counterblast to Tobacco* proclaiming the many harmful consequences of smoking. A heavy import duty was also placed on tobacco in order to deter its use. Smoking was later banned by Pope Urban VIII (1623-44) with offenders excommunicated, while in Russia Tsar Michael I (1596-1645) went as far as to order the execution of second offenders. Both measures failed to influence its use.

Tobacco is now cultivated in Russia for the alkaloid *nicotine* for use as an insecticide; the cured leaves containing up to 10% nicotine. Smoking tobacco was also cultivated in tropical America from before the arrival of Columbus and it too is not known in the wild, with *N. otophora*, *N. sylvestris* and *N. tomentosiformis* as its pugitive ancestors. The cured leaves contain 1.5-4% nicotine (Baker, 1964; Brouk, 1975; Mabberley, 1997).

Tobacco quality depends very much on the weather, variety, soil and curing. Four types of tobacco are recognised according to the method of curing the mature leaves: (1) *Flue cured* using artificial heat for drying in specially constructed barns and producing Virginia and Amarelo tobaccos; (2) *Air cured* by drying on racks in the shade at ambient temperatures for Burley (for cigarettes, pipe and chewing tobaccos), Maryland and cigar tobaccos; (3) *Sun cured*, i.e. sun dried, for Turkish and other oriental varieties; and (4) *Fire cured* by direct drying over a fire to produce Kentucky tobacco.

The cured leaves are then fermented in heaps for 4-6 weeks, a process that results in the disintegration of any remaining chlorophyll and the removal of any volatile nicotine, leaving only the fixed nicotine. Liquorice paste, honey, sugar, molasses, rum and the fragrant seeds (which contain 2-3% coumarin) of *Dipteryx odorata* (Tonka beans) cured in rum, may be used to flavour the tobacco. Buyers select according to leaf size, thickness, colour, texture, flavour/aroma, rate of burn and processing qualities.

Tobacco is mainly used in cigarettes, cigars and for pipe tobacco; small quantities of chewing and snuff tobaccos are also being produced. For cigarettes old leaves are known as *fillers* and the young leaves as *leaf tip*; the cigarette papers are made mainly from flax fibres. Heavy bodied and aromatic *cigar-fillers* form the core of cigars, and are bound together by finer textured and more elastic cigar-*binders*, with an outer layer of the elite, thin and silky *cigar-wrappers*. While tobacco consumption has declined in the West because of an increased awareness of the health risk, it has increased in the East in pace with increasing populations (Brouk, 1975; Purseglove, 1987; Robbins, 1995).

1.2   Masticatories

Masticatories, from Late Latin *masticare*, Greek *mastikhë*, to grind the teeth, refer to substances that are chewed to produce salivation. They have a long history. The resin of *Pistacia lentiscus* (mastic), for example, has been used as a masticory at least since the time of Theophrastus (ca. 370-287 BC), as its Greek vernacular name testifies (Long, 1987; Mabberley, 1997).

Chewing gum from the latex of *Manilkara zapota* (chicle, chiku, naseberry, sapodilla [plum], beef apple) of Mexico and Central America was being chewed by the Mayan people of Guatemala long before the sweetened, flavoured version was commercialised in the late 19th century. The chicle is tasteless and contains no alkaloids, the flavours are added later. It became famous during World II following the arrival of US troops in Europe. The latex contains 20-40% of a gutta-percha-like material, which is boiled and evaporated for chewing gum. The latex from other *Manilkara* species are now used as a substitute for that of *M. zapota*. Chewing gum also contains 85% of a latex from the Malaysian tree *Dyera costulata* (jelutong); during World War II the latex of *Couma macrocarpa* (sorva, lechi-caspi) from the Upper Amazon was used as a substitute (Brouk, 1975; Brücher, 1989).

Through much of southern Asia to Oceania the half ripe and cured or fully ripe seeds *Areca catechu* (betal palm) are chewed as a masticatory. The seeds are normally sliced and mixed with various spices, such as cinnamon, cloves, cardamom (*Elettaria cardamomum*) and nutmeg and wrapped in the leaf of *Piper betel* (betel pepper) smeared with lime. Alternatively the betel may be mixed with tobacco, with or without additives, before wrapping. It is a mild stimulant, blackening the teeth and giving a red stain to saliva as well as sweetening the breath. Its use has been known from antiquity, being mentioned by Herodotus in 340 BC. Betel contains the alkaloid *arecoline*, a mild narcotic, producing a sense of well-being and dulling the appetite.

The seeds (cola nuts) of *Cola* species (cola) from tropical Africa are also chewed. They contain 2-4% *caffeine*, with traces of *theobromine* and the glucoside *kolanin*, which acts as a heart stimulant. The use of the nuts in cola drinks has now been largely supplanted by synthetics.

In the eastern Andes the local inhabitants daily chew the dried, powdered leaves of *Erythroxylum coca* (coca) mixed with unslaked lime and the alkaline ash of *Chenopodium quinoa*. The users are able to maintain their blood glucose levels despite poor diets. The leaves of *E. novagranatense* from lower altitudes are similarly used. The leaves are slightly narcotic, containing 1% *cocaine* and are the source of the drug cocaine. Cocaine is a debilitating, addictive narcotic, causing euphoria, indifference to pain and tiredness, increased alertness and sexual desire (Brouk, 1975; Mabberley, 1997). The presence of both cocaine and nicotine in Egyptian mummies is still an ethnobotanical mystery and suggests an early unrecorded trade between the Old and New Worlds. Tests for these drugs in the ancient dead from India and the Far East may indicate a possible route.

Although commonly referred to as a narcotic, *Duboisia hopwoodii* (pituri) is best regarded as a masticatory. It is chewed by the Aboriginal tribes of Central Australia as a *stimulant*, to promote excitement, and was used before fighting or other important events, and on long and difficult walks to reduce fatigue. Indigenous species of *Nicotiana* are similarly used (M. Lazarides, pers. comm., 1998).

2.   NARCOTICS

Narcotics are toxic drugs that dull the senses, induce sleep and, with prolonged use, become addictive. When used medicinally to induce sleep such drugs are referred to as *hypnotics*. It is the actual dosage that distinguishes between being a poison, a medicine or a narcotic (*'Sola dosis facit venenum'* of Paracelsus). In modern society the use of narcotics is largely recreational, while in more primitive societies their use may include medicinal, ritual, religious and recreational purposes (Schultes and Hofmann, 1992; Macpherson, 1995).

2.1   Hallucinogens and Psychoactive Drugs

Hallucinogens are compounds characterised by their ability to produce distortions of perception, emotional changes, depersonalisation and a variety of effects on memory and learned behaviour. They may also produce *psychosis*, a serious disorder of the mind, amounting to insanity. Some 150 plants are known to be used for their hallucinogenic properties, with many more with hallucinogenic properties that are apparently not used as such. Surprisingly, the aboriginal populations of Australia and New Zealand and the peoples of the Pacific Islands do not use hallucinogens, although plants containing hallucinogenic principles are certainly known to occur in Polynesia. The majority of plant hallucinogens contain alkaloids as their active principles. The major exception is *Cannabis*, where the active principles are the non-nitrogenous *cannabinoids* present in a resinous oil. See Chapter 19 for a discussion on hallucinogenic fungi (Schultes and Hofmann, 1992; Macpherson, 1995).

2.1.1   Cannabis

Cannabis sativa subsp. *indica* (Indian hemp) is believed to be native to Central Asia, possibly China. The plant has been cultivated and used for medicinal purposes in India since 900-800 BC, and in North Africa since medieval times. Its cultivation has now spread throughout the world, apart from the Arctic regions and the tropical rain forest. It is reputed to be the US's most lucrative cash crop, worth $32 billion. The more northerly distributed subsp. *sativa* (hemp) is cultivated for its fibre and is considerably less potent, despite which its cultivation in the UK became illegal 1951. Cannabis-free cultivars are now available and production was reintroduced in 1993.

The short and much branched cultivars are the source of the volatile psychoactive cannabis resin exuded from glandular hairs. The resin may be smoked as a hallucination inducing fumitory, creating a pleasurable state of mind, albeit damaging to the health. It is also used as a snuff or chewed.

There are wide variations in the psychoactive effects of cannabis preparations, depending on the type of plant used, preparation, method of administration, dosage, personality of the user, and social and cultural background. While the use of cannabis is reputed not to cause physical dependence, its abuse leads to passivity, apathy and inertia, effects that were first described in 2736 BC in the pharmacopoeia of the Chinese herbalist, Emperor Shen Nung.

The crude resin known as *charas* is mainly produced in central Asia and is extracted by rubbing the tops of the plants with the hands, or beating with a cloth. The purified resin known as *marihuana, marijuana* (Mexico), *pot* (US), *dagga* (South Africa), *kif* (Morocco), *hashish* (Turkey), is obtained from the dried flower heads of female plants. *Ganja* (India) is obtained from the resin-rich pressed and dried unfertilised flower heads of female plants. It is usually smoked, often with tobacco. A relatively mild preparation known as *bhang* (Hindustani), is the pounded paste of spices and the dried leaves and flowering heads of both male and female plants. It is consumed either as candy or as a tisane (Kirby, 1963; Brouk, 1975; Purseglove, 1987; Schultes and Hofmann, 1992; Macpherson, 1995; Robbins, 1995; Mabberley, 1997).

The active component of cannabis resin is Δ^1*-tetrahydrocannabinol* (Δ*-9-THC*). The member of the active fraction present in most cases is the Δ^1-3,4-*trans* isomer; the isomeric Δ^6-3,4-*trans* isomer also occurs naturally and is similar to the Δ^1 isomer in pharmaceutical activity. It has now been demonstrated that the cannabinoids are centrally acting analgesics with a distinct mode of action, which could lead to new classes of non-opioid analgesics.

The Δ-9-THC has proved lethal to some insects. However, in trials with the larvae of *Arctia caja* (tiger moth) and the grasshopper *Zonocerus elegans*, some insects died while others survived by storing the THC in their bodies (Harborne, 1988; Sharp, 1990; Schultes and Hofmann, 1992; Meng *et al.*, 1998).

2.1.2   Opium

The second major hallucinogenic drug from the Old World is opium. From among the *ca.* 25 alkaloids present the principal alkaloid of opium is *morphine* which, with its salts, are very valuable albeit highly addictive analgesic drugs. *Opium* is produced from the dried latex harvested from the lanced immature capsules of the cultigen *Papaver somniferum* subsp. *somniferum* (opium poppy), a process that has remained unchanged since it was described by Pliny the Elder (23-79 AD) in his *Historia Natuuralis*. Opium has been used for millennia as a medicine and recreational drug. Cultivated in the Neolithic, it is mentioned in the Sumerian writings, and dispensed by the pharmacies of ancient Egypt and Persia.

In China the opium addicts smoke powder balls; in the Western World addicts either inject liquid morphine or the synthetic alkaloid based on morphine known as *diamorphine hydrochloride* (*heroin*), which can also taken as a snuff. The tincture of opium known as *laudanum*, obtained by dissolving opium in alcohol, was first prepared in the mid-17th century by Thomas Sydenham, he is also credited with prescribing Peruvian bark, i.e. quinine for treating malaria. Laudanum was, until modern times, a popular soporific and analgesic drug (Renfrew, 1973; Brouk, 1975; Sharp, 1990; Duin and Sutcliffe, 1992; Mabberley, 1997).

2.1.3   Peyote

Mexico is undoubtedly the world's richest area of plant diversity and use of hallucinogens among aboriginal societies, with South America a close second. In Mexico the best known hallucinogenic drug is *peyote*, obtained from the spineless woolly cactus *Lophophora williamsii* (peyote cactus). The ritual use of peyote has now spread northwards through the USA to the Amerindians of Canada. The fresh or dried aerial parts, *mescal buttons*, are either chewed or used a tisane prepared by boiling in water. The active principles are non-volatile so that the buttons loose none of their potency during storage. Some 30 alkaloids are present, of which the active ingredient is the alkaloid *mescaline (β-(3,4,5-trimethoxyphenyl)-ethylamine)*, which is closely related to a neurotransmitter, the brain-hormone norepinephrine. Mescaline acts as a depressant of the central nervous system, producing kaleidoscopic hallucinations and a feeling of weightlessness. It has even been suggested that the use of mescal by the shamans stimulated the strikingly bizarre artistic designs of the early American civilisations (Brouk, 1975; Sharp, 1990; Schultes and Hofmann, 1992; Mabberley, 1997).

2.1.4   Other Hallucinogens

In South America the indigenous peoples use hallucinogens obtained from a number of species, including the seeds of *Anadenanthera*, the bark of *Banisteriopsis*, and the bast resin of *Virola*; even the leaves of *Erythroxylum coca* can produce hallucinations if sufficient quantities are masticated (Prance, 1984; Schultes and Hofmann, 1992).

Ott (1998) discusses the effect of a number of toxic honeys, in which bees have sequested naturally occurring secondary metabolites from floral and extrafloral nectaries. For example, the Mayans deliberately exploited the psychoactive honey from *Turbina corymbosa* (ololiuqui) for the ritual mead, *balché*, which was used as a shamanic inebriant for hallucinations and visions. Elsewhere in southern Mexico the seeds are used in Aztec ceremonies as a hallucinogenic intoxicant with reputed analgesic properties. The active principles are ergotine alkaloids and lysergic acid derivatives. These were previously known from the cereal fungus *Claviceps purpurea* (ergot). Ergotism, from eating flour contaminated by ergots, resulted in intense pain and hallucinations, a condition known as St Anthony's fire. Although Linnaeus (1763) in his *De Raphania*, wrongly attributed ergotism to the seeds of the common black turnip-radish mixed in the grain (Schultes and Hofmann, 1992; Mabberley, 1997).

3.   SOAPS, COSMETICS AND FRAGRANCES

3.1   Soaps

Soaps are cleansing agents manufactured in bars, granules, flakes, or liquid form from the sodium and potassium salts of fatty acids, particularly stearic, palmitic and oleic acids. The vegetable oils and fats from which many soaps are prepared consist essentially of the glyceryl esters of these acids. During manufacture vats containing the oils or fats are heated with dilute NaOH (less frequently KOH) solution. When hydrolysis is completed the soap is 'salted out' or precipitated with NaCl. The soap is then treated as required with perfumes and made into tablets. For example, Castile soap is manufactured from olive oil, transparent soap from decolorised vegetable oils, and liquid green soap from KOH and vegetable oils.

Traditional cleansing agents are based on *saponins*, i.e. water-soluble glycosides that foam when agitated. They occur in a wide variety of plants, where their presence is believed to deter predation. Typical examples of such soap substitutes include the leaves of *Agave* spp., *Atriplex polycarpa* (all scale), *Carica papaya* (pawpaw) and *Saponaria officinalis* (soapwort), and the fruits of *Sapindus* spp. (soapberry); among the species used for hair shampoos are the seeds of *Simmondsia chinensis*, the leaves of *Vaseyanthus* sp. and the young shoots of *Gynerium sagittatum* (uva grass) (Sharp, 1990; Cotton, 1996; Mabberley, 1997).

3.2   Cosmetics

Cosmetics are preparations applied to either beautify or decorate the body (from French *cosmétique*, from adjective 'of adornment', from Greek *kosmëtikos*, skilled in arranging, from *kosmëtos*, well ordered, from *kosmein*, to arrange, from *kosmos*, order). Their use dates back into antiquity. The powdered leaves of *Lawsonia inermis* (henna) are believed to have been used in Pharaonic Egypt for colouring the finger nails red, and of *Isatis tinctoria* (woad) for colouring the bodies of the Ancient Britons blue. Henna is still used today, as is the orange colouring obtained from the testa of *Bixa orellana* (annatto), the original Amerindian body paint. The Seri of Baja California use a range of face paint materials; they also tattoo their bodies with the mashed leaves of *Condalia globosa* mixed with the ashes of *Olneya tesota* (ironwood). In some cultures it is customary to blacken the teeth using wood tar obtained from such species as *Eugenia tumida*, *Fagraea racemosa*, *Tamarindus indica*, etc., or the juice from *Rothmannia macrophylla*. First used by the inhabitants of the Sonoran Desert for dressing the hair, the liquid wax from the seeds of *Simmiondsia chinensis* (jojoba) is now widely used in a number of commercial cosmetic preparations (Manatee, 1990; Cotton, 1990; Fleeter and Mossier, 1985; Lemons *et al.*, 1991; Mabberley, 1997).

3.3   Fragrances

Fragrances based on essential oils (see Chapter 15) have been widely used from the earliest times. Today they are added to such products for personal use as perfumes, deodorants, shampoos, bath lotions, toilet soaps, toothpastes and mouth washes, and industrially for laundry soaps, detergents, cleaning agents, air freshness, etc. (Coupon, 1995).

The Ancient Egyptians were famous for their scents and perfumes. Because the distillation of alcohol was not known until the 4th century BC, the essential oil had to be extracted by steeping plants, flowers or shavings of fragrant wood in oil, which is then added to other oils or fat. The oils used included balanas oil from *Balanites aegyptiaca*, moringa oil from *Moringa peregrina*, syn. *M. aptera* (the *M. oleifera*, syn. *M. pterygosperma* cited by Manniche (1989) is a native of northern India and Pakistan, and is not known from the wild in Egypt *fide* Verdcourt (1985)), olive oil from *Olea europaea*, almond oil from *Prunus dulcis*, etc. According to Theophrastus, *Concerning Odours*, balanos was the preferred oil because it was the least viscous, followed by fresh raw olive oil and almond oil. One of the more famous of the Egyptian perfumes was made in the Delta city of Mendes, consisting of balanos oil, myrrh and resin from *Commiphora* spp. Interestingly, the order in which the ingredients were added to the oil was extremely important since the last one to be added imparted the most pungent scent. See Manniche (1989) for other recipes.

The origins of the present-day industry began in 1367 with the creation by the Queen Elizabeth of Hungary of an alcoholic solution of fragrant herbs known as 'Queen of Hungary Water'. This was followed in 1690 by the development of 'Eau de Cologne' by Jean-Antoine Farina. During the 18th century commercial perfumery houses began to appear, with producers and compounders providing the essential oils and creating fragrance compounds for the perfumery houses. In the 19th century developments in organic chemistry led to the synthesis of the first organic products, such as vanillin and benzyl aldehyde. Despite such advances, the essential oil industry is a conservative one which does not readily lend itself to synthetic substitutes for the better class of perfumes (Brud, 1995; Coppen, 1995).

The essential oils are the most widely used source of perfume by indigenous societies. An example of an alternative mode is in the Sudan where the smoke from *Combretum adenogonium*, syn. *C. fragrans* and *Terminalia brownii* are used by women to scent their bodies. In West Africa sachets containing the rhizomes and tubers of *Cyperus* spp. and *Kyllinga* spp., or their smoke, are used for scent. As an antithesis of fragrance, in Nubian Egypt the leaves of *Lawsonia inermis* placed in the hollows of the arms act as a deodorant (Burkill, 1985; Manniche, 1989).

4. CONTRACEPTIVES AND ABORTIFACIENTS

Plant sources of contraceptives as means of avoiding pregnancy despite sexual activity are based on the hormone **progesterone** to govern the growth and development of the uterus during pregnancy. Progesterone can be manufactured from the animal sterol **cholesterol** and from certain plant steroid sapogenins, such as **diosgenin** from the fruits of *Balanites aegyptiaca* and *Trigonella foenum-graecum* (fenugreek), the fleshy tubers of *Dioscorea elephantipes* (elephant's-foot, hotentot-bread), *D. opposita* (Chinese yam), and from **stigmasterol**. Stigmasterol was first extracted from the seeds of *Physostigma venenosum* (Calabar bean), it is now readily isolated from many plant sources, often in the presence of **sitosterol**, from which it is difficult to separate. Toxic alkaloids too can cause abortion, including the fungus *Claviceps purpurea* (ergot), and even from tobacco smoke (Abu-Al-Futuh, 1989; Lapinskas, 1989; Okigbo, 1989; Tyler, 1989; Blackwell, 1990; Sharp, 1990).

Plants containing volatile oils have been widely utilised in traditional medicines to induce or stimulate menstrual flow, i.e. **emmenagogues**, and are known somewhat less euphemistically as **abortifacients**. They bring about artificial abortion by producing pelvic congestion through intestinal irritation; toxic doses of the actual volatile oil being more effective than the plant material (Tyler, 1989). The latter include the leaves of *Ruta graveolens* (rue), the leaves and tips of *Chrysanthemum vulgare* (tansy), *Hedeoma pulegioides* (American pennyroyal) and *Mentha pulegium* (European pennyroyal), the tops of *Juniperus sabina* (savin), the fruits of *Juniperus communis* (juniper) and *Petroselinum crispum* (parsley), and the oleoresin of *Pinus palustris* (longleaf pine).

Manniche (1989) suggests that the seeds of *Apium graveolens* (celery) and not parsley were used as an emmenagogue by the Ancient Egyptians. While Lawless (1992) also reports emmenagogue properties for celery, other literature sources, e.g. Launert (1981), Chiej (1984) and Mabey (1988) only mention parsley. This example serves to illustrates how information is often handed down without the recipient checking the primary references and plant identities, and thereby providing a major source of conflicting information.

5. PLANTS OF RITUAL OR RELIGIOUS SIGNIFICANCE

Tree worship and the use of leaves, flowers, etc. in religious ceremonies almost certainly date back to prehistoric times. That useful plants should be venerated is understandable, less understandable are the very many plants without apparent usefulness to a community that are also associated with myths and traditions. Their use is presumably due to their association with religious beliefs, or perhaps because of their resemblance to the emblem of a particular deity or even the name of a sage associated with them, and thereby making the plant sacred. Various plant parts are traditionally used to counteract witchcraft or the evil eye, or burnt to drive away mosquitoes and other pests. Scents and perfumes are used to appease the gods, while others are believed to restore fertility, etc. (Gupta, 1971).

The Ancient Greeks associated some plants with particular gods and their godly attributes, e.g. purification, fertility and growth. Thus wheat is sacred to Dementer, who taught agriculture to man, and the vine associated with Dionysus, the god of wine and ecstasy. Victorious athletes were awarded garlands of wild olive leaves (*Olea europaea*) at the Olympic Games, bay leaves (*Laurus nobilis*) at the Pythian Games, and wild celery (*Apium graveolens*) at the Nemean Games.

Similarly, in India the traditional Hindu almanac, the Panchang, which is based on an astrological concept of the movements of the Sun through the constellations, has a reigning deity for each constellation together with an associated sacred tree to be worshipped. Plants are also believed to influence body functions, ailments and disease, and have been linked with reputed medical properties of plants to counteract that influence. The root bark of *Calotropis gigantea*, for example, is associated with the constellation Sravena and the diety Vishnu, and is used to treat intermittent fever. While in Bali the Hindu burial ceremonies include the yellow variety of the bamboo *Schizostachyum brachycladium*.

In Europe the 17th century some philosophers and herbalists still maintained that every plant and every illness was governed by a constellation or planet and that a disease caused by one planet could be cured by a plant belonging to an opposing planet, or conversely, by a herb belonging to the planet responsible for the disease. Probably the best known of these astrological botanists was Nicolas Culpepper and his *Physicall Directory* published in 1649 and his fanciful linking of herbs with astrology. Fortunately for medicine other herbals were free of such fancies (Vickery, 1995). See Chapter 16 for further information on herbals.

The hallucinogenic drugs may also have a ritual or religious significance in addition to their purely narcotic usage, their hallucinogenic properties being associated with evil spirits, particularly by the New World shamans. In medieval Europe the three notorious hexing herbs were *Atropa belladonna* (deadly nightshade), *Hyoscyamus niger* (henbane) and *Mandragora officinarum* (mandrake). All three contain relatively high concentrations of tropane alkaloids, chiefly **atropine**, **hyoscyamine** and **hyoscine** (**scopolamine**), with the hyoscine producing the hallucinogenic effects considered responsible for their magical attributes (Bennet *et al.*, 1992; Schultes and Hofmann, 1992; Dransfield and Widjaja, 1995; Rose and Dietrich, 1996).

In some mythologies forked roots are attributed with human properties. Thus the fanciful resemblance to the human form of the forked roots of the mandrake were regarded as a talisman by the ancient Assyrians; the roots were used to ward off evil spirits. In early western mythology the mandrake was reputed to emit screams when pulled from the ground. Dioscorides describes how a dog is tied by the neck to the plant and a piece of meat thrown to the dog, causing it to lunge and uproot the mandrake. The shrieks and groans during the uprooting plus the foul odour brought about the demise of the dog, while the master's ears are stopped against the sound, which would otherwise drive him mad. Mandrake was also known as a 'gallows man' because it was believed to grow beneath a gallows, fertilised by the urine or semen of the hanged man. Such gallows men were attributed with greater powers than mandrakes found elsewhere. The foul-scented root of *Ferula assa-foetida* (Satan's faeces) was also used as a talisman to ward off evil (Embooden, 1974; Blackwell, 1990; Schultes and Hofmann, 1992).

In Africa there are numerous myths concerning *Adansonia digitata* (baobab). Many cultures believe the tree to be inhabited by spirits, and in Senegal the Sérères make use of any hollow baobabs for bodies denied burial, i.e. those of the griot cast - poets, musicians, sorcerers, drummers and buffoons. They are suspended for mummification so that their bodies will not pollute the earth. In East Africa, Resa, Lord of Rain, is said to live in a great baobab that holds up the sky. Elsewhere prominent trees may be regarded as the traditional meeting place for the elders, others are worshipped as fertility symbols, etc. (Wickens, 1982).

In Australia myths relate how careless land owners will become blind unless certain lands are protected from burning. As a result fragments of the rain forest are protected from fire and valuable fire-sensitive species, e.g. *Dioscorea* spp. (yams) are conserved. While such a protective practice has a practical function in conserving the yams as a food resource, some argue that the advantageous consequences of a social practice does not necessarily explain its existence (Cotton, 1996).

Europe too has its fair share of religious and ritual plants. Floral tributes at funerals were probably first used to mask the odour of rotting flesh. In Italy white chrysanthemums are a favourite funeral flower and as such are never used for indoor decoration. In various parts of the UK *Sambucus nigra* (elder) was often planted near habitation for protection against witches; sometimes it was also associated with fairies and good luck (Vickery, 1995).

In Ireland ands Wales a sprig of *Ulex europaeus* (gorse) was traditionally hung over the doorway or brought into the house on May Day to ward off witches and fairies (Grigson, 1958; Lucas. 1960). On the other hand Vickery (1995) provides examples from Ireland and the Channel Islands where it is considered unlucky to bring gorse into the house, and from Hampshire where dragons were believed to either live or were born in gorse flowers.

The tradition of well-dressing in the Peak District of Derbyshire is considered by some to have its origins in pagan times when springs and wells were decorated with green branches and flowers. With the coming of Christianity such veneration was banned but many of the wells were purged of their pagan associations and rededicated to the Blessed Virgin Mary or a saint and the pagan rite continued under the banner of Christianity. While at Tissington well-dressing is claimed to be in thanksgiving for the village escaping the ravages of the Black Death. What ever their origin, well-dressing is certainly a tourist attraction. (Vickery, 1995, who should be consulted for further examples). The importance of such religious and ritual use of plants to communities cannot be ignored. People are entitled to their beliefs and administrators and developers should be aware of such uses and beliefs and respect them.

6.   SEASONS AND WEATHER

Most people are familiar with the changing seasons and their association with the life cycle of wild and cultivated plants. Apart from biologists relatively few among the industrialised nations today correlate the changing life cycle with other biological events. In the UK there are a number of old country sayings that still reflect such observations. Thus, when the lilac (*Syringa vulgaris*) is in flower it is a bad time to buy calves. This is because flowering coincides with lush pastures and rich milk causing the calves to flower.

The so called 'primitive societies' may even rely on certain plants, referred to as 'calendar plants', for their hunting and survival. Often with conspicuous flowers, they serve to indicate certain important seasonal occurrences that could otherwise be difficult to assess, such as when an animal food resource is at its best. For example, when *Brachychiton paradoxus* is flowering the Arnhem Land Aborigines know that there are plenty of baby sharks in the sea, scrub fowl have laid their eggs, and mudcrabs contain eggs and are at their best (Yunupinu *et al.*, 1995).

Plants may also be used to forecast the weather. In Wales rapidly opening lilac blossoms indicates rain falling soon; if they quickly droop and fade it is a sign of a warm summer. Also in the UK some people hang a piece of dried seaweed near the door, when it gets damp it is a sign of rain (Vickery, 1995). As a child I was taught that rain would follow when the wind blowing through trees and shrubs reveals the underside of leaves.

Chapter 22

At the Start of the 21st Century

The world population is now 6 billion, of whom 0.8 billion are malnourished and 1.2 billion live on a daily income of less than US$1.00. The population is continuing to rise and is expected to reach 11.5-12 billion by 2050. Even now many of the undeveloped nations are doubling their populations within 20-25 years, expanding at a faster rate than sustainable agriculture and forestry can be maintained. The effects of this demographic explosion is likely to be compounded by the necessity of adapting to the yet unknown full effects of global warming. The balance between the need to conserve the environment and the ever increasing demand for land and water resources will become more difficult to maintain.

Sustainable productivity, whether from natural resources, agriculture (crops and livestock), forestry, etc. without damaging the environment, is the dream of all conservationists. In theory sustainable productivity is probably attainable although in practice it is economically unattainable. This is because the developed countries continue to seek ever higher living standards, for which they require cheap imports from the less developed countries on their own terms, regardless of the environmental and human consequences for the developing countries. For facts and figures for North America see Prescott-Allen and Prescott-Allen (1986).

1.   DEVELOPED AND DEVELOPING COUNTRIES

While it is theoretically possible to feed the world's population, present-day economics and politics make such a possibility unlikely. Food security is, and is likely to continue to be, a major global problem, especially for the developing countries where crippling debt repayments, low agricultural inputs, erosion, desertification and deforestation make it difficult, if not impossible, to grow sufficient food for local needs. Yet, in order to obtain foreign currency, these countries are compelled to supply non-essential products, such as. out of season salad crops, cut flowers and

415

newspaper pulp, to the developed countries. On the other hand, the developed countries have the capacity to provide food for export yet fail to do so because crop surpluses would mean lower prices for their farmers. Even if surplus food was made available to those countries in need, the transport costs, etc. are prohibitively high. For example, during the Ethiopian famine in the 1980s 1 ton of wheat costing *ca.* £100 at the port of lading had doubled in value by the time it had reached the port of unloading, to which distribution costs had to be added!.

A very wide range of under-utilised crops and other plant resources have been identified and are available to suit the needs of both the developed and less developed nations of the world, although I fear that the economics of production will always control what product reaches the world markets. For example, it is largely the economics of jojoba oil production that currently restricts the crop to a relatively small scale acreage for the lucrative cosmetics industry, whereas large scale cropping would be required to provide a cheap substitute for the oil from the endangered sperm whale. Growers would have to accept a substantial drop in sale price to meet that demand.

From the very beginning of human life on earth plants have been utilised by man, and many poor people in the developing countries continue to rely on non-wood forest products (NWFP) for their survival and/or income, while the developed countries have tended to regard NWFP solely as objects of commerce. It is only during the past two decades that the international organisations have fully appreciated the scale and importance of NWFP for the rural populations of developing countries.

Throughout history wild plants have been brought into cultivation to meet an increasing demand for guaranteed quantity, quality and price for food and industry, accompanied by a shift from a domestic to a plantation or industrial economy. Where these demands cannot be met an often more profitable development in recent years has been to produce synthetic substitutes, especially for pharmaceuticals. Plantation and industrial production have also resulted in the rural producers loosing their competitive edge in the national and international markets and having to increasingly rely on supplying rural households and local markets. (Wickens, 1991; Killman, 1999).

No single country can be expected to produce the range of plant products that it requires from its own internal natural resources, yet many countries still fail to fully utilise what they have available. India's Green Revolution was a good example of what can be done to provide greater self sufficiency and a better trade balance, although the swing to monoculture and high agrochemical inputs is rapidly becoming a cause for concern. Trade, of course, is essential, but it should be for the mutual advantage of both the exporting and importing countries, narrowing the currently widening monetary gap between the developed and less developed nations. As Mohandas Ghandhi has wisely stated *"The earth has enough for everyone's needs but not for somebody's greed."*

The United Nations Convention on Biological Diversity adopted at Nairobi in 1992 encouraged *"co-operation between government authorities and the private sector in developing methods for sustainable use of biological resources."* It also called for an equitable sharing of the benefits arising from the use of biotechnology by apportioning access to genetic resources, the transfer of relevant technologies and suitable reimbursement for the rights over resources and technologies. Among the many ethical and legal issues now recognised regarding NWFP, especially between the developing countries and the multinationals, are: (1) Who is the rightful owner? (2) What are the benefits for the rural population/indigenous group/country of origin? (3) Who monitors their interests? (4) What are the mediation/negotiation processes available? and (5) How successful have they been?

The pharmaceutical industry provides a good example as to how the Rio Convention can be implemented. Basically, the global concentration of plant diversity and traditional healers are in the species-rich tropical forests, while the technological infrastructure, pharmaceutical expertise and financial resources are mainly in the developed countries of the temperate regions. Thus, plants collected in the tropics are being mainly investigated and developed by academic institutions and private companies in Europe, USA and Japan. The imbalance between the accessors and owners of indigenous knowledge raises a wide range of ethnic and political issues, a number of which have been addressed by the Nairobi Convention. The recent interest by pharmaceutical companies in the ethnobotanical approach to herbal remedies discussed in Chapter 16. has meant a new approach to research protocols. Thus, Shaman Pharmaceuticals have developed a working collaborative relationship with local communities, traditional healers and local scientific institutions, providing a working model as to how a large pharmaceutical company can create an equitable and potentially lucrative partnership within the Nairobi Convention. In the long term Shaman will return a percentage of its profits to all the indigenous communities and countries with which it has worked (Carlson *et al.*, 1997b). It is to be hoped that other industries will develop a similar philosophy and partnerships.

2. GENETIC ENGINEERING

World trade is increasingly being controlled by multinational organisations motivated by profit. This is exemplified by the development of genetically modified crops whose seeds not only respond to the seed and agrochemical industry's own brands of herbicides/pesticides but could include the introduction of a terminator gene to prevent the germination of future generations, thereby forcing the farmer to buy new seed each year. Obviously the terminator gene cannot be used for all cereals, the brewing, distilling and starch industries still require cereals capable of germinating! In early October 1999 Monsanto bowed to popular outrage, particularly from Europe, and have ceased production involving the terminator gene, at least for the time being.

Perhaps the terminator gene technology could be beneficially applied to *Cannabis sativa* subsp. *indica* to control the psychotropic drugs industry? However, should the terminator gene escape the effects would be disastrous.

There are extravagant claims of GM crops being able to solve the world's food problem. This would be possible if the GM crops were suitable for the developing countries and in the unlikely event of the subsistence farmers in these developing countries being able to afford the cost of the seeds and chemical inputs. Even so, there could be potential benefits for the developing countries if, for example, suitable low input GM crops could be developed to solve their problems of hunger and nutrition by using the technology now available to increase the calorific content of starch-storage organs of the major starch crops and the vitamin content, etc. of others. This would undoubtedly be beneficial not only to the undernourished people of the developing countries but also help reduce the indignity of famine relief. Perhaps the savings made by the developed countries in costly relief could be used finance such developments? It should be remembered that there are ethical problems regarding the introduction of animal genes into food plants for those whose religions forbid the eating of animals.

Both pharmaceuticals and GM foods undergo rigorous health investigations for any injurious effects before they are considered safe. Pharmaceuticals, unlike food, are usually consumed in small quantities for relatively short periods. Foods are consumed daily and in large quantities, presenting the possibility, however remote, of a build up of injurious substances. I am unaware of any long-term studies on the consumption of GM foods, especially on pregnant women and their offspring. Until such studies have been completed to the satisfaction of nutritionists and other concerned bodies, there must remain a shadow of uncertainty regarding their safety.

A more serious danger is the long-term effect of GM crops on the environment, which has never been properly assessed. This includes the 'scorched earth' action of the agrochemical input on birds, insects and soil organisms, and the inevitable escape of genes from the GM crops into the environment. If highly trained nuclear scientists can still have the occasional 'accident', then the chances of farmers having more frequent 'accidents' is even greater. There is already the example of a seed merchant supplying oilseed rape accidentally contaminated with GM rape and the seeds being widely planted in Europe and the crop being destroyed prematurely as unmarketable.

The US Corn Belt, for which the GM technology was first developed, with vast cultivated fields and the minimum of headlands, hedges and other wildlife habitats, certainly does not provide the most suitable environment for such investigations. It was left to the more environmentally conscious Europe to demonstrate that there are dangers to some species of wildlife. The adverse effects on a single insect species can have unseen consequences elsewhere in the food chain. Darwin (1859) was the first to provide simple examples of such food chains, e.g. *Trifolium pratense* (red clover) relying on pollination by the bumble bee, whose nests are destroyed by field mice who, in turn, are eaten by cats, to which I might add the malicious rumour, hopefully untrue, that the cats are sometimes curried and served up in oriental restaurants. A failure in one link will have repercussions further down the chain. But, as Wilson (1992) has pointed out, this is an over-simplification since each link in the chain is linked to other chains, forming a veritable mesh. Other chains will link the red clover with the fungus *Sclerotinia trifoliorum* (clover rot) and the eelworm *Anguillulina dipsaci* responsible for 'clover sickness' (Robinson, 1947). The monitoring of the environmental effects is clearly a highly complex and long-term investigation, especially in the developing countries and in the species-rich tropics.

MAFF (1996) has already reported the sobering fact that 41% of the fruit and vegetables analysed in the UK contain pesticide residues. Insecticide and pesticide pollution of the environment are already well documented; could there be any additional risk from growing GM crops? There are already indications that the increase in use of glyphosate herbicides in association with glyphosate-resistant GM crops could have an adverse effect on wild life. Despite the manufacturer's claims that glyphosate has a half-life of 25 days, the results from independent scientists range between 40 and 150 days. It has also been demonstrated that glyphosate has adverse effects on N-fixing and other soil organisms, stunt the growth of worms and change the microbial balance of the soil. Although the GM crops require fewer applications than non-GM crops, they do require insecticides/pesticides designated by the breeder, the accumulative effect of successive applications of which on the environment have yet to be investigated. There is no possibility of using more environmentally friendly agrochemicals (Anonymous, 1998). How long will it take before the pests and diseases build up a resistance to the agrochemicals. It has happened with non-GM crops and there is no reason why the GM crops should be the exception.

The escape of genes from GM crops into related non-GM crops and wild species is now recognised. Cross-pollination is the obvious culprit. The ability to identify and destroy every GM escapee is an impossible task and will become an increasing problem as the growing of GM crops becomes more widespread. Since the honeybee forages over a radius of 5 km (Butler, 1959), pollen transfer by bees can be expected to occur over a similar range. Isolating wind pollinated GM crops is clearly an impossible task

As the growing of GM cops increases in area short distance wind dispersal of contaminated pollen will become more widespread (see Chapter 6 for details). What is even more worrying is the possibility of wind dispersal over even greater distances. The best authenticated long distance pollen transport is 800 km for pollen carried by the harmattan winds of the southern Sahara and the Sahel (Maley, 1972). The danger may be minuscule but it still exists. Such risks are shared by non-GM crops but with the difference that the GM crops are still very much an unknown risk.

While there is considerable consumer resistance, especially in Europe, to GM foods and the manner in which the multinationals appear to impose their products into the food chain, the technology should not be condemned. Less controversial are the attempts by organisations such as the International Crop Research Institute for the Semi-arid Tropics (ICRISAT) to modify the physiological resistance of crops to drought, salinity, etc. intended for the benefit of the less developed nations. There are considerable technical problems involved since such GM modifications involve several genes, unlike the more common single gene manipulation for food crops. There are also definite benefits to be obtained from producing GM pharmaceutical and medical products, hopefully under rigorous security control to prevent any abuse.

3.   ECONOMIC BOTANY AND THE UK

During the 20th century the UK's involvement in economic botany has passed through several stages: (1) The colonial period, developing new crops and introducing them to the various colonies in order to provide a balanced supply of raw materials within the British Empire. The Royal Botanic Gardens, Kew (RBG) played an active role in recommending, collecting and introducing new plants and training economic botanists for the colonies; (2) The doldrums of post colonialism, with the former colonies working independently and trying to adopt a veneer of western civilisation regardless of the costs. The individual countries were unable to maintain an internal balanced flow of raw materials, and their economies deteriorated. Research at the RBG on economic plants virtually ceased; and (3) The reawakening. As part of the EU's Common Agricultural Policy there has been increased interest during the past two decades in alternative crops for the UK. Amongst other activities a new unit, the Alternative Crops Unit, was established within MAFF to investigate alternative crops for the UK, and research on biomass as a source of energy is being funded by the Department of Trade and Industry (Dover, 1985; Carruthers, 1986; Chisholm, 1994; MAFF, 1994a).

With the help of funding by OXFAM, research on economic plants at the RBG recommenced in 1981 with the setting up of the Survey of Economic Plants for Arid and Semi-arid Tropics (SEPASAT) which, in 1985, became the Survey of Economic Plants for Arid and Semi-arid Lands (SEPASAL). SEPASAL, together with other units and the Museum collections now form Kew's Centre for Economic Botany, recreating with modern technology Sir William Hooker's idea of *"a collection that would render great service, not only to the scientific botanist, but to the merchant, the manufacturer, the physician, ..."* (Wickens, 1993). The UK Chapter of the US parent Society for Economic Botany was founded in 1991 and goes from strength to strength. The UK's first MSc course in Ethnobotany at the University of Canterbury, held in conjunction with the RBG, commenced in 1998. The UK economic botanists, whatever their disciplines, are once again active and united in a common cause.

4. FINIS

There are several lines of research that are likely to become increasingly important in the future: (1) An increased search for novel biochemicals and pharmaceuticals, especially among the marine algae; (2) An increased use of biofuels and their by-products. As a renewable resource the plants offer a suitable alternative to the world's ever diminishing reserves of fossil fuels. According to Morris and Ahmed (1992) not only is it possible for plants to recapture their share of the industrial materials market that they enjoyed in the 1920s, they could also replace at least one-third of all such materials derived from petroleum-based stocks. The potential is there but it will be the economics that decide; (3) A greater use of the Cyanobacteria as a source of food, especially in the developing countries of the arid and semi-arid tropics; (4) An increase in the use of genetic engineering; and (5) Greater emphasis on reclamation and maintenance of the environment, especially in the realm of pollutants and, in the case of heavy metals, their sequestering and recycling.

Whatever the future globally, it is ultimately cost, supply and demand that will control what plants will be used where and for which purposes; the more cynical may even believe that profit alone will be the sole deciding factor.

Bibliography

Abayo, G.O., English, T., Eplee, R.F., Kanampiu, F.K., Ransom, J.K. and Gressel, J. (1998) Control of parasitic witchweeds (*Striga* spp.) on corn (*Zea mays*) resistant to acetolactate synthase inhibitors, *Weed Science* **46**, 459-466.

Abu-Al-Futuh (1989) Study on the processing of *Balanites aegyptiaca* fruits for drug, food and feed, in G.E. Wickens, N. Haq, P. Day (eds.), *New Crops for Food and Industry,* Chapman and Hall, London, pp. 272-279.

Adams, R.P. (1991) Cedar wood oil - analyses and properties, in H.F. Liskens and J.F. Jackson (eds.), *Essential Oils and Waxes. Modern methods of plant analysis,* New Series vol 12. Springer-Verlag, New York, pp. 159-173.

Aderkas, P. von (1984) Economic history of ostrich fern, *Matteuccia struthiopteris,* the edible fiddlehead, *Economic Botany* **38**, 14-23.

Ainsworth, J. (1994) *From Vine to Wine,* Part II, Direct Wines (Windsor), Reading.

Alados, C.A., Barroso, F.G., Aguirre, A. and Escós, J. (1996) Effects of early season defoliation on further herbivore attack on *Anthyllis cyisoides* (a Mediterranean browse species), *Journal of Arid Environments* **34**, 455-463.

Allen, G.O. (1950) *British stoneworts (Charophyta),* Buncle & Co., Arbroath, Scotland.

Allen, O.N. and Allen, E.K. (1981) *The Leguminosae,* University of Wisconsin Press, Madison, and Macmillan, London.

Alpino, P. (1592) *De plantis aegypti liber,* Venice.

Altschul, S. von Reis (1973) Drugs and Foods from Little Known Plants: Notes in Harvard University Herbarium., Harvard University Press, Cambridge, MA.

Altschul, S. von Reis (1977) Exploring the herbaria, *Scientific American,* May, 36-104.

Alvarez-Buylla Roces, M.A., Lazos Chavero, E. and Garcia-Barrios, J.R. (1989) Home gardens of a humid tropical region in south east Mexico. An example of an agroforestry cropping system in a recently established community, *Agroforestry Systems* **8**, 133-156.

Anderson, D.M.W. (1985) Gums and resins, and factors influencing their economic development, in G.E. Wickens, J.R. Goodin and D.V. Field (eds.), *Plants for Arid Lands,* Allen & Unwin, London, pp. 343-356.

Anderson, D.M.W. (1990) Commercial prospects for gum exudates from the *Acacia senegal* complex, *Nitrogen Fixing Tree Research Reports* **8**, 91-92.

Anderson, D.M.W. (1993) Some factors influencing the demand for gum arabic (*Acacia senegal* (L.) Willd.) and other water-soluble tree exudates, *Forest Ecology Management* **58**, 1-18.

Anderson, D.M.W. and Dea, I.C.M. (1968) Structural studies of some unusual forms of the gum from *Acacia senegal, Carbohydrate Research* **6**, 104-110.

Andrade-Lima, D. (1981) The caatingas dominium, *Revista Brasilera de Botânica* **4**, 149-153.

Anjaria, J. (1996) Ethnoveterinary pharmacology in India: past, present and future, in C.M. McCorckle, E. Mathias. and T.W. Schillhorn van Veen (eds.), *Ethnoveterinary Research and Development*, Intermediate Technology Publications, London, pp. 137-147.

Anonymous (1895) Some new ideas. The plants cultivated by aboriginal people and how used in primitive commerce, *The [Daily] Evening Telegraph, Philadelphia*, Thursday, 5 December, Vol. 64, No. 134, p. 2.

Anonymous (1906) Guayule in Deutsch-Ostafrika, *Der Tropenpflanzer* **10**, 397.

Anonymous (1959) *Chung-Kuo Thu Nung Yao Chih* (Repertorium of Plants used in Chinese Agricultural Chemistry), Kho-Hsüeh, Peking.

Anonymous (1971) *The Colour Index*, 6 vols, 3rd edn, Society of Dyers and Colourists, Bradford..

Anonymous (1978) International Convention for the Protection of New Varieties of Plants of 2 December 1961 as revised at Geneva on 10 November 1972 and on 23 October 1978, *UPOV Newsletter* No. 15.

Anonymous (1980) Biotechnology: Report of a Joint Working Party, HMSO, London.

Anonymous (1987) *Mode of Operation of the Turbidity Test*, Institut Technique des Céréales et des Fourrages, Station Experimentale, Boigneville, France.

Anonymous (1989) Gondwanan chemistry aids taxonomy, *Jodrell Newsletter, Royal Botanic Gardens, Kew* **4**, 1.

Anonymous (1993) The tools of rice biotechnology, *Tropical Agriculture Association UK Newsletter* **13** (4), 18.

Anonymous (1994) The use of vegetable oils to fuel diesel engines, *Tropical Agriculture Association UK Newsletter* **14** (2), 30.

Anonymous (1997a) Revolutionary rice sprouts greener shoots, *New Scientist* **154** (2085), 12.

Anonymous (1997b) Continuing US-EU labelling row, *The Environment Digest* **6/7**, 17-18.

Anonymous (1997c) Reducing pesticides, *The Garden* **122**, 768.

Anonymous (1998) Say 'no' to GMOs, *Gardening Which?* June, 196-197.

Appelqvist, L.A. (1989) The chemical nature of vegetable oils, in G. Röbbelen, R.K. Downey and Amram Ashri (eds.), *Oil Crops of the World*, McGraw-Hill, New York, pp. 22-37.

Appleyard, H.M. and Wildman, A.B. (1969) Fibres of archaeological interest: their examination and identification, in D. Brothwell (ed.), *Science in Archaeology. A survey of progress and research*, 2nd edn., Thames & Hudson, London, pp. 624-633.

Arber, A. (1965) The Gramineae. A Study of Cereal, Bamboo and Grass, J. Cramer, Weinheim.

Archer, T.C. (1853) Popular Economic Botany; or the botanical and commercial characters of the principal articles of vegetable origin, used for food, clothing, tanning, dyeing, building, medicine, perfumery, etc., Reeve and Co., London.

Archer, T.C. (1865) Profitable Plants: description of the principal articles of vegetable origin used food, clothing, tanning, dyeing, building, medicine, perfumery, etc., Routledge, Warne and Routledge, London.

Archibold, O.W. (1995) *Ecology of World Vegetation*, Chapman & Hall, London.

Armisen, R. and Galantas, F. (1987) Production, properties and uses of agar, in D.J. McHugh (ed.), *Production and Utilization of Products from Commercial Seaweeds*, Fisheries Technical Paper 288. FAO, Rome, pp. 1-57.

Armstrong, D.G. (1985) The general implications of biotechnology in the agricultural industry, in L.G. Copping and P. Rodgers (eds.), *Biotechnology and its Application to Agriculture*, Monograph No. 32, British Crop Protection Council, Croydon, pp. 3-11.

Arnold, T.H., Wells, M.J. and Wehmeyer, A.S. (1985) Khoisian food plants: taxa with potential for future economic exploitation, in G.E. Wickens, J.R. Goodin and D.V.Field, (eds.), *Plants for Arid Lands*, Allen & Unwin, London, pp. 69-86.

Arnon, I. (1992) Agriculture in Dry Lands: Principles and Practice, Developments in Agricultural and Managed-Forest Ecology 26, Elsevier Science Publishers, Amsterdam.

Aronson, J. (1965) Economnic halophytes - a global review, in G.E. Wickens, J.R. Goodin and D.V. Field (eds.), *Plants for Arid Lands,* Allen & Unwin, London, pp. 177-188.

Aronson, J.A. (1989) *HALOPH. A Data Base of Salt Tolerant Plants of the World,* Office of Arid Lands Studies, University of Arizona, Tucson.

Aronson, J., Wickens, G.E. and Birmbaum, E. (1990) An experimental technique for the long distance transport of evergreen or deciduous cuttings under tropical conditions, *FAO/IBPGR Plant Genetic Resources Newsletter* **81/82,** 47-48.

Askew, K. and Mitchell, A.S. (1978) The Fodder Trees and Shrubs of the Northern Territory, Extension Bulletin No. 16, Division of Primary Industry, Darwin.

Atchley, J. and Cox, P.A. (1984) Breadfruit fermentation in Micronesia, *Economic Botany* **39,** 326-325.

Atkin, T. (1996) *From Vine to Wine,* Direct Wines (Windsor), Reading.

Aubrecht, E., Biacs, P., Lajos, J. and Léder, F.-Né (1998) Buckwheat: systematization, name and cultivation of buckwheat, in P.S. Belton, and J.R.N. Taylor (eds.), *Increasing the Utilisation of Sorghum, Buckwheat, Grain Amaranth and Quinoa for Improved Nutrition. Selected Papers from the International Association for Cereal Science and Technology Cereals Conference Symposium "Challenges in Speciality Crops" Held at the 16th ICC Cereals Conference, Vienna, Austria 9 May 1998,* Institute of Food Research, Norwich, pp. 22-48.

Baerheim Svendsen, A. (1984) Biogene Arzneistoffe - heute noch oder heute wieder? in F.-C. Czygan (ed.), *Biogene Arzneistoffe.* Friedr, Vieweg und Sohn, Braunschweig/ Wiesbaden, pp. 27-44.

Bagnouls, F. and Gaussen (1957) Climats biologiques et leur classification, *Annals de Géographie* **355,** 193-220.

Bailey, J. (1999) *Penguin Dictionary of Plant Sciences,* 2nd edn., Penguin Books, London.

Baillon, H.E. (1859) *Monographie des Buxaces,* Paris.

Baker, E.G. (1929) *Leguminosae of Tropical Africa,* Part 2, Unitas Press, Ostende.

Baker, H.G. (1964) *Plants and Civilization,* Macmillan & Co., London.

Balick, M.J. (1988) Jessenia and Oenocarpus: neotropical oil palms worthy of domestication, FAO Plant Production and Protection Paper 88, FAO, Rome.

Baranov, A.l. (1962) On the economic use of wild plants in N.E. China, *Quarterly Journal of the Taiwan Museum* **5** (1-2), 107-115.

Barnes, D. and Barnes, S. (1978) *Instant Home Brewing,* Rigby, Australia.

Barnett, J.A., Payne, R.W. and Yarrow, D. (1990) *Yeasts: characteristics and identification,* 2nd edn., Cambridge University Press, Cambridge.

Barrau, J. (1971) L'ethnobotanique au carrefour des sciences naturelles et des sciences humaines, *Bulletin de la Société Botanique de France* **118,** 237-248.

Barrau, J.F. (1989) The possible contribution of ethnobotany to the search for new crops for food and industry, in G.E. Wickens, N. Haq and P. Day (eds.), *New Crops for Food and Industry,* Chapman & Hall, London, pp. 402-410.

Barron, G.L. (1977) *The nematode-destroying fungi. Topics in Mycobiology No. 1,* Canadian Biological Publications, Guelph, Ontario.

Barton, G.M. (1979) *Chemicals from trees, outlook for the future. Technical Report No. 4.* Presented at the Eighth World Forestry Congress, Jakarta, Indonesia, October 16 to 28, 1978.

Barz, W. and Ellis, B.E. (1981) Potential of plant cell cultures for pharmaceutical production, in J.J. Beal and E. Reinhard (eds.), *Natural Products as Medicinal Agents,* Hippokrates Verlag, Stuttgart, pp. 471-507.

Batanouny, K.H. (1999) *Wild Medicinal plants in Egypt. An Inventory to Support Conservation and Sustainable Use,* Academy of Scientific Research and Technology, Cairo and International Union for Conservation, Morges.

Batra, L.R. and Millner, P.D. (1974) Some Asian fermented foods and beverages, and associated fungi, *Mycologia* **66,** 942-950.

Baum, B.R. (1978) *The Genus Tamarix,* Israel Academy of Sciences and Humanities, Jerusalem.

Baumer, M. (1983) Notes on Trees and Shrubs in Arid and Semi-arid Regions. EMASAR Phase II, FAO, Rome.

Bean, W.J. (1970) *Trees & Shrubs Hardy in the British Isles*, 8th edn. revised, vol. 1, A-C,. John Murray, London.

Beckerman, S. (1977) The use of palms by the Bari Indians of the Maracaibo Basin, *Principes*, **21** (4), 143-154.

Beech, F.W. and Pollard, A. (1970) *Winemaking and Brewing*, The Amateur Winemaker, Andover.

Bell, E.A. (1965) Preface, in G.E. Wickens, J.R. Goodon and D.V.Field (eds.), *Plants for Arid Lands*, Allen & Unwin, London, pp.vi-vii.

Ben-Amotz, A. and Avran, M. (1983a) Accumulation of metabolites by halotolerant alga and its industrial potential, *Annual Review of Microbiology* **37**, 95-119.

Ben-Amotz, A. and Avron, M. (1983b) On the factors which determine massive β-carotene accumulation in the halotolerant alga *Dunaliella bardawil*, *Plant Physiology* **72**, 593-597.

Bennet, S.S.R., Gupta, P.C. and Vijendra Rao, R. (1992) *Venerated Plants*, Indian Council of Forestry Research and Education, Dehra Dun.

Bentham, G. and Hooker, J.D. (1862-83) *Genera Plantarum*, 3 vols., A. Black (vol. 1), William Pamplin (id.), Lovell Reeve & Co., Williams & Norgate, London.

Berghofer, E. and Schoenlechner, R. (1998) Grain-amaranths: Nutritive value and potential uses in the food industry, in P.S. Belton and J.R.N. Taylor (eds.), *Increasing the Utilisation of Sorghum, Buckwheat, Grain Amaranth and Quinoa for Improved Nutrition. Selected Papers from the International Association for Cereal Science and Technology Cereals Conference Symposium "Challenges in Speciality Crops" Held at the 16th ICC Cereals Conference, Vienna, Austria 9 May 1998*, Institute of Food Research, Norwich, pp. 65-80.

Bhat, K.K.S. (1998) Medical plant information databases, in, G. Bodeker, K.K.S. Bhat, J. Burley and P. Vantomme (eds.), *Medicinal Plants for Forest Conservation and Health Care*, Non-wood Forest Products 11, FAO, Rome, pp. 60-77.

Bianchini, F. and Corbetta, F. (1975) *The Fruits of the Earth*, Bloomsberry Books, London.

Billings, W.D. (1957) Physiological ecology, *Annual Review of Plant Physiology* **8**, 375-392.

Björk, I. (1996) Starch: nutritional aspects, in A.-C. Eliasson (ed.), *Carbohydrates in Food*, Marcel Dekker, New York, pp. 505-553.

Blackwell, W.H. (1990) *Poisonous and Medicinal Plants,* Prentice Hall, Englewood Cliffs, NJ.

Blakeney, A.B. (1996) Rice, in R.J. Henry and P.S. Kettlewell (eds.), *Cereal Grain Quality*, Chapman & Hall, London, pp. 55-76.

Blau, P.A., Feeney, P., Contardo, L. and Robson, D.S. (1978) Allyglucosinolate and herbicorous caterpillars: a contrast in toxicity and tolerance, *Science* **200**, 1296-298.

Bloomfield, F. (1985) *Jojoba and Yucca*, Century Publishing, London.

Blundell, T. (1996) New scientific opportunities for the agriculture of developing countries, *Tropical Agriculture Association UK Newsletter* **16** (1), 22-26.

Blunden, G. (1991) Agricultural uses of seaweeds and seaweed extracts, in M.D. Guiry and G. Blunden (eds.), *Seaweed Resources in Europe: Uses and Potentia*, John Wiley & Sons, Chichester, pp. 63-81.

Böcher, T.W. (1960) Infraspecific differentiation, Coreferate, *Botanica Medica* **8** , 224-225.

Bodson, L. (1996) Veterinary medicine, in S. Hornblower and A. Spawforth (eds.), *The Oxford Classical Dictionary*, 3rd edn., Oxford University Press, Oxford and New York, pp. 1592-1593.

Boom, B.M. (1985) Ethnopteridology of the Chácobo indians in Amazonian Bolivia, *American Fern Journal* **75**, 19-21.

Booth, F.E.M. and Wickens, G.E. (1988) Non-timber Uses of Selected Arid Zone Trees and Shrubs in Africa. FAO Conservation Guide 19, FAO, Rome.

BOSTID (1980) *Firewood Crops: Shrub and Tree Species for Energy Production*, National Academy of Science, Washington, DC.

Böttcher, H.M. (1963) *Miracle Drugs. A History of Antibiotics*, Heinemann, London (English translation of the 1959 German edition).

Boulger, G.S. (1889) The Uses of Plants. A manual of economic botany with special reference to vegetable products introduced during the last fifty years, Roper and Drowley, London.

Braendle, R. and Crawford, R.M.M. (1987) Rhizome anoxia tolerance and habitat specialization in wetland plants, in R.M.M. Crawford (ed.), *Plant life in Aquatic and Amphibious Habitats. British Ecological Society Publication No. 5*, Blackwell, Oxford, pp. 397-410.

Bramwell, M. (ed.) (1982) *The International Book of Wood*, AH, London.

Brand, J.C. and Cherikoff, V. (1985) The nutritional composition of Australian aboriginal food plants of the desert regions, in G.E. Wickens, J.R. Goodin and D.V. Field (eds.), *Plants for Arid Lands*, Allen and Unwin, London, pp. 53-68.

Bressani, R. and Elias, L.G. (1980) Nutritional value of legume crops for humans and animals, in R.J. Summerfield and A.H. Bunting (eds.), *Advances in Legume Science*, Royal Botanic Gardens, Kew, pp. 135-155.

Brian, P.W., Elson, G.W., Hemming, H.G. and Wright, J.M. (1952) The phytotoxic properties of alternaric acid in relation to the etiology of plant diseases caused by *Alternaria solani* (Ell. & Mart.) Jones & Grout, *Annals of Applied Biology* **39**, 308-321.

Briand, X. (1991) *Seaweed harvesting in Europe*, in M.D. Guiry and G. Blunden (eds.), *Seaweed Resources in Europe: Uses and Potentia*, John Wiley & Sons, Chichester, pp. 259-308.

Brickell, C.D. (ed.) (1980) *International Code of Nomenclature for Cultivated Plants*, 7th edn., *Regnum vegetabile* **104**, Bohn, Scheltema & Holkema, Utrecht.

Briggs, M.K. (1996) Riparian Ecosystem Recovery in Arid Lands. Strategies and References, University of Arizona Press, Tucson.

Brightman, F.H. and Nicholson, B.E. (1966) *The Oxford Book of Flowerless Plants*, Oxford University Press, London.

Brisebarre, A.-M. (1996) Tradition and modernity: French shepherds' use of medicinal bouquets, in C.M. McCorckle, E. Mathias and T.W. Schillhorn van Veen (eds.), *Ethnoveterinary Research and Development*, Intermediate Technology Publications, London, pp. 76-90.

British Pharmacopoeia Commission (1988) *British Pharmacopoeia*, HMSO, London.

British Pharmacopoeia Commission (1993) *British Pharmacopoeia*, HMSO, London.

Brokensha, D. and Riley, W.R. (1986) Changes in uses of plants in Mbeere, Kenya, *Journal of Arid Environments* **11**, 75-80.

Brooks, W.H. (1978) Jojoba - a North American desert shrub; its ecology, possible commercialization, and potential as an introduction into other arid regions, *Journal of Arid Environments* **1**, 227-246.

Brouk, B. (1976) *Plants Consumed by Man*, Academic Press, London.

Brown, J.A.C. (1977) *Pears Medical Encylopaedia*. Revised by A.M. Hastin-Bennett, Sphere Books Ltd, London.

Brücher, H. (1989) Useful Plants of Neotropical Origin and Their Wild Relatives, Springer-Verlag, Berlin, Heidelberg.

Brud, W.S. (1995) Formulation and evaluation of fragrance for perfumery cosmetics and related products, in K. Tuley De Silva (ed.), *A Manual on the Essential Oil Industry*, UNIDO, Vienna, pp. 179-201.

Brummittt, R.K. (1992) *Vascular Plant Families and Genera*, Royal Botanic Gardens, Kew.

Brummitt, R.K. and Powell, C.E. (eds.) (1992) *Authors of Plant Names*, Royal Botanic Gardens, Kew.

Bruneau de Miré, P. and Quézel, P. (1959) Sur la présence de la bruyère en arbre (*Erica arborea* L.) sur les sommets de l'Emi Koussi (Massif du Tibesti), *Compte rendu sommaire des seances de la Societe de Biogéographie* **315**, 66-70.

BSI (1979) Glossary of Paper. board, pulp and related terms (ISO title: Paper, board, pulp and related terms-vocabulary). BS 3203: 1979, ISO 4046-1978, British Standards Institute, London.

Buchanan, R.E. and Gibbons, N.E. (eds.) (1974) *Bergey's Manual of Determinative Bacteriology*, 8th edn., Williams and Wilkins, Baltimore.

Burkill, H.M. (1985, 1994, 1995, 1997, in press) *The Useful Plants of West Tropical Africa*, vol. 1 Families A-D), vol. 2 (Families E-I), vol 3 (Families J-L), vol. 4 (Families M-R), vol. 5 (Families S-Z, Addenda Families A-R, and Cryptograms)., Royal Botanic Gardens, Kew.

Burton, W.G. (1989) *The Potato*, 3rd edn., Longman Scientific & Technical, Harlow, Essex.

Butler, C.G. (1959) *The World of the Honeybee*, Readers Union/Collins, London.

Callaghan, T.V., Abdelnour, H. and Lindley, D.K. (1988) The environmental crisis in the Sudan: the effect of water-absorbing synthetic polymers on tree germination and early survival, *Journal of Arid Environments* **14**, 301-317.

Candolle, A.L.P. de (1882) *L'Origine des Plantes Cultivées*, Germer Balliére et Cie, Paris.

Candolle, A.P. (1819) Théorie élémentaire de la botanique, ou exposition des principes dela classification naturelle et de l'art de décrire et d'étudier les végétaux, 2nd. edn., Deterville, Paris.

Carlson, T.J., Cooper, R., King, S.R. and Rozhon, E.J. (1997a) Modern science and traditional healing, in *Special Publication 200 (Phytochemical Diversity)*, Royal Society of Chemistry, London, pp. 84-95.

Carlson, T.J., Iwu, M.M., King, S.R., Obialor, C. and Ozloko, A. (1997b) Medicinal plant research in Nigeria: an approach for compliance with the Convention on Biological Diversity, *Diversity* 13 (1), 29-33.

Carlsson, R. (1989) Green biomass of native plants and new, cultivated crops for multiple use: food, fodder, fuel, fibre for industry, phytochemical products and medicine, in G.E. Wickens, N. Haq and P. Day (eds.), *New Crops for Food and Industry*, Chapman & Hall, London, 101-107.

Carruthers, S.P. (ed.) (1986) *Alternative Enterprise for Agriculture in the UK. CAS Report 11*, Centre for Agriculture Strategy, University of Reading, Reading.

Carruthers, S.P. (1994) Fibres, in S.P. Carruthers, F.A. Miller and C.M.A. Vaughan (eds.), *Crops for Industry and Energy. CAS Report No. 15*, Centre for Agricultural Strategy, University of Reading, Reading, pp. 92-108.

Caulton, E., Keddie, S. and Dyer, A.F. (1995) The evidence of airborne spores of bracken, *Pteridium aquilinum* (L.) Kuhn, in the rooftop airstream over Edinburgh, Scotland, in R.T. Smith and J.A. Taylor (eds.), *Bracken: an Environmental Issue. Special Publication no. 2*, International Bracken Group, Aberystwyth.

Chalk, L. and Akpalu, J.D. (1963) Possible relation between the anatomy of the wood and buttressing, *Commonwealth Forestry Review* **42**, 53-58.

Chamberlain, D. and Stewart, C.N. (1999) Transgene escape and transplastomics, *Nature Biotechnology* **17**, 331-332.

Chapman, V.J. and Chapman, D.J. (1980) *Seaweeds and their Uses*, 3rd edn., Chapman & Hall, New York.

Champ, M. (1994) Definition, analysis, physical and chemical characterization of starch, in *EURESTA report, FLAIR programme*, European Commission DGXII, Brussels, pp. 1-14.

Chase, M. (1999) New classification of flowering plants, *Kew Scientist* **15**, 4-5.

Chèvre, A.-M., Baranger, F.E.A. and Renard, M. (1997) Gene flow from transgenic crops, *Nature* **289**, 924.

Chiej, R. (1984) *The Macdonald Encyclopedia of Medicinal Plants*, Macdonald, London.

Chew, F.S. (1988) Searching for defensive chemistry in the Cruciferae, or, do glucosinolates always control interactions of Cruciferae and their potential herbivores and symbionts? No! in K.C. Spencer (ed.), *Chemical Mediation of Coevolution*, Academic Press, San Diego, CA, pp. 81-112.

China National Corporation of Traditional and Herbal Medicine (1994-1995) *Chinese Medicinal Resources*, 6 vols, Science Publishers, Beijing.

Chisholm, C.J. (ed.) (1994) *Towards a UK Research Strategy for Alternative Crops*, Silsoe Research Institute, Silsoe, Beds.

Christensen, H. (1997). Uses of ferns in two indigenous communities in Sarawak, Malaysia, in R.J. Johns (ed.), *Holtum Memorial Volume*, Royal Botanic Gardens, Kew, pp. 177-192.

Chudnoff, M. (1979) *Tropical Timbers of the World*, 2 vols., Forest Products Laboratory, Forestry Service, USDA, Madison, WI (Reproduced in 1980 by the National Technical Information Service, US Department of Commerce, Springfield, VA.

Ciferri, O. (1983) *Spirulina*, the edible microorganism, *Microbiological Reviews* **47**. 551-578.

Ciferri, O. and Tiboni, O. (1985) The biochemistry and industrial potential of *Spirulina*, *Annual Revue of Microbiology* **39**, 503-520.

Clayton, W.D. and Renvoize, S.A. (1986) *Genera Gramineum. Grasses of the World*, Kew Bulletin Additional Series No. 43, HMSO, London.

Clements, F.E. (1907) *Plant Physiology and Ecology*, Constable, London.

Clements, M.A., Muir, H,J. and Cribb, P.J. (1986) A preliminary report on the symbiotic germination of European terrestrial orchids, *Kew Bulletin* **41**, 437-445.

Cloudsley-Thompson, J.L. (1969) *The Zoology of Tropical Africa*, Weidenfeld & Nicolson, London.

Cloudsley-Thompson, J.L. (1977) *Man and the Biology of Arid Zones*, Edward Arnold, London.

Cole, J.N. (1979) *Amaranth from the Past to the Future*, Rodale Press, Emmaus, PA.

Cole, M.D. (1992) The significance of the terpenoides in the Labiatae, in R.M. Harley and T. Reynolds (eds.), *Advances in Labiate Science*, Royal Botanic Gardens, Kew, pp. 315-324.

Coles, S. (1970) *The Neolithic Revolution*, British Museum (Natural History), London.

Committee on Forest Development in the Tropics (1985) *Tropical Forest Action Plan*, FAO, Rome.

Cook, F.E.M. (1995) *Economic Botany Data Collection Standard. Prepared for the International Working Group on Taxonomic Databases for Plant Science (TDWG)*, Royal Botanic Gardens, Kew.

Cooke, G.B. (1961) *Cork and the Cork Tree,* Pergamon Press, Oxford.

Copeland, E.B. (1942) Edible ferns, *American Fern Journal* **32**, 121-126.

Coppen, J.J.W. (1995) *Flavours and Fragrances of Plant Origin*. Non-Wood Forest Products 1, FAO, Rome.

Coppen, J.J.W. and Hone, G.A. (1995) *Gum Naval Stores: Turpentine and Rosin from Pine resin*. Non-Wood Forest Products 2, FAO, Rome.

Cooper, M.R. and Johnson, A.W. (1998) *Poisonous Plants and Fungi in Britain. Animal and Human Poisoning*, 2nd edn., HMSO, London.

Corley, R.H.V. (1988) Breeding and propagation of tropical plantation crops, *Tropical Agriculture Association UK Newsletter* **8** (3), 18, 22-24.

Corner, E.J.C. (1964) *The Life of Plants*, Weidenfeld and Nicolson, London.

Corner, E.J.H. (1966) *The Natural History of Palms*, Weidenfeld and Nicolson, London.

Cortella, A.R. and Pochettino, M.L. (1994) Starch grain analysis as a microscopic diagnostic feature in the identification of plant material, *Economic Botany* **48**, 171-181.

Cossalter, C. (1991) *Acacia senegal*: gum tree with promise for agroforestry. *NFT Highlights, NFTA 91-02*, Nitrogen Fixing Tree Association, Hawaii.

Cotton, C.M. (1996) *Ethnobotany: Principles and Applications*, John Wiley & Sons, Chichester.

Coupland, R.T. (1992) Approach and generalizations, in R.T. Coupland (ed.), *Natural Grasslands. Introduction and Western Hemisphere. Ecosystems of the World 8A,* Elsevier, Amsterdam, pp. 1-6.

Coward, P. (1997) Vegetable brush fibres, *Tropical Agriculture Association UK Newsletter* **17** (1), 10-14.

Cox, P.A. and Ballick, M.J. (1994). The ethnobotanical approach to drug discovery, *Scientific American* **271**, 82-87.

Crane, E. (1990) *Bees and Beekeeping: Science, Practice and World Resources*, Heinemann Newnes, Oxford.

Crane, E. and Walker, P. (1984) *Pollination Directory for World Crops*, International Bee Research Association, London.

Crane, E., Walker, P. and Day, R. (1984) *Directory of Important World Honey Sources*, International Bee Research Association, London.

Crawford, R.M.M., Studer, C. and Studer, K. (1989) Deprivation indifference as a survival strategy in competition advantages and disadvantages of anoxia tolerance in wetland vegetation, *Flora* **182**, 189-201.

Crawley, M.J. (1990) Biocontrol and biotechnology, in *Brighton Crop Protection Conference: Weeds -1989*, vol. 3, British Crop Protection Council, Brighton, pp. 969-978.

Croft, J.R. and Leach, D.N. (1985) New Guinea salt fern (*Asplenium acrobryum* complex): identity, distribution, and chemical composition of its salt, *Economic Botany* **39**, 139-149.

Crompton, C.W., McNeill, J., Stahevitch, A.E. and Wojtas, W.A. (1988) *Preliminary inventory of Canadian weeds. Technical Bulletin 1988-9E*, Biosystematics Research Centre, Ottawa.

Cross, C.F. and Bevan, E.J. (1900) *A Text-Book of Paper-Making*, 2nd edn., E. & F. Spon, London; Spon & Chamberlain, New York.

Cross, H. (ed.) (1984) *Grow Your Own Energy*, Basil Blackwell, Oxford.

Curtis, O.F. and Clark, D.G. (1950) *An Introduction to Plant Physiology*, McGraw-Hill, New York.

Cutler, D.F. and Richardson, I.B.K. (1989) *Tree Roots and Buildings*, 2nd edn., Longman, Harlow, Essex.

D'Mello, J.P.F. (1991) Toxic amino-acids, in J.P.F. D'Mello, C.M. Duffus and J.H. Duffus, (eds.), *Toxic Substances in Crop Plants*, Royal Society of Chemistry, Cambridge, pp. 21-48.

Dagne, E. (1998) Integration of traditional phytotherapy into general health care: an Ethiopian perspective, in H.D.V. Prendergast, N.L. Etkin, D.R. Harris and P.J. Houghton (eds.), *Plants for Food and Medicine*, Royal Botanic Gardens, Kew, pp. 47-55.

Dalal-Clayton, D.B. (1981) *Black's Agricultural Dictionary*, A. & C. Black, London.

Dalla Torre, C.G. de and Harms, H. (1900-1907) *Genera Siphonogamarum ad Systema Englerianum Conscripta*, G. Engelmann, Leipzig.

Dales, M.J. (1992) A Revue of Plant Material Used for Controlling Insect Pests of Stored Products, Bulletin 65, Natural Resources Institute, Chatham.

Dando, T.R. and Young, J.E. (1990) *Catalogue of Strains*, The National Collection of Industrial and Marine Bacteria Ltd., Aberdeen.

Danin, A. (1983) *Desert Vegetation of Israel and Sinai*, Cana Publishing House, Jerusalem.

Danin, A. (1996) *Plants of Desert Dunes*, Adaptations of Desert Organisms, (C.T. Cloudsley-Thompson, ed.), Springer-Verlag, Berlin, Heidelberg.

Darwin, C. (1859) *The Origin of Species by Means of Natural Selection or the Preservation of Favoured Races in the Struggle for Life*, John Murray, London.

Darwinkel, A. (1996) *Secale cereale* L., in G.H.J. Grubben and S. Partohardjono (eds.), *Cereals. Plant Resources of South-East Asia No. 10*, Backhuys Publishers, Leiden, pp. 90-95

David, E. (1978) *English Bread and Yeast Cookery*, Book Club Associates, London.

Davis, J.H.C. (1993) Recent developments in plant breeding - a new green revolution? *Tropical Agriculture Association UK Newsletter* **13** (4), 5-9, 20.

Davis, M. and Eberhard, A.A. (1991) Combustion characterisitics of fuelwoods, *South African Forestry Journal* **199**, 17-22.

Davis, P.H. (ed.) (1965) *Flora of Turkey and the East Aegean Islands*, Vol. 1, University Press, Edinburgh.

Davis, P.H. and Heywood, V.H. (1963) *Principles of Angiosperm Taxonomy*, Oliver & Boyd, Edinburgh and London.

Davison, M.W. (ed.) (1994) *Field Guide to Trees and Shrubs of Britain*, Reader's Digest, London.

Day, M.H. (1969) *Fossil Man*, Hamlyn Publishing Group, London.

Dayton, W.A. (1948) Poisonous plants, in A. Stefferud (ed.), *Grass. The Yearbook of Agriculture 1948*, U.S. Department of Agriculture, Washington, DC, pp. 729-734.

DeBuch, I. (ed.) (1964) *Biological Control of Insect Pests and Weeds*, Chapman & Hall, London.

de Wet, J.M.J. (1981) Grasses and the culture history of man, *Annals of the Missouri Botanical Garden* **68**, 87-104.

de Wet, J.M.J. (1990) Origin, evolution and systematics of minor cereals, in E. Seetharam, K.W. Riley and G. Harinarayana (eds.), *Small Millets in Global Agriculture. Proceedings of the First International Small Millets Workshop, Bangalore, India, October 29-November 2, 1986*, Aspect Publishing, London, pp. 19-30.

de Wet, J.M.J. (1992) The three phases of cereal domestication, in G.P. Chapman (ed.), *Grass Evolution and Domestication*, Cambridge University Press, Cambridge, pp. 176-198.

Devitt, J. (1986) A taste for honey: Aborigines and the collection of ants associated with mulga in central Australia, in P.S. Sattler (ed.), *The Mulga Lands*, Royal Society of Queensland, Brisbane, pp. 40-44i

Devitt, J. (1988) *Contemporary Aboriginal Women and Subsistence in Remote, Arid Australia*, Ph.D Thesis, Department of Anthropology and Sociology, University of Queensland.

Devitt, J. (1989) Honeyants: a desert delicacy, *Australian Natural History* **22**, 588-595.

Devitt, J. (1992) Acacias: a traditional Aborigininal food source in central Australia, in A.P.N. House and C.E. Harwood (eds.), *Australian Dry-zone Acacias for Human Food*, Australian Tree Seed Centre, Canberra, pp. 37-53.

Dickinson, C.I. (1963) *British Seaweeds*, Eyre & Spottiswoode, London.

Dikshit, A. and Husain, A. (1984) Antifungal action of some essential oils against animal pathogens, *Fitoterapia* **55**, 171-176.

Dimbleby, G. (1978) *Plants and Archaeology. The Archaeology of the Soil*, Granada Publishing, St Albans.

Dirar, H.A. (1984) Kawal, meat substitute from fermented *Cassia obtusifolia* leaves, *Economic Botany* **38**, 342-249.

Dirar, H.A. (1993) *The Indigenous Fermented Foods of the Sudan. A Study in African Food, and Nutrition*, CAB International, Wallingford, UK.

Dirar, H.A., Harper, D.B. and Collins, M.A. (1985) Biochemical and microbiological studies on kawal, a meat substitute derived by the fermentation of *Cassia obtusifolia* leaves,. *Journal of the Science of Food and Agriculture* **36**, 881-892.

Döbereiner, J., Day, J.M. and Dart, P.J. (1972) Nitrogenase activity and oxygen sensitivity of *Paspalum notatum - Azotobacter paspali* association, *Journal of General Microbiology*, **71** 103-116.

Doggett, H. (1970) *Sorghum*, Longmans, Green and Co., London.

Don, G. (1837) *A General System of Gardening and Botany*, vol. 3, London.

Dover, P.A. (1985) *A Guide to Alternative Combinable Crops*, Royal Agricultural Society of England, London and Agricultural Development & Advisory Service, Stoneleigh, Warwickshire.

Dowling, R.M. and McKenzie, A.A. (1993) *Poisonous Plants. A Field Guide*, Department of Primary Industries, Brisbane.

Dransfield, J. (1986) *Flora of Tropical East Africa: Palmae*, Balkema, Rotterdam.

Dransfield, J. and Manokaran, J. (eds.) (1993) *Rattans. Plant Resources of South-East Asia No. 6*, Purdoc Scientific Publishers, Wageningen.

Dransfield, S. and Widjaja, E.A. (1995) Introduction, in S. Dransfield and E.A. Widjaja (eds.), *Bamboos. Plant Resources of South-East Asia No. 7*, Backhuys Publishers, Leiden, pp. 15-49.

Drar, M. (1970) *A Botanic Expedition to the Sudan in 1938*, Edited after author's death with introductory notes by Vivi Täckholm, Publication of the Cairo University Herbarium No. 3.

Drechsler, C. (1934) Organs of capture in some fungi preying on nematodes, *Mycologia*, **26** 135-144.

Drew, M.C. and Stolzy, L.H. (1991) Growth under oxygen stress, in Y. Waisel, A. Eshel and U. Kafkafi (eds.), *Plant Roots: the Hidden Half*, Dekker, New York, pp. 331-350.

Driscoll, C.J. (1990) *Plant Sciences: Production, Genetics and Breeding*, Ellis Horwood, Chichester.

Duffus, C.M. and Duffus, J.H. (1991) Introduction and overview. in J.P.F. D'Mello, C.M. Duffus and J.H. Duffus (eds,), *Toxic Substances in Crop Plants*, Royal Society of Chemistry, Cambridge, pp. 1-20.

Dufour, D.L. (1987) Insects as food: a case study from the northwest Amazon, *American Anthropologist* **89**, 383-397.

Duin, N. and Sutcliffe, J. (1992) *A History of Medicine from Prehistoric to the Year 2020*, Simon and Schuster, London.

Duke, J.A. (1992) Tropical botanical extracts, in N. Plotkin and L. Famolare (eds.), *Sustainable Harvesting and Marketing*, Island Press for Conservation International, Covelo, CA, and Washington, DC, pp. 53-62.

Duke, J.A. and Ayensu, E.S. (1985) *Medicinal Plants of China*, 2 vols., Reference Publications, Algonac, MI.

Duvick, D.N. (1984) Genetic diversity in major farm crops on the farm and in reserve, *Economic Botany* **38**, 161-178.

Eames, A.J. and MacDaniels, L.H. (1947) *An Introduction to Plant Anatomy*, 2nd edn., McGraw-Hill, New York.

Eberhard, A.A. (1990) Fuelwood calorific values in South Africa, *South African Forestry Journal* **152**, 17-22.

Eckardt, F.E. (1965) F.E. (1965) Remarques préliminaires concernant la méthodologie de l' éco-physiologie végétale et l'organisation du colloque de Montpellier, in F.E. Eckardt (ed.), *Methodology of Plant Eco-Physiology. Proceedings of the Montpellier Symposium. Arid Zone Research 23*, UNESCO, Paris, pp. 1-10.

Eckhoff, S.R. and Paulsen, M.R. (1996) Maize, in R.J. Henry and P.S. Kettlewell (eds.), *Cereal Grain Quality*, Chapman & Hall, London, pp. 77-112.

Edney, M.J. (1996) Barley, in R.J. Henry and P.S. Kettlewell (eds.), *Cereal Grain Quality*, Chapman & Hall, London, pp. 113-151.

Edwards, G. and Walker, D.A. (1983) C_3, C_4 *Mechanisms, and Cellular and Environmental Regulation of Photosynthesis*, Blackwell, Oxford.

Edwards, P.J. (1984). The use of fire as a management tool, in P. de V. Booysen and N.M. Tainton (eds.), *Ecological Effects of Fire in South African Ecosystems*, Springer-Verlag, Berlin, Heidelberg, pp. 349-362.

Ehrlich, P.R. and Raven, P.H. (1964) Butterflies and plants: a study in co-evolution, *Evolution* **18**, 586-608.

Eilert, U., Wolters, B. and Nahrstedt, A. (1980) Antibiotic properties of seeds of *Moringa oleifera*, *Planta Medica* **39**, 235.

Eilert, U., Wolters, B. and Nahrstedt, A. (1981) Antibiotic principle of seeds of *Moringa oleifera* and *Moringa stenopetala*, *Planta Medica* **42**, 55-61.

Elfaki, A.E., Dirar, H.A., Collins, M.A. and Harper, D.B. (1991) Biochemical and microbiological investigations of sigda - A Sudanese fermented food derived from sesame oilseed cake, *Journal of the Science of Food and Agriculture* **57**, 351-365.

El-Ghonemy, A.A., El-Din, H.K. and El-Razzak, H.A. (1974) *Thymelaea hirsuta* (L.) Endl. A possible cellulose raw material for paperpulp industry in Egypt, *Proceedings of the Egyptian Academy of Science*, **27** 51-66.

Embooden, W.A. (1974) *Bizarre Plants, Magical, Monstrous, Mythical*, Studies Vista, London.

Emrich, W. (1985) *Handbook of Charcoal-making*. Series E. Energy from Biomass, vol.7. D,. Reidel, Dordtrecht.

Engler, A. and Diels, L. (1936) *Syllabus der Pflanzenfamilien*, Aufl. 112, Gebruder Borntaeger, Berlin (revised by H. Melchior, 2 vols., 1954, 1964).

Ensminger, A.H., Ensminger, M.E., Konlande, J.E. and Robson, J.R.K. (1994) *Food and Nutrition Encyclopaedia*, 2nd edn, 2 vols., CRC Press, Boca Ratan, FL.

Esau, K. (1953) *Plant Anatomy*, John Wiley & Sons, New York and Chapman & Hall, London.

Evans, L.T. (1973) The effect of light on plant growth, development and yield, in R.O. Slatyer (ed.), *Plant Responses to Climatic Factors. Proceedings of the Uppsala Symposium, 1970*. Ecology and Conservation No. 5, UNESCO, Paris, pp. 21-35.

Everist, S.L. (1972) *Poisonous Plants of Australia*, Australian Natural Science Library, Angus & Robertson Publishers, London.

Everist, S.L. (1981) *Poisonous Plants of Australia*, revised edn., Angus & Robertson, London.

Ewen, S.W.B. and Pusztal, A. (1999) Effect of diets containing genetically modified potatoes expressing *Galanthus nivalis* lectin on rat small intestine, *The Lancet* **354**, 1353-1354.

Faegri, K. and Iversen, J. (1964) *Textbook of Pollen Analysis*, Blackwell Scientific Publications, Oxford.

Fagg, C.W. and Stewart, J.L. (1994) The value of *Acacia* and *Prosopis* in arid and semi-arid environments, *Journal of Arid Environments* **27**, 3-25.

FAO (!962) *Charcoal for domestic and industrial use*, FAO, Rome.

FAO (1973) *Guide for planing pulp and paper enterprises.* Forestry and Forest Products Studies No. 18, FAO, Rome.

FAO (1982) *Fruit-bearing Forest Trees.* FAO Forestry Paper 34, FAO, Rome.

FAO (1983) *Food and Fruit-bearing Forest Species. 1: Examples from Eastern Africa.* FAO Forestry Paper 44/1, FAO, Rome.

FAO (1984a) *Food and Fruit-bearing Forest Species. 2: Examples from Southeastern Asia.* FAO Forestry Paper 44/2, FAO, Rome.

FAO (1984b) *Traditional (Indigenous) Systems of Veterinary Medicine for Small Farmers in India,* FAO-RAPA, Bangkok.

FAO (1984c) *Traditional (Indigenous) Systems of Veterinary Medicine for Small Farmers in Thailand,* FAO-RAPA, Bangkok.

FAO (1984d) Traditional (Indigenous) Systems of Veterinary Medicine for Small Farmers in Nepal, FAO-RAPA, Bangkok.

FAO (1986a) *Food and Fruit-bearing Forest Species. 3: Examples from Latin America.* FAO Forestry Paper 44/3, FAO, Rome.

FAO (1986b) *Traditional (Indigenous) Systems of Veterinary Medicine for Small Farmers in Pakistan,* FAO-RAPA, Bangkok.

FAO (1988a) *Traditional Food Plants. A resource book for promoting the exploitation and consumption of food plants in arid, semi-arid and sub-humid lands of Eastern Africa.* FAO Food and Nutritional Paper 42, FAO, Rome.

FAO (1988b) *The Eucalypt Dilemma,* FAO, Rome.

FAO (1989) *Plant Genetic Resources,* FAO, Rome.

FAO (1990) Specifications for identity and purity, in *Specifications for identity and purity of certain food additives. Emulsifiers, enzyme preparations, flavouring agents, food colours, thickening agents, miscellaneous food additives.* FAO Food Nutrition Paper 49, FAO, Rome, pp. 23-25.

FAO (1991a) *TraditionalVeterinary Medicine in Sri Lanka,* FAO-RAPA, Bangkok.

FAO (1991b) *Traditional Veterinary Medicine in Indonesia,* FAO-RAPA, Bangkok.

FAO (1992) *Traditional Veterinary Medicine in the Philippines,* FAO-RAPA, Bangkok.

FAO (1993a) *FAO Trade Yearbook Vol. 46, 1992,* FAO, Rome.

FAO (1993b) *A decade of wood energy activities within the Nairobi Programme of Action.* FAO Forestry Paper 108, FAO, Rome.

FAO (1994) *Production 1993. FAO Yearbook vol. 47,* FAO Statistics Series No. 117, FAO, Rome.

FAO-UNESCO (1985) *Soil Map of the World. Revised Legend,* FAO, Rome.

Faria, S.M. de, Lewis, G.P., Sprent, J.J. and Sutherland, J.M. (1989) Occurrence of nodulation in the Leguminosae, *New Phytologist* **111**, 607-619.

Farjon, A. (1998) *World Checklist and Bibliography of Conifers,* Royal Botanic Gardens, Kew.

Farnsworth, N.R. (1984) The role of medicinal plants in drug development, in P. Krogsgaard-Larsen, C. Brøgger and H. Kofod (eds.), *Natural Products and Drug Development,* Munksgaard, Copenhagen, pp. 17-30.

Farnsworth, N.R. and Soejarto, D.D. (1985) Potential consequences of plant extinction in the United States on the current and future availability of prescription drugs, *Economic Botany* **39**, 231-240.

Farrelly, D. (1984) *The Book of Bamboo,* Thames and Hudson, London.

Feeney, P. (1975) Biochemical co-evolution between plants and their insect herbivores, in L.E. Gilbert and P.S. Raven (eds.), *Co-evolution of Animals and* Plants, University of Texas Press, Austin, TX, pp. 3-19.

Felger, R.S. and Moser, M.B. (1985) *People of the Desert and Sea. Ethnobotany of the Seri Indians,* University of Arizona Press, Tucson.

Fenical, W. (1983) Marine plants: a unique and unexplored resource, in *Plants: The Potentials for Extracting Protein, Medicines, and Other Useful Chemicals - Workshop Proceedings.* US Congress Office of Technology Assessment, Washington, DC, pp. 147-153.

Fenton, B., Stanley, K., Fenton, S. and Bolton-Smith, C. (1999) Differential binding of the insecticidal lectin BNA to human blood cells, *The Lancet* **354**, 1354-1355.

Ferrando, A. (1981) *Traditional and Non-Traditional Foods*, FAO Food and Nutrition Series No. 2, FAO, Rome.

Ferri, M.G. (1961) Problems of water relations of some Brazilian vegetation types, with special consideration of the concepts of xeromorphy and xerophytism, in *Plant-Water Relation-ships in Arid and Semi-Arid Condition. Proceedings of the Madrid Symposium. Arid Zone Research 16*, UNESCO, Paris, pp. 191-197.

Fielding, D. (1998) Ethnoveterinary medicine in the tropics - key issues and the way forward? *Tropical Agriculture Association UK Newsletter* **18**,2, 17-19.

Fitter, A.H. and Hay, R.K.M. (1985) *Environmental Physiology of Plants*, 2nd edn., Academic Press, London.

Flach, M. and Rumawas, F. (eds.) *Plants Yielding Non-Seed Carbohydrates*. Plant Resources of South-East Asia No. 9, Backhuys Publishers, Leiden.

Flach, M. and Rumawas, F. (1996) Introduction, in M. Flach and F. Rumawas (eds.), *Plants Yielding Non-Seed Carbohydrates. Plant Resources of South-East Asia No. 9*, Backhuys Publishers, Leiden, pp. 15-42.

Flowerdew, B. (1995) *Bob Flowerdew's Complete Book of Companion Gardening*, Kyle Cathy, London.

Food Standards Agency (2000) *The Food Standards Agency - Functions and Structure*, Fact Sheet, Food Standards Agency, London..

Ford, R. I. (1978) Ethnobotany: historical diversity and synthesis, in R.I. Ford (ed.), *The Nature and status of Ethnobotany*, Anthropological Papers Vol. 67, Museum of Anthropology, University of Michigan, Ann Arbor, pp. 33-49.

Forestry Authority (1994) *Plant Health Import and Export Controls: Wood and Bark. A Guide for Importers and Exporters Trading Directly with Countries outside the European Community.* Plant Health Leaflet No 1 (revised July 1984), The Forestry Authority, Edinburgh.

Fourie, T.G., Swart, E. and Snyckers, F.O. (1992) Folk medine: a viable starting point for pharmaceutical research, *South African Journal of Science* **88**, 190-192.

Fowles, G. (1977) *Straight-Forward Liqueur Making*, Fowles, Reading.

Frodin, D.G. (1984) *Guide to Standard Floras of the World*, Cambridge University Press, Cambridge.

Franz, J.C. (1984) *Jojoba - a new cash crop for arid lands*, Gesellschaft für Regionalforschung und Angewandte Geographie, Nürnberg.

Fryxell, P.A. (1965) Stages in the evolution of *Gossypium*, *Advances in Frontiers of Plant Science* **10**, 31-56.

Gabányi, S. (1997) *Whisk(e)y*, Abbeville Press, New York.

Gade, D. (1970) Ethnobotany of canihua (*Chenopodium nudicaule*), rustic seed crop of the Altiplano, *Economic Botany* **24**, 55-61.

Gaussen, H. (1955) Expression des milieux par des formules écologiques; leur représentation cartographique, *Colloques internationaux de Centre national de la Recherche scientifique* **59**, 257-269.

Gentil, l. (1907) *Liste des plantes cultivées dans serres chaudes et coloniales du Jardin Botanique de l'Etat à Bruxelles*, Weissenbruch, Bruxelles.

Gentry, H.S. (1958) The natural history of jojoba (*Simmondsia chinensis*) and its cultural aspects, *Economic Botany* **12**, 261-295.

Gentry, H.S. (1972) Supplement to the natural history of jojoba, in E.F. Hause and W.G. McGinnies (eds.), *Jojoba and its Uses - An International Conference*, Office of Arid Land Studies, Tucson, pp. 11-12.

Ghisalberti, E.L. (1979) Propolis: a review, *Bee World* **60** (2), 59-84.

Gibbs Russell, G.E. (1983) The taxonomic position of C_3 and C_4 *Alloteropsis semialata* (Poaceae) in southern Africa. *Bothalia* **14** (2), 20 5-213.

Gilbert, M.G. (1986) Notes on East African Vernonieae (Compositae). A revision of the *Vernonia galamensis* complex, *Kew Bulletin* **41**, 19-35.

Gindel, I. (1970) The nocturnal behaviour of xerophytes grown under arid conditions, *New Phytologist* **69**, 399-404.

Gintzburg, C. (1963) Some anatomical features of splitting of desert shrubs, *Phytomorph*ology **13**, 92-97.

Gledhill, M. and McGrath, P. (1997) Monsanto's cotton gets the Mississippi blues, *New Scientist* **2106**, 4-5.

Glenn, E.P., Brown. J.J. and Blumwald, E. (1999) Salt tolerance and crop potential of halophytes, *Critical Reviews in Crop Science* **18** (2), 227-255.

Glenn, E.P., O'Leary, J.W., Watson, M.C., Thompson, T.L. and Kuehl, R.O. (1991) *Salicornia bigelovii* Torr.: an oilseed halophyte for seawater irrigation, *Science* **251**, 1065-1067.

Glime, J.M. and Saxena, D. (1991) *Uses of Bryophytes*, Today and Tomorrow's Printers and Publishers, New Delhi.

Glossary Revision Special Committee (1989) *A Glossary of Terms used in Range Management*, Society for Range Management, Denver, CO.

Glover, J. and Gwynne, M.D. (1962) Light rainfall and plant survival in E. Africa. I, Maize, *Journal of Ecology* **50**, 111-118.

Glover, P.E. Glover, J. and Gwynne, M.D. (1962) Light rainfall and plant survival in E. Africa. II, Dry grassland vegetation, *Journal of Ecology* **50**, 199-206.

Godon, B. and Petit, L. (1971) Les mais grain: prestockages, séchage et qualité, v. propriétes des protéines, *Annals de Zootechnie* **20**, 641-644.

Göhl, B. (1981) *Tropical feeds. Feed information summaries and nutritive values.* FAO Animal Production and Health Series No. 12, FAO, Rome.

Gowen, S.R. (1998) Biocontrol of plant parasitic nematodes. *Tropical Agriculture Associatio, Newsletter* **18** (3), 11

Grabow, W., Slabbert, J.L., Morgan, W.S.G. and Jahn, S. Al A. (1985) Toxicity and mutagenicity evaluation of water coagulated with *Moringa oleifera* seed preparations using fish, protozoan, bacterial, coliphage, enzyme and Ames *Salmonella* assays, *Water SA* (Pretoria) **11** (1), 9-14.

Grami, B. (1998) Gaz of Khunsar: the manna of Persia, *Economic Botany* **52**, 183-191.

Grant-Downton, R. (1998) Dry Lazarus, *The Garden* **123**, 656-659.

Grattan, S.R. and Grieve, C.M. (1993) Mineral nutrient acquisition and response by plants grown in saline environments, in M. Pessarakli (ed.), *Handbook of Plant and Crop Stress*, M. Dekker, New York, pp. 203-226.

Green, C.L. (1995) *Natural colourants and dyestuffs*, Non-Wood Forest Products No. 4, FAO, Rome.

Green, P.S. and Wickens, G.E. (1989) The *Olea europaea* complex, in Kit Kan (ed.), *The Davis & Hedge Festschrift*, Edinburgh University Press, Edinburgh, pp. 287-289.

Green, T. (1816-24) *The Universal Herbal; or, Botanical, Medical, and Agricultural Dictionary*, 2nd edn, 2 vols, Caxton Press, London.

Greenland, D.J. (1996) Brown rice, *Tropical Agriculture Association UK Newsletter* **16** (2), 19.

Greenwood, E.A.N. (1986) Water use by trees and shrubs for lowering saline groundwater, in E.G. Barrett-Leonard, C.V. Malcolm, W.R. Stern and S.M.Wilkins (eds.), *Forage and Fuel Production from Salt Affected Wasteland. Proceedings of a Seminar held at Cunderdin, Western Australia, 19-27 May 1984*, Elsevier, Amsterdam, 423-434 (reprinted from *Reclamation and Revegetation Research* **5**, 423-434).

Gressel, J., Kleifeld, Y. and Joel, D.M. (1994) Genetic engineering can help control parasitic weeds, in A.H. Pieterse, J.A.C. Verkleij and S.J.ter Borg (eds.), *Biology and Management of Orobanche. Proceedings of the Third International Workshop on Orobanche and Related Striga Research*, Royal Tropical Institute, Amsterdam, pp. 406-418.

Greuter, W., Barrie, F.R., Burdet, H.M., Chaloner, W.G., Demoulin, V., Hawksworth, D.L., Jorgensen, P.M., Nicholson, D.H., Silva, P.C., Trehane, P. and McNeil, J. (eds.) (1994) *International Code of Botanical Nomenclature (Tokyo Code), Regnum vegetabile* **131**,. Koeltz Scientific Books, Köningstein.

Greuter, W., Hawksworth, D.L., McNeil, J., Mayo, M.A., Minelli, A. Sneath, P.H.A., Tindall, B.J., Trehane, P. Tubbs, P. (eds.) (1996) Draft BioCode: The prospective international rules for the scientific names of organisms, *Taxon* **45**, 349-372.

Grigson, G. (1958) *The Englishman's Flora*, Phoenix House, London.

Grove, A.T. (1985) The arid environment, in G.E. Wickens, J.R. Goodin and D.V. Field (eds.), *Plants for Arid lands*, Allen & Unwin, London, pp. 9-18.

Grubben, G.J.H. and Siemonsma, J.S. (1996) *Fagopyrum esculentum* Moench, in G.J.H. Grubben and S. Partohardjono (eds.), *Cereals. Plant Resources of South-East Asia No. 10*, Backhuys Publishers, Leiden, pp. 95-99.

Grubben, G.J.H., Partohardjono, S. and van der Hoek, H.N. (1996) Introduction, in G.J.H. Grubben and S. Partohardjono (eds.), *Cereals. Plant Resources of South-East Asia No. 10*, Backhuys Publishers, Leiden, pp. 15-72.

Guiry, M.D. and Blunden, G. (1991) Conclusions and outlook, in M.D. Guiry and G. Blunden (eds.), *Seaweed Resources in Europe: Uses and Potentia*, John Wiley & Sons, Chichester, pp. 409-413.

Gupta, S.M. (1971) *Plant Myths and Indians in India*, E.J. Brill, London.

Gutale, F.S. and Ahmed, M.A. (1984) *Cordeauxia edulis* pigment, cordeauxiaquinone is deposited on bones and may stimulate hemopoiensis in rats, *Rivista Tossicologia Sperimentale e Clinico 14*, 57-62.

Gutterman, Y. (1993) *Seed Germination in Desert Plants*, Adaptations of Desert Organisms (ed. J.L. Cloudsley-Thompson), Springer-Verlag, Berlin, Heidelberg.

Hagerman, A.E., Robbins, C.T., Weerasuriya, Y., Wilson, T.C. and McArthur, C. (1992) Tannin chemistry in relation to digestion, *Journal of Range Management* 45, 57-62.

Hale, M.E. Jnr. (1983) *The Biology of Lichens,* 3rd edn., Edward Arnold, London.

Hall, I.M. (1964) Use of macro-organisms in biological control, in I. DeBuch (ed.) *Biological Control of Insect Pests and Weeds*, Chapman & Hall, London, pp. 610-628.

Hall, I.R., Lyon, A.J.E., Wang, Y. and Sinclair, L. (1998a) Ectomycorrhizal fungi with edible fruiting bodies, 2. *Boletus edulis, Economic Botany*, **52**, 44-56.

Hall, I.R., Zambonelli, A. and Primavera, F. (1998b) Ectomycorrhizal fungi with edible fruiting bodies, 3. *Tuber magnatum* (Tuberaceae), *Economic Botany* 52, 192-200.

Hall, J.B. and Walker, D.H. (1991) *Balanites aegyptiaca. A Monograph*, School of Agriculture and Forest Sciences Publication No. 3, University of Wales, Bangor.

Hall, N., Boden, R.W., Christian, C.S., Condon, R.W., Dale, F.A., Hart, A.J., Leigh, J.H., Marshall, J.K., McArthur, A.G. Russell, V. and Turnbull, J.W. (eds.) (1972) *The Use of Trees and Shrubs in the Dry Country of Australia*, Australian Government Publishing Service, Canberra.

Halligan, P. (1975) Toxic terpenes from *Artemesia californica, Ecology* **56**, 999-1003.

Handa, S.S. (1998) The integration of food and medicine in India, in H.D.V. Prendergast, N.L. Etkin, D.R. Harris and P.J. Houghton (eds.), *Plants for Food and Medicine*, Royal Botanic Gardens, Kew, pp. 57-68.

Harborne, J.B. (1988) *Introduction to Ecological Biochemistry*, 3rd edn., Academic Press, London.

Harder, R., Schumacher, W., Firbas, F. and von Denffer, D. (1965) *Strasburger's Textbook of Botany*, 28th edn. (English translation by P. Bell, P. and D. Coombe), Longmans, Green & Co., London.

Harlan, J.R. (1989) Wild-grass seed harvesting in the Sahara and Sub-sahara of Africa, in D.R. Harris and G.C. Hillman (eds.), *Foraging and Farming: The Evolution of Plant Exploitation*, One World Archaeology-B, Unwin Hyman, London, pp. 79-98.

Harris, D.R. (1998) Introduction: the multidisciplinary study of cross-cultural plant exchange, in H.D.V. Prendergast, N.L. Etkin, D.R. Harris and P.J. Houghtin (eds.), *Plants for Food and Medicine*, Royal Botanic Gardens, Kew, pp. 85-91.

Harshberger, J.W. (1896) The purposes of ethno-botany, *Botanical Gazette* (Chicago) **21**, 146-154 (reprinted in *American Antiquarian* **17** (2), 73-81).

Hart, C. (1991) *Practical Forestry for the Agent and Surveryor*, 3rd edn., A. Sutton, Stroud, Glos.

Hartley, C.W.S. (1967) *The Oil Palm*. 2nd edn., Longmans, Green & Co., London.

Hattersley, P.W. and Watson, L. (1992) Diversification of photosynthesis, in G.P. Chapman, (ed.), *Grass Evolution and Domestication*, Cambridge University Press, Cambridge, pp. 38-116.

Hawkes, J.G. (1980) *Crop Genetic Resources Field Collection Manual for Seed Crops, Root and Tuber Crops, Tree Fruit Crops and Related Wild Species*, IBPGR, Rome and Eucarpa, Wageningen.

Hawkes, J.G. (1983) *The Diversity of Crop Plants*, Harvard University Press, Cambridge, MA.

Hawkes, J.G. (1991) The importance of genetic resources in plant breeding, *Biological Journal of the Linnean Society* **43**, 3-10.

Hawkes, J.G. (1998) The introduction of New World crops into Europe after 1492, in H.D.V. Prendergast, N.I. Etkin, D.R. Harris and P.J. Houghton (eds.), *Plants for Food and Medicine*, Royal Botanic Gardens, Kew, pp. 147-159.

Hawksworth, D.L., Kirk, P.M., Sutton, B.C. and Pegler, D.N. (eds.) (1995) *Ainsworth's & Bisby's Dictionary of the Fungi*, 8th edn., CAB International, Wallingford.

Hazen, J. (1993) *Mustard. Making your Own Gourmet Mustard*, Chronicle Books, San Francisco.

Hecht, S.B., Anderson, A.B. and May, P. (1988) The subsidy from nature: shifting cultivation, successional palm forests and rural development, *Human Organization* **47** (1), 25-35.

Hedrick, V.P. (ed.) (1972) *Sturtevant's Edible Plants of the World*, Dover Publications, New York. (originally published as Sturtevant, E.L. (1919) *Sturtevant's Notes on Edible Plant,*. New York Department of Agriculture 27th Annual Report, vol. 2, Part II, J.B. Lyon Co., Albany, NY).

Hehn, V. (1885) *The wanderings of plants and animals from their first home*, Swan Sonnenschein, London.

Heiser, C.B. (1985) Ethnobotany of the naranjilla (*Solanum quitoesnse*) and its relatives, *Economic Botany* **39**, 4-11.

Heiser, C.B. (1986) Economic botany: past and future, *Economic Botany* **40**, 261-266.

Heller, J. (1996) *Physic nut (Jatropha curcas* L.), Promoting the Conservation and Use of Underutilized and Neglected Crops 1, IBPGR, Rome.

Hemingway, J.S. (1995) The mustard species: condiment and food ingredient use and potential as oilseed crops, in D.S. Kimber and D.I. McGreggor (eds.), *Brassica Oilseeds*, C.A.B. International, Wallingford.

Henderson, L. (1983) Barrier plants in South Africa, *Bothalia* **14**, 635-639.

Hepper, F.N. (1963, 1968) *Flora of West Tropical Africa*, vols 2 and 3, Crown Agents, London.

Heywood, V.H. (ed.) (1971) *The Biology and Chemistry of the Umbelliferae*, Linnean Society, London.

Heywood, V. (1990) Conservation of germplasm of wild plant species, in O.T. Sandlund, K. Hinder and A.H.D. Brown (eds.), *Conservation of Biodiversity for Sustainable Development*, Scandinavian University Press, Oslo, pp. 189-203.

Heywood, V.H., Harborne, J.B., Turner, B.L. (eds.) (1977) *The Biology and Chemistry of the Compositae*, 2 vols., Academic Press, New York.

Hidalgo, O. (1974) *Bambu su Cultivo y Applicación Arquitecture, Ingeniería Artesanía,* Estudios Techices Colombianos Limitada, Cali, Colombia.

Hill. A.F. (1937) *Economic Botany. A Textbook of Useful Plants and Plant Products*, McGraw-Hill, New York.

Hill, A.F. (1952) *Economic Botany. A Textbook of Useful Plants and Plant Products*, 2nd edn., McGraw-Hill Book Company, New York.

Hillis, W.T. (ed.) (1962) *Wood Extractives and Their Significance to the Pulp and Paper Industries*, Academic Press, London.

Hizukuri, S. (1996) Starch - analytical aspects, in A.-C. Eliasson (ed.), *Carbohydrates in Food*, Marcel Dekker, New York, pp. 347-429.

Hodgkinson, K.C. and Griffin, G.F. (1982) Adaptation of shrub species to fires in the arid zone, in W.R. Barker and P.J.M. Greenslade (eds.) *Evolution of the Flora and Fauna of Arid Australia,* Peacock Publications, Frewville, NSW, pp. 145-152.

Hoffmann, J.J. and McLaughlin, S.P. (1986) *Grindelia camporum*: potential cash crop for the arid Southwest, *Economic Botany* **40**, 162-169.

Hoffman J., A.E. (1989) *Cactaceas En la flora silvestre de Chile*, Fundacion Claudio Gay, Santiago de Chile.

Hoizey, D. and Hoizey, M.-J. (1993) *A History of Chinese Merdicine*, translated from the French by P. Bailey, Edinburgh University Press, Edinburgh.

Holland, P. (1678) *The new world of English Words, or, a general dictionary*, 4th edn., London.

Holgren, P.K., Keuken, W. and Schoffield, E.K. (eds.) (1981) *Index herbariorum: a guide to the location and contents of the world's public herbaria. Pt. 1, The herbaria of the world*, 7th edn., *Regnum vegetabile* **106**, Bohn, Scheltema & Holhema, Utrecht.

Holliday, P. (1990) *A Dictionary of Plant Pathology*, Cambridge University Press, Cambridge.

Holness, B.L. (1995) Implications for third world produce, *Tropical Agriculture Association UK Newsletter* **15** (4), 29-30.

Holt, J.G. (ed. in chief) (1984-1989) *Bergey's Manual of Systematic Bacteriology*, 2nd ed., 4 vols., Williams & Wilkins, Baltimore, MA.

Holt, J.G., Krieg, N.R., Sneath, P.H.A., Staleg, J.T. and Williams, C.T. (eds.) (1994) *Bergey's Manual of Determinative Bacteriology*, 9th edn., Williams and Wilkins, Baltimore.

Horrobin, D.F. (ed.) (1990a) *Omega-6 Essential Fatty Acids. Pathophysiology and Roles in Clinical Medicine*, Wiley-Liss, New York.

Horrobin, D.F. (1990b) Gamna linolenic acid: an intermediate in essential fatty acid metabolism with potential as an ethnic pharmaceutical and as a food, *Reviews in Contemporary Pharmacotherapy* **1**, 1-45.

Horrobin, D.F. and Lapinskas, P. (1998) The commercial development of food plants used in medicines, in H.D.V. Prendergast, N.L. Etkin, D.R. Harris and P. Houghton (eds.), *Plants for Food and Medicine*, Royal Botanic Gardens, Kew, pp. 75-81.

Horton, R. (1999) Genetically modified foods: "absurd" concern or welcome dialogue, *The Lancet* **354**, 1314-1315.

Hostettmann, K. and Marston, A. (1987) Plant molluscicide research - an update, in K.E. Mott (ed.), *Plant Molluscicides*, John Wiley & Sons, Chichester, pp. 299-320.

Hough, C. (1998) Reaping willows, *Green Futures* **11**, 37.

House, W.A. and Welch, R.M. (1984) Effects of nutritionally occurring antinutrients in the nutritive value of cereal grains, potato tubers and legume seeds, in R.M. Welch and W.H. Gabelman (eds.), *Crops as Sources of Nutrients for Human,*. American Society of Agronomy, Special Publications No. 48, pp. 9-35.

Howes, F.N. (1948) *Nuts. Their Production and Everyday Use*, Faber & Faber, London.

Howes, F.N. (1949) Sources of poisonous honey, *Kew Bulletin* **4**, 167-171.

Hoyt, E. (1988) *Conserving the Wild Relatives of Crops*, IBPGR, Rome, IUCN and WWF, Gland, Switzerland.

Hudson, N. (ed.) (1954) *Hortus sanitatis* (Facsimile edition of 1485), Quaritel, London.

Hughes, D. (1994) The growing interest in herbs and spices, *Tropical Agriculture Association UK Newsletter* **14** (3), 3-4.

Hunter, D. (1947) *Papermaking: The History and Technique of an Ancient Craft*, 2nd edn., Knopf, New York.

Hutchinson, J. (1926, 1934) *The Families of Flowering Plants*, Vol. 1, *Dicotyledons*, Vol. 2, *Monocotyledons*, Oxford University Press, London.

Hutchinson, J.B., Silow, R.A. and Stephens, S.G. (1947) *The Evolution of Gossypium*, Oxford University Press, London.

Hutchinson, P. (1973) *The Complete Home Carpenter*, Marshall Cavendish, London.

Ibrahim, K. (1975) *Glossary of Terms Used in Pasture and Range Survey Research*, Ecology and Management, FAO, Rome.

Ibrahim, M.A. (1996) Ethno-toxicology among Nigerian agropastoralists, in C.M. McCorckle, E. Mathias and T.W. Schillhorn van Veen (eds.), *Ethnoveterinary Research and Development*, Intermediate Technology Publications, London, pp. 54-59.

Indergaard, M. and Østgaard (1991). Polysaccharides for food and pharmaceutical uses, in M.D. Guiry and G. Blunden (eds.), *Seaweed Resources in Europe: Uses and Potentia*, John Wiley and Sons, Chichester, pp. 169-183.

Indrbhakdi, S. (1989) Traditional paper making: northern Thailand, *Biomass Users Network, Network News* **3** (4) 1, 4-5.

Inman, R.B., Dunlop, P. and Jackson, J.F. (1991). Oils and waxes of eucalypts. Vacuum distillation method for essential oils, in H.F. Linskens and J.F. Jackson (eds.), *Essential Oils and Waxes.* Modern methods of plant analysis, New Series vol. 12, Springer-Verlag, New York, pp. 195-203.

Innamorati, T.Fossi (1973) Notizie di medicina populare african nell 'Erbario tropicale di forenze, *Webbia* **28**, 81-134 (with English summary).

Irvine, F.R. (1952) Supplementary and emergency food plants of West Africa, *Economic Botany* **6**, 23-40

Irvine, F.R. (1953) *A Text-Book of West African Agriculure. Soils and Crops*, 2nd edn., Oxford University Press, London.

ISO (1981) *Mustard Seed Specification*, I.S.O. 1237-1981(E), 2nd edn., International Organization for Standardization, Geneva, pp. 1-11.

ISO (1997) *ISO in brief*, International Standards Organisation, Geneva.

Jackson, B.D. (1893-95) *Index Kewensis Plantarum Phaerogamarum*, 4 vols., Clarendon Press, Oxford (with 19 Supplements, 1921-1991), Also available on CD-ROM from Oxford University Press, Corby.

Jackson, D.L. and Jacobs, S.W.L. (1985) *Australian Agricultural Botany*, Sydney University Press, Sydney.

Jacobson, K.M. (1997) Moisture and substrate stability determine VA-mycorrhizal fungal community distribution and structure in arid grassland, *Journal of Arid Environments* **35**, 59-75.

Jacobson, M. (1982) Plants, insects, and man - their interrelationships, *Economic Botany* **36**, 346-354.

Jackson, P.J., Unkefer, P.J., Delhaize, E. and Robinson, N.J. (1990) Mechanisms of trace metal tolerance in plants, in F. Katterman (ed.), *Environmental Injury in Plants*, Academic Press, San Diego, pp. 231-255.

Jahn, S. Al A. (1981) *Traditional water purification in tropical developing countries*, GTZ, Eschborn.

Jahn, S.Al A. and Dirar, H. (1979) Studies on natural water coagulants in the Sudan with special reference to *Moringa oleifera* seeds, *Water SA* (Pretoria) **5** (2), 90-97.

Jahn, S. Al A.Musnad, H.A. and Burgstaller, H. (1986) The trees that purifies water: cultivating multipurpose Moringaceae in the Sudan, *Unasylva* **38** (152), 23-29.

Jain, S.K. and Sutarno, J. (1996) *Amaranthus* L. (grain amaranth)., in G.J.H. Grubben and S. Partohardjono (eds.), *Cereals. Plant Resources of South-East Asia No. 10*, Backhuys Publishers, Leiden, pp. 75-79.

James, L.F., Nielsen, D.B. and Panter, K.E. (1992) Impact of poisonous plants on the livestock industry, *Journal of Range Management* **45**, 3-8.

Janick, J. (1986) *Horticultural Sciences*, 4th edn., W.H. Freeman and Co., New York.

Jansen, P.C.M., van der Wilk, C. and Hetterscheid, W.L.A. (1996) *Amorphophalus*, in M. Flach and F. Rumawas (eds.), *Plants Yielding Non-Seed Carbohydrates. Plant Resources of South-East Asia No. 9*, Backhuys Publishers, Leiden, pp. 45-50.

Jardin, C. (1967) *List of Foods Used in Africa*, FAO, Rome.

Jarman, C. (1998) *Plant Fibre Processing. A Handbook*, Small-Scale Textiles. Intermediate Technology Publications, London.

Jeffrey, C. (1968) Systematic categories for cultivated plants, *Taxon* **17**, 109-114.

Jeffrey, C. (1982) *An Introduction to Plant Taxonomy*, 2nd edn., Cambridge University Press, Cambridge.

Jeffrey, C. (1989) *Biological Nomenclature*, 3rd edn., Edward Arnold, London.

Jerrard, H.G. and McNeill, D.B. (1992) *Dictionary of Scientific Units*, 6th edn., Chapman & Hall, London.

Joel, D.M., Klelfeid, Y., Losner-Goshen, D., Herzlinger, G. and Gressel, J. (1995) Transgenic crops against herbicides, *Nature* **374**, 220-221.

Johnson, D.V. (1985) Present and potential economic usage of palms in arid and semi-arid areas, in G.E. Wickens, J.R. Goodin and D.V. Field (eds.), *Plants for Arid Lands*, Allen & Unwin, London, pp. 189-202.

Johnson, M.S. (1984) Effects of soluble salts on water absorption by gel-forming soil conditioners, *Journal of the Science of Food and Agriculture* **35**, 1063-1066.

Johnson, S. (1755) *Dictionary of the English Language*, London.

Johnston, M.C. (1972) *Flora of Tropical East Africa: Rhamnaceae*, Crown Agents for Overseas Governments and Administrations, London.

Jones, V.H. (1941) The nature and state of ethnobotany, *Chronica Botanica* **6** (10), 219-221.

Joseleau, J.P. and Ullmann, G. (1985) A relation between starch metabolism and the synthesis of gum arabic, *Bulletin of the International Group for the Study of Mimosoideae* **13**, 46-54.

Jowitt, R. (1989) *A Classification of Foods and Physical Properties*, Food Science Publishers, Hornchurch, Essex.

Joy, E.T. (1962) *English Furniture A.D. 43-1950*, B.T. Batsford, London.

Kain, J.M. (1991) Cultivation of attached seaweeds, in M.D. Guiry and G. Blunden (eds.), *Seaweed Resources in Europe: Uses and Potentia*, John Wiley & Sons, Chichester, pp. 309-377.

Kalkman, C. (1989) Economic botany in South-East Asia, in J.S. Siemonsma and N. Wulijarni-Soetjipto (eds.), *Proceedings of the First PROSEA International Symposium May 22-25, 1989, Jakarta, Indonesia*, Pudoc, Wageningen, pp. 48-56.

Kassam, A.H., Kowal, J.M. and Sarraf, S. (1996) *Climatic Adaptability of Crops. Consultant's Report.* Agroecological Project, FAO, Rome.

Kearney, T. and Peebles, R.H. (1951) *Arizona Flora*, University of California Press, Berkeley and Los Angeles.

Keating, W.G. (1980) Utilization of mixed species through grouping and standards. *Australian Forestry* **43**, 233-244.

Keating, W.G. and Bolza, E. (1982) *Characteristics, Properties and Uses of Timbers.* Vol.1 South-East Asia, Northern Australia and the Pacific, Inkata Press, Melbourne.

Keay, R.W.J.(1954, 1958) *Flora of West Tropical Africa*, vol. 1, Crown Agents, London.

Keeley, J.E. and Keeley, S.C. (1988) Chaparral. In: Barbour, M.G. and Billings, W.D. (eds.), *North American Terrestrial Vegetation*, Cambridge University Press, New York, pp. 165-207.

Kellogg, A. (1859) *Simmondsia pabulosa. Proceedings of the California Academy of Natural Sciences* **2**, 21-22.

Killmann, W. (1999) Editorial, *Non-Wood News* [FAO, Rome] **6**, 1-2.

Kingsbury, J.M. (1964) *Poisonous Plants of the United States and Canada*, Prentice-Hill, Englewood Cliffs, NJ.

Kirby, R.H. (1963) *Vegetable Fibres. Botany, Cultivation, and Utilization*, Leonard Hill (Books), London; Interscience Publications, New York.

Kiritsakis, A. and Markakis, P. (1991) Olive oil analysis, in H.F. Linskens and J.F. Jackson, (eds.), *Essential Oils and Waxes.* Modern Methods of Plant Analysis, New Series vol 12,. Springer-Verlag, New York, pp. 1-20.

Kirston, B.E. and Kurtzman, C.P. (eds.) (1988) *Yeasts (Living Resources for Biotechnology,.* Cambridge University Press, Cambridge.

Kloos, H. and McCullough, F.S. (1987) Plants with recognized molluscicidal activity, in K.E. Mott (ed.), *Plant Molluscicides*, John Wiley & Sons, Chichester, pp. 45-108.

Knight, J. (1997) White, with a touch of potato, *New Scientist* 1 Nov. **2106**, 7.

Ko, Swan Djien (1982) Indigenouis fermented foods, in A.H. Rose (ed.), *Fermented Food*, Academic Press, London.

Kobayashi, A. and Kawakami, (1991) Analysis of essential oils of tea, in H.K. Linskens and J.F. Jackson (eds.), *Essential Oils and Waxes.* Modern Methods of Plant Analysis, New Series vol 12, Springer-Verlag, New York, pp. 21-40.

Kochhar, S.I. and Singh, B.M. (1989) Plant resources for AD 2001, in M.S. Swaminathan and S.L. Kochhar (eds.), *Plants and Society*, Macmillan Publishers, London and Basingstoke, pp. 556-617.

Köppen, W. and Geiger, R. (1936) *Handbuch der Klimatologie*, Borntraeger, Berlin.

Kreig, M.B. (1965) *Green Medicine, the Search for Plants that Heal*, Harrap, London.

Kronenberg, H.J. (1984) Reduction of incubation time for tempeh fermentation by use of pregerminated inoculum, *Economic Botany* **38**, 433-438.

Kruger, F.J. and Bigalke, R.C. (1984) Fire in fynbos, in P. de V. Booysen and N.M. Tainton, (eds.), *Ecological Effects of Fire in South African Ecosystems*, Ecological Studies 48, Springer-Verlag, Berlin, Heidelberg, pp. 67-114.

Kubitzki, K. (1990-) *The Families and Genera of Vascular Plants*, 4 vols to date, Springer-Verlag, Berlin.

Kulper, H.A., Noteborn, H.P.J.M. and Peijnenburg, A.A.C.M. (1999) Adequacy of methods for testing the safety of genetically modicified foods., *The Lancet* **354**, 1315-1316.

Kummerow, J. (1973) Comparative anatomy of sclerophylls of Mediterrabean climatic areas, in F. di Castri and H.A. Mooney (eds.), *Mediterranean-type Ecosystems: Origin and Structure*, Springer-Verlag, New York, pp. 157-167.

Kunkel, G. (1984) *Plants for Human Consumption*, Koeltz Scientific Books, Köenigatein.

Kurz, S. (1876) Bamboo and its use, *Indian Forester* **1**, 219-269, 335-362.

LaFleur, J.R. (1992) *Marketing of Brazil Nuts*, FAO, Rome.

Lamb, R. (1995) *Forests, Fuels and the Future. Wood energy for sustainable development*. Forestry Topics Report No. 5, FAO, Rome.

Land, S. and Farquhar, I. (1998) *Smart Plants. A Farmers' Guide to: Genetically Modified Organisms (GMOs) in Arable Agriculture*, Farmers' Link, Thetford, Norfolk.

Lane, J.A. (1992) Developments in *Striga* control, *Tropical Agriculture Association UK Newsletter* **12** (2), 5-6.

Lange, M. and Hora, F.B. (1963) *Collins Guide to Mushrooms & Toadstools*, Collins, London.

Langer, R.H.M. and Hill, G.D. (1982) *Agricultural Plants,* Cambridge University Press, Cambridge.

Lapinskas, P. (1989) Commercial exploitation of alternative crops, with special reference to evening primrose, in G.E. Wickens, N. Haq and P. Day (eds.), *New Crops for Food and Industry*, Chapman and Hall, London, pp. 216-221.

Latie, M.A., Dasgupta, S.R. and De, B.C. (1987) Preservative treatment of bamboo and low cost housing, Bulletin: *Wood Preservation Series*. **3**, 1-6 (*Bamboo Abstracts*, **2** (1), no. 890135, 1989).

Launert, E. (1981) *Edible and Medicinal Plants of Britain and Northern Europe*, Country Life Books, London.

Lauremberg, P. (1631) *Horticultura,* M. Merianim Fracofurti ad Moenum.

Law, C.N. (1988) Recent developments in plant breeding methods, *Tropical Agriculture Association UK Newsletter* **8** (4), 8.

Lawless, J. (1992) *The Encyclopaedia of Essential oils. A Complete Guide to the Use of Aromatics in Aromatherapy, Herbalism, Health and Well-Being*, Element Books, Ltd., Shaftesbury, Hants.

Lawrence, B.M. (1995) The isolation of aromatic materials from natural plant products, in K. Tuley de Silva (ed.), *A Manual on the Essential Oil Industry*, UNIDO, Vienna.

Lawrence, G.H.M. (1951) *Taxonomy of Vascular Plants*, Macmillan, New York.

Lazarides, M. and Hince, B. (1993) *CSIRO Handbook of Economic Plants of Australia,* CSIRO Publications, Victoria.

Leaf, M.J. (1983) The green recolution and cultural change in a Punjab village (1965-1978), *Economic Development and Cultural Change* **31** (2), 227-270.

Leakey, R.R.B. and Last, F.T. (1980) Biology and potential of *Prosopis* species in arid environments, with particular reference to *P. cineraria*. *Journal of Arid Environments* **3**, 9-24.

Le Houérou, H.N. (1984) Rain use efficiency: a unifying concept in arid-land ecology, *Journal of Arid Environments* **7**, 212-247.

Le Houérou, H.N. (1986) The desert and arid zones of northern Africa, in M. Evenari, J. Noy-Meir and D.W. Goodall (eds.), *Hot deserts and arid shrublands,* Ecosystems of the World, vol. 12B, Elsevier, Amsterdam, pp. 117-141.

Le Houérou, H.N. (1996) Climate change, drought and desertification, *Journal of Arid Environments* **34**, 133-185.

Le Houérou, H.N., Bingham, R.L. and Skerbeck, W. (1988) Relationship between the variability of primary production and the variability of annual precipitation in world arid lands, *Journal of Arid Environments* **15**, 1-18.

Le Houérou, H.N., Popov, G.F. and See, L. (1993) *Agro-Bioclimatic Classification of Africa.* Agrometeorology
Series Working Paper No. 6, FAO, Rome.

Lemmens, R.H.M.J. and Wulijarni-Soetjipto, R.P. (eds.) (1991) *Dye and Tannin-Producing Plants. Plant Resources of South-East Asia No. 3*, Purdoc, Wageningen.

Lemmens, R.H.M.J., Wulijarni-Soetjipto, R.P., van der Zwan and Parren, M. (1991) Introduction, in R.H.N.J. Lemmens and R.P. Wulijarni-Soetjipto (eds.), *Dye and Tannin-Producing Plants. Plant Resources of South-East Asia No. 3*, Purdoc, Wageningen, pp. 15-34.

Léonard, J. and Compère, P. (1967) *Spirulina platensis* (Gom.) Geitl algue bleu des grande valeur alimentaire par sa richesse en protèines, *Bulletin du Jardin Botanique National de Belgique* 37 (1), Suppl. 1-23.

Levitt, J. (1980) *Responses of Plants to Environmental Stress*, 2nd edn., 2 vols., Academic Press, New York.

Levitt, J., Sullivan, C.Y. and Krull, E. (1960) Some problems in drought resistance, *Bulletin of the Research Council, Israel*, Sect. D **8**, 173-180.

Lewington, A. (1990) *Plants for People*, Natural History Museum, London and Royal Botanic Gardens, Kew.

Liener, I.E. (1980) Heat-labile antinutritional factors, in R.J. Summerfield and A.H. Bunting, (eds.), *Advances in Legume Science*, Royal Botanic Gardens, Kew, pp. 157-170.

Lin, J.H. and Panzer, R. (1994) Use of Chinese herbal medicine in veterinary science: history and perspectives, *Revue Scientifique et Technique de le Office International des Epizooties* **13**, 425-432.

Lindley, J. (1846) *The vegetable kingdom; or the structure, classification and uses of plants*, 1st edn., Bradbury and Evans, London.

Lindley, J. (1849) *Medicinal and Oeconomical Botany*, Bradbury and Evans, London.

Link, J.H.F. (1821-1822) *Enumeratio plantarum horti regii botanici berolinensis altera*, vol. 2., Berlin.

Linnaeus, C. (1737) *Flora lapponica*, S. Schouten, Amsterdam.

Linnaeus, C. (1745) *Flora svecica*, C. and G.J. Wishoff, Lugduni Batavorum.

Linnaeus, C. (1748) *Flora economica*, Uppsala.

Linnaeus, C. (1751) *Philosophia botanica*, Kiesewetter, Stockholm/Chatelain, Amsterdam.

Linnaeus, C. (1752) *Plantae esculentae patriae*, Uppsala.

Linnaeus, C. (1753) *Species plantarum*, Stockholm.

Linnaeus, C. (1763) *De Raphania*, Uppsala

Linskens, H.-F. and Jackson, J.F. (eds.) (1991a) *Essential Oils and Waxes*, Modern Methods of Plant Analysis, New Series vol. 12, Springer-Verlag, New York.

Linskens, H.-F. and Jackson, J.F. (1991b) Introduction, in H.F. Linskens and J.F. Jackson. (eds.), *Essential Oils and Waxes*. Modern Methods of Plant Analysis, New Series vol 12, Springer-Verlag, New York, pp. v-viii.

Linskens, H.-F. and Jorde, W. (1997) Pollen as food and medicine - a review, *Economic Botany* **51**, 78-87.

Lisansky, S.G. (1985) Microbial insecticides, in L.G. Copping and P. Rodgers (eds.), *Biotechnology and its Application to Agriculture. Monography No. 32*, British Crop Protection Council, Croydon, pp. 145-151.

Lister, J.H., Eugster, C.H. and Karrer, P. (1955) Cordeauxiachinon ein Blattfarbstoff aus *Cordeauxia edulis*, *Helvetica Chimica Acta* **38**, 212-222.

Lock, G.W. (1962) *Sisal*, Tropical Science Series, Longman, London.

Long, T.H. (etym. ed.) (1994) *Reader's Digest Universal Dictionary*, Reader's Digest Association, London.

Loscher, U.(1986) Variety descriptions according to Plant Breeder's Rights, in L.J.G. van der Maessen (ed.), *First International Symposium on Taxonomy of Cultivated Plants, Acta Horticultura* **182**, 59-62.

Losey, J.E., Rayor, L.S. and Carter, M.E. (1999) Transgenic pollen harms monarch larvae, *Nature* **399**, 214.

Low, D. (1847) *Elements of Practical Agriculure*, Longman, Brown, Green and Longmans, London and A. & C. Black, Edinburgh.

Lucas, A.T. (1960) *Furze. A Survey and History of its Uses in Ireland*, The Stationery Office, Dublin.

Lucas, G.Ll. (1971) The baobab project, in H. Merxmuller (ed.), Proceedings of the Seventh Plenary Meeting of the AETFAT, Mitteilungen der Botanischen Staatssammlung, München **10**, 162-164.

Lugt, C.B. (1987) Feasibility of growth and production of molluscicidal plants, in K.E. Mott (ed.), *Plant Molluscicides*, John Wiley & Sons, Chichester, pp. 231-244.

Lumpkin, T.A. and Plucknett, D.L. (1982) *Azolla as a Green Manure. Use and Management in Crop Production*, Westview Press, Boulder, CO.

Mabey, R. (1972) *Food for Free*, Collins, London.

Mabey, R. (ed.) (1988) *The Complete New Herbal*, Penguin Books, London.

Mabey, R. (1996) *Flora Britanica. The Definitive New Guide to Wild Flowers, Plants and Trees*, Sinclair-Stevenson, London.

Mabberley, D.J. (1973) Evolution in the giant groundsels, *Kew Bulletin* **28**, 61-96.

Mabberley, D.J. (1997) *The Plant Book. A Portable Dictionary of the Vascular Plants*, 2nd edn., Cambridge University Press, Cambridge.

Macksad, A., Schoental, R. and Coady, A. (1970) The hepatoxic action of a traditional Bedu plant remedy "ramram", *Journal of the Kuweit Medical Association* **4**, 297-299.

Maconochie, J.R. (1982) Regeneration of arid zone plants: a floristic survey, in W.R. Barker and P.J.M. Greenslade (eds.), *Evolution of the Flora and Fauna of Arid Australia*, Peacock Publications, Frewville, NSW, pp. 141-144.

Macdonald, J. (1995) *The Complete Book of Gardening Tips*, Carnell, London.

Macpherson, G. (ed.) (1995) *Black's Medical Dictionary*, 38th edn., A. & C. Black, London.

Maessen, L.J.G. van der (ed.) (1986) First International Symposium on Taxonomy of Cultivated Plants, *Acta Horticultura* **182**, 1-436.

MAFF (1971) *Bees for Fruit Pollination*, Advisory Leaflet 328, Ministry of Agriculture, Fisheries and Food (Publications), Pinner, Middlesex

MAFF (1994a) *Alternative Crops, New Markets*, MAFF, London.

MAFF (1994b) *The Plant Health Guide for Importers*, MAFF Publications, London.

MAFF (1994c) *Chemicals in Food. Managing the Risks.* Leaflet PB1695, MAFF, London.

MAFF (1996) *Annual Report of the Working Party on Pesticide Residues*, HMSO, London.

MAFF (1997a) *Guide to the Plant Varieties Act 1997*, Plant Variety Rights Office, Cambridge.

MAFF (1997b) *Genetic Modification and Food (revised)*, Food Safety Directorate, MAFF, London.

MAFF (1997c) *About Food Additives.* Leaflet PB0552 (revised), MAFF, London.

MAFF (1998a) *UK Plant Breeders' Rights Handbook*, Plant Variety Rights Office, Cambridge.

MAFF (1998b) *Guide to National Listings of Varieties of Agricultural and Vegetable Crops in the UK*, MAFF, London.

MAFF (1998c) *Understanding Radioactivity in Food.* Leaflet PB1212, MAFF, London.

MAFF (1998d) *Your Guide to Additives.* Leaflet PB0552/1 (revised), MAFF, London.

MAFF (1999) *Fact Sheet: Genetic Modification of Crops and Food*, Additives and Novel Foods Division, MAFF, London.

Maksoud, S.A., El Hadidi, M.N. and Amer, W.M. (1994) Beer from the early dynasties (3500-3400 cal B.C.) of Upper Egypt, detected by archaeochemical methods, *Vegetation History and Archaeobotany* **3**, 219-224.

Maley, J. (1972) La sédimentation pollinique actuelle dans la zone du lac Tchad (Afrique centrale), *Pollen et Spores* **14**, 263-307.

Malleshi, N.G. (1990) Processing of small millets for food and industrial use, in E. Seetharam, K.W. Riley and G. Harinarayana (eds.), *Small Millets in Global Agriculture. Proceedings of the First International Small Millets Workshop, Bangalore, India, October 29-November 2, 1986*, Aspect Publishing, London, pp. 325-339.

Manniche, L. (1989) *An Ancient Egyptian Herbal*, British Museum Publications, London.

Margulis, L. and Schwartz, K.V. (1998) *Five Kingdoms. An Illustrated Guide to the Phyta of Life on Earth*, 3rd edn., W.H. Freeman, New York.

Marticorena, C. and Quezada, M. (1985) Catálogo de la flora vascular de Chile, *Gayana Botanica* **42**, 1-2, 3-157.

Martin, L.C. (1983) *Wildflower Folklore*, Eastwood Press, Charlotte, NC.

Martinez-Meza, E. and Whitford, W.G. (1996) Stemflow, throughfall and channelization of stemflow by roots in three Chihuahuan Desert shrubs, *Journal of Arid Environments* **32**, 271-287.

Maslin, B.R. (1981) Should Acacia be divided? *Bulletin of the International Group for the Study of the Mimosoideae* **16**, 56-76.

Mason, H.L. (1950) Taxonomy, systematic botany and biosystematics *Madroño* **10**, 193-208.

Mastebroek, H.D., van Soest, L.J.M. and Siemonsma, J.S. (1996) *Chenopodium* L. (grain chenopod), in G.J.H. Grubben and S. Partohardjono (eds.), *Cereals. Plant Resources of South-East Asia No. 10*, Backhuys Publishers, Leiden, pp. 79-83.

Mathias-Mundy, E. and McCorkle, C.M. (1989) *Ethnoveterinary medicine: an annotated bibliography. Bibliographies in Technology and Social Change, No. 6*, Technology and Social Change Program, Iowa State University, Ames, IO.

Matthei, O. (1986) El genero *Bromus* L. (Poaceae) en Chile, *Gayana Botanica* **43** (1-4), 47-110.

Matthews, J.D. (1989) The changing picture of forest management, in M.S. Swaminathan and S.L. Kochhar (eds.), *Plants and Society*, Macmillan Publishers, London and Basingstoke, pp. 491-522.

May, L.W. (1978) The economic uses and associated folklore of ferns and fern allies, *Botanical Review* **44**, 491-528.

May, P., Anderson, A.B., Balick, M.j. and Frazão, J.M.F. (1985) Subsistence benefits from the babassu palm (*Orbignya martiana*), *Economic Botany* **39**, 113-129.

May, V. (1978) Areas of recurrence of toxic algae within Burrinjuck Dam, New South Wales, Australia. *Telopea*, **1** (5), 295-313.

Maydell, H.-J. von (1986) *Trees and Shrubs of the Sahel: their Characteristics and Uses*, GTZ, Eschborn.

Maydell, H.-J. von, Budowski, G., Le Houérou, H.N., Lundgren, B. and Steppler, H.A. (1982) What is agroforestry? *Agrofprestry Systems* **1** (1), 7-12.

McCorkle, C.M. and Mathias-Mundy, E. (1992) Ethnoveterinary medicine in Africa, *Journal of the International African Institute* **62**, 59-93.

McCorckle, C.M., Mathias, E. and T.W. Schillhorn van Veen, T.W. (eds.) *Ethnoveterinary Research and Development*, Intermediate Technology Publications, London.

McDougall, G.J., Morrison, I.M., Stewart, D., Weyers, J.D.B. and Hillman, J.R. (1993) Plant fibres: botany, chemistry and processing for industrial use, *Journal of the Science of Food and Agriculture* **62**, 1-20.

McHugh, D.J. (1987) Production, properties and uses of alginates, in D.J. McHugh (ed.), *Production and Utilization of Products from Commercial Seaweeds. Fisheries Technical Paper 288*, FAO, Rome, pp. 58-115.

McNeil, J. (1996) The BioCode: integrated biological nomenclature for the 21st century? in J.L. Reveal, J.L. (ed.), *Proceedings of a Mini-Symposium on Biological Nomenclature in the 21st Century*, University of Maryland, College Park, MD, pp. 1-10.

McWhirter, A. and Clasen, L. ((eds.) (1986) *Foods that Harm, Foods that Heal*, Reader's Digest Association, London.

Meadley, J. (1989) The commercial implications of new crops, in G.E. Wickens, N. Haq and P. Day (eds.), *New Crops for Food and Industry*, Chapman & Hall, London, pp. 23-28.

Meikle, R.D. (1985) *Flora of Cyprus*, vol. 2, Royal Botanic Gardens, Kew.

Meidner, H. and Sheriff, D.W. (1976) Seasonal dimorphism of foliage in Californian coastal sage scrub, *Oecologia* (Berlin) **51**, 385-388.

Mendelsohn, R. and Balick, M.J. (1995) The value of undiscovered pharmaceuticals in tropical forests, *Economic Botany* **49**, 223-228.

Mendelsohn, R. and Balick, M.J. (1997) Valuing undiscovered pharmaceuticals in tropical forests, *Economic Botany* **51**, 328.

Metcalfe, C.R. (1983) Ecological anatomy and morphology general survey, in C.R. Metcalfe and L. Chalk (eds.), *Anatomy of the Dicotyledons,* 2nd edn., vol. 2., Claredon Press, Oxford, pp. 127-152.

Melville, R. (1947) The nutritive value of nuts, *Chemistry and Industry* **22**, 304-306.

Meng, I.D., Menning, B.H., Martin, W.J. and Fields, H.L. (1998) An analgesia circuit activated by cannabinoids, *Nature* **395**, 381-383.

Menninger, E.A. (1977) *Edible Nuts of the World*, Horticultural Books, Stuart, FL.

Miller, D.L. (1975) Annual crops: a renewable source for cellulose, *Applied Polymer Symposia, No. 28*, New York, pp. 21-28.

Miller, W.C. and West, G.P. (1956) *Black's Veterinary Dictionary*, 4th edn., A. & C. Black, London.

Minson, D.J. (1988) The chemical composition and nutritive value of tropical legumes, in P.J. Skerman, D.G. Cameron and F. Riveros (eds.), *Tropical Forage Legumes*, 2nd edn., FAO Plant Production and Protection Series no. 2, FAO, Rome, pp. 185-193.

Minson, D.J. (1990) The chemical composition and nutritive value of tropical grasses, in P.J.: Skerman and F. Riveros (eds.), *Tropical Grasses*, 2nd edn. *FAO Plant Production and Protection Series no. 23*, FAO, Rome, pp. 163-180.

Mirov, N.T. (1954) Composition of turpentines of Mexican pines, *Unasylva* **8**, 167-173.

Mitchell, M.E. and Guiry, M.D. (1983) Carrageen: a local habitation or a name? *Journal of Ethnopharmacology* **9**, 347-351.

Moldenke, H.N. and Moldenke, A.L. (1952) *Plants of the Bible, Chronica Botanica* **28**, Ronald Press, New York.

Molisch, H. (1937) *Der Einfluss einer Pflanze auf die andere-allelopathie*, Fischer-Verlag, Jena.

Molyneux, R.J. and Ralphs, M.H. (1992) Plant toxins and palatability to herbivores, *Journal of Range Management* **45**, 13-18.

Monk, L.S., Fagerstedt, K. V. and Crawford, R.M.M. (1989) Oxygen toxicity and superoxide dismutase as an antioxidant in physiological stress, *Physiologia Plantarum* **76**, 456-459.

Monod, T. (1970) Sur des endocarpes de *Celtis* du gisement néolithique d'Amekni (Ahaggar), *Bulletin de l'Institut Français d'Afrique Noire* **32**, sér. A, 585-593.

Montagné, P. (1977) *New Larousse Gastronnomique. The World's Greatest Cookery Reference Book*, Book Club Associates, London.

Montalembert, M.R. de and Clément, J. (1983) *Fuelwood supplies in the developing countries*, FAO Forestry Paper 42, FAO, Rome.

Moore, P.D. (1998) Getting to the roots of tubers, *Nature*, **395**, 330-331.

Moore, R.M. (1993) Grasslands of Australia, in R.T. Coupland (ed.), *Natural Grasslands. Eastern Hemisphere and Résumé. Ecosystems of the World 8B*, Elsevier, Amsterdam, pp. 315-360.

Mori, S.A. and Prance, G.T. (1990) *Flora Neotropica: Lecythidaceae*, Part 2, New York Botanical Garden, New York.

Morris, C.F. and Rose, S.P. (1996) Wheat, in R.J. Henry and P.S. Kettlewell (eds.), *Cereal Grain Quality*, Chapman & Hall, London, pp. 3-54.

Morris, D. and Ahmed, I. (1992) *The Carbohydrate Economy*, Institute for Local Self-Reliance. Washington, DC.

Morris, J.W. and Manders, R. (1981) Information available with the PRECIS data bank of the National Herbarium, Pretoria, with examples of uses to which it may be put, *Bothalia* **13** (3-4), 473-485.

Morton, J.F. and Voss, G.L. (1987) The argan tree (*Argania sideroxylon*, Sapotaceae), a desert source of edible oil, *Economic Botany* **42**, 221-233.

Motley, T.J. (1994) The ethnobotany of sweet flag, *Acorus calamus* (Araceae), *Journal of Economic Botany* **48**, 397-412.

Mott, K.E. (ed.) (1987) *Plant Molluscicides*, John Wiley & Sons, Chichester.

Mouret, M. (1985) Gummosis of acacias: current histological research, *Bulletin of the International Group for the Study of the Mimosoideae* **13**, 38-45.

Muchaili, J.H. (1989) Management problems in the introduction of new crops: the Zambian experience, in G.E.. Wickens, N. Haq and P. Day (eds.), *New Crops for Food and Industry*, Chapman & Hall, London, pp. 29-35.

Muller, C.H. and Chou, C.-H. (1972) Phytotoxins: an ecological phase of phytochemistry, in J.B. Harborne (ed.), *Phytochemical Ecology,* Academic Press, London, pp. 201-216.

Müller [Argoviensis], J.L. (1864-69) Buxaceae, in A. De Candolle (ed.), *Prodromus Systematis Naturalis Regni Vegetabilis* **16** (1), V. Masson et filli, Paris, p.9.

Multon, J.L. (1991) Basics of moisture measurement in grain, in L.D. Hill (ed.), *Uniformity by 2000. An International Workshop of Corn and Soybean Quality*, Scherer Communications, Urbana, IL.

Muñoz-Pizarro, C. (1944) Sobre la localidad-tipo de *Bromus mango* Desv., *Agricultura Técnica, Chile* **4** (1), 98-101.

Muñoz-Pizarro, C. (1948) Cinco especies nuevas de plantas para Chile, *Agricultura Técnica, Chile* **8** (2), 83-85.

Mueller, F.J.H. von (1885) *Select Plants Readily Eligible for Industrial Use in Extra-Tropical Countries*, Government Printer, Melbourne.

Nabhan, G.P. (1985a) Native crop diversity in Aridoamerica: conservation of regional gene pools, *Economic Botany* **39**, 387-399.

Nabhan, G.P. (1985b) *Gathering the Desert*, University of Arizona Press, Tucson.

Nabhan, G.P. and Sheridan, T.E. (1977) Living hedgerows of the Rio San Miguel, Sonora, Mexico: traditional technology for floodplain management, *Human Ecology* **5**, 97-111.

Nandi, O.I., Chase, M.W. & Endress, P.K. (1998) A combined cladistic analyses of Angiosperms using *rbc*L and non-molecular data sets, *Annals of the Missouri Botanical Garden* **85**, 132-212.

National Academy of Sciences (1980) *Firewood Crops*, National Academy of Sciences, Washington, DC.

National Research Council (1975) *Underexploited Tropical Plants with Promising Economic Value*, National Academy of Sciences, Washington, DC.

National Research Council (1989) *Lost Crops of the Incas: Little-known Plants of the Andes with Promise for Worldwide Cultivation*, National Academy Press, Washington, DC.

National Research Council (1992) *Neem. A Tree for Solving Global Problems*, National Academy Press, Washington, DC.

National Research Council (1996) *Lost Crops of Africa, Volume 1: Grains*, National Academy Press, Washington, DC.

Needham, J. (1986) *Science and Civilisation in China*, Vol. 6, *Biology and Biological Technology*, Part I: Botany, Cambridge University Press, Cambridge.

Nerd, A., Irijimovich, V. and Mizrahi, Y. (1998) Phemology, breeding system and fruit development of argan (*Argania spinosa*, Sapotaceae) cultivated in Israel, *Economic Botany* **52**, 161-167.

Noble, J.C. (1982) The significance of fire in the biology and evolutionary ecology of mallee *Eucalyptus* populations, in W.R. Barker and P.J.M. Greenslade (eds.), *Evolution of the Flora and Fauna of Arid Australia*, Peacock Publications, Frewville, NSW, pp. 153-159.

Noble, P.S. (1991a) Achievable productivity of certain CAM plants; basis for high values compared with C_3 and C_4 plants, *New Phytologist* **119**, 183-205.

Noble, P.S. (1991b) *Physiochemical and Environmental Plant Physiology*, Academic Press, London.

Noble, P.S. (1996) Responses of some North American CAM plants to freezing temperatures and doubled CO_2 concentrations: implications of global climatic change for extending cultivation, *Journal of Arid Environments* **34**, 187-196.

Noggle, G.R. and Fritz, G.J. (1983) *Introductory Plant Physiology*, 2nd edn., Prentice-Hall, Englewood Cliffs, NJ.

Nord, E.C. and Countryman, C.M. (1972) Fire protection, in C.M. McKell, J.P. Blaisdell and J.R. Gooding (eds.) *Wildland shrubs - their iology and utilization*, USDA Forest Service Federal Technical Report INT-1, 1972, Ogden, UT, pp. 88-97.

Nowacki, E. (1980) Heat-stable antinutritional factors in leguminous plants, in R.J. Summerfield and A.H. Bunting (eds.), *Advances in Legume Science*, Royal Botanic Gardens, Kew, pp. 171-177.

Nuttall, T. (1844) On *Simmondsia*, a new genus of plants from California, *Hooker's London Journal of Botany* **3**, 400-401.

Obeid Mubarak, M., Bari, E.A., Wickens, G.E. and Williams, M.A.J. (1982) The vegetation of the central Sudan, in M.A.J. Williams and D.A. Adamson (eds.), *A Land between Two Niles*, Balkema, Rotterdam, pp. 143-164.

O'Brien, J.P. (1997) Pulp fibres, *Tropical Agriculture Association UK Newsletter* **17** (1), 8-10.

Office of Technology Assessment (1983) *Plants. The Potentials for Extracting Protein, Medicines, and other Useful Chemicals. Workshop Proceeding.* Office of Technology Assessment, Congress of the United States, Washington, DC.

Ogle, B.M. and Grivetti, L.E. (1985) Legacy of the chameleon: Edible wild plants in the Kingdom of Swaziland, Southern Africa: A cultural, ecological, nutritional study, *Ecology of Food and Nutrition* **16** (3), 193-208; **17** (1) 1-64.

Ohnishi, O. (1998) Search for the wild ancestors of buckwheat III. The wild ancestor of cultivated common buckwheat and of tertiary buckwheat, *Economic Botany* **52**, 123-133.

Okafor, J.C. (1973) Development of forest tree crops for food supplies in Nigeria, *Forest Ecology and Management* **1**, 235-247.

Okafor, J.C. (1980a) Edible indigenous woody plants in the rural economy of the Nigerian forest zone, *Forest Ecololgy and Management* **3**, 45-55.

Okafor, J.C. (1980b) Trees for food and fodder on the savanna areas of Nigeria, *International Tree Crops Journal* **1**, 131-141

Okigbo, B.N. (1989) New crops for food and industry: the roots and tubers in tropical Africa, in G.E. Wickens, N. Haq, P. Day (eds.), *New Crops for Food and Industry*, Chapman & Hall, London, pp. 123-134.

Oliver, B. (1960) *Medicinal Plants in Nigeria*, Nigerian College of Arts, Science and Technology, Ibadan.

Oliver-Bever, B. (1986) *Medicinal Plants in Tropical West Africa*, Cambridge University Press, Cambridge.

Olofsdotter, M., Navarez, P. and Rebulanan, M. (1997) Rice allelopathy - where are we and how far can we get? in *1957 Brighton Crop Protection conference - Weeds*. Vol. 1, British Crop Protection Council, Farnham, pp. 99-104.

Oppenheimer, H.R. (1960) Adaptations to drought: xerophytism, in *Plant-Water Relationships in Arid and Semi-Arid Conditions*. Reviews of Research. Arid Zone Research 15, UNESCO, Paris, pp. 105-138.

Ott, J. (1998) The Delphic Bees: bees and toxic honeys as pointers to psychoactive and other medicinal plants, *Economic Botany* **52**, 260-266.

Oubré, A.Y., Carlson, T.J., King, S.R. and Reaven, G.M. (1997) From plant to patient: an ethnomedical approach to the identification of new drugs for the treatment on NIDDM, *Diabetologia* **40**, 614-617.

Owadally, A.W. (1979) The dodo and the tambalacoque tree, *Science* **203**, 1363-1264.

Oyen, L.P.A. and Andrews, D.J. (1996) *Pennisetum glaucum* (L.) R.Br., in G.J.H. Grubben, and S. Partohardjono (eds.), *Cereals. Plant Resources of South-East Asia No. 10*, Backhuys Publishers, Leiden, pp. 119-123.

Ozenda, P. (1977) *Flora du Sahara*. 2nd edn, CNRS, Paris.

Palmer, E. and Pitman, N. (1972) *Trees of Southern Africa covering all known indigenous species in the Republic of South Africa, South-West Africa, Botswana, Lesotho and Swaziland*, vol. 1, A.A. Balkema, Cape Town.

Parker, J. (1968) Drought-resistance mechanisms, in T.T. Kozlowski (ed.), *Water Deficits and Plant Growth*, vol. 1, Academic Press, New York, pp. 195-234.

Parodi, L.R. (1950) Las gramineas tóxicas para el ganado en le Repúblico Argentina, *Revista Argentina de Agronomía* **17** (3) 163-229.

Parry, M. (1990) *Climate Change and World Agriculture*, Earthscan Publications, London.

Passmore, R., Nicol, B.M. and Rao, M.N. (1974) *Handbook on Human Nutritional Requirements*. FAO Food and Nutrition Series No. 4, FAO Nutritional Studies No. 8, and WHO Monograph Series No. 61, FAO, Rome.

Pathak, M.D. and Saxena, R.C. (1979) Insect resistance, in J. Sneep and A.J.T. Hendricksen, (eds.), *Plant Breeding Perspectives*, Centre for Agricultural Publishing and Documentation, Wageningen, pp. 270-285.

Pauly, G., Yani, A., Piovetti, L. and Bernard-Dagan, C. (1983). Volatile constituents of the leaves of *Cupressus dupreziana* and *Cupressus sempervirens*, *Phytochemistry* **22**, 957-959.

Pauw, C.A. and Linder, H.P. (1997) Tropical African cedars *Widdringtonia*, Cupressaceae): systematics, ecology and conservation status, *Botanical Journal of the Linnean Society* **123**, 297-319.

Pedley, L. (1981) Classification of acacias, *Bulletin of the International Group for the Study of the Mimosoideae* **9**, 42-48.

Pegler, D. (1990) *Field Guide to the Mushrooms and Toadstools of Britain and Europe*. Kingfisher Books, London.

Pelletier, S.W. (1983) The nature and definition of an alkaloid, in S.W. Pelletier (ed.), *Alkaloids: Chemical and Biological Perspectives*, John Wiley & Sons, New York, pp. 1-31.

Perdue, R.E., Carlson, K.D. and Gilbert, M.G. (1986) *Vernonia galamensis*, potential new crop source of epoxy acid, *Economic Botany* **40**, 54-68.

Perdue, R.E., Jones, E. and Nyati, C.T. (1989) *Vernonia galamensis*: a promising new industrial crop for the semi-arid tropics and subtropics, in G.E. Wickens, J.R. Goodin and D.V. Field (eds.), *Plants for Arid Lands*, Chapman & Hall, London, pp. 197-207

Perez-Llano, G.A. (1944) Lichens: their biological and economic significance, *Botanical Review* **10** (1), 1-65.

Perry, F. (1961) *Water Gardening*, Country Life, London.

Pessarakli, M. (ed.) (1993) *Handbook of Plant and Crop Stress*, Marcel Dekker, New York.

Petch, T. (1930) Buttress roots, *Annals of the Royal Botanic Gardens, Peradeniya* **11**, 277-285.

Petterson, D.S., Harris, D.J. and Allen, D.G. (1991) Alkaloids, in J.P.F. D'Mello, C.M. Duffus and J.H. Duffus (eds.), *Toxic Substances in Crop Plants*, Royal Society of Chemistry, Cambridge, pp. 148-179.

Philips, C. (1993) Trees for fire protection, in D. Race (ed.), *Agroforestry. Trees for Productive Farming*, Agmedia, Victoria, pp. 223-238.

Philips, E. (1678) *The New World of English Words*, 4th edn., London.

Pietroni, A. (1999) Gathered wild food plants in the upper valley of the Serchio River (Garfagnana), Central Italy, *Economic Botany* **53**, 327-341.

Pietroni, P.C. (ed.) (1994) *Reader's Digest Family Guide to Alternative Medicine*, Reader's Digest Association, London.

Pickersgill, B. (1999) Crop introductions and the development of secondary areas of diversity, in H.D.V. Prendergast, N.I. Etkin, D.R. Harris and P.J. Houghton (eds.) *Plants for Food and Medicine*, Royal Botanic Gardens, Kew, pp. 93-105.

Pincock, S. (1997) Spud gun targets disease, *New Scientist* **154** (2092), 18.

Plucknett, D.L., Smith, N.J.H., Williams, J.T. and Anishetty, N.M. (1987) *Gene Banks and the World's Food*, Princetown University Press, Princetown, N.J.

Polhill, R.M. (1968) Miscellaneous notes on African species of *Crotalaria* L. : II., *Kew Bulletin* **22**, 169-348.

Polhill, R.M. (1981) Tribe 29: *Crotalarieae* (Benth.)Hutch., in R.M. Polhill and P. Raven. (eds.), *Advances in Legume Systematics Part 1*, Royal Botanic Gardens, Kew, pp. 399-402.

Polhill, R.M. and Wiens, D. (1998) *Mistletoes of Africa*, Royal Botanic Gardens, Kew.

Poljakoff-Mayber, A. and Lerner, H.R. (1993) Plants in saline environments, in M. Pessarakli (ed.), *Handbook of Plant and Crop Stress*, Marcel Dekker, New York, pp. 65-96.

Pollak, G. and Waisel, Y. (1970) Salt secretion in *Aleuropus litoralis* (Willd.) Parl., *Annals of Botany* **34**, 879-888.

Posey, D.A. (1984) A preliminary report on diversified management of tropical forest by the Kayapó indians of the Brazilian Amazon, in G.T. Prance and J.A. Kallunki (eds.), *Ethnobotany in the Neotropics*.

Proceedings of a Symposium at Oxford, Ohio. Advances in Economic Botany, vol.1, New York Botanic Garden, Bronx, NY, pp. 112-126.

Postgate, J. (1992) *Microbes and Man*, 3rd edn., Cambridge University Press, Cambridge.

Poulter, N. (1988) Tropical root crops in African: traditional processing and improvement, *Tropical Agriculture Association UK Newsletter* **8** (2), 1-6.

Powers, S. (1875) Aboriginal botany, *Proceedings of the California Academy of Sciences* **5**, 373-379.

Prance, G.T. (1984) The use of edible fungi by Amazonian Indians, in G.T.Prance and J.A. Kallunki (eds.), *Ethnobotany in the Neotropics. Proceedings of a Symposium at Oxford, Ohio. Advances in Economic Botany, vol. 1*, New York Botanical Garden, Bronx, pp. 127-139.

Prance, G.T. (1991) What is ethnobotany today? *Journal of Ethnopharmacology* **32**, 209-216.

Prance, G.T. and Mori, S.A. (1979) *Lecythidaceae Part 1. Flora Neotropica Monograph* **21**, New York Botanical Garden, New York.

Prescott-Allen, C. and Prescott-Allen, R. (1986*) The First Resource. Wild Species in the North American Economy*, Yale University Press, New Haven, CT.

Prescott-Allen, R. and Prescotte-Allen, C. (1983) *Genes from the Wild. Using Wild Genetic Resources for Food and Raw Materials*, Earthscan Publications, London.

Prescott-Allen, R. and Prescott-Allen, C. (1990) How many plants feed the world? *Conservation Biology* **4**, 365-374.

Price, R.S. (1911) The roots of some North African desert grasses, *New Phytologist* **10**, 328-339.

Princen, L.H. (1983) New oilseed crops on the horizon, *Economic Botany*, **37** 478-492.

Purseglove, J.W. (1985) *Tropical Crops. Monocotyledons*, Revised and Updated, Longman Group Ltd., Harlow, Essex.

Purseglove, J.W. (1987) *Tropical Crops. Dicotyledons*, Longmans Scientific and Technical, Harlow, Essex.

Rambelli, R. (1985) *Manual on Mushroom Cultivation*, FAO Plant Production and Protection Paper 43. FAO, Rome.

Ramsbottom, J. (1949) *A Handbook of the Larger British Fungi*, British Museum (Natural History), London.

Ramsbottom, J. (1960) *Mushrooms and Toadstools. A Study of the Activities of Fungi*, The New Naturalist Series. Collins, London.

Rasaonaivo, P. (1990) Rain forests of Madagascar: sources of industrial and medicinal plants, *Ambio* **19**, 421-423.

Rauh, W. (1985) The Peruvian-Chilean deserts, in M. Evenari, J. Noy-Meir and D.W. Goodall (eds.), *Hot Deserts and Arid Shrublands. Ecosystems of the World, vol. 12A*, Elsevier, Amsterdam, pp. 239-267.

Ranklior, P.R. (1986) *International Directory of Geotextiles and Related Products*, Manstock Geotechnical Consultancy Services, Manchester.

Raynal, J. (1973) Notes cypérologiques. Contribution no. 19: la classification de la sous-famille des Cyperoideae, *Adansonia* Sér. 2 **13**, 145-171.

Read, B.E. and Yü-thien, L. (1933) *Chinese Medicinal Plants from the 'Pên Tshao Kang Mu' A.D. 1596. A Botanical Chronical and Pharmacological Reference List*, Publication of the Peking Natural History Bulletin, Peking [Beijing].

Rees, A.R. (1972) *The Growth of Bulbs: Applied Aspects of the Physiology of Ornamental Bulbous Crop Plants*, Academic Press, London, New York.

Reis, Siri von and Lipp F.J. Jnr. (1982) *New Plant Sources for Drugs and Foods from the New York Botanical Garden Herbarium*, Harvard University Press, Cambridge, MA

Renfrew, J.M. (1973) *Palaeoethnobotany. The Prehistoric Food Plants of the Near East and Europe*, Colombia University Press, London.

Renvoize, S.A., Cope, T.A., Cook, F.E.M., Clayton, W.D. and Wickens, G.E. (1992) Distribution and utilization of grasses of arid and semi-arid regions, in G.P. Chapman (ed.), *Desertified Grasslands: Their Biology and Management*, Academic Press, London, pp. 3-16, 323-332.

Reveal, J.l. and Hoogland, R.D. (1990) Simmondsiaceae (Muell. Arg.) Validation of five family names in the Magnoliophyta, *Bulletin du Museum Naturelle, section B, Adansonia* **2**, 205-208.

Rexen, F. and Munck, L. (1984) *Cereal Crops for Industrial Use in Europe. Report Prepared for: The Commission of the European Communities*, EUR 9617 EN, Carlsberg Research Center, Copenhagen.

Rice, E.L. (1975) Allelopathy - an update, *The Botanical Revue* **45**, 15-109.

Rice, E.L. (1984) *Allelopathy*, Academic Press, New York.

Rice, E.L. (1992) Allelopathic effects on nitrogen cycling, in S.J.H. Rizvi, S.J.H. and V. Rizvi, (eds.), *Allelopathy: Basic and Applied Aspects*, Chapman & Hall, London, pp. 31-58.

Richards, P.W. (1957) *The Tropical Rain Forest. An Ecological Study,* Cambridge University Press, Cambridge.

Riddle, J.R. (1985) *Dioscorides on Pharmacy and Medicine*, University of Texas Press, Austin.

Ridley, M. (1973) *Treasures of China*, The Dolphin Press, Christchurch, Hampshire.

Riley, W.R. and Brokensa, D. (1988) *The Mbeere in Kenya*, 2 vols., University Press of America, Lanham, MD.

Risi, J. and Galwey, N.W. (1989) Chenopodium grains of the Andes: a crop for temperate lands, in G.E. Wickens, N. Haq and P. Day (eds.) *New Crops for Food and Industry*, Chapman & Hall, London, pp. 222-234.

Rivett, D.E., Jones, G.P. and Tucker, D.J. (1989) *Santalum accuminatum* fruit: a prospect for horticultural development, in G.E. Wickens N. Haq and P. Day (eds.), *New Crops for Food and Industry,* Chapman & Hall, London, pp. 208-215.

Rivinus, A.Q. (1690) *Introductio generalis in rem harbariam*, G. Guntheri, Leipzig.

Rizvi, S.J.H. and Rizvi, V. (1992) Exploitation of allelochemicals in improving crop productivity, in S.J.H. Rizvi and V. Rizvi (eds.), *Allelopathy: Basic and Applied Aspects*, Chapman & Hall, London, pp. 459-469.

Rizvi, S.J.H., Haque, H., Singh, V.K. and Rizvi, V. (1992) A description called allelopathy, in S.J.H. Rizvi and V. Rizvi (eds.), *Allelopathy Basic and Applied Aspects*, Chapman & Hall, London, pp. 1-10.

Robbins, P. (1995) *Tropical Commodities and their Markets. A Guide and Directory*, Kogan Page, London.

Robertson, P. (1994) *The New Shell Book of Firsts*, Headline Book Publishing, London.

Robinson, D.H. (1947) *Leguminous Forage Plants*, 2nd edn., Edward Arnold & Co., London.

Rochebrune, A.T. de (1879) Recherches d'ethnographie botanique sur la flore des sépultures péruviennes d'Ancon, *Actes de Société linnéenne de Bordeaux* **33**, 343-358.

Rohde, E.S. (1922) *The Old English Herbals*, Longmans, Green & Co., London.

Rooney, L.W. (1996) Sorghum and millets, in R.J. Henry and P.S. Kettlewell (eds.), *Cereal Grain Quality,* Chapman & Hall, London, pp. 153-177.

Rose, H.J. and Dietrich, B.C. (1996) Plants, sacred, in S. Hornblower and A. Spawforth (eds.), *The Oxford Classical Dictionary*, 3rd edn., Oxford University Press, Oxford and New York, p. 1189.

Rosengarten, F. Jr (1984) *The Book of Edible Nuts*, Walker and Company, New York.

Rosillo-Calle, F., Rezende, M.A.A. de, Furtado, P. and Hall, D.O. (1996) *The Charcoal Dilema. Finding Sustainable Solution for Brazilian industry*, Intermediate Technology Publications, London.

Rotherham, E.R., Blaxell, D.F., Briggs, B.G. and Carolin, R.C. (eds.) (1975) *Flowers and Plants of New South Wales and Southern Queensland*, A.H. & A.W. Reed, Sydney.

Roughan, P.G. (1989) *Spirulina*: a source of dietary gamma linolenic acid? *Journal of the Science of Food and Agriculture* **47**, 85-93.

Royon, R. (1986) Cultivated variety denominations and trade marks, in L.J.G. van der Maessen (ed.), *First International Symposium on Taxonomy of Cultivated Plants*, Acta Horticultura **182**, 273-275.

Ruales, J. (1998) Quinoa (*Chenopodium quinoa* Willd.): nutritive quality and technical aspects as human food, in P.S. Belton and J.R.N. Taylor (eds.), *Increasing the Utilisation of Sorghum, Buckwheat, Grain Amaranth and Quinoa for Improved Nutrition. Selected Papers from the International Association for Cereal Science and Technology Cereals Conference Symposium "Challenges in Speciality Crops" Held* at the 16th ICC Cereals Conference, Vienna, Austria 9 May 1998, Institute of Food Research, Norwich, pp. 49-64.

Ruddle, K. (1973) The human use of insects: examples from the Yupa, *Biotropica* **5** (2), 94-101.

Rudgley, R. (1998) *Lost Civilisations of the Stone Age*, Century, London.

Rumphius, G.E. (1741-1755) *Herbarium amboinense*. 7 vols., Amsterdam.

Rundell, P.W. (1982) Water uptake by organs other than roots, in O.L. Lange, P.S. Noble, C.B. Osmond and H.Zeigler (eds.), *Physiological Plant Ecology. II Water Relations and Carbon Assimilation*, Springer-Verlag, Berlin, Heidelberg, New York, pp.111-134.

Russell, E. J. (1950) *Soil Conditions and Plant Growth*, 8th edn. (Recast and rewritten by E.W. Russell), Longmans, Green & Co., London.

Russell, E.W. (1959) *The World of the Soil*, Collins, London.

Saint-Pierre, C. and Ou, B. (1994) Lac host-trees and the balance of agroecosystems in South Yunnan, China, *Economic Botany* **48**, 21-28.

Salaman, R.N. (1949) *The History and Social Influence of the Potato*, Cambridge University Press, Cambridge.

Salisbury, E. (1961) *Weeds & Aliens*, Collins, London.

Santelices, B. (1987) The wild harvest and culture of the economically important species of *Gelidium* in Chile, in M.S. Doty, J.F. Caddy and B. Santelices (eds.), *Case Studies of Some Commercial Seaweed Resources*. Fisheries Technical Paper No. 281, FAO, Rome, pp. 109, 205.

Saunders, C.F. (1976) *Edible and Useful Plants of the United States and Canada*, Dover Publications, New York (formerly published in 1934 as *Useful Plants of the United States and Canada*, R.M. McBride & Co., New York).

Saunders, J.H. (1961) *The Wild Species of Gossypium and their Evolutionary History*, Oxford University Press, London.

Scarborough, J. (1996) Pharmacology, in S. Hornblower and A. Spawforth (eds.), *The Oxford Classical Dictionary*, 3rd edn., Oxford University Press, Oxford and New York, pp. 1154-1155.

Schillhorn van Veen, T.W. (1996) Sense or nonsense? Traditional methods of animal disease prevention and control in the African savannah, in C.M. McCorckle, E. Mathias and T.W. Schillhorn van Veen (eds.), *Ethnoveterinary Research and Development*, Intermediate Technology Publications, London, pp. 25-36.

Schimper, A.F.W. (1898) *Pflanzen-Geograaphie auf physiologischer Grundlage*, Jena (translated by Fisher, W.R. (1903), revised and edited by Groom, P. and Balfour, I.B. (1958) as *Plant-Geography upon a Physiological Basis*, Clarendon Press, Oxford).

Schmidt, J. and Stavisky, N. (1983) Uses of *Thymelaea hirsuta* (mitnan) with emphasis on hand papermaking, *Economic Botany* **37**, 310-321.

Schneider, C.K. (1907) *Illustriertes Handbuch des Laubholzkunde*, vol. 2, Gustav Fischer, Jena.

Schramm, W. (1991) Cultivation of unattached seaweeds, in M.D. Guiry and G. Blunden,. (eds.), *Seaweed Resources in Europe: Uses and Potentia*, John Wiley & Sons, Chichester, pp. 379-408.

Schultes, R.E. (1960) Tapping our heritage of ethnobotanical lore, *Economic Botany* **14**, 257-252.

Schultes, R.E. and Hofmann, A. (1992) *Plants of the Gods: Their Sacred, Healing and Halucinogenic Powers*, Healing Arts Press, Rochester, VE.

Schwabe, C.W. (1996) Ancient and modern veterinary beliefs, practices and practitioners among Nile Valley peoples, in C.M. McCorckle, E. Mathias and T.W. Schillhorn van Veen (eds.) *Ethnoveterinary Research and Development*, Intermediate Technology Publications, London, pp.37-45.

Schuiling, D.L. and Jong, F.S. (1996) *Metroxylon sagu* Rottboell, in M. Flach and F. Rumawas (eds.), *Plants Yielding Non-Seed Carbohydrates. Plant Resources of South-East Asia No. 9*, Backhys Publishers, Leiden, pp. 121-126.

Schultes, R.E. (1980) The plant kingdom as a source of new medicines, in L.H. Chesson (ed.), *The 1980 Longwood Program Seminars Volume*, University of Delaware, Newark, pp. 2-8.

Scott, S.E. and Wilkinson, M.J. (1999) Low probability of chloroplast movement from oilseed rape (*Brassica napus*) into wild *Brassica rapa*, *Nature Biotechnology* **17**, 390-392.

Seaward, M.R.D. and Williams, D. (1976) An interpretation of mosses found in recent archaeological excavations, *Journal of Archaeological Science* **3**, 173-177.

Seetharam, E., Riley, K.W. and Harinarayana, G. (eds.) (1990) *Small Millets in Global Agriculture. Proceedings of the First International Small Millets Workshop, Bangalore, India, October 29-November 2, 1986*, Aspect Publishing, London.

Sen, D,N. and Mohammed, S. (1993) General aspects of salinity and the biology of saline plants, in M. Pessaraki (ed.), *Handbook of Plant and Crop Stress*, Marcel Dekker, New York, pp. 125-145.

Shaltout (1992) Dimension analysis of *Thymaea hirsuta* (L.) Endl. fibres, *Feddes Repertorium* **103** (3-4), 93-106.

Sharma, S.D. and Shastry, S.V.S. (1965) Taxonomic studies in genus *Oryza* L. III *O. rufipogon* Griff. *sensu stricto* and *O. nivara* Sharma et Shastry *nom. nov.*, *Indian Journal of Genetics and Plant Breeding* **25**, 157-167.

Sharp, D.W.A.(ed.) (1990) *The Penguin Dictionary of Chemistry*, 2nd edn., Penguin Books Ltd., London.

Shay, E.G. (1993) Diesel fuel from vegetable oils: status and opportunities. *Biomass and Bioenergy* **4** (4), 227-242.

Sheldon, J.W., Balick, M.J. and Laird, S. (1997) *Medicinal Plants: Can Utilization and Conservation Coexist?* New York Botanical Garden, Bronx.

Shery, R.W. (1972) *Plants for Man*, 2nd edn., Prentice-Hall, Englewood Cliffs, NJ.

Shinohara, K., Zhao, Y. and Sato, G.H. (1989) Food production by selective algal biomass, in G.E. Wickens, N. Haq and P. Day (eds.), *New Crops for Food and Industry*, Chapman & Hall, London, pp. 333-340.

Simmonds, N.W. (1959) *Bananas*, Longmans, Green & Co., London.

Simmonds, N.W. (1962) *The Evoluition of the Bananas*, Longman, London.

Simmonds, N.W. (1979) *Principles of Plant Improvement*, Longman, Harlow, Essex.

Simmonds, N.W. (1996) Profits, projects and plant breeding, *Tropical Agriculture Association UK Newsletter* **16** (1), 31-32.

Simmonds, N.W. (1997) Breeding perennial crops, *Tropical Agriculture Association UK Newsletter* **17**, 4, 27-29.

Simmonds, N.W. and Shepherd, K. (1955) The taxonomy and origins of the cultivated banana, *Journal Linnean Society of Botany* **55**, 302-312.

Simon, J. (1994) *Discovering Wine*, Reed International Books Ltd., London.

Simpson, J.A. and Weiner, E.S.C. (1989) *Oxford English Dictionary*, 2nd edn., Clarendon Press, Oxford.

Sims, A. (198-) Has desert shrub a link with cancer? *Kkaleej Times Magazine*, pp. 14-16.

Singer, C. (1927) The herbal in antiquity, *Journal of Hellenic Studies* **47**, 1-52.

Siopongco, J.O. and Munandar, E.M. (1987) Advantages and disadvantages of bamboo for construction, in *Technology Manual on Bamboo as Building Material*, Regional Network in Asia for Low-Cost Building Materials Technologies and Construction Systems, Manila, Philippines, pp. 24-25 (*Bamboo Abstracts*, **2** (1), no. 890122, 1989).

Sjkåk-Bræk, G. and Martinsen, A. (1991) Applications of some algal polysaccharides in biotechnology, in M.D. Guiry and G. Blunden (eds.), *Seaweed Resources in Europe: Uses and Potentia*, John Wiley & Sons, Chichester, pp. 219-257.

Skerman, P.J. and Riveros, F. (1990) *Tropical Grasses*, FAO Plant Production and Protection Series No. 13, FAO, Rome.

Skerman, P.J., Cameron, D.G. and Riveros, F. (1988) *Tropical Forage Legumes*, 2nd edn. *FAO Plant Production and Protection Series No. 2*, FAO, Rome.

Skorupa, L.A. and Assis, M.C. (1998) Collecting and conserving ipecac (*Psychotria ipecacuanha*, Rubiaceae) germplasm in Brazil, *Economic Botany* **52**, 209-210.

Slatyer, R.O. (1973) The effect of internal water status on plant growth, development and yield, in R.O. Slatyer (ed.), *Plant Response to Climatic Factors. Proceedings of the Uppsala Symposium, 1970,* UNESCO, Paris, pp. 177-191.

Slee, R.W. (1991) The potential of small woodlands in Britain for edible mushroom production, *Journal of the Royal Scottish Forestry Society* **45**, 3-12.

Smartt, J. (1990) *Grain Legumes: Evolution and Genetic Resources,* Cambridge University Press, Cambridge.

Smidsrød, O. and Christensen, B.E. (1991) Molecular structure and physical behaviour of seaweed colloids as compared with microbial polysaccharides, in M.D. Guiry and G. Blunden (eds.), *Seaweed Resources in Europe: Uses and Potentia,* John Wiley & Sons, Chichester, pp. 185-217.

Smith, F. (1976) *The Early History of Veterinary Literature and its British Development. Vol. 1. From the Earliest Period to A.D. 1700,* J.A. Allen & Co., London (Reprinted with minor corrections from the Journal of Comparative Pathology and Therapeutics, 1912-1918).

Smith, J. (1949) *Distribution of Tree Species in the Sudan in Relation to Rainfall and Soil Texture.* Forests Department Bulletin No. 4, Ministry of Agriculture, Khartoum.

Smith, R.D. (1985) Seed banks: a useful tool in conservative plant evaluation and exploitation, in G.E. Wickens, J.R. Goodin and D.V. Field (eds.), *Plants for Arid Lands,* Allen & Unwin, London, pp. 321-331.

Smith, S.D., Monson, R. and Anderson, J.E. (1997) *Physiological Ecology of North American Plants.* Adaptations of Desert Organisms (ed. J.L. Cloudsley-Thompson), Springer-Verlag, Berlin, Heidelberg.

Snyder, E.M. (1984) Industrial microscopy of starches, in R.L. Whistler, J.N. BeMiller and E.F. Paschall (eds.), *Starch Chemistry and Technology,* 2nd edn., Academic Press, San Diego, pp. 661-673.

Soerianegari, I. and Lemmens, R.H.M.J. (ed.) (1993) *Timber Trees. Major Commercial Timbers. Plant Resources of South-East Asia No. 5(1),* Purdoc Scientific Publishers, Wagenningen.

Soil Survey Staff (1937) *Soil Survey Manual,* US Department of Agriculture, Miscellaneous Publications No. 274, US Department of Agriculture, Washington, DC.

Soil Survey Staff (1951) *Soil Survey Manual,* US Department of Agriculture Handbook No. 18, US Department of Agriculture, Washington, DC.

Soil Survey Staff (1975) *Soil Taxonomy,* US Department of Agriculture Handbook No. 426, US Department of Agriculture, Washington, DC.

Spindler, K. (1994) *The Man in the Ice. The Preserved Body of a Neolithic Man Reveals the Secrets of the Stone Age,* Weindenfeld and Nicolson, London.

Sprent, J.I. (1979) *The Biology of Nitrogen-fixing Compounds,* McGraw-Hill, London.

Sprent, J.I. (1985) Nitrogen fixation in arid environments, in G.E. Wickens, J.R. Goodin and D.V. Field (eds.), *Plants for Arid Lands,* Chapman & Hall, London, pp. 215-229.

Sprent, J.I. and Sprent, P. (1990) *Nitrogen Fixing Organisms: Pure and Applied Aspects,* 2nd edn., Chapman & Hall, London.

Staples, R. (1981) Summary and discussions of environmental tolerance, in J.T. Manassah and E.J. Briskey (eds.), *Advances in Food-producing Systems for Arid and Semiarid Lands,* Academic Press, New York, pp. 533-540.

Stearn, W.T. (1951) Mapping the distribution of species, in J.E. Lousley (ed.), *The Study of the Distribution of British Plants,* Botanical Society of the British Isles, Oxford, pp. 48-64.

Stearn, W.T. (1976) From Theophrastus and Dioscorides to Sibthorp and Smith: the background and origin of the Flora Graeca, *Biological Journal of the Linnean Society* **8**, 285-298.

Stearn, W.T. (1992) *Botanical Latin,* 4th edn., David and Charles, London.

Steinkraus, K.H. (ed.) (1995) *Handbook of Indigenous Fermented Foods,* 2nd edn., Marcel Decker, New York.

Stover, R.H. and Simmonds, N.W. (1987) *Bananas,* 3rd. edn., Longman, Harlow.

Straughan, R. and Reiss, M. (1996) *Ethics, Morality and Crop Biotechnology,* Biotechnology and Biological Sciences Research Council, Swindon.

Street, H.E. and Öpik, H. (1984) *The Physiology of Flowering Plants: Their Growth and Development*, 3rd edn., Arnold, London.

Subba Rao, N.S. and Kaushik, B.D. (1989) Microbes and economic prosperity, in M.S.: Swaminathan and S.L. Kochhar (eds.), *Plants and Society*, Macmillan Publishers, London and Basingstoke, pp. 367-402.

Subbarao, G.V. and Johansen, C. (1993) Strategies and scope for improving salinity tolerance in crop plants, in M. Pessarakli (ed.), *Handbook of Plant and Crop Stress*, Marcel Dekker, New York, pp. 559-595.

Sunberg, M.D. (1985) Trends in distribution of stomata in desert plants, *Desert Plants* **7**, 154-157.

Suzuki, O. (1988) Production of γ-linolenic acid by fungi and its industrialisation, in T.H.: Applewhite (ed.), *World Conference on Biotechnology for the Fats and Oil Industry*, American Oil Chemists Society, Chicago, pp. 277-287.

Suzuki, T. and Hasegawa (1974a) Lipid molecular species of *Lipomyces starkeyi*, *Agricultural Biology and Chemistry* **38**, 1371-1376.

Suzuki, T. and Hasegawa (1974b) Variation in lipid compositions and in the lipid molecular species of *Lipomyces starkeyi* cultivated in the glucose sufficient and the glucose deficient media, *Agricultural Biology and Chemistry* **38**, 1485-1492.

Syed, R.A. (1979) Studies on oil palm pollination by insects, *Bulletin of Entomological Research* **69**, 213-224.

Szabolcs, I. (1993) Soils and salinization, in M. Pessarakli (ed.), *Handbook of Plant and Crop Stress*, Marcel Dekker, New York, pp. 3-11.

Täckholm, V. (1970) *A Botanic Expedition to the Sudan in 1938 by Mohammed Drar. Edited after Author's Death with Introductory Notes by Vivi Täckholm*, Publication of the Cairo Herbarium No. 3. Cairo University Press, Cairo.

Täckholm, V., Täckholm, G. and Drar, M. (1941) *Flora of Egypt*, vol. I, Cairo University Press, Cairo.

Tatham, A., Maya Pandya and Shewry, P. (1998) Pseudocereals: methodologies in physicochemical analysis, in P.S. Belton and J.R.N. Taylor (eds.) *Increasing the Utilisation of Sorghum, Buckwheat, Grain Amaranth and Quinoa for Improved Nutrition. Selected Papers from the International Association for Cereal Science and Technology Cereals Conference Symposium "Challenges in Speciality Crops" Held at the 16th ICC Cereals Conference, Vienna, Austria 9 May 1998*, Institute of Food Research, Norwich, pp. 99-106.

Taylor, J. (1998) Enhancing food security in Africa: through the development of sorghum food technologies, in P.S. Belton and J.R.N. Taylor (eds.), *Increasing the Utilisation of Sorghum, Buckwheat, Grain Amaranth and Quinoa for Improved Nutrition. Selected Papers from the International Association for Cereal Science and Technology Cereals Conference Symposium "Challenges in Speciality Crops" Held at the 16th ICC Cereals Conference, Vienna, Austria 9 May 1998*, Institute of Food Research, Norwich, pp. 81-98.

Temple, S.A. (1977) Plant-animal mutalism: coevolution with dodo leads to near extinction of plant, *Science* **197**, 885-886.

Tenore, M. (1845) *Catalogo dell'Orta Royale di Napoli*, ex Typographia Diarii encyclopedici, Neapoli.

Terrell, E.E. and Batra, L.R. (1982) *Zizania latifolia* and *Ustilago esculenta*, a grass-fungus association, *Economic Botany* **36**, 274-285.

The Crucible Group (1994) *People, Plants, and Patents*, International Development Research Centre, Ottawa.

Thomas, E. and Wernicke, W. (1978) Morphogenesis in herbaceous crop plants, in T.A. Thorpe (ed.) *Frontiers of Plant Tissue Culture*, International Association Plant Tissue Culture, Calgary, pp. 403-410.

Thomson, L. (1992) Australia's subtropical dry-zone *Acacia* species with human food potential, in A.P.N. House and C.E. Harwood (eds.), *Australian Dry-zone Acacias for Human Food*, Australia Tree Seed Centre, Canberra, pp. 3-36.

Thornthwaite, C.W. (1948) An approach towards a rational classification of climate, *Geographical Review* **38**, 55-94.

Tieghem, P. van (1898) Sur le genre *Simmondsie* considéré comme type d'une famille distincte, les Simmondsiacées, *Journal de botanique* (Morot) **12**, 103-112.

Timmermann, B.N. and Hoffmann, J.J. (1988) The potential for the commercial utilization of resins from *Grindelia camporum*, in E.E. Whitehead, C.E. Hutchinson, B.N. Timmermann, and R.G. Varady (eds.), *Arid Lands: Today and Tomorrow*, Westview Press, CO and Belview Press, London, pp. 1321-1339.

Tootill, E. (1984) *The Penguin Dictionary of Botany*, Penguin Books, London.

Tralau, H. (ed.) (1961-81) *Index Holmensis. A World Index of Plant Distribution Maps*, 5 vols, Scientific Publications, Zurich.

Trehane, P., Brickell. C.D., Baum, B.R., Hetterscheid, W.L.A., Leslie, A.C., McNeill, J., Sponberg, S.A. and Vrugtman, F. (eds.) (1995) *International Code of Nomenclature for Cultivated Plants. Regnum vegetabile* **133**, Quarterjack Publishing, Wimborne, UK.

Troll, C. (1965) Seasonal climates of the earth, in E. Rodenwalt and H. Jusatz (eds.), *World Maps of Climatology*, Springer-Verlag, Berlin.

Tuley De Silva, K. (ed.) (1995) *A Manual on the Essential Oil Industry*, UNIDO, Vienna.

Turner, J.S. and Picker, M.D. (1993) Thermal coupling of an embedded dwarf succulent from southern Africa (*Lithops* spp., Mesembryanthaceae), *Journal of Arid Environments* **24**, 361-385.

Turner, R.M., Bowers, J.E. and Burgess, T.L. (1995) *Sonoran Desert Plants. An Ecological Atlas*, University of Arizona Press, Tucson, AZ.

Turner, S. and Skiöld, B. (1983) *Handmade Paper Today. A Worldwide Survey of Mills, Papers, Techniques and Uses*, Lund Humpries, London.

Tutin, T.G., Heywood, V.H., Burges, N.A., Valentine, D.H., Walters, S.M. and Webb, D.A. (eds.) (1964) *Flora Europaea*. Vol. 1, University Press, Cambridge.

Tyler, V.E. (1986) Plant drugs in the twenty-first century, *Economic Botany* **40**, 279-288.

Tyler, V.E. (1989) Potential hazards of herbal medications, in M.S. Swaminathan and S.l. Kochhar (eds.), *Plants and Society*, Macmillan Publishers, London and Basingstoke.

UPOV (1984) *Nomenclature. Records of a Symposium Held to the Occasion of the Seventeenth Ordinary Session of the Council of the International Union for the Protection of New Varieties*, Geneva 12 October 1983, UPOV Publication No. 341 (R).

van Emden, H.F. (1989) Recent advantages in the control of insect pests in tropical crops, *Tropical Agriculture Association UK Newsletter* **9** (3), 2-5.

van Ginkel, M. and Villareal, R.L. (1996) *Triticum* L., in G.J.H. Grubben and S. Partohardjono (eds.), *Cereals. Plant Resources of South-East Asia No. 10*, Backhuys Publishers, Leiden, pp. 137-143.

van Gorkom, K.W. (1880-1881) *De Oost-Indische cultures*, 2 vols, De Bussy, Amsterdam.

Vaughan, J.G. and Geissler, C.A. (1997) *The New Oxford Book of Food Plants*, Oxford University Press, Oxford.

Vavilov, N.I. (1926) Studies on the origin of cultivated plants, *Bulletin of Applied Botany, Genetics and Plant Breeding* **16**, 1-248.

Vavilov, N.I. (1951) The origin, variation, immunity and breeding of cultivated plants, *Chronica Botanica* **13**, 1-366. [Translation from Russian by K. Starr Chester].

Vavilov, N.I. (1992) *Origin and Geography of Cultivated Plants*. Translation by D. Löve of *Proiskhozdenie i geografiia kulturnykh rasteni*, Cambridge University Press, Cambridge.

Verdcourt, B. (1985) A synopsis of the Moringaceae. *Kew Bulletin*, **40**, 1-23.

Verdcourt, B. and Halliday, P. (1978) A revision of *Psophocarpus* (Leguminosae - Papilionoideae - Phaseoleae), *Kew Bulletin* **33**, 191-227.

Verdcourt, B. and Trump, E.C. (1969) *Common Poisonous Plants of East Africa*, Collins, London.

Verheij, E.W.M. and Coronel, R.E. (eds.) (1991) *Edible Fruits and Nuts. Plant Resources of South-East Asia No. 2*, Pudoc, Wageningen.

Vickery, R. (1995) *A Dictionary of Plant Lore*, Oxford University Press, Oxford.

Voisin, A. (1959) *Grass Productivity*, Crosby Lockwood & Son, London.

Volkens, G. (1887) *Die Flora der Aegyptisch-Arabischen Wüste auf Grundlage anatonisch-physiolgisher Forschungen*, Borsträger, Berlin.

von Willert, D.J., Eller, B.M., Werger, M.J.A., Brinckmann, E. and Ihlenfeldt, H.-D. (1992) *Life Strategies of Succulents in Deserts*, Cambridge Studies in Ecology, Cambridge University Press, Cambridge.

Vriende, H.J. de (ed.) (1984) *The Old English Herbarium and Medicina de Quadrupedibus*, Oxford University Press, London.

Waisel, Y., Eshel, A. and Kafkafi, U. (eds.) (1991) *Plant Roots: The Hidden Half*, Marcel Dekker, New York.

Walker, P.M.B. (ed.) (1991) *Chambers Science and Technology Dictionary*, W. & R. Chambers Ltd., Edinburgh.

Wallis, E.S., Wood, I.M. and Byth, D.E. (1989) New crops: a suggested framework for their selection, evaluation and commercial development, in G.E. Wickens, N. Haq, N. and P. Day (eds.), *New Crops for Food and Industry*, Chapman & Hall, London, pp. 36-52.

Walter, H. (1963) Climatic diagrams as a means to comprehend the various climatic types for ecological and agricultural purposes, in R.J. Rutter and F.H. Whitehead (eds.), *The Water Relations of Plants*, Blackwell Scientific Publishers, Oxford.

Walter, H. and Breckle, S.-W. (1985) *Ecological Systems of the Geobiosphere. 1 Ecological Principles in Global Perspective*, Springer-Verlag, Berlin.

Walter, H. and Breckle, S.-W. (1986) *Ecological Systems of the Geobiosphere. 2 Tropical and Subtropical Zonobiomes*, Springer-Verlag, Berlin, Heidelberg.

Walter, H. and Breckle, S.-W. (1989) *Ecological Systems of the Geobiosphere. 3. Temperate and Polar Zonobiomes of Northern Eurasia*, Springer-Verlag, Berlin.

Walter, H. and Lieth, H. (1960-67) *Klimadiagramm-Weltatlas*, G. Fischer, Jena.

Wang, S. and Huffman, J.B. (1981) Botanochemicals: supplements to petroleum, *Economic Botany* **35**, 369-352.

Wang, S., Huffman. J.B. and Rockwood, D.L. (1982) Qualitative evaluation of fuelwood in Florida - a summary report, *Economic Botany* **36**, 381-388.

Wang, Y., Hall, I.R. and Lynley, A.E. (1997) Ectomycorrhyizal fungi with edible fruiting bodies, 1. *Tricholoma matsutake* and related fungi, *Economic Botany* **51**, 311-327.

Waterman, P.G. (1989) Bioactive phytochemicals - the search for new sources, in G.E. Wickens, N. Haq and P. Day (eds.), *New Crops for Food and Industry*, Chapman & Hall, London, pp. 378-390.

Watling, R. and Seaward, M.R.D. (1976) Some observations on puff-balls from British archaeological sites, *Journal of Archaeological Science* **3**, 165-172.

Watson, A.K. (1989) Current advances in bioherbicide research, in *Brighton Crop Protection Conference - Weeds*, British Crop Protection Council, Farnham, pp. 987-996.

Watt, J.M. and Breyer-Brandwijk,M.G. (1962) *The Medicinal and Poisonous Plants of Southern and Eastern Africa*. 2nd edn., E. & S. Livingstone, Edinburgh.

Webster, C.C. and Watson, G.A. (1988) Tree crops of the humid tropics, *Tropical Agriculture Association UK Newsletter* **8** (4), 11-19.

Webster, F.H. (1996) Oats, in R.J. Henry and P.S. Kettlewell (eds.), *Cereal Grain Quality*, Chapman & Hall, London, pp. 153-177.

Webster, J.M. (1972) *Economic Nematology*, Academic Press, London.

Weier, K.I. (1976) *Nitrogen Economy of Pastures,* CSIRO Australia, Division of Tropical Crops and Pastures Report 1975-76, CSIRO, Melbourne.

Weinstein, A. (1991) *Introduction of New Forest Species in Israel*, Agriculture Research Organization, Volcani Center, Bet Dagan, Israel.

Weipert, D. (1996) Rye and triticale, in R.J. Henry and P.S. Kettlewell (eds.), *Cereal Grain Quality*, Chapman & Hall, London, pp. 205-224.

Went, F.W. (1971) Parallel evolution, *Taxon* **20**, 197-226.

Westman, W.E. (1981) Seasonal distribution of foliage in Californian coastal sage scrub, *Oecologia* (Berlin) **51**, 385-388.

Whistler, R.L. and Corbett, W.M. (1957) Polysaccharides, in W. Pigman (ed.), *The Carbohydrates:Chemistry, Biochemistry, Physiology*, Academic Press, New York, pp. 641-208.

White, F. (1962) Geographic variation and speciation in Africa with particular reference to *Diospyros*, in D. Nichols (ed.), *Taxonomy and Geography*, The Systematics Association, London, pp. 47-103.

White, F. (1976) The underground forests of Africa: a preliminary review, *Gardens' Bulletin* **29**, 57-71.

Whitford, W.G., Anderson, J. and Rice, P.M. (1997) Stemflow contribution to the 'fertile island' effect in creosote bush, *Larrea tridentata*, *Journal of Arid Environments* **35**, 451-457.

Whitworth, J.W. and Whitehead, E.E. (eds.) (1991) *Guayule Natural Rubber. A Technical Publication with Emphasis on Recent Findings*, Office of Arid Land Studies, Tucson.

WHO (1976) *WHO Handbook on Basic Health policy: Alternative Approaches to Meeting the Basic Health Needs of Developing Countries*. SEA/RC.29/Min.5, World Health Organization, Geneva.

WHO (1990a) Technological evaluation of certain food additives and contaminants. *Food Additive Series* **26**, WHO, Geneva, pp. 77-79.

WHO (1990b) Evaluation of certain food additives and contaminants, *Technical Report Series* **798**, WHO, Geneva. p.24.

Whyte, R.O. (1947) Anglo-Eyptian Sudan, in *The Use and Misuse of Shrubs and Trees as Fodder*, Imperial Agricultural Bureau Joint Publication No. 10, London, pp. 94-109

Wickens, G.E. (1969) A study of *Acacia albida* Del. (Mimosoideae), *Kew Bulletin* **23**, 181-202.

Wickens, G.E. (1982) The baobab - Africa's upside-down tree, *Kew Bulletin* **37**, 173-209.

Wickens, G.E. (1987) Ecosystem data and geographic distribution of molluscicides, in K.E. Mott (ed.), *Plant Molluscicides: Papers Presented at a Meeting of the Scientific Working Group on Plant Molluscicides*, UNDP/World Bank/WHO Special Programme for Research and Training in Tropical Diseases, John Wiley and Sons, Chichester, pp. 205-230.

Wickens, G.E. (1990) What is economic botany? *Economic Botany* **44**, 12-28.

Wickens, G.E. (1991) *Non-wood forest products: the way ahead*, FAO Forestry Paper 97, FAO, Rome.

Wickens, G.E. (1992) Arid and semiarid ecosystems, in W.A. Nierenberg (ed.), *Encyclopedia of Earth System Science*, vol. 1, Academic Press, San Diego, pp. 113-118.

Wickens, G.E. (1993) Two centuries of economic botanists at Kew, *The Kew Magazine* **10**, 84-94, 132-138.

Wickens, G.E. (1995a) *Edible nuts. Non-Wood Forest Products 5*, FAO, Rome.

Wickens, G.E. (1995b) *Role of Acacia species in the rural economy of dry Africa and the Near East*, FAO Conservation Guide 27, FAO, Rome.

Wickens, G.E. (1998) *Ecophysiology of Economic Plants of Arid and Semi-Arid Lands,.Adaptations of Desert Organisms*, (ed. J.L. Cloudsley-Thompson), Springer-Verlag, Berlin, Heidelberg.

Wickens, G.E., Goodin, J.R. and Field, D.V. (eds.), (1985) *Plants for Arid Lands*, Allen & Unwin, London.

Wigmore, S. (2000) The woad ahead, *Reading Reading* **30**, 9-10.

Williams, J.T. (1989) Plant germplasm preservation: a global perspective, in L. Knutson and A.K. Stoner (eds.), *Germplasm Preservation, Global Imperatives*, Kluwer Academic Publishers, Dordrecht, Boston and London, pp. 81-96.

Williams, Jr., S.B. and Weber, K.A. (1989) Interlectual property protection and plants, in L. Knutson and A.K. Stoner (eds.), *Intellectual Property Rights Associated with Plants, American Society of Agronomy Special Publication no. 52,* Crop Science Society of America, American Society of Agronomy, Soil Science Society of America, Madison, WI, pp. 91-107.

Willis, J.C. (1973) *A Dictionary of the Flowering Plants and Ferns*, 8th edn. (revised by H.K. Airy Shaw), Cambridge University Press, Cambridge.

Wilson. E.O. (1992) *The Diversity of Life*, Belknap Press of Harvard.University Press, Cambridge, MA.

Wilson, J.M. and Witcombe, J.R. (1985) Crops for arid lands, in G.E. Wickens, J.R. Gooding, and D.V. Field (eds.), *Plants for Arid Lands*, Allen & Unwin, London, pp. 35-52.

Wilson, R.T. (1978) The 'gizu' winter grazing in the south Libyan Desert, *Journal of Arid Environments* **1**, 327-344.

Winter, K. (1981) C_4 plants of high biomass in arid regions of Asia - occurrence of C_4 photosynthesis in Chenopodiaceae and Polygonaceae from the Middle East and USSR, *Oecologia* (Berlin) **48**, 100-106.

Withers, L.A. (1991) *In vitro* conservation, *Biology Journal of the Linnean Society* **43**, 31-42.

Woodroof, J.G. (1979) *Tree Nuts: production, processing, products*, 2nd edn., Avi Publishing Company, Westport CT.

Woods, A. (1985) The potential for the *in vitro* propagation of a number of economically important plants for arid areas, in G.E Wickens, J.R. Goodin and D.V. Field (eds.), *Plants for Arid lands*. Allen & Unwin, London, pp. 333-342.

Woodward, M. (1985) *Gerard's Herball. The Essence thereof distilled by Marcus Woodward from the Edition of Th. Johnson, 1636*, Studio Editions, London.

Wormersley, J. (1981) *Plant Collecting and Herbarium Development. A Manual*, Plant Production and Protection Paper 33, FAO, Rome.

Wullstein, L.H. (1991) Variation in N_2 fixation (C_2H_2 reduction) associated with rhizosheaths of Indian ricegrass (*Stipa hymenoides*), *American Midland Naturalist* **126**, 76-81.

Wullstein, L.H., Bruening, M.L. and Bollen, W.B. (1979) Nitrogen fixation associated with sand grain root sheaths (rhizosheaths) of certain xeric grasses, *Physiologia Plantarum* **46**, 1-4.

Xia, B. and Abbott, I.A. (1987) Edible seaweeds and their place in the Chinese diet, *Economic Botany* **41**, 341-353.

Xia, P.G. and Peng, Y. (1998) Ethnopharmacology and research on medicinal plants in China, in H.D.V. Prendergast, N.L. Etkin, D.R. Harris and P.A. Houghton (eds.), *Plants for Food and Medicine. Proceedings of the Joint Conference of the Society for Economic Botany and the International Society for Ethnopharmacology London, 1-6 July 1996*, Royal Botanic Gardens, Kew, pp. 31-39.

Yanovsky, E. (1936) *Food Plants of the North American Indians*, USDA Miscellaneous Publication No. 237, Washington, DC.

Yunupinu, B., Yunupinu-Marika, L., Marika, D., Marika,B, Marika, B., Marika, R. and Wightman, G. (1995) Rirratjinu ethnobotany: aboriginal plant use from yirrkala, Arnhem Land, Australia, *Northern Territories Botany Bulletin* **21**.

Zin, J. and Weiss, C. (1980) *La Salud por Medico de las Plantas Medicinales*, 6th edn., Editorial Salesiana, Santiago, Chile.

Taxonomic Index

Taxa arranged according to the Five Kingdoms classification

Families and genera of the vascular plants (Brummitt, 1992) and authors of plant names (Brummitt and Powell, 1992) are according to those recognised at the Royal Botanic Gardens, Kew. Accepted names are in bold, synonyms noted in the cited literature are in *italics*.

**CONIFEROPHYTA, CYCADOPHYTA,
GINKGOPHYTA AND GNETOPHYTA**
(Gymnosperms)
Nomenclature of conifers according to Farjon
(1998)

FILICINOPHYTA, LYCOPHYTA, PSILOPHYTA AND SPHENOPHYTA (Ferns and Fern Allies)

Chemical Index

Nomenclature according to Sharp (!990). Major references in bold.

Subject Index

(Major references in bold)